D1130687

An

Introduction to

ULTRATHIN
ORGANIC FILMS

From

Langmuir–Blodgett to Self-Assembly

An

Introduction to

ULTRATHIN ORGANIC FILMS

From

Langmuir–Blodgett to Self-Assembly

Abraham Ulman

Corporate Research Laboratories
Eastman Kodak Company
Rochester, New York

ACADEMIC PRESS, INC.
Harcourt Brace Jovanovich, Publishers
Boston   San Diego   New York
London   Sydney   Tokyo   Toronto

This book is printed on acid-free paper. ⊚

Copyright © 1991 by Academic Press, Inc.
All rights reserved.
No part of this publication may be reproduced or
transmitted in any form or by any means, electronic
or mechanical, including photocopy, recording, or
any information storage and retrieval system, without
permission in writing from the publisher.

ACADEMIC PRESS, INC.
1250 Sixth Avenue, San Diego, CA 92101

United Kingdom Edition published by
ACADEMIC PRESS LIMITED
24–28 Oval Road, London NW1 7DX

Library of Congress Cataloging-in-Publication Data

Ulman, Abraham, date.
    An introduction to ultrathin organic films : from Langmuir
  –Blodgett to self-assembly / Abraham Ulman.
       p. cm.
    Includes bibliographical references (p.   ) and index.
    ISBN 0-12-708230-1 (alk. paper)
    1. Organic compounds. 2. Thin films—Surfaces. 3. Monomolecular
  films. 4. Surface chemistry. I. Title.
  QC176.9.073U44 1991
  530.4′175—dc20   90-26668
                                                                   CIP

Printed in the United States of America
91 92 93 94   9 8 7 6 5 4 3 2 1

To my wife Hanna
and my children Tzipor, Amihai, and Matan.

"Today...I propose to tell you of a real two-dimensional world in which phenomana occur that are analogous to those described in 'Flatland.' I plan to tell you about the behavior of molecules and atoms that are held at the surface of solids and liquids."

— I. Langmuir, *Science* **1936**, *84*, 379.

CONTENTS

PREFACE

About 70 years ago, Langmuir published his first work on the study of two-dimensional systems of molecular films at the gas–liquid interface [1]. However, it was only about 10 years ago that interest in this area started to grow at an impressive pace; and in about the same time, a self-assembled (SA) monolayer of octadecyltrichlorosilane ($C_{18}H_{37}SiCl_3$, OTS) was introduced as a possible alternative to the Langmuir–Blodgett (LB) system [2], and, recently, a relatively thick (~0.1 μm) multilayer film was formed by self-assembly of methyl 23-trichlorosilyltricosanoate [3].

That Langmuir–Blodgett and SA films are important is apparent from the exponential growth in the number of publications in these areas. I feel that at least part of this interest results from the fact that optoelectronics [4–10] and molecular electronics [11–19] have become areas at the frontier of materials science. In both cases, there are limitations to what inorganic materials can provide, and therefore ordered organic materials are likely to become increasingly important. In both areas, it is believed that LB and SA monolayers may provide the desired control on the order *at the molecular level* and thus should be considered potential techniques for the construction of future organic materials.

Although nonlinear optics (NLO) and molecular electronics represent distinct areas of investigation, we see that they share a common requirement for the development of thermally stable, ordered organic (or organometallic) molecular systems. It is important that organic materials be developed in which interesting properties can be tailored by incorporating appropriate chromophores or other functional groups. Furthermore, a high degree of control over the orientation of these groups also is desirable.

Thus, it can be seen that the advent of materials science requires research and development in what may be termed *materials engineering*. The goal of such research is to attain the capability of *assembling individual molecules into highly ordered architectures*. Although attainment of such a goal until recently could be considered practically impossible, the development of oriented organic mono-molecular layers by the LB and SA techniques opens the way to the realization of this goal. Analogues of nature's lipid bilayers, these two-dimensional structures allow scientists to construct films with a high degree of orientation within layers, and to construct *molecular architectures* by varying the constituents of adjacent monolayers.

One of the exciting aspects of thin organic films, in my view, is that it is an interdisciplinary area, forming bridges between the areas of physics, chemistry, and biology. With all this in mind, this area has not yet been accepted as a legitimate research interest in many chemistry departments in this country.

In 1965, George L. Gaines, Jr., wrote a book entitled *Insoluble Monolayers at Liquid–Gas Interfaces* [20]. This work, which has provided guidance for many of us in the area, still is one of the most used and cited references. However, since that book was published, there has been no other book on this subject. Recent developments in the area of ultrathin organic films have prompted me to write this book, with the idea of starting where Gaines finished, and complementing his work. There are excellent works in related areas, and they are listed as well. I refer the reader to discussions in other books when appropriate.

The continually growing contribution of LB and SA systems to the chemistry and physics of thin organic films, of course, is widely recognized. Equally well-known is the difficulty in keeping up-to-date with the multidisciplinary research in this area that seems to be spawned at an ever-increasing rate. Indeed, when I started the literature search for this book, it became clear very quickly that it would be impossible to cover the entire richness of this area in only one book. The original plan was to have a book of no more than 300 pages, but the book you hold in your hands is considerably longer. This is a result of my desire to cover as many areas—and to provide as many references—as possible. Most of these references are taken from before December 1989; however, in some cases, references from later dates have been included.

The development of surface analytical tools in the last decade makes it possible to address structural issues of monomolecular layers in great detail, and with unprecedented precision. Therefore, I start the book with Part One, "Analytical Tools," with the hope that it will help the reader understand the structure and properties of monolayers and films. Part Two is dedicated to Langmuir–Blodgett films. Part Three describes the preparation and properties of self-assembled monolayers and films. In Part Four, we discuss the modeling of LB and SA monolayers. I decided to write this part because I believe modeling will become an obvious part of any materials science activity in the 1990s and beyond. Finally, in Part Five, we discuss applications of LB and SA films.

One remark has to be made regarding the literature on this subject. This literature contains hundreds of patents and papers in Japanese for which no English version has been available. Therefore, I could not refer to them in a precise way, and, unfortunately, did not even mention most of them in the list of references.

This book is an overview, with a strong emphasis on materials, and has been written for scientists and graduate students; I hope that both will find it useful as an introduction to the field. Let me take this opportunity to invite readers to bring to my attention errors and/or obscurities. I thank everyone in advance for his or her effort.

It is impossible for me to finish this introduction without emphasizing the encouragement, support, and cooperation of my wife, Hanna, throughout my scientific adventures, and during the preparation of this book. To her I dedicate this book with great love.

Abraham Ulman
Rochester, New York

References

1. Langmuir, I. *Trans. Faraday Soc.* **1920**, *15*, 62.
2 Sagiv, J. *J. Am. Chem. Soc.* **1980**, *102*, 92.
3. Tillman, N.; Ulman, A.; Penner, T.L. *Langmuir* **1989**, *5*, 101.
4. Bloembergen, N.; Ducuing and, J.; Pershan, P. S. *Phys.Rev.* **1962**, *127*, 1918.
5. Bloembergen, N.; Pershan, P. S. *Phys.Rev* **1962**, *62*, 606 .
6. Yariv, A. *Quantum Electronics,* 3ed, Wiley: New York (1986).
7. Byer, R. L. in *Nonlinear Optics,* Harper, P. G., and Wherrett, B. S.,Eds., Academic Press: New York (1979).
8. Kurtz, S. K. in *Quantum Electronics: A Treatise,* **1-B**; Rabin, H. and Tang C. L., Eds., Academic Press: New York (1975); Chapter 3.
9. Zyss, J. in *Current Trends in Optics,* F. T. Arecchi and F. R. Aussenegg, Eds., pp. 122-133, Taylor and Francis: London (1981).
10. *ACS Symposium Series* **233**; Williams, D. J., Ed., Washington, D.C. (1983).
11. Hopfield, J. J.; Onuchic, J. N.; Beratan, D. N. *Science* **1988**, *241*, 817.
12. *Molecular Electronic Devices,* Carter, F. L. ed.; Marcel Dekker: New York, 1982.
13. *Molecular Electronic Devices II* , F. L. Carter, ed.; Marcel Dekker: New York, 1987.
14. Aviram, A. J. *J. Am. Chem. Soc.* **1988**, *110*, 5687.
15. Haddon,R. C.; Lamola, A. A. *Proc. Nat. Acad. Sci. USA* **1985**, *82*, 1874.
16. Milch, J. R. *Computers Based on Molecular Implementations of Cellular Automata, Third International Symposium on Molecular Electronic Devices,* North Holland, in press.
17. Van Brunt, J. *Biotechnology* **1985**, *3*, 209.
18. Birge, R. R., Lawrence, A. F.; Findsen, L. A. *Proceeding of the 1986 International Congress on Technology and Technology Exchange* ; Pittsburgh, PA, 1986.
19. Palacin S.; Ruaudel-Teixier, A.; Barraud, A. *Mol. Cryst. Liq. Cryst.* , **1988**, *156*, 331.
20. Gaines, G. L., Jr. *Insoluble Monolayers at Liquid–Gas Interfaces,* Interscience: New York (1966).

ACKNOWLEDGEMENTS

I am grateful to those who collaborated with me in the investigation of self-assembled monolayers, for their work, help, discussions, and encouragement. They have contributed directly to my work, and thus indirectly to this book. Many thanks to Dr. Tomas L. Penner and Dr. Jay Schildkraute, both of Eastman Kodak Company, for reading parts of the manuscript and for their comments, and to Nancy J. Armstrong of Eastman Kodak Company for providing some of the drawings in Part Two.

Special acknowledgment and thanks to Dr. Yizhak Shnidman of Eastman Kodak Company for advice and help in writing Part Four of the book.

I am indebted to the management of the Corporate Research Laboratories, Eastman Kodak Company, for their encouragement, and for their help in the preparation of this manuscript.

Grateful acknowledgment is made to all who extended their permission to reproduce material from their work. These resources are mentioned in the text or in the captions, as appropriate.

SUGGESTIONS FOR FURTHER READING

Books

1. Gaines, G. L., Jr. *Insoluble Monolayers at Liquid–Gas Interfaces;* Interscience: New York, 1966.
2. Adamson, A. W. *Physical Chemistry of Surfaces* ; Wiley: New York, 1982.
3. Griffiths, P. R.; de Haseth, J. A. *Fourier Transform Infrared Spectroscopy,* Wiley: New York, 1986.
4. Harrick, N. J. *Internal Reflection Spectroscopy*; Harrick: Ossing, 1979.
5. Czanderna, A. W., Ed. *Methods of Surface Analysis*; Elsevier: Amsterdam, 1975.
6. Rosen, M. J. *Surfactants and Interfacial Phenomena*; Wiley: New York, 1978.
7. Tadros, Th. F., Ed. *Surfactants*; Academic Press: New York, 1984.
8. Hiemenz, P. C., *Principles of Colloid and Surface Chemistry* ; Marcel Dekker: New York, 1986.
9. Davies, J. T.; Rideal, E. K., *Interfacial Phenomena*, Academic Press: New York, 1961.
10. Johnson, R. E., Jr.; Dettree, R. H., in *Surfaces and Colloid Science*, Matijevi´c, E., Ed., Wiley: New York, 1969, Vol. 2.
11. Neumann, A. W.; Good, R. J., in *Surfaces and Colloid Science*, Plenum: New York, 1979, Vol. 11.
12. Kitaigorodski, A. T., *Organic Chemical Crystallography* , Consultants Bureau: New York, 1961.
13. Fendler, J. H. *Membrane Mimic Chemistry* , Wiley: New York, 1982.
14. Israelachvili, J. N. *Intermolecular and Surface Forces*, Academic Press: London, 1989.
15. Drexler, K. E. *Engines of Creation, the Coming Era of Nanotechnology* , Anchor Press: New York, 1986.

16. Allen, M. P.; Tildesley, D. J. *Computer Simulation of Liquids*, Clarendon Press: Oxford, 1987
17. Ciccotti, G.; Frenkel, D.; McDonald, I. R. *Simulation of Liquids and Solids*, North Holland, 1987.
18. Ciccotti, G.; Hoover, W. G., Eds.; *Molecular Dynamics Simulations of Statistical Mechanical Systems*, Proc. of the International School of Physics "Enrico Fermi", Course XCVII, North Holland, 1985.
19. Binder, K. *Monte Carlo Methods in Statistical Physics* (2nd Ed.). *Topics in Current Physics* Vol. 7, Springer: Berlin, 1986.
20. Prasad, P. N.; Williams, D. J. *Introduction to Nonlinear Optical Effects in Organic Molecules and Polymers*, Wiley: New York, 1990.

Review Articles

1. Kuhn H. *Thin Organic Films* **1989**, *178*, 1.
2. Laschewsky, A. *Angew. Chem. Int. Ed. Engl. Adv. Mater.* **1989**, *28*, 1574.
3. Peterson, I. R. *Spec. Publ. R. Soc. Chem.* **1989**, *69*, 317.
4. Vandevyver, M.; Barraud, A. *J. Mol. Elect.* **1989**, *4*, 207.
5. Roberts, G. G. *Adv. Chem. Ser.* **1988**, *218*, 225.
6. Vandevyver, M. *Thin Solid Films* **1988**, *159*, 243.
7. Morizumi, T. *Thin Solid Films* **1988**, *160*, 413.
8. Swalen, J. D. *Thin Solid Films* **1988**, *160*, 197.
9. Blinov, L. M. *Sov. Phys. Usp.* **1988**, *31*, 623.
10. Ringsdorf, H.; Schlarb, B.; Venzmer, J. *Angew. Chem. Int. Ed. Engl.* **1988**, *27*, 113.
11. Peterson, I. R. *J. Chim. Phys. Phys.-Chim. Biol.* **1988**, *85*, 997.
12. Peterson, I. R. *J. de. Chim. Phys.* **1988**, *85*, 997.
13. Moriizumi, T. *Thin Solid Films* **1988**, *160*, 413.
14. Peterson, I. R. *J. Mol. Electron.* **1987**, *3*, 103.
15. Biddle, M. B.; Rickert, S. E. *Ferroelectrics* **1987**, *76*, 133.
16. Sugi, M. *Thin Solid Films* **1987**, *152*, 305.
17. Tredgold, R. H. *Thin Solid Films* **1987**, *152*, 223.
18. Ringsdorf, H.; Schmidt, G.; Schneider, J. *Thin Solid Films* **1987**, *152*, 207.
19. Pomerantz, M. *Thin Solid Films* **1987**, *152*, 165.
20. Khanarian, G. *Thin Solid Films* **1987**, *152*, 265.
21. Swalen, J. D. *Thin Solid Films* **1987**, *152*, 151.
22. Reichert, W. M.; Bruckner, C. J.; Joseph, J. *Thin Solid Films* **1987**, *152*, 345.
23. Swalen, J. D. *J. Mol. Electron.* **1986**, *2*, 155.
24. Barraud, A.; Vandevyver, M. in *"Nonlinear Opt. Prop. Org. Mol. Cryst.."*, Chemla, D. S.; Zyss, J., Eds., Academic Press: Orlando, (1987), Vol. 1, 357.
25. Honeybourne, C. L. *J. Phys. Chem. Solids* **1987**, *48*, 109.
26. Ertl, G. *Langmuir* **1987**, *3*, 4.
27. Sugi, M. *J. Mol. Electron.* **1985**, *1*, 3.
28. Miyano, K. *Jpn. J. Appl. Phys. I* **1985**, *24*, 1379.
29. Roberts, G. G. *Adv. Phys.* **1985**, *34*, 475.
30. Roberts, G. G. *Contemp. Phys.* **1984**, *25*, 109.
31. Berendsen, H. J. C.; van Gunsteren, E. F. in *Molecular Liquids, Dynamics and Interaction*, Barness, E. J.; Thomas, W. J.; Yarwood, J., Eds.; NATO ASI series C135, p 475, Reidel: New York, 1984.
32. Roberts, G. G. *Sens. Actuators* **1983**, *4*, 131.

33. Peterson, I. R. *IEE Proc. Part I* **1983**, *130*, 252.
34. Petty, M. C. *Endeavour* **1983**, *7*, 65.
35. Berton, M. *J. Macromol. Sci. Rev. Macromol. Chem.* **1981**, *C2*, 61.
36. Vincett, P. S.; Roberts, G. G. *Thin Solid Films* **1980**, *68*, 135.
37. Pitt, C. W.; Walpita, L. M. *Thin Solid Films* **1980**, *68*, 101.
38. Gaines, G. L., Jr. *Thin Solid Films* **1980**, *68*, 1.
39. Kuhn, H. *J. Photochem.* **1979**, *10*, 111.
40. Blake, M. *Prog. Surf. Membr. Sci.* **1979**, *13*, 87.
41. Agarwal, V. K. *Electrocomponent Sci. Tech.* **1975**, *2*, 1.
42. Blake, M. *Tech. Surface Colloid Chem. Phys.* **1972**, *1*, 41.

PART ONE

ANALYTICAL TOOLS

In the study of monolayers and thin films, we are interested in both their surface and bulk properties. Ellipsometry is used to measure the thickness and uniformity of freshly prepared films. Contact angles with different liquids are measured to evaluate wetting properties, surface-free energy, and uniformity, and to get information on surface order. Fourier transform infrared (FTIR) spectroscopy, in both grazing-angle and attenuated total reflection (ATR) modes, is used to learn about the direction of transition dipoles, and to evaluate dichroic ratios, molecular orientation, packing, and coverage. We also use surface potential measurements to get information on the coherence of a film at the water–air interface and on metal surfaces, electron spectroscopy for chemical analysis (ESCA) to study surface composition and monolayer structure, and surface imaging techniques to learn about surface topography.

The development of analytical tools for the study of thin organic films has been dramatic in the last decade. Therefore, it is the intent of this part of the book to introduce the reader to these techniques, and to provide the tools for better understanding of the models and applications treated in subsequent chapters. However, it is beyond the scope of this book to give a detailed physical background to every analytical tool; therefore, a brief explanation of the physical principles is given, followed by the respective literature in each case. I have listed as many examples as possible, since I feel that readers will find it useful to have references relevant to their specific areas of interest.

1.1. Analysis of Film Properties

A. Thickness Measurements

The estimation of film thickness usually is the first characterization method we use for a newly made film. Here, it is proper to mention the classic work of Katherine Blodgett on optical interference of thick films to determine their thickness and refractive index [1]. Today, however, *ellipsometry* is the common optical technique for the determination of the thickness and refractive index of thin homogeneous films [2–5]. When a plane-polarized light interacts with a surface at some angle, it is resolved into its parallel and perpendicular components (*s*- and *p*-polarized, respectively). These components are reflected from the surface in a different way; i.e., the amplitude and phase of both components are changed. When the *s*- and *p*-polarized reflected light beams are combined, the result is elliptically polarized light. Ellipsometry uses this phenomenon to estimate the thickness of a transition region between the surface and air by measuring the ratio r between r_p and r_s, the reflection coefficients of the *p*- and *s*-polarized light, respectively.

In a typical ellipsometer (Figure 1.1) [6], monochromatic light (He–Ne laser) is plane polarized (p = angle of polarization) and impinges on the surface. A compensator changes the reflected beam that is elliptically polarized to plane-polarized (a = angle of polarization). The analyzer then determines the angle a, by which the compensator polarized the beam.

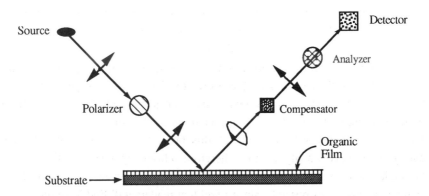

Figure 1.1. A schematic description of an ellipsometer, where ➤➤ denotes unpolarized light, ◄—► denotes plane-polarized light, and ⬭ denotes elliptically polarized light. (Adapted from Gaines [6], © 1965, Wiley Interscience.)

2

These two angles (p and a) give the phase shift between the parallel and perpendicular components (Δ), and the change in the ratio of the amplitudes of the two components ($\tan\psi$), as

$$e^{i\Delta}\tan\psi = \left\{\frac{\left(\dfrac{E_{\text{reflected }(p)}}{E_{\text{reflected }(s)}}\right)}{\left(\dfrac{E_{\text{incident }(p)}}{E_{\text{incident }(s)}}\right)}\right\}, \qquad 1.1$$

where $\Delta = 2p + \pi/2$ and $\psi = a$. For a clean surface, Δ and ψ are directly related to the complex index of refraction of the surface,

$$\hat{n}^s = n^s(1 - ik^s), \qquad 1.2$$

where n^s is the ordinary refraction index and k^s is the extinction coefficient.

Once a film having a different index of refraction n_f is coated on the surface, Δ and ψ are related to the complex indices of both the film and the substrate, and to the film thickness. Typically, the compensator is set at an azimuth angle of $70°$ and the experimental data are expressed as

$$\delta\Delta = \Delta_0 - \Delta, \qquad 1.3$$

$$\delta\psi = \psi_0 - \psi, \qquad 1.4$$

where Δ_0 and ψ_0 are the ellipsometric angles characteristic of the clean substrate surface, and Δ and ψ those measured for the substrate with the film.

Practically, we have to estimate the refractive index of the organic film (although in principle for thickness >50 Å, ellipsometry can determine both the thickness and the refractive index). Usually, we use the value of 1.50 as the refractive index for organic monolayers [7]. This value is suggested based on the assumption that the monolayer is crystalline, similar to polyethylene, and its refractive index thus should be in the region 1.49–1.55 [8]. However, recently, Wasserman et al. found from x-ray studies that electron density in alkyltrichloro-silane monolayers is similar to that of bulk paraffins, rather than to that of crystalline paraffins, and suggested that the value of 1.45 should be more adequate for the refractive index of the organic monolayer [9]. We note that this can be correct only under the assumption that there was no electronic resonance in the measurement. On the other hand, Tillman et al. measured a value of $n_f = 1.50$ for a sample of 15 monolayers of methyl 23-trichlorosilyltricosanoate [10], and calculated that an increase of 0.05 in the refractive index translates to a decrease in the thickness by ~1 Å [11]. Porter et al. [12] found that if the value of

3

n_f = 1.50 is used (as for polyethylene) for a monolayer of docosanethiol (n = 21), a thickness of 31.6 Å is measured. On the other hand, if the value of n_f = 1.45 is used, the thickness is estimated as ~33 Å. The calculated maximum thickness for all-*trans* chain with 25° tilt is ~29 Å. On the other hand, Rabe *et al.* reported a value of n_f = 1.55 for a sample of 40 monolayers of cadmium arachidate deposited on a (111) silicon wafer with its natural oxide [13]. The reader should not be confused by the differences in refractive index values, since the reported studies were made on different films. We suggest that the value of n_f = 1.50 be used for monolayers of simple alkyl chains of, for example, alkanethiols on gold (Part Three), with chains of C_{10} and longer. (See also the discussion on surface potential.) The introduction of metal ions increases the polarity of the material, and therefore, a higher refractive index (n_f > 1.50) should be used. In cases of monolayers of alkyl chains of C_9 and shorter, values of n_f < 1.50 (e.g., 1.45) should be used. We do not suggest that the refractive index changes sharply, but note that it may change considerably with chain length, as is apparent from studies of surface potential of alkanethiol monolayers on gold. (See Section 1.1.D).

Of course, other values of n_f should be used for different molecular structures; e.g., a value of n_f ~ 1.32–1.35 is used for perfluorinated materials. In general, it is recommended that the refractive indices of model compounds be used to estimate the value of n_f for a film [14].

Rabe and Knoll reported accuracy of 2 Å in detecting thickness differences with a lateral resolution of about the size of a laser spot, thus making possible the investigation of large domains in a film, laterally homogeneous in thickness and refractive index [15].

Another optical method to obtain thickness data is *Nomarski microscopy* [16–18]. Here, working with reflected light and a metal-coated sample, accuracy of ~30 Å with lateral resolution of ~1 μm can be obtained. This, of course, is not a method sensitive enough to study monolayers, but may be convenient for investigating steps in the optical thickness of a film (having a different number of monolayers).

The use of *plasmon-surface polarization* (PSP) microscopy for the study of film thickness was discussed by Rothenhausler *et al.* [19]. The sensitivity in this case is ~1 Å with lateral resolution of ~1 μm. This method, therefore, is very sensitive to lateral inhomogeneities that may result, for example, from domains with different tilt angles of the alkyl chains.

The most detailed and accurate information on thickness comes from *x-ray reflectivity* studies. Here, a highly collimated monochromatic beam is used to record peaks very close (<1°) to the total reflection angle (the critical angle of the film) [20, 21]. Recently, Wasserman *et al.* used this method to study complete and partial monolayers of alkyltrichlorosilanes on silicon surfaces [9]. This method also provides information on the complex refractive index and the roughness of the film.

The use of *x-ray standing waves* is another very accurate technique that was introduced by Nakagiri *et al.* [22]. The estimated accuracy in these experiments is ± 0.5 Å with a detection limit of $\sim 10^{-2}$ monolayer coverage of the surface (10^{13} atoms per cm^2).

Cuypers *et al.* described the first ellipsometer for the study of film thickness at the solid–liquid interface [23]. In a second paper, they studied adsorption of protein on Cr surfaces, and Cr covered with stacked monolayers of phospholipids [24]. This technique is especially valuable for the study of adsorption at the solid–liquid interface. For example, controlling the adsorption of proteins and their kinetics is very important in the development of polymeric materials for prostates, as well as for membranes for bioseparation. Studies of adsorption kinetics in real time, and with control on solution parameters, can be achieved with this technique.

The first IR ellipsometry study of oriented molecular monolayers was described by Benferhat *et al.* [25]. They looked at one to 19 monolayer films of behenic acid using *spectroscopic phase-modulated ellipsometry* (SPME). With this highly sensitive method, they observed a structural transition when the film thickness increased from one to more than nine monolayers. Moreover, they could show anisotropic orientation in single-monolayer samples. Ishino and Ishida determined the ordinary and extraordinary optical constants of LB monolayers in the IR region [26]. Based on these constants, they calculated internal and external reflection spectra of LB films on silicon and poly(methyl methacrylate) (PMMA) surfaces. Note that the advantage of using IR is mainly where the material under study absorbs the visible light.

In summary, ellipsometry is the most useful (and simple) optical technique for the determination of the thickness and refractive index of monolayer and multilayer films on solid substrates. The choice of refractive index has to be made with caution, but for films of simple alkyl chains, the value of $n_f = 1.50$ may be used. Other techniques (e.g., x-ray reflectivity) are not straightforward, and usually require a more skilful worker.

5

B. Fourier Transform Infrared (FTIR) Spectroscopy

Infrared (IR) spectroscopy is our everyday tool for the study of molecular packing and orientation in ultrathin organic films. We use FTIR in two main spectroscopy modes: *attenuated total reflection* (ATR) and *grazing-angle* (GA). Let us begin with a discussion on ATR FTIR spectroscopy.

1. Attenuated Total Reflection (ATR) Spectroscopy

A typical setup for an ATR FTIR experiment is described in Figure 1.2. A Brewster-angle rotating Ge polarizer is placed in the IR beam, in front of the *C* face of an ATR crystal that can be made of Si, Ge, or ZnSe. Separate background spectra are recorded for the *s*- and *p*-polarizations; and the crystal then is used as a substrate either for LB or SA monolayer samples. The cross-section area of the IR beam is ~1 cm^2 and thus is sufficient to cover the entire area of the *C* faces of the ATR crystal. As a result, complete coverage of faces *A* by the IR radiation is achieved. Hence, the measured absorbance should be proportional to the fraction of the total area of faces *A* and *C* covered by the monolayer. All six faces of the crystal are covered with the monolayer film. However, only the *A* faces contribute to the measured signal via internal reflection. This is because the area of the *C* faces is only ~7% of the total area (*A* + *C*) in a typical ATR crystal, and the differences between the transition mode (in the *C* faces) and the ATR mode (in the *A* faces) are not very large. Thus, it was suggested by Maoz and Sagiv that no corrections for this effect are needed [27].

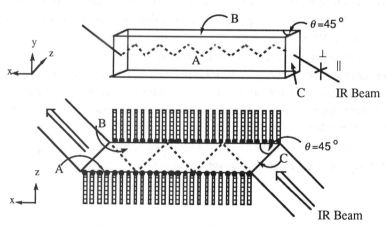

Figure 1.2. A schematic description of an optical setup for ATR measurements.

6

The theory for analysis of ATR data was developed by Harrick [28, 29], and applied by Haller and Rice [30]. The following relations were predicted by Maoz and Sagiv for the ATR experiment described in Figure 1.2 [27].

(a) For a perfect all-*trans* molecular orientation perpendicular to the surface, with all vibrations parallel to the surface of the substrate (e.g., $-CH_2-$ stretching vibrations, Figure 1.10), the following equation can be written,

$$D = \frac{A_s}{A_p} = \frac{0.42d + 5d_{e\perp y}\cos 45^0}{0.42d + 5d_{e||x}\cos 45^0}, \qquad 1.5$$

where D is the dichroic ratio, and is defined as the ratio between the absorbance recorded with perpendicular (A_s), and parallel (A_p) polarizations, respectively. Note that parallel and perpendicular polarizations are defined in respect to the xz plane in Figure 1.2. d is the thickness of the film, $d_{e\perp y}$ is the effective thickness of the film for the perpendicular polarization (electrical field parallel to the y axis in Figure 1.2), and $d_{e||x}$ is the effective thickness of the film for the x component of the parallel polarization. 0.42 and 5 are the areas (in cm^2) of faces C and A, respectively (for a 50 x 10 x 3 mm ATR crystal), and cos 45° corrects the expression given by Harrick [28] for the sampling area factor, which now is explicitly accounted for in Equation 1.5. This equation takes into account the contribution at faces A, as well as that of the direct transmission through faces C (for which $A_p = A_s$). If we insert the refractive indices for the crystal and organic film into Equation 1.5 — and N$_2$ (n_1(Ge) = 4.03, n_1(Si) = 3.42, n_1(ZnSe) = 2.42, n_2(organic film) = 1.5, and n_3(N$_2$) = 1) — we get A_s / A_p(Ge) = 1.061, A_s / A_p(Si) = 1.094, and A_s / A_p (ZnSe) = 1.245. If we now omit the contribution from faces C, we get A_s / A_p(Ge) = 1.071, A_s / A_p(Si) = 1.103, and A_s / A_p(ZnSe) = 1.260. For absorbance measured with nonpolarized IR radiation ($A = 1/2A_s + 1/2A_p$), we get A(Ge) / A(Si) / A (ZnSe) = 0.598 / 0.704 / 1.

(b) For random molecular orientation, where vibrations are distributed randomly in space, the following equation can be written:

$$D = \frac{0.42d + 5d_{e\perp y}\cos 45^0}{0.42d + 5d_{e||}\cos 45^0}, \qquad 1.6$$

where all notations are as previously described and $d_{e||} = d_{e||x} + d_{e||z}$. Using the same refractive indices, we get A_s / A_p(Ge) = 0.879, A_s / A_p(Si) = 0.897, A_s / A_p(ZnSe) = 0.969, and A(Ge) / A(Si) / A(ZnSe) = 0.584 / 0.690 / 1.

(c) For perfect molecular orientation, with all vibrations perpendicular to the surface, $A_s = 0$ and $A_p \propto d_{e||z}$. Vibrations that are perpendicular to the surface

7

are more difficult to detect by ATR. (Usually, we use grazing-angle spectroscopy for this purpose.) This is because in the ATR mode, $d_{e||z} < d_{e||x}$. For example, the A_p values predicted for transitions perpendicular to the surface are smaller than those predicted for parallel transitions of the same intrinsic intensity (transition moment dipoles) by factors of 2.22 (Ge), 2.10 (Si), and 1.67 (ZnSe). Therefore, it is recommended that ZnSe be used when possible for analysis of molecules with both parallel and perpendicular transitions, and where grazing-angle spectroscopy is not available.

Most of the monolayer and multilayer films that usually are investigated fall within the limits of cases (a) and (b). The symmetric stretching of the carboxylate group, and the symmetric stretching of the methyl group (Figure 1.6) in well-oriented monolayers, are two examples of vibrations belonging to case (c). When going from case (a) to case (c) without changing the number of molecules per unit film area, we get the following relations:

(a) $A_{s\,(oriented)} / A_{s\,(random)} = 1.50$ for all materials,
(b) $A_{p\,(oriented)} / A_{p\,(random)} = 1.223$ (Ge), 1.211 (Si), 1.153 (ZnSe),
(c) $A_{(oriented)} / A_{(random)} = 1.352$ (Ge), 1.346 (Si), 1.324 (ZnSe).

These differences in the magnitude of the measured absorbances are due to the redistribution of the transition dipoles in space, from a situation where all are in the x, y plane (e.g., $\nu_s(CH_2)$, $\nu_a(CH_2)$ in an all-*trans* perpendicular orientation, as in case (a), to a uniform distribution of the transition dipoles along the x, y, and z directions, as in case (c).

It is important to emphasize that the preceding discussion does not take into account the optical anisotropy of the oriented films [31], and assumes the same refractive index for all the samples. However, if the films have the same molecular density, such as, for the first approximation, complete monolayers [31–33], these constraints are not critical. On the other hand, these considerations may be important in partial monolayers, since significant deviations may arise when going down in surface concentration. The refractive index of the film, n_2, is a monotonic function of the molecular density; it varies from 1.00 in the limit of zero surface coverage to about 1.50 in a complete closely packed monolayer. The effective film thickness parameters, $d_{e\perp y}$ and $d_{e||x}$, are proportional to n_2, while $d_{e||z}$ is proportional to $1/n_2^3$ [28]. Therefore, measured absorbances in submonolayer coverages may not be expected to vary linearly with the surface coverage. The reader is directed to Reference 27 for more detailed discussion.

To evaluate quantitatively the orientation of vibrational modes from the dichroic ratio in molecular films, we assume a uniaxial distribution of transition dipole moments with respect to the surface normal (z axis in Figure 1.2). This assumption is reasonable for a crystalline-like, regularly ordered monolayer assembly. An alternative, although more complex, model is to assume uniaxial symmetry of transition dipole moments about the molecular axis, which itself is tilted (and uniaxially symmetric) with respect to the z axis. As monolayers become more liquid-like, this latter model may become a progressively more valid model [34, 35]. We define ϕ as the angle between the transition dipole moment M and the surface normal. (Note that $0° \leq \phi \leq 90°$.) The absorbance due to the components E_x, E_y, and E_z of the electric field of the evanescent wave [28, 36] in the ATR experiment [34] is given by

$$A_z = \frac{1}{2}M^2 E_z^2 \sin^2\phi \ , \qquad\qquad 1.7$$

$$A_x = \frac{1}{2}M^2 E_x^2 \sin^2\phi \ , \qquad\qquad 1.8$$

$$A_y = \frac{1}{2}M^2 E_y^2 \sin^2\phi \ . \qquad\qquad 1.9$$

Light polarized parallel to the incident plane (p-polarized) has components in the x and z directions, while light polarized perpendicular to the incident plane (s-polarized) has only a y component. Therefore, the dichroic ratio D is given by

$$D = \frac{A_s}{A_p} = \frac{A_y}{A_x + A_z} \ . \qquad\qquad 1.10$$

This equation is written in terms of A_s and A_p (defined earlier), omitting the contribution of the entrance and exit faces (C in Figure 1.2). The observed dichroic ratio, D_{obsd}, should, in principle, include a contribution A_T from the transmission absorbance from the entrance and exit faces, as in

$$D_{obsd} = \frac{A_s + A_T}{A_p + A_T} \ . \qquad\qquad 1.11$$

It was suggested above to neglect the contribution A_T on the basis of the small area of the entrance and exit faces [27]. We examined this suggestion in a transmission IR experiment with a monolayer of molecule 1 on Si (Part Three) [11]. This experiment produced a very weak spectrum, with an absorbance, at 2920 cm^{-1}, of only 6% of the absorbance in the ATR mode (using p-polarized light). This would affect D by less than 0.01 at values close to $D = 1.0$, and is in agreement, therefore, with setting $D \approx D_{obsd}$.

The expressions for the electric field components E_x, E_y, and E_z were evaluated by Haller and Rice [30], based on a theory developed by Harrick [28].

We inserted these expressions into Equations 1.7–1.9, and 1.10 [assuming case (a) from earlier, although it can be done as well for (b)], calculated a plot of ϕ vs. D (Figure 1.3), and estimated the unknown angle ϕ for a given monolayer from this plot.

Kopp et al. studied the rearrangement of LB layers of tripalmitin using ATR FTIR spectroscopy [37]. They reported that these LB layers undergo a spontaneous transition from a liquid–crystalline state to a microcrystalline one. This transition is accompanied by a shift of the CH_2-bending band from 1469 to 1473 cm^{-1}, and of the ester group motion from 1169 to 1182 cm^{-1}. Ohnishi et al. studied monolayers and multilayers of cadmium arachidate on glass, and reported a regular perpendicular alignment of the alkyl chains [38].

Hartstein et al. reported that thin metal overlayers or underlayers enhanced the absorption from monolayers by a factor of 20 [39]. The total enhancement, including contributions from the ATR geometry, was almost 10^4.

Okamura et al. studied monolayers of 1,2-dipalmitoyl-3-sn-phosphatidyl-choline (DPPC) on a Ge plate [40]. They reported that the alkyl chains in these layers were oriented vertically to the Ge surface, with an all-trans configuration, irrespective of the surface pressure at the film transfer. This result suggested that throughout the surface pressure examined, there were islands or surface micelles on the surface.

Kimura et al. studied LB films of stearic acid with one to nine monolayers deposited on Ge [41]. They examined the CH_2 scissoring band and suggested that the alkyl chains of the stearic acid in the first monolayer are in a hexagonal or pseudohexagonal subcell packing. These chains are perpendicular to the surface and, therefore, are free to rotate around their molecular axis.

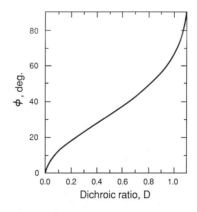

Figure 1.3. Plot of orientation angle ϕ vs. dichroic ratio D. (From Tillman et al. [11], © 1988, Am. Chem. Soc.)

Table 1.1. Infrared spectra of fluorocarbon chains. (Constructed from Naselli *et al.* [46], © 1989, Am. Inst. Phys.)

Absorption (cm^{-1})	Polarization[a]	Assignment[b]
1300	⊥	ν_a(CC), δ(CCC)
1242	⊥	ν(CF2), r(CF2)
1213	∥	δ(CCC), ν(CC)
1153	⊥	ν_s(CF2), δ(CF2)
638	∥	ω(CF2), r(CF2)
553	⊥	δ(CF2)
503	∥	ω(CF2)

[a]Indicates the direction with respect to helix axis. [b]Note the notations: ν - vibration, δ - rotation, r - rocking , and ω - wegging.

On the other hand, in LB films thicker than two monolayers, the alkyl chains in monolayers other than the first one crystallized in a monoclinic form, and were packed alternately with a tilt of ~30° with respect to the surface normal.

Maoz and Sagiv used ATR FTIR to investigate the oxidation reaction of double bonds with penetrated permanganate ions (MnO_4^-) in self-assembled monolayers [42].

Davies and Yarwood studied ATR FTIR of LB films of ω-tricosenoic acid on SiO_2/Si substrates [43]. They reported that the integrated intensity increased roughly linearly with increasing thickness, and that there was a degree of random variation in the alkyl chain tilt.

Kamata *et al.* studied the enhancement in ATR IR spectra of LB films of stearic acid overcoated with gold and silver films, and reported that for the same metal thickness, gold was more effective than silver [44]. In another paper, the same authors expanded the study to other metals and reported the order Au > Ag > Pt ≈ Cu > Ni > Al for the enhancement of the asymmetric CH_3 stretching in one stearic acid LB layer overcoated samples [45]. They explained the origin of the enhancement order in terms of the complex dielectric constant of the metal and its oxidation potential.

Recently, Naselli *et al.* studied LB films of semifluorinated fatty acid $[C_8F_{17}(CH_2)_{10}CO_2H]$ [46]. They reported that in this case, unlike its hydrocarbon analogue, the chains are inclined due to packing constraints dictated by the considerably larger cross section of the fluorocarbon chains. In their paper, they present a detailed discussion on IR spectra. We show these numbers here as a reference for LB and SA films of fluorocarbon amphiphiles (Table 1.1).

11

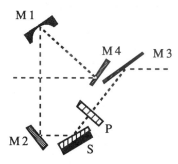

Figure 1.4. An optical arrangement for measuring polarized reflection–absorption spectra of a monolayer or thin film on the surface of a mirror S at near grazing incidence. P is a polarizer, M1, M2, M3, and M4 are mirrors. (Adapted from Ishitani *et al.* [47], © 1982, Am. Chem. Soc.)

In summary, ATR spectroscopy is an important analytical tool for the study of molecular packing and orientation in monolayers and multilayers, and should be used routinely. The reader should note, however, that the slope of the curve describing the angle between the transition dipole moment and the surface normal (ϕ) as the function of the dichroic ratio (D) is not uniform (Figure 1.3). Thus, for angles close to 90° (small chain tilts), small changes in D yield very large changes in ϕ. Therefore, it is difficult to determine accurately tilt angles of 0°–15°, and many measurements, on statistically different samples, should be made to achieve reliable orientation information.

2. Grazing-Angle Spectroscopy

Reflection–absorption (RA) or *grazing-angle* spectroscopy is a very useful technique that gives information about the direction of transition dipoles in a sample. Figure 1.4 presents an optical setup for a grazing-angle experiment [47].

Theoretical consideration of the IR spectroscopy of monolayers adsorbed on a metal surface showed that the reflection–absorption spectrum is measured most efficiently at high angles of incidence, and that only the component of incident light that is parallel to the plane of incidence gives measurable absorption [48]. Figure 1.5 presents a schematic description of a monomolecular film on a mirror, with the incident light and direction of the polarization, and Figure 1.6 presents, in detail, an alkyl thiol molecule on a metal surface.

Note the direction of the different transition dipoles. Thus, while both the symmetric and asymmetric methylene vibrations (v_s and v_a, respectively), are parallel to the metal plane, the symmetric vibration and both asymmetric vibrations of the methyl group have components that are perpendicular to the surface. Therefore, the methylene groups in a perpendicular, all-*trans* alkyl chain will not be picked up by the *p*-polarized light. Of course, once the alkyl chain tilts from the normal to the plane, the symmetric and asymmetric vibrations of the methylene groups no longer are parallel to the surface and thus will appear in the grazing-angle spectra.

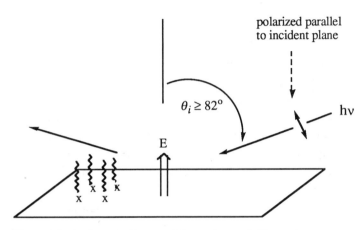

Figure 1.5. A schematic diagram of the grazing-angle IR experiments.

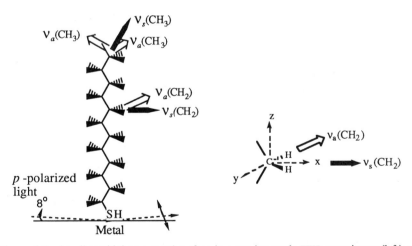

Figure 1.6. An alkanethiol on a metal surface in a grazing-angle FTIR experiment (left), and an in-plane diagram of the CH_2 group and its transition dipoles (right). Note that the CH_2-plane is parallel to that of the substrate, and that both $v_s(CH_2)$ and $v_a(CH_2)$ are in that plane, where $v_s(CH_2)$ and $v_a(CH_2)$ are orthogonal to each other.

13

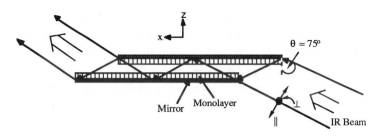

Figure 1.7. An optical setup for the study of monolayers in a reflection–absorption mode. (Adapted from Gaines [6], © 1965, Wiley Interscience.)

The intensity of the methylene vibrations in the spectra is a direct function of the tilt angle of the alkyl chain. This unique property of the grazing-angle experiment allows the calculation of molecular orientation from FTIR spectra.

A modification of the simple grazing-angle experiment is *multiple reflection – absorption spectroscopy*, which was described first by Gaines (Figure 1.7) [6]. It has a unique value for the detection of order–disorder transformations in the structure of a monolayer, resulting from its interaction with an external physical or chemical agent.

Two parallel Al–glass mirrors, separated by a 3-mm-thick, frame-shaped spacer are used in this experiment. These are then mounted on an ATR attachment, with the mirrors set at a 75° angle of incident. At this geometry, most of the incident beam radiation is used in the measurement. The signal can be enhance by using only the parallel polarization (∥). The multiple reflections make this technique more sensitive than the regular grazing-angle, which is only a one-reflection experiment.

A pioneering work in the area of grazing-angle FTIR spectroscopy in monolayers was made by Allara and Swalen [49]. The characterization of alkanethiol monolayers on gold by grazing-angle FTIR was made by Porter *et al.* [12]. They carried out a very detailed study of the structure as a function of the chain length. A part of Table I in their paper is presented in Table 1.2. Figure 1.8 presents the C–H stretching region of the IR spectrum of an SA monolayer of $CH_3(CH_2)_{21}SH$ on gold with the peak assignments [50–54]. The assignment of the bands in the C–H stretching region is important, since these peaks appear in any IR spectra regardless of the assembling technique. The band at 2965 cm^{-1} is the asymmetric in-plane C–H stretching mode of the CH_3 group, $v_a(CH_3,ip)$. The bands at 2973 and 2879 cm^{-1} are the symmetric C–H stretching modes of the CH_3 group, $v_s(CH_3, FR)$. The abbreviation FR denotes the Fermi resonance

14

interaction, which causes the splitting of this band [52]. The bands at 2917 and 2850 cm^{-1} are the asymmetric (v_a) and symmetric (v_s) C–H stretching modes of the CH$_2$ group. Porter *et al.* used $v_a(CH_3)$ and $v_s(CH_2)$ for structural interpretation simply because there is minimal overlap between them. A careful inspection of the data in Table 1.2 and Figure 1.8 suggests that (*a*) the formation of the two-dimensional assembly does not alter the peak positions as compared to the bulk spectra, and (*b*) both peak positions and intensities are the function of the alkyl chain length n.

Table 1.2. Peak positions for CH$_3$–(CH$_2$)$_n$–SH C–H stretching modes in crystalline and liquid states and adsorbed on gold. (A part of Table I from Porter *et al.* [12], © 1987, Am. Chem. Soc.)

Structural Group	C–H Stretching Mode	Peak Position [a] for Crystalline and Liquid State, cm^{-1}		peak positions [a]			
		Crystalline [b]	Liquid [c]	$n = 21$	$n = 15$	$n = 9$	$n = 5$
–CH$_2$–	v_a	2918	2924	2918	2918	2920	2921
	v_s	2851	2855	2850	2850	2851	2852
CH$_3$–	v_a(ip)	*d*	*d*	2965	2965	2966	2966
	v_a(op)	2956	2957	*e*	*e*	*e*	*e*
	v_s(FR)	*f*	*f*	2937	2938	2938	2939
	v_s(FR)	*f*	*f*	2879	2879	2878	2878

a. ±1 cm^{-1}. *b*. Crystalline-state positions determined for CH$_3$–(CH$_2$)$_{21}$–SH in KBr. *c*. Liquid-state positions determined for CH$_3$(CH$_2$)$_7$SH. *d*. The v_a(ip) is masked by the strong v_a(op) in the crystalline- and liquid-state spectra. *e*. The position of v_a(op) could not be determined because of low signal-to-noise ratio. This is the result of the orientation of this mode with respect to the surface. *f*. Both v_s(Fr) bands are masked by the strong v_a(CH$_2$) band.

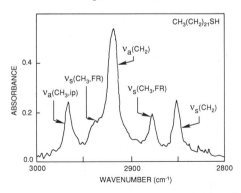

Figure 1.8. The C–H region in the IR spectrum of CH$_3$–(CH$_2$)$_{21}$–SH on gold. (From Porter *et al.* [12], © 1987, Am. Chem. Soc.)

15

It is useful, at this point, to refer to the work of Snyder *et al.* [50, 51]. The trend toward higher peak frequencies as the length of the alkyl chain decreases in the monolayer is in agreement with their findings for polyethylene. Thus, they reported 2920 cm^{-1} for the $v_a(CH_2)$ mode in crystalline polyethylene, which is 8 cm^{-1} lower than the peak position in the liquid state (2928 cm^{-1}). They also reported 2850 cm^{-1} for $v_s(CH_2)$ in crystalline polyethylene, which is 6 cm^{-1} lower than the peak value in the liquid state (2856 cm^{-1}). The same trend in peak position is reported in Table 1.4 for the solid and liquid phases, where changing from the solid to the liquid phase shifts the aforementioned modes by +6 and +4 cm^{-1}, respectively. Let us compare the frequencies for the monolayers with alkyl chains ($n > 15$). Here, $v_a(CH_2) = 2918$ cm^{-1}, the same frequency as in the KBr spectrum of the solid $CH_3(CH_2)_{21}SH$. On the other hand, this mode appeared at 2921 cm^{-1} in the $CH_3(CH_2)_5SH$ ($n = 5$) monolayer, approaching 2924 cm^{-1}, which is the value in the pure liquid ($n = 7$). Thus, it is recommended that the FTIR spectrum of a monolayer be compared with the spectrum of the pure material in solution (or neat for a liquid), and in KBr (for a solid), to establish the range of frequencies for the different modes. This is especially true where the molecule exhibits a richer spectrum due to different functionalities. Such an FTIR study may help to establish, for example, that there are different degrees of packing (e.g., liquid-like and solid-like) in different parts of the monolayer, due to different types of interactions between the different functional groups.

Golden *et al.* studied the grazing-angle spectra of LB and self-assembled arachidate monolayers on aluminum surfaces [54]. Rabolt *et al.* studied cadmium arachidate monolayers on AgBr for transmission, and on Ag for grazing-angle experiments. They found that, independent of the substrate, the chains of the fatty acid salt are oriented within a few degrees of the normal to the surface of the substrate [55].

Bonnerot *et al.* studied LB layers of docosanoic and ω-tricosenoic acids on aluminum surfaces [56]. They observed a structural transition when the thickness increased from one to > 7 layers. The hexagonal subcell became orthorhombic, and the axis of the chain, perpendicular to the substrate for the first layer, tilted progressively to reach a final limit of 23° for behenic acid, and 18° for ω-tricosenoic acid.

Naselli *et al.* used grazing-angle FTIR spectroscopy to study thermally induced order–disorder transitions in LB films of cadmium arachidate [57]. They suggested a pretransitional disordering prior to melting. The introduction of an aromatic group into the molecule (e.g., tetradecylbenzoic acid) improved the high temperature stability of the films, probably due to enhanced intermolecular interactions between the head groups.

16

Hallmark *et al.* compared self-assembled monolayers of octadecyltrichloro-silane ($C_{18}H_{37}SiCl_3$, OTS) and LB monolayers of cadmium arachidate, and concluded that the chain axis in the OTS monolayer has some degree of tilt, while that in the arachidate salt monolayer is oriented normal to the substrate [58].

Dote and Mowery studied the orientation of LB monolayers of stearic acid and perdeuterated stearic acid [$CD_3(CD_2)_{16}COOH$] on gold and native-oxide aluminum (Al_2O_3/Al) surfaces [59]. This is an important work, since it was observed that monolayers on aluminum aged with time and realigned to a more perpendicular orientation. In another paper, the authors assigned bands in deuterated alkyl chains [60]. The $v_a(CD_2)$ and $v_s(CD_2)$ appear at 2194 and 2086 cm^{-1}, respectively, while the $v_a(CD_3)$ appears at 2212 and 2221 cm^{-1} for the in-plane and out-of-plane modes, respectively, and at 2103 and 2076 cm^{-1} appear $v_s(CD_2)$, and $v_s(CD_3)$, respectively.

In summarizing the discussion on FTIR spectroscopy, we note that both ATR and GA measurements are straightforward and can be carried out after relatively little training. Today's instruments are menu-driven, and have excellent laser stability and the sensitivity required for studies of partial monolayers. Data analysis is done with the aid of advanced software, which does baseline corrections, mathematical subtraction of spectra, etc. From the author's experience, it seems that a vacuum-bench FTIR spectrometer should be the preferred instrument. This is because purging the sample compartment with dry nitrogen may take as long as two hours, and during this time the instrument is idle. There are some who use mathematical subtraction of water and CO_2 spectra (or parts thereof) to avoid long purging time. This is not a good practice and should be avoided.

It usually is accurate enough to use 4 cm^{-1} resolution and to collect 2000 scans for routine studies of single monolayers. However, one should use, if possible, 2 cm^{-1} resolution and a good signal-to-noise ratio when calculating molecular packing and orientation.

In conclusion, grazing-angle FTIR spectroscopy is a useful technique that gives information about the direction of transition dipoles in an organic thin film. Note that small changes in the angle of incidence result in large changes in the IR spectra. Thus, a comparison between samples can be done only if their spectra were recorded using the same angle of incidence.

C. Raman Spectroscopy

Resonance Raman and IR spectroscopies have the same set of selection rules at the surface; i.e., transition dipoles that are parallel to the surface are not detected by *p*-polarized light. However, while symmetrical bonds (e.g., $N = N$) appear very weak in IR, they are detected easily by resonance Raman.

The first account of Raman spectroscopy in multilayer films is that of Cipriani *et al.*, who studied LB multilayers of barium stearate in a total reflection mode [61]. Lis *et al.* studied Raman spectra of LB multilayers of behenic acid, barium behenate, and barium *cis*-13-erucate [62]. They compared the ratio I_{2890}/I_{2850} (methylene C–H stretching) of these multilayers with that of samples of crystalline hydrocarbon chains, and found that fatty acid multilayers are less ordered than the crystalline sample.

Surface-enhanced Raman scattering (SERS) first was observed for samples of pyridine adsorbed on roughened silver surfaces [63–65]. Sandroff and Herschbach studied SERS of organic sulfides adsorbed on silver [66]. They reported that room-temperature adsorption of disulfides and sulfides (RSSR and RSR, respectively, where R = C_6H_5— and C_6H_5–CH_2—), decompose on the silver surface to give RS^-–Ag^+, thus implying facile cleavage of S–S and C–S bonds. From ring-stretching frequencies, they concluded that while phenyl species are lying flat on the surface, benzyl species are sticking up. Note that the surfaces used in these studies were rough, and therefore, conclusions about molecular orientation should be considered with caution.

Rabolt *et al.* demonstrated for the first time the use of integrated optics in Raman scattering studies [67, 68]. Here, the material was made into an asymmetric slab waveguide, or a composite waveguide structure in which both the optical field intensity of the in-coupled laser source and the scattering volume of the sample were increased significantly. Using this technique, they obtained Raman spectrum of a single-dye monolayer (27 Å).

Sandroff *et al.* observed an SERS spectrum from 1-hexadecanethiol ($C_{16}H_{33}SH$) adsorbed on silver island films [69]. This monolayer has a solid-like all-*trans* confirmation; however, on contact with chloroform, there was a decrease in the number of *trans* bonds (i.e., an increase in the number of *gauche* bonds). This is an important experiment, since it demonstrates the fact that monolayers can rearrange, thus decreasing both bulk and surface order. SERS spectra of benzenethiol and propane thiol (C_6H_5SH and C_3H_7SH, respectively) were reported by Joo *et al.* [70]. In both cases, the results indicated that the thiol was chemisorbed dissociatively on the silver surface by rupture of the S–H bond, thus forming the thiolate salt.

Evidence for the electromagnetic mechanism in SERS is given by Kovacs *et al.*, where LB films of arachidic acid were used as a spacer layer to control the separation distance between evaporated indium and silver metal island films, and LB monolayers of substituted phthalocyanine [71]. Their results rule out the short-range chemical mechanism, i.e., an interaction between metal atoms and the phthalocyanine molecules. Cotton *et al.* looked at SERS spectra of dye-containing monolayers [72]; These monolayers were spaced from the surface by accurately known increments by deposition of inert spacer monolayers. They observed SERS spectra from monolayers spaced ~160 Å from the silver surface, thus concluding that in this case, the enhancement is due purely to electromagnetic mechanism.

Evans *et al.* studied SERS in a different way [73]. They adsorbed alkanethiol derivatives (see preceding) on hydrophilic gold surfaces and evaporated thin (~60 Å) silver films on top of the monolayers. A marked difference could be observed as the length of the alkyl chain "above" the sulfone group increased from C_4 to C_{12}. Thus, while for the former, good enhancement was observed, in the latter, it was hardly possible to recognize a spectrum in the noise. This is clear evidence for a short-range chemical mechanism, where the silver atoms can penetrate the liquid-like, disordered C_4, but not the crystalline, ordered C_{12} chain.

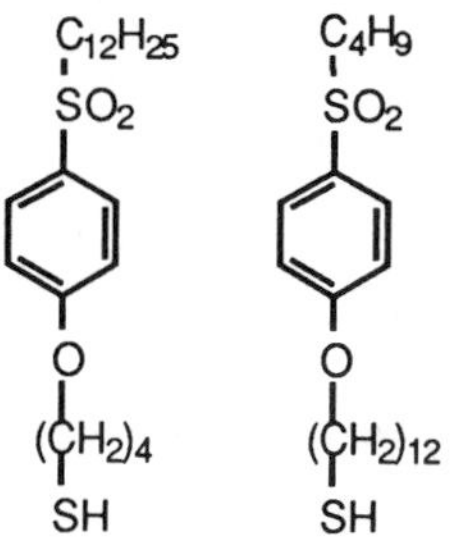

The question is, how was a spectrum of a dye spaced ~160 Å from a silver surface enhanced, while that of the alkoxyphenylsulfonyl derivative spaced ~16 Å from the silver film was not? A possible answer to this problem is given in the mechanism of the *electromagnetic enhancement theory* [74, 75]. This theory predicts that the influence of the optical properties of the surrounding medium on the signal intensity depend on adsorbate surface coverage [76, 77]. Moreover, it is predicted that the response of the system can be quite complex when the electronic states of the adsorbate are close to those of the substrate (resonance) [76, 78].

Recently, Kim *et al.* studied SERS spectra in monolayers of the cyanine dye (S-120) [79]. They found that the maximum SERS enhancement was observed at

500 nm, close to the Ag surface plasmon resonance. The electronic spectra of S–120 both in methanol solution and in monolayers show peaks at this region. Thus, interaction between the surface plasmon and the adsorbate molecules is possible. On the other hand, the alkoxyphenylsulfonyl group absorbs in the uv, and no such interaction is possible.

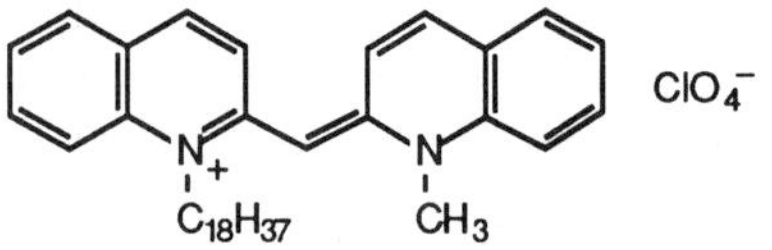

1-methyl-1'-octadecyl-2,2'-cyanine perchlorate (S-120)

Note that SERS in not useful for studies of molecular orientation, since different bonds interact with silver atoms (electromagnetically or chemically) in a different way, and thus, their Raman absorptions can not provide quantitative analyses of tilt angles or *trans*-to-*gauche* isomerization. On the other hand, SERS is useful for studies of packing effects on peak positions, and for investigations of electron transfer and other reactions in thin organic films.

Schlotter *et al.* published the first papers on Raman spectroscopy of LB monolayers [80–82]. They looked at LB films of cadmium stearate and cadmium arachidate with thicknesses from one to 27 layers, using an intensified optical multichannel array detector. The two main peaks they detected are at 2840 cm^{-1} and 2880 cm^{-1}, corresponding to $v_s(CH_2)$ and $v_a(CH_2)$, respectively. They reported significant disruption of lateral packing and conformational order in the first few layers.

FT Raman spectroscopy recently has been applied by Zimba *et al.* for the study of organic thin films and dye-containing monolayers [83]. The advantage of this technique is the lack of resonance enhancement, and therefore, the possible immediate interpretation of the results in terms of molecular orientation.

Rothenhäusler *et al.* introduced plasmon surface polariton field-enhanced Raman spectroscopy [19, 84–89]. A *plasmon surface polariton* (PSP) is a coupled excitation state of a surface plasmon and a photon at a metal–dielectric interface. Since the surface plasmon does not interact with the adsorbed molecules, but with the *p*-polarized light, it does not belong to the same resonance enhancement phenomena as SERS. Therefore, it is possible to use PSP Raman scattering spectra to get information on molecular orientation in monolayer assemblies. Figure 1.9 presents PSP Raman spectra in the C–H stretching range of cadmium arachidate (CdA) LB multilayer assemblies [19].

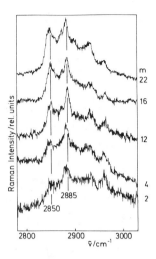

Figure 1.9. PSP field-enhanced Raman spectra in the C–H scattering range of CdA LB multilayers of varying thicknesses. The number of monolayers deposited are denoted by *m*. The two prominent peaks at 2850 cm^{-1} and 2885 cm^{-1} are marked. (From Rothenhäusler *et al.* [89], © 1988, Elsevier Science Publishers.)

In general, Raman spectroscopy does not require expertise much different than FTIR, but in practice, it is done by spectroscopists. Further development of Raman spectroscopy as a tool for quantitative analysis of molecular orientation (e.g., *trans–gauche* isomerization) will contribute to the understanding of self-assembly processes, order–disorder transitions, and other molecular events in thin organic films.

D. Surface Potential

The *surface potential* or *Volta potential differences* method is an important tool for analysis of the electrical structure of surfaces, i.e., the distribution of polar vs. nonpolar groups (For example, OH vs. CH_3 groups, respectively). Let us divide the discussion into the measurements of surface potential on the water–air interface, and then solid substrates.

On the water–air interface, we measure the Volta potential between the surface of a liquid and that of a metal probe immersed in the liquid. Figure 1.10 presents one available procedure. This potential is influenced by the pressure of the monolayer film on the water surface. Thus, the practice usually is to measure the difference between the potential of the clean surface and that of a monolayer-covered subphase (ΔV), rather than try to determine a single-surface potential. The potential difference is given by the Helmholtz formula,

$$\Delta V = \frac{4\pi\sigma d}{D} ,$$

1.12

where σ is the charge density, d is the distance between the electrodes (the thickness of the film), and D is the dielectric constant. A much simpler way to illustrate the situation is presented in Figure 1.11.

Let us assume that in this distance, there are polar molecules, i.e., molecules with effective dipole moment μ. Now, if there are N molecules per cm^2, then $\sigma d = Ned = N\mu$, and if we assume that $D = 1$ and that $\mu = \mu \cos\theta$, where θ is the tilt angle of the molecular dipole from the normal, we get

$$\Delta V = \frac{4\pi\mu N}{D} = 4\pi\mu N \cos\theta .$$

1.13

We note that surface potential obtained by this method is rather empirical, since it is not known if the structure of water layers near the monolayer is not altered by the polar hydrophilic groups (i.e., if the potential of the clean water surface is a true reference) [90].

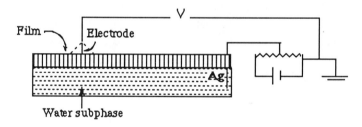

Figure 1.10. An electrode method for measuring surface potential; V, voltmeter.

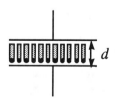

Figure 1.11. An organic monolayer between two metal surfaces.

22

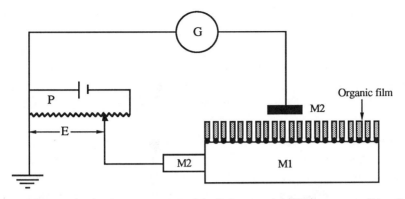

Figure 1.12. A setup for the measurement of the Volta potential difference on a solid surface. G is a galvanometer, P is a potentiometer, M1 is the substrate metal, M2 is the electrode.

In the case of a monolayer on a substrate we use a slightly different setup, which is illustrated in Figure 1.12 [91]. Here we use a metal as the substrate (M1) and another metal (M2) as the electrode, which is a very short distance from the surface. A piece of M2 is also connected to the substrate. The potentiometer is adjusted so that no current flows in the system. In this arrangement, we get

$$E_0 = \Delta \Psi_{M1,M2} ,\qquad\qquad 1.14$$

where $\Delta \Psi_{M1,M2}$ is the difference in the Volta potential between M1 and M2. The reader may find a more detailed discussion on this system in References 92 and 93. If the clean M1 is replaced by M1 coated with a monolayer, and the same procedure is repeated, a new value is established on the potentiometer,

$$E_1 = \Psi_{M1(monolayer)} - \Psi_{M2} .\qquad\qquad 1.15$$

The difference between these two values is the change in surface potential,

$$\Delta V = E_1 - E_0 .\qquad\qquad 1.16$$

We have seen in Equation 1.13 that the surface potential basically is the sum of the molecular dipoles. This is correct especially in organic films, where there is no ionic double layer to contribute to the surface potential. Thus, if a surface reaction changes the molecular dipole, the change in the surface potential may provide information on the change in molecular dipole as a result of this reaction.

$$\Delta V = N \Delta \mu.\qquad\qquad 1.17$$

There is a large body of literature on surface potential. Starting in the mid-1960s, surface potential measurements were used to investigate phospholipid monolayers [94–120]; the interaction of such monolayers with ions and charged macromolecules [121–132]; monolayers of fatty acids; and sulphate, sulfonate, phosphate, and phosphonate derivatives [133–151]. Surface potential studies of polymeric monolayers [152–154], nonionized systems (e.g., long-chain alcohols) [155, 156], and polypeptides [167–161] also were carried out. Other systems that were studied are monolayers of porphyrins [162, 163], ionophores [164–167], and enzymes [168].

Frommer et al. studied the interaction of DNA with monolayers using surface potential [169], and Sahay et al. studied $KMnO_4$ oxidation of cyanine dyes at the water–air interface [170]. Ter-Minassian-Saraga et al. studied elastic and viscoelastic monolayers [171], and Dragcevic developed a technique for simultaneous measurements of dynamic surface tension and surface potential of films at the water–air interface [172]. Finally, Oliveira et al. published results on the estimation of group dipole moments from surface potential measurements of monolayers containing sulfoxide groups [173]. In this work, they reevaluated the model suggested previously by Demchak and Fort [155], and report that this model may not be appropriate for monolayers with long aliphatic chains. They suggested new values for local permittivities that are more applicable for such compounds.

Evans et al. studied surface potential as the function of alkyl-chain length in alkanethiol monolayers on gold [174]. They envisaged the monolayer as a dipole sheet (Figure 1.13), with a sheet of negative charges residing close to the metal–monolayer interface, and a sheet of positive charges closer to the monolayer–air interface. The direction of the polarization (dipole sheet) was inferred from the positive sign of the change in the surface potential with respect to the clean gold reference electrode. Further conformation of this assignment was obtained by looking at a monolayer in which the upper part of the alkyl chain was perfluorinated [73]. This yielded a large negative surface potential value due to the high electronegativity of the fluorine atoms.

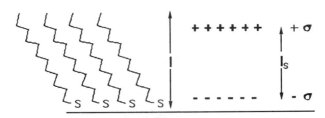

Figure 1.13. The representation of an alkanethiol monolayer as a dipole sheet.

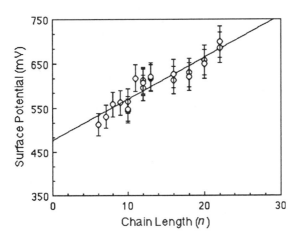

Figure 1.14. The variation of surface potential with chain length (n) in alkanethiol monolayers on gold. (From Evans and Ulman [174], © 1990, Elsevier Science Publishers.)

In molecular terms, this implies that the effective R^+–S^- dipole (where R = C_nH_{2n+1}) must be larger than the Au^+–S^- dipole, which seems reasonable since the Au^+ could be screened within a very short distance by the electrons within the metal, whereas the charges in the monolayer cannot.

If σ (Figure 1.13) represents the charge per unit area, and l_s the separation of the charged planes, then the surface potential, ΔV, is directly proportional to σl_s. Assuming that l_s varies linearly with increasing chain length n, then one would also expect ΔV to vary linearly with n. Note that l_s is not necessarily the same as the monolayer thickness l, though they both are expected to vary linearly with n. In actuality, however, one would expect the surface potential to exhibit a gradual decrease in its dependence on chain length and reach some limiting value, beyond which it would be unaffected by further increases in chain length.

This saturation-type effect should occur when the number of methylene units in the molecule is sufficient to effectively screen, by induced dipole effects, the net charge on the sulfur. It is evident from Figure 1.14 that ΔV does vary linearly with n. A straight-line fit to the data yields a change in ΔV of 9.3 mV per CH_2 unit, which is in close agreement with the average value of 9.0 mV per CH_2 unit reported by Kuchhal *et al.* for monolayers of straight-chain alcohols at the water–air interface (25°C, $\pi = 40$ mN/m) [175].

However, Harkins and Fischer found a variation of 22 mV per CH_2 unit for fatty acid monolayers at the water–air interface (25°C, area per molecule = 22 Å²) [176], a value that is approximately twice as large as that found by Evans *et al.*

25

(and may be associated with the fact that there are two oxygens per molecule contributing to the net polarization). It would seem apparent from the data that the number of carbons in the aliphatic chains in this study was not sufficient to see the suggested saturation effects.

It is of interest to determine what information about the monolayers can be gained from such surface potential studies. Perhaps the most obvious is that one gains knowledge about the size and direction of the net dipole of the monolayer, as indicated earlier. We already have mentioned that in monolayers of alkanethiols on gold, the net dipole moment, perpendicular to the monolayer surface, is oriented such that its positive end is toward the monolayer–air interface and its negative end is close to the gold surface. In addition, it has been shown that the net perpendicular component of the dipole moment increased linearly with increasing n. It has been suggested in the literature that the net dipole can be considered to be comprised of several parts, which may be combined in an additive manner,

$$\mu_{net}/\varepsilon_{net} = (\mu_1/\varepsilon_1) + (\mu_2/\varepsilon_2) , \qquad 1.18$$

where $\mu_{net}/\varepsilon_{net}$ represents the net *apparent* dipole moment for the monolayer, (μ_1/ε_1) represents the *apparent* dipole of the head-group region, and (μ_2/ε_2) is that of the hydrocarbon tail region. See Figure 1.13 [155, 173, 177]. The ε values represent the permittivity or dielectric values in the various regions. Assuming that all the films exhibit the same uniform coverage, then the first term on the right of Equation 1.18, (μ_1/ε_1), should be constant for all monolayers and can be equated with the intercept value of $\mu_{net}/\varepsilon_{net}$ for $n = 0$. Thus, the variation of 1.18 with n can be equated with either μ_2 or ε_2, or both, varying as the carbon number n is increased. From the molecular point of view, it would seem apparent that μ_2 should remain constant, since each additional methylene unit should not contribute any additional net dipole moment to that already existing in the molecule. Equation 1.18 may thus be rewritten as

$$\varepsilon_2 = \mu_2/[(\mu_{net}/\varepsilon_{net}) - (\mu_1/\varepsilon_1)] . \qquad 1.19$$

For simplicity, Evans *et al.* let μ_2 take the value 0.288 which corresponds to that for the $n = 22$ molecule, assuming $\varepsilon_2 = 2.25$, i.e., the bulk value for close-packed methylene chains. Keeping μ_2 constant, they found that the value of ε_2 varied greatly with the length of the hydrocarbon chain (Figure 1.15).

In the limit of $n \rightarrow \infty$, ε_2 approaches the bulk value for the permittivity of a close-packed methylene chain. Such a variation of ε_2 with n often is ignored, especially in determinations of monolayer thicknesses using ellipsometry. That such a variation is intuitively correct can be seen from the evaluation of ε from the Clausius–Mossotti relationship, which though not strictly valid for such a monolayer arrangement, does display the correct qualitative dependence of ε on

$$\frac{\varepsilon - 1}{\varepsilon + 2} = \frac{4\pi\alpha}{3v} .$$ 1.20

The normalized values of ε determined from Equation 1.20, using calculated values for the polarizabilities for a thiol, and from the experimental values ε_2, are shown in Figure 1.15. It is clear that the n dependence of ε_2 is, at least qualitatively, as expected. Thus, the variation of the *apparent* dipole moment with n may be associated with the changing nature of the dielectric or, in molecular terms, the increase in the molecular polarizability of the molecules, with increasing chain length. The fact that the permittivity of the monolayer may change so significantly for short alkane chains must be kept in mind when estimating thicknesses of these monolayers using techniques such as ellipsometry. There, the input of the refractive index (n_f) of the monolayer is required, and since $\varepsilon \approx n_f^2$, using the same value of n_f for short ($n < 10$), and for long ($n > 10$) alkyl chains may result in measurements that deviate from the true film thickness. (See also the discussion earlier on thickness measurements.) As the permittivity of the monolayer approaches that for a bulk alkane, surface potential should become independent of chain length, and the bulk n_f value can be used.

Another interesting fact is that (μ_1 / ε_1) does not equal zero for $n = 0$, which implies that there must be some net polarity within the head-group region; i.e., the gold–sulfur bond cannot be purely covalent. The sign of the intercept implies that the residual moment is acting in the same direction as that found in the monolayer cases; however, without further data for the short alkyl chain region, where $n < 6$, such an extrapolation may be unwarranted. In principle, knowing the intercept value should allow one to determine the net charge associated with each sulfur atom,

$$|q| = |\mu_l| / l ,$$ 1.21

where l is the charge separation in the dipole. To compute $|q|$, one must assume a knowledge of l and ε_1, since the intercept equals (μ_1 / ε_1). Taking ε_1 to be ~ 6.4 and $l = 2$ Å, they estimated that the size of the residual charge on the sulfur is approximately one sixth of electronic charge (i.e., $\sim 0.26 \times 10^{-19}$ C). Again, although such values should be treated with a little caution, they do indicate that the gold–sulfur bond is not of a purely covalent or purely ionic nature, and can be described best as polar covalent.

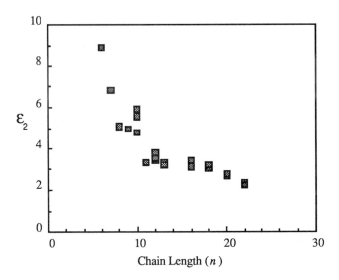

Figure 1.15. The variation of the dielectric constant of the hydrocarbon portion of a monolayer, as a function of chain length. Calculated assuming that for $n = 22$, ε_2 approaches the bulk value for an alkane chain and that μ_2 is constant. (From Evans and Ulman [174], © 1990, Elsevier Science Publishers.)

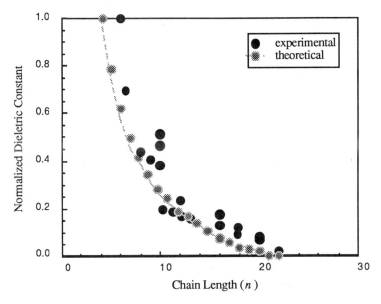

Figure 1.16. Comparison between the variation of ε_2 found experimentally to that calculated theoretically using Equation 1.20. Both sets of results are normalized. (From Evans and Ulman [174], © 1990, Elsevier Science Publishers.)

E. Surface Viscosity

The rheology of LB films has a very important effect on their transfer from the water–air interface to a solid substrate. Surface viscosity is a function of the molecular weight of the substrate, and its packing and order in the two-dimensional film. Thus, as the film becomes more close-packed and crystalline, its viscosity increases and it becomes more difficult for it to be transferred onto a solid substrate. Therefore, surface viscosity measurements of LB films will help to establish a protocol for the transfer process. A schematic description of a damped oscillator for the measurement of surface viscosity is presented in Figure 1.17.

In this experiment, the successive amplitude of the damped oscillator with a vane of length l (l is the perimeter), and a disc with a moment of inertia (I) on an LB film is measured, and its $\log_{10}$ is denoted as λ. λ_0 is the same parameter, but for a clean surface. τ is the period of the oscillation with the monolayer, while τ_0 is the same parameter for a clean surface.

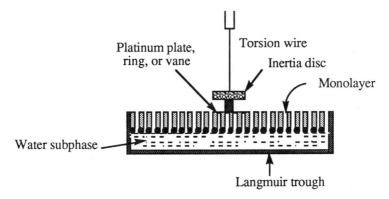

Figure 1.17. A study of surface viscosity by the damped oscillator method.

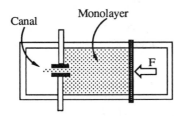

Figure 1.18. A top view of the canal-type viscometer.

The surface viscosity η_s can be calculated according to

$$\eta^s = \frac{9.2l}{l^2}\left(\frac{\lambda}{\tau} - \frac{\lambda_0}{\tau_0}\right). \qquad\qquad 1.22$$

The canal-type viscometer provides a method for surface viscosity measurement [178–180]. Figure 1.18 presents a top view of this viscometer. Myers and Harkins developed the following equation for the surface viscosity in this case [181],

$$\eta^s = \frac{\Delta\gamma a^3}{12l\,(dA\,/dt\,)}, \qquad\qquad 1.23$$

where $\Delta\gamma$ is the applied surface pressure, a is the width and l the length of the channel, and dA/dt is the rate of emergence of the monolayer from the channel. However, Harkins and Kirkwood later added a term that takes into account the dragging of a subphase of viscosity η [182],

$$\eta^s = \frac{\Delta\gamma a^3}{12l\,(dA\,/dt\,)} - \frac{a\eta}{\pi}. \qquad\qquad 1.24$$

Values of surface viscosities are of the order of 10^{-2} to 10^{-4} g/sec or surface poises (sp). For further discussion, the reader is referred to Adamson's book [183].

Two interesting studies of surface viscosity are those of Malcolm [184,185] and Daniel and Hart [186]. Without getting into the details of their experiments, the general conclusion of these studies, which used completely different materials (synthetic polypeptides in the α-helix conformation and ω-tricosenoic acid, respectively), is the same. That is, there is a deformation of the film in the middle due to the shear forces, while at the two ends — at the fixed and at the moveable barriers — there is no deformation (Figure 1.19). These results emphasize how sensitive the deposition process may be to the two-dimensional film rheology. It also suggests that enough relaxation time should be allowed before the actual deposition takes place, and that deposition should be carried out near the fixed barrier.

Buhaenko *et al.* studied the influence of shear viscosity (η^s) on maximum possible deposition rate (V_{max}) as a function of chain length in fatty acids, and in two isomers of tricosenoic acid [187]. Figure 1.20 shows the shear viscosity η^s measured at $\pi \approx 10$ mN m^{-1} for n-alkanoic acids ($C_{n-1}H_{2n-1}COOH$) as a function of n. In both cases, there is a rapid increase in surface viscosity with the increasing number of carbon atoms in the alkyl chain, with no apparent odd–even effect. Figure 1.21 presents their results for the compression moduli

κ_2 (defined as

$$\kappa_2 = -A(\partial \pi / \partial A)_T ,\qquad\qquad 1.25$$

and measured at ≈ 30 mN m^{-1}).

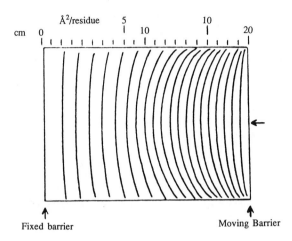

Figure 1.19. Influence of shear forces on compressed Langmuir film. (From Malcolm [184], © 1985, Academic Press.)

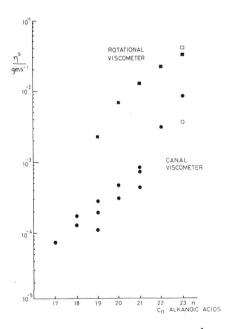

Figure 1.20. Shear viscosity η^S measured at $\pi \sim 10$ mN m^{-1} for n-alkanoic acids as a function of n. (From Buhaenko *et al.* [187], © 1985, Elsevier Science Publishers.)

31

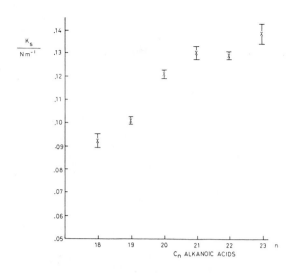

$$\frac{\kappa_s}{N\,m^{-1}}$$

Figure 1.21. Compression moduli κ_2 for n-alkanoic acids ($C_{n-1}H_{2n-1}COOH$), as a function of n. (From Buhaenko *et al.* [187], © 1985, Elsevier Science Publishers.)

These results show an increase in κ_2 with the increase in chain length. The authors of that study suggest that this trend indicates an increase in intermolecular repulsive interactions with the increasing number of carbon atoms. This explanation is not reasonable, since the van der Waals interactions (or hydrophobic bonding) increase with the increasing molecular weight, which explains why the two-dimensional order and crystallinity increase with the increasing number of carbon atoms in the alkyl chain.

Recently, Halperin *et al.* published a study of the mechanical behavior of surface monolayers using orthogonal Wilhelmy plates [188]. They report that both octadecanoic acid (C_{18}) and tetracosanoic acid (C_{24}) on water deform as elastic solids at $\pi \sim 30$ mN m^{-1}, with different relaxation times attributed by the authors to different water solubility of the C_{18} and C_{24} acids. No difference in behavior was detected at low surface-pressure ($\pi < 20$ dyn/cm), and in this region, no relaxation was detected. Therefore, since Buhaenko *et al.* have measured η^s at ~ 10 mN m^{-1}, it is safe to suggest that any comparison between η^s and V_{max} differences in their η^s values is simply a result of differences in the molecular weight of the different substrates. On the other hand, they report their V_{max} results at ~ 30 mN m^{-1}, where a difference in relaxation time was observed. These results show that water entrainment is a function of the water solubility of the substrate, and not of its head group only.

The information on surface rheology in the literature is mainly on simple compounds, probably due to the fact that surface rheology is still not in the mainstream of rheology. However, there is no doubt that a better understanding of molecular structure–surface rheology relationships is essential for the development of successful transfer conditions that lead to reproducible LB film properties.

F. Electron Paramagnetic Resonance (EPR)

Electron Paramagnetic Resonance (EPR) is a very sensitive technique that allows detection of a very small number of unpaired spins in an organic film, provided that these spins do not interact with other molecules in the film. EPR is used mainly to determine the angular distribution of the molecules in the film. This is possible because in the EPR experiment, contrary to spectroscopic methods (e.g., optical absorption, Raman, etc.), the resonance frequency of the unpaired electron depends on the orientation of the molecule.

The first EPR measurement on LB films was described by Messier and Marc [189]. Most of the EPR work can be divided into three main parts, the first being on the characterization of porphyrin and phthalocyanine derivatives. Janzen *et al.* reported on the first light-induced EPR signal in chlorophyll monolayers containing chloranil as the electron acceptor [190]. Other investigators used EPR to elucidate the order of copper tetraphenylporphyrin in behenic acid films [191], and to determine the structure in pure monolayers of copper phthalocyanine derivatives [192–194]. Palacin *et al.* studied a tetraoctadecyl derivative of copper tetrapyridino-porphyrazine in LB films [195]. EPR studies of monolayers, and alternate layers built from mixtures of the porphyrazine with ω-tricosenoic acid, suggest that these LB films are well-ordered, and that the porphyrazine ring lies flat on the substrate. In a recent work, Palacin and Barraud studied a cobalt (II) complex of the tetramethyl tetrapyridinoporphyrazine derivative [196]. Using ion exchange at an ω-tricosenoic acid monolayer–water interface, they could build and transfer monolayers of this semi-amphiphilic molecule. EPR spectra suggest that as in the previous case, where the alkyl chains were substituted directly on the macrocycle [195], the porphyrazine rings lie flat on the substrate. This is an important result, since it suggests that ion exchange may be used as a general route to well-ordered LB films with other ionic molecules. Barraud *et al.* pioneered the work on electrically conducting LB films, and used EPR to study the $TCNQ^-$, and the TTF^+ derivatives [197–201]. The recent interest in conducting LB film in Japan has resulted in increasing activity in EPR [202–207].

33

Merocyanine dyes also were studied using EPR. These experiments included both pure dye monolayers and those where the dyes were diluted with fatty acids, monolayers where intermolecular electron transfer to different donor molecules occurred, and where the EPR was light-induced [208–212]. Studies of *J*-aggregates, their thermal stability, and the in-plane and out-of-plane molecular orientation of the merocyanine dye molecules within the LB films also were studied using EPR [213, 214]. Another use of EPR is in studies of monolayer structure, where small concentrations of spin-label materials (e.g., a nitroxide derivative) in the LB film are used. The experiments may give information on molecular motion and orientation within the film [215–218].

In an interesting work, Bonosi *et al.* recently prepared a series of nitroxide-containing strearic acid derivatives, where the position of the five-member paramagnetic unit in the chain was changed systematically [219]. The bulky paramagnetic ring has a cross-section area larger than that of the alkyl chain, and therefore introduces free volume and disorder into the monolayer, which was indicated by the EPR spectra.

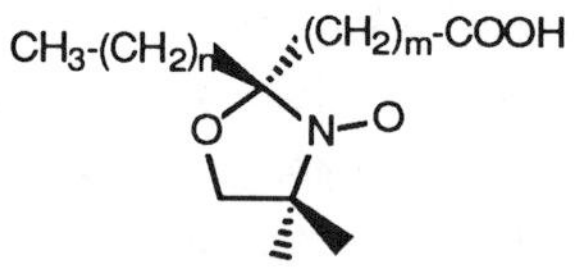

Furthermore, it was shown that the dipole–dipole effects and spin-exchange interactions were a function of the position of the paramagnetic group along the chain. The exchange interaction increased with the increased distance of the NO group from the COOH head group, which reflects both the decrease in interlayer distance of the paramagnetic units, and the increased mobility of the molecules in the two-dimensional assembly. The dipole–dipole effects show the same trend, i.e., to increase with the decreasing interlayer distance among the paramagnetic moieties.

EPR also was used to study the structure of LB films of diacetylene derivatives [$CH_3–(CH_2)_{17}–C{\equiv}C–C{\equiv}C–CO_2R$, where R = H, 0.5 Cu^{2+}], suggesting a molecular tilt angle of 45° from the surface normal [220]. Nakahara *et al.* used EPR among other analytical techniques to study LB films of long-chain derivatives of ferrocene, diferrocene, and oxidized ferrocenium [221], and Yoshino *et al.* used EPR to study the effect of doping (insulator–metal transition) in poly(3,4-dialkylthiophene) [222]. Finally, it is interesting to mention that the interaction of radicals such as NO_2 and similar gases can be studied with EPR [223].

G. X-ray Diffraction

X-ray diffraction is a powerful analytical tool for obtaining and investigating the structure of thin organic films. The angular directions of the diffraction peaks give the dimensions of the unit cell, while the intensities of these peaks give the electron density in the unit cell. A three-dimensional picture of electron density can be obtained from experiments using different angles of incidence. However, x-rays are very penetrating and one would not expect to obtain diffraction from LB monolayers due to the small amount of material. By working close to the angle for total internal reflection, such diffractions can be obtained due to the high intensity of the interfering beams. An example of the use of different x-ray techniques can be found in the work of Rieutord *et al.*, who used three x-ray experiments (synchrotron radiation) to determine the structure of an LB film. Thus, diffraction experiments in the transmission geometry yielded the interlamellar ordering, small-angle glancing diffraction yielded the interlamellar organization, and a critical reflection experiment provided the number of molecules per unit cell [224].

The first studies of x-ray diffraction from LB films were done by Holley and Bernstein [225,226]. In the early studies, LB films of heavy metal salts (e.g., lead, barium, cadmium, silver, etc.) of fatty acids were studied, mainly with the idea that these heavy atoms would help establish the layer structure of the film. With the development of better diffractometers and computer software, it became possible to extend structural studies beyond those films that contained heavy atoms, and higher-order Bragg peaks could be obtained. Today, a resolution of better than 2 Å is possible, and therefore, valuable structural information can be obtained. When this information is compared with information obtained from other techniques, a full interpretation of the x-ray diffraction becomes possible. In this section, a general review of the literature is presented, while a more rigorous discussion on the structure of LB films will occur in Part Two.

Pomerantz and Segmüller were the first to obtain data from monolayers of manganese stearate $[(C_{18}H_{37}CO_2)_2Mn]$ when they used highly collimated, highly monochromatic x-ray beams [227]. Prakash *et al.* applied a technique developed by Marra *et al.* [228] to study the in-plane structure of lead stearate monolayers using glancing-angle x-ray diffraction. Their conclusion is that the two-dimensional structure of the alkyl chains is similar in its static structure to an ideal smectic-B liquid crystal. Comparing the structure of LB films of lead myristate $[(C_{13}H_{27}CO_2)_2Pb]$, lead stearate $[(C_{17}H_{35}CO_2)_2Pb]$, and lead behenate $[(C_{21}H_{43}CO_2)_2Pb]$, they concluded that all the three types of films have the same in-plane structure [229, 230]. Belbeoch *et al.* studied x-ray diffraction from LB films of behenic acid, and of its silver (Ag$^+$) salt [231]. Their results reveal that while behenic acid films are orthorhombic, the symmetry of the silver salt films is triclinic. Allain *et al.* studied multilayer films of behenic acid and concluded that the last deposited monolayer is poorly filled and quite disordered [232]. Rieutord *et al.* found a new, nontilted structure of behenic acid monolayers [233]. Bosio *et al.* reported the first density profile of LB monolayers of lead stearate at the water–air interface [234].

Outka *et al.* compared calcium arachidate (CaA) and cadmium arachidate (CdA) monolayers on Si(111) substrates using near-edge x-ray absorption fine-structure (NEXAFS), and reported that while the alkyl chains in the CaA monolayers are normal to the surface ($0 \pm 15°$), those in the CdA monolayers are tilted by $33 \pm 5°$ [235]. In later papers, they report that monolayers of arachidic acid on Si(111) surfaces are not ordered at all [236.237].

Mizushima *et al.* studied the structure of cadmium stearate [238], and then used an extension of the ordinary kinematical theory to analyze Bragg and non-Bragg reflections for LB films of cadmium stearate [239]. They were able to show that the film crystallizes with time. Fromherz *et al.* used medium-angle x-ray scattering to calculate film thickness in LB bilayers of cadmium salts of a series of fatty acids [240]. They reported that film thickness increased linearly with chain length, with an inclination of the chain of about 15°.

Jark *et al.* studied LB films of cadmium arachidate by means of soft x-ray diffraction [241]. They calculated the reflectivity profile from the specular reflection and the chemical composition of the films, and demonstrated excellent agreement between experimental and calculated interference patterns. Thus, using soft x-rays, they achieved a significant enhancement in the spatial resolution, and could measure film thickness from 20 to 2500 Å (with uncertainty of ~1%).

Kajiyama *et al.* deposited LB films (200 monolayers) of barium salts of C_{18}, C_{20}, and C_{22} fatty acids on polystyrene substrates and used x-rays to study structural defects in these samples [242]. Their most important result was that the crystallite size increased from 77–108 Å lateral and 468 Å longitudinal for the C_{18}Ba sample, to 115–145 Å lateral and 594 longitudinal for the increasing chain length sample of C_{22}Ba. This result suggests the coherence length in a two-dimensional crystal is a function of the alkyl chain length, i.e., of the intermolecular van der Waals interactions, and thus as these interactions increase, so does the coherence length.

Skita *et al.* studied two sets of multilayer films: in the first, finite sequences of arachidic acid ($C_{19}H_{39}CO_2H$) and myristic acid monolayers, and in the second, finite sequences of arachidic acid and 10,12-pentacosadiynoic acid

$[CH_3-(CH_2)_{11}-C\equiv C-C\equiv C-(CH_2)_8-CO_2H]$. Their results revealed that the surface monolayer, i.e., the outmost layer at the multilayer–air interface, is disordered, and that ordering of the surface can be induced by the deposition of an additional monolayer [243, 244]. Although the last bilayer deposited symmetrically in the aforementioned studies displayed the most significant asymmetry, Fischetti et al. showed in a later work that the interior arachidic-acid bilayers retained asymmetry [245].

Dutta et al. reported on x-ray studies of monolayers of lead stearate and then lignoceric acid $(C_{23}H_{47}CO_2H)$ on the water surface [246]. They reported an area per molecule of 17.8 Å in a triangular lattice, and a first-order melting transition with increasing area. Later, they compared lead stearate, tetracosamoinc acid, and 1-eicosanol [247]. In an interesting, detailed work, Barton et al. studied LB monolayers of 1-heneicosanol $(C_{21}H_{43}OH)$ on water [248]. Their results are very important for the understanding of packing and lateral order in assemblies of alkyl chains. (See Part Two.) In a subsequent paper, Lin et al. studied the kinetics of structural phase transition in LB monolayers of heneicosanol on water [249]. Here, they reported on the first observation of the microscopic relaxation mechanism, where immediately after compression, the structure was a uniaxially disordered structure, and was transformed to an undistorted hexagonal structure, at a rate that was strongly temperature-dependent. This is an interesting observation, since it means that the system began in a pseudohexagonal phase, where the CCC planes of the molecules are partially oriented, and was transformed slowly to the hexagonal phase, where these planes are randomly oriented about their chain axes. For this to happen, the chains also should become less tilted; however, the authors do not discuss tilt angles in this paper.

Grundy et al. studied the structure of docosanoic acid $(C_{21}H_{43}COOH)$ monolayers at the water–air interface, and found that in the condensed phase of cadmium docosanoate, the molecules are perpendicular to the surface [250]. Upon reduction of surface pressure, the film breaks up into islands (> 0.2 μm) of close-packed, perpendicular molecules. However, if the pH also is reduced (pH ≈ 3) and the cadmium ions are displaced, the molecules in the expanded film are 30° tilted relative to the surface normal. Buhaneko et al. used x-rays to study the temperature behavior of spread and deposited monolayers of docosanoic acid and its cadmium salt [251]. There are two interesting results in this study: (a) The structure and temperature behavior of the films depend crucially on the subphase, and (b.) Si(111) (with the surface oxide, SiO_2) is a preferred substrate and yields better films than hydrophobicized (silanized) glass.

Kjaer et al. and then Möhwald compared LB films of arachidic acid and the phospholipid dimyristoylphosphatidic acid (DMPA) [252, 253]. They found that the thickness of the C_{20} monolayers on water surface and on solid support was similar. However, the coherence length on water was 22 lattice spacings, while on SiO_2/Si substrate, it was only ≈ 8 spacings, indicating that surface roughness may have an effect on this parameter. Note that since the structure of an LB film on a solid substrate may not be identical to that on the water surface, one should be careful in assigning differences in lattice spacing only to surface roughness. For monolayers of DMPA, they found that x-ray data complement their fluorescence microscopy experiments, indicating coexistence of a fluid and a denser gel phase in the isobar region of the π–A isotherm, where the coherence length changed from 10 in the mixed region to ≈ 100 in the purely gel phase.

Richardson and Blasie used high-resolution x-ray diffraction to study the melting of arachidic acid monolayers [254]. In this experiment, they built a heterogeneous structure in which one, three, and 10 bilayers of arachidic acid nested between two polymer bilayers of $[C_{12}H_{25}-C\equiv C-C\equiv C-(CH_2)_{16}-CO_2H]$. The results of this study were that the average in-plane molecular chain density in an individual monolayer, and the mean tilt angle for the straight all-*trans* chains from the surface normal, were strongly related. Upon heating, a two-stage process occurred, consisting of a slowly evolving, continuous untilting of the chains, thus extending the profile thickness of the monolayer, followed by a continuous decrease in the profile thickness due to chain melting via the formation of kinks and jogs.

L'vov *et al.* studied alternate-layer films of tricosanoic acid ($C_{22}H_{45}COOH$) / 1-docosylamine ($C_{22}H_{45}NH_2$), and of docosanoic acid/1-docosylamine [255]. They showed that these films consist of asymmetric bilayers with a smaller tilt of the alkyl chain than that observed in the corresponding films of the pure compounds. This immediately suggested that the interlayer-attractive electrostatic interactions reduce the repulsion among the head groups, and thus allow the molecules to pack more closely.

Pomerantz *et al.* were the first to study an OTS SA monolayer , and an SA three-layer sample of methyl 23-trichlorosilyltricosanoate, $Cl_3Si-(CH_2)_{22}-COOCH_3$ [256]. Wasserman *et al.* used x-ray reflectivity in a comparison with ellipsometry to study SA monolayers of alkyltrichlorosilanes [257]. This study helped in establishing the structure of partial monolayers of OTS on silicon. (See earlier discussion.) Both studies are discussed further in Part Three.

Grayer Wolf *et al.* used grazing-angle x-ray diffraction and reported on the first study of amino acids-containing monolayers, i.e., a palmitoyl-(R)-lysine derivative, $[CH_3(CH_2)_{14}-CO-NH-(CH_2)_3-CH_2-CH(NH_3^+)COO^-]$, on the water surface [258]. The coherence length was 500 Å, and the packing arrangement of the α-amino acid head groups was very similar to that found in the crystal structure of the α form of glycine. The same authors investigated the phase transition in the LB monolayer of a perfluorododecyl-1H,1H,2H,2H derivative of glutamic acid, (S) $[CF_3-(CF_2)_9-(CH_2)_2-OCO-CH_2-CH(NH_3^+)-$ $-COO^-]$ [259]. The grazing-angle x-ray diffraction results indicated that there was a solid–solid transition from a hexagonal lattice, where the chains are normal to the surface, to a distorted-hexagonal lattice, where the chains have a tilt of $26 \pm 7°$. At 70 mN/m, the coherence length exceeded 1000 Å. Landau et al. used x-ray structure analysis of LB films for crystal nucleation [260].

Day and Lando studied the morphology of crystalline diacetylene monolayers polymerized at the water–air interface [261]. Tredgold et al. used x-ray diffraction to study the structure of porphyrin and polymer LB films [262]. Tredgold et al. also studied alternating layers of mesoporphyrin IX and protoporphyrin IX derivatives with various fatty acids [263]. They reported that the best film-forming porphyrin was the mesoporphyrin IX-diol, and that the use of esters leads to segregation of the two materials in the plane of the film. This segregation suggests that molecules diffuse after deposition over distances of the order of 1 μm. Oyanagi et al. used EXAFS (edge x-ray absorption fine structure) to study LB films of a trisubstituted Cu(II) phthalocyanine derivative [264]. Their results suggest that the phthalocyanine plane faces the dipping direction, but is not perpendicular to the substrate surface. Fujiki and Tabei studied LB films of several phthalocyanines containing tert-butyl, isopropyl and cyano groups, both in the free base and Ni(II), Cu(II), and Pb(II) complexes [265]. While they could not suggest a packing arrangement in these films, their results suggested an edge-on configuration of the macrocycle at the water–air interface. Dent et al. reported on x-ray reflectivity results from three phthalocyanine derivatives [266].

Oyanagi et al. studied LB films of merocyanine dyes using synchrotron-polarized x-ray radiation. Their major conclusions are that while the chromophores are highly oriented at room temperature, they lost this preferential orientation when heat treated [267]. Duschl et al. [268], Kim et al. [269], and Nakano et al. [270] published papers on pure and mixed LB films of different J-band-forming cyanine dyes. Imazeki et al. reported on a memory effect of chromophore orientation in merocyanine LB films [271]. Tredgold et al. used x-ray diffraction to study monolayer films of azobenzene derivatives before and after annealing [272, 273]. The study of electron density profiles revealed that the annealing process results in a phase change (from smectic-A liquid crystal structure to a tilted phase).

An interesting example of structure properties relationships being established is that of the LB film of N-docosylpyridinium tetracyanoquinodimethane (TCNQ) [274]. It was reported that while the diffraction pattern of 12-layer film showed only one Bragg peak, at least five more diffractions developed upon aging, and the more ordered material behaved as a semiconductor. Otsuka and Saito reported on x-ray diffraction from a 66-layer film of a charge-transfer complex between TCNQ and a new TTF derivative [275].

In the area of polymerizable monolayers, Hirota et $al.$ studied the polymerization of mixed LB films of β-parinaric acid ($trans$ C_2H_5–$(CH=CH)_4$–$(CH_2)_7$–COOH) and stearic acid [276].

Banerjie and Lando reported on the irradiation-induced solid-state polymerization of oriented LB films of octadecyl acrylamide [277].

There is one report on x-ray studies of LB films of perfluorocarboxylic acids. Nakahara et $al.$ reported on LB films of up to 50 monolayers of perfluoroundecanoic acid ($C_{10}F_{21}$COOH) deposited from the Al^{3+} subphase [278]. Their results indicate that these Y-type multilayers are well-organized with thicknesses of 16.1 Å. They do not indicate if the chains are tilted, although from their area per molecule of 33.2 Å^2, it can be suggested that the chains have a tilt of $\approx 30°$, based on the value of 27.2 Å^2 for polytetrafluoroethylene (PTFE) [279].

Yang et $al.$ studied the structures of poly(3-n-hexadecylpyrrole) and poly(3-n-octadecylpyrrole) LB films using near-edge x-ray absorption fine structure study (NEXAFS) [280]. Their results indicated that the alkyl chains are tilted in the monolayers, they are normal to the platinum substrate in multilayers.

Watanabe et $al.$ studied mixed multilayers containing stearic acid and various poly(3-alkyl thiophenes) [281]. They reported that these films exhibit well-defined layer spacing, e.g., 49 Å bilayer repeat distance in a 30-monolayer sample of poly(3-n-octyl thiophene) and stearic acid in 1.4:1 molar ratio, respectively.

In concluding the discussion on x-ray diffraction, we note that diffraction experiments using synchrotron radiation are complicated, require lengthy preparation and data collection, and usually are caried out by crystallographers. Data collection from a single monolayer on a substrate is rather difficult, and the preferred method in these cases is electron diffraction. (See later.) Diffraction from thick films is a relatively easy experiment, and the analysis is similar to that for organic crystals.

H. Electron Diffraction

Electron diffraction is a useful technique for the determination of the in-plane structure of thin organic films. Compared to x-ray or neutron diffraction, the interaction with matter is stronger, and the electron beam can be focused down to 5 Å, which makes rapid scanning of small details (defects, crystallites, etc.) possible. The disadvantage, however, is that samples have to be stable *in vacuo* to sustain the electron beam energies.

Havinga and de Wael [282] and Germer and Storks [283] studied LB films of cadmium stearate, but since there was no theoretical analysis available at that time, these studies, although correct, are not complete. However, Russell *et al.* reinvestigated these monolayers and reported that the alkyl chains in LB films of cadmium stearate are normal to the monolayer plane, and in an orthorhombic configuration [284]. Bonnerot *et al.* [285] and Garoff *et al.* [286] published two important papers on the structure of monolayers. Both groups studied electron diffraction from LB films of fatty acids, and concluded that the first monolayer deposited in hexagonal packing, and that over the next few layers, there existed a transition leading to the final orthorhombic or monoclinic structure (depending on the fatty acid).

Vogel and Wöll used low-energy electron diffraction (LEED) to study two-dimensional crystal structures of single fatty acid monolayers on metal single crystals (Cu, Ag, Au) [287, 288]. Their remarkable result is that monolayer crystals have a coherence length over the whole monolayer (20 mm in diameter).

Recently, Riegler used electron diffraction to study thermal behavior of cadmium stearate, arachidate, and behenate [289]. He reported a sharp decrease in diffraction peak intensities at temperatures below the bulk melting points, indicating a pretransition disordering. These temperatures were ~35° for cadmium stearate, ~55° for cadmium arachidate, and ~75° for cadmium behenate. His conclusion was that hexagonal geometry was preserved, but that the thermally induced random tilt orientational disorder was the cause for the decrease in intensity.

Peterson and Russell studied LB films of ω-tricosenoic acid ($CH_2=CH-(CH_2)_{20}-CO_2H$) [290]. They reported that the structure of these LB films varied with the plane of the crystal on which it was coated. For example, orthorhombic packing occurred on the (100) plane of InP, while on the (111) face on the same crystal, the alkyl chains were in a triclinic packing. Both orthorhombic and triclinic packing arrangements had been observed over a wide range of deposition conditions; however, in the latter, the inclination of the chains to the substrate normal gave anisotropic deposition. In another contribution, Peterson and Russell found that the structure of ω-tricosenoic acid films varied with subphase conditions [291]. Thus, the acid gave films with tilted packing, while in LB films of its cadmium salt, the alkyl chains were more perpendicular. Jones *et al.* used *reflection high-energy electron diffraction* (RHEED) to study LB films of ω-tricosenoic acid deposited on hydrophilic silicon substrates [292]. They reported that the structure of the first few monolayers is a granular one, with the molecules equally tilted by ~20° with respect to the substrate normal, but where the azimuthal angle of the tilt can take any value. On the other hand, in thicker films, subsequent layers developed a single tilt in the upward direction with respect to the substrate normal. Recently, Peterson *et al.* used RHEED to investigate surface pressure dependence of molecular tilt in LB films of ω-tricosenoic acid [293]. Their results indicate that the orientation at which molecules pack in these films varies continuously with the surface pressure at deposition (25 to 45 mN m^{-1}). Their explanation for this behavior was that films do not consist of pure acid, but incorporate an amount of water depending on the deposition surface pressure. It seems, however, that there may be an alternative explanation, and we shall discuss it in Part Three.

Lieser *et al.* studied the order and packing in LB films of cadmium salts of long-chain diacetylene carboxylic acids [294]. They reported that UV irradiation of the LB films resulted in polymerization of the diacetylene without destruction of the films. A more detailed discussion on monolayers of diacetylenes can be found in Part Two. In recent work, Sotnikov *et al.* studied LB films of tetraynoic molecules, R–(C≡C)$_2$–(CH$_2$)$_4$–(C≡C)$_2$–(CH$_2$)$_8$–COOH [295]. Their results indicated that upon irradiation and polymerization, there was a change in the molecular packing of the film. The number of methylene units between the two diacetylene groups (four), as well as between the diacetylene and carboxylic head group (eight), is definitely not optimal. It seems, however, that a systematic molecular engineering approach may lead to a molecule capable of topochemical polymerization in LB films. Naegele *et al.* used electron diffraction to study the polymerization of cadmium octadecylfumarate in LB films [296]. Duda *et al.* reported on the structure of LB monolayers and multilayers of two quite different preformed polymers [297]. Deposited multilayers of a copolymer of octadecyl methacrylate and dodecyl methacrylate were studied and it was found that the alkyl chains are perpendicular to the substrate normal, indicating interpenetration of alkyl chains. Using the same principle, they reported that a noncomplete substitution of γ-methyl ester groups by long-chain alkyl groups in poly(γ-methyl-L-glutamate) resulted in liquid-like, transferable monolayers. The results indicated that the α-helices were oriented with the helix axis in the flow direction on the substrate, and that interdigitation of alkyl chains occurred. The directionality reported for the helix axis probably is a result of the shear forces in the meniscus during deposition.

Neal *et al.* obtained RHEED from monolayers of 4-heptadecylamido-4'-nitrostilbene on both (111) and (100) faces of silicon [298]. They reported that the stilbene molecules form two-dimensional crystallites, where the π-systems aligned parallel to each other and were tilted at an angle of 55° to the substrate normal.

$$C_{17}H_{35}-\overset{\overset{\textstyle O}{\|}}{C}-NH-\hspace{-4pt}\langle\bigcirc\rangle\hspace{-4pt}=\hspace{-4pt}\langle\bigcirc\rangle\hspace{-4pt}-NO_2$$

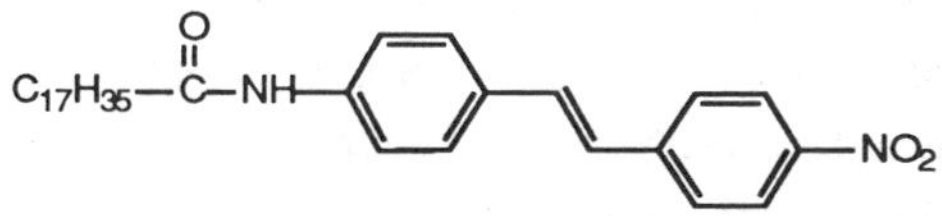

Heard *et al.* used the same stilbene derivatives and demonstrated that high-quality diffractions can be obtained using transmission electron diffraction (TED) from LB films [299]. The results indicated that the stilbene derivative aligned well when deposited either as a monolayer or in alternate multilayer structures with ω-tricosenoic acid.

Earls *et al.* studied both reflection and transmission electron diffraction from LB films of an alkyl hemicyanine derivative [300].

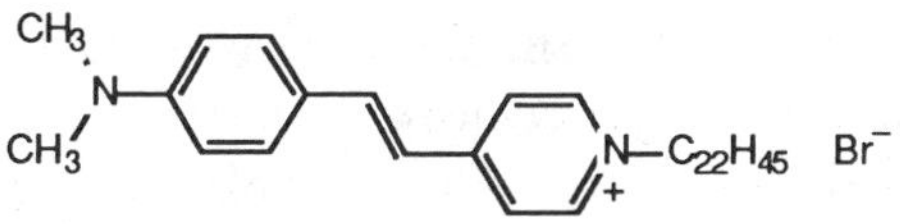

$$\underset{CH_3}{\overset{CH_3}{\diagdown}}N-\hspace{-4pt}\langle\bigcirc\rangle\hspace{-4pt}=\hspace{-4pt}\langle\bigcirc\rangle\hspace{-4pt}\overset{+}{N}-C_{22}H_{45}\quad Br^-$$

Here, they reported that the molecules (both chromophore and alkyl chain), essentially are parallel to each other in a quasi-crystalline molecular packing, and tilted 40° to the substrate normal. Duschl *et al.* reported on the verification of the brickstone arrangement model, suggested by Kuhn [301], for the chromophore head groups in *J*-aggregates [302].

Luk *et al.* used electron diffraction to study LB films of mesoporphyrin IX dimethyl ester ididium chloride, and reported on sharp hexagonal diffraction patterns, indicating a high degree of crystallinity in these domain-type structures [303].

Fryer *et al.* reported on a diffraction from a single monolayer of copper tetra-*t*-butylphthalocyanine [304]. The diffraction indicated that the film is not uniform, that the domains are small and highly disordered, and that the macrocycles are lying in columns in the plane of the film with an intermolecular separation of 3.3 Å, and an inter-columnar separation of 19 Å. Troitskiy *et al.* studied different alternating sequences of LB films containing Cu^{2+} or $(VO)^{2+}$

phthalocyanine derivative and barium behenate [305]. Electron diffraction of films with various sequences of bilayer alternation shows that considerable azimuthal disorder appeared in the mutual orientation of crystallites in barium behenate bilayers upon introducing phthalocyanine layers between them. Frisby and Bonnerot studied the electron diffraction from LB films of the charge transfer salt N-docosylpyridinium 7,7,8,8-tetracyanoquino-dimethene [306].

Rahman et al. used electron diffraction to study LB films of poly(3-n-hexadecyl pyrrole) [307]. They reported that these films are substantially different from bulk polymerized films, and that the two-dimensional organization and polymerization yield a smooth coverage over a large area of the substrate.

Strong and Whitesides reported on the only electron-diffraction studies of self-assembled alkanethiol monolayers on gold [308]. This paper will be discussed in Part Three.

I. Neutron Diffraction

There are four interesting properties of *neutron diffraction* studies:

(*a*) Neutrons have no coulombic interaction with matter.

(*b*) In a reflection experiment, close to the critical angle, neutron reflectivity is very sensitive to changes in the refractive index of the material. Thus, since the refractive index is different for protonated and deuterated materials (e.g., CH_2 vs. CD_2), neutron diffraction studies of deuterated molecules may provide additional structural information.

(*c*) Neutrons are scattered by the electron magnetic moment, therefore giving both structural and magnetic information.

(*d*) Neutron scattering by nuclei is not relayed to the chemical element. However, an isotope effect may have an important effect on the neutron diffraction, and may be revealing about the structure.

The first neutron diffraction from LB films was reported by Nicklow et al. [309]. They studied neutron diffraction ($\lambda = 4.15$ Å) from films as thin as three layers of manganese stearate. Hayter et al. reported neutron diffraction ($\lambda = 15$ Å) from three layers of deuterated calcium stearate deposited on glass [310]. In this study, they could answer the question whether water exists in LB films as a molecular layer in the hydrophilic region, since exchanging H_2O for D_2O showed a considerable change in the calculated reflectivity profile. Highfield et al. obtained the density profile of deuterated cadmium arachidate multilayers deposited on glass [311]. From this data they could calculate molecular thickness of 24.8 ± 0.2 Å, indicating that this thickness decreases with the reduction of the total number of monolayers. Grundy *et al.* used neutron diffraction to study monolayers of docosanoic acid and of cadmium docosanoate at the water–air interface [250], and Buhaneko *et al.* studied the effect of temperature on the structure of spread monolayers and deposited films of docosanoic acid and cadmium salt [151]. Dent *et al.* reported on neutron diffraction from spread monolayers of T15, a cyanoterphenyl molecule that exhibits a nematic liquid crystal at elevated temperatures [266].

Two reports on new reflectometers appeared in recent years. Penfold *et al.* reported on a time-of-flight neutron reflectometer for the study of both liquid and solid surfaces and interfaces [312]. Their results indicated that this design allows neutron reflectivities approaching 10^{-6} to be measured. Stamm *et al.* reported on a new design of neutron reflectometer for the study of solid and liquid interfaces [313].

Finally, we note that neutron diffraction may be used in the study of assemblies of molecules that contain bulky groups in the alkyl chain. (See the beginning of Part Four for details.) Here, neutron diffraction studies of partially deuterated chains (either above or below the bulky group) may provide additional structural information that, together with electron and x-ray diffraction, can help us understand the contribution of different molecular parts to the overall packing and orientation scheme.

J. High-Resolution Electron Energy Loss Spectroscopy

In *high-resolution electron energy loss spectroscopy* (HREELS), the atomic process is an inelastic scattering of incident electrons from the surface [314, 315]. The energy of the scattered electrons is measured, and from the energy difference (loss), the nature of adsorbed species and their vibrational and electronic spectra can be revealed.

HREELS offers several advantages: (*a*) A single instrument gives both vibrational and electronic spectra; (*b*) the method is highly sensitive, to the low level of < 0.1% of a monolayer; (*c*) it can be applied to various surface topologies; and (*d*) it is sensitive to molecular orientation. The selection rule in HREELS is similar to that in grazing-angle IR spectroscopy; i.e., only normal vibrations that are orthogonal to the substrate surface will be accessed [316]. The first HREELS of LB monolayers of fatty acids on polycrystalline substrates was demonstrated by Wandass and Gardella [317]. They deposited LB monolayers of four fatty acids (octadecanoic, docosanoic, *cis*-9-octadecenoic, and octatridecane

-9,12,15-trienoic) on hydrophilic Ag/AgO surfaces and studied their spectra. (The resolution in this experiment was ±1 meV or 8.066 cm⁻¹.) A typical HREELS of a monolayer of stearic acid is presented in Figure 1.22.

Note that while in most single crystal/small molecule systems (e.g., NO, N$_2$, CO, etc.) the relative loss feature/elastic peak ratios are low (≤1%), these ratios in all spectra reported in this study are much higher (6–10%). This may be due either to the very high scattering cross section at the surface, or a large absorbance density of the ordered assembly on the surface (or both). High-resolution spectra give detailed information on vibration spectra. The C–H stretch appears at 359 meV (2896 cm⁻¹), while in the octatridecane-9,12,15-trienoic spectrum, the peak at 372 meV (~3001 cm⁻¹) is assigned as a shifted olefinic C–H stretch.

The C=O stretch, which is weak due to its ~20 Å distance from the scattering electrons, appeared at 203 meV (1637 cm⁻¹). This position of the C=O peak is different from the 213 meV (1718 cm⁻¹) measured for monomeric acids, and is explained by the charge delocalization, as well as the back donation of electron density from the Ag d orbitals to the C=O π* orbitals. The C–O–H in-plane bend appeared at 173.8 meV (~1402 cm⁻¹), the C–O stretch of the dimer at 159.3 meV (1285 cm⁻¹), and the CH$_2$ rock peak at 91.5 meV (738 cm⁻¹). A higher energy peak at 520 neV (4194 cm⁻¹) is assigned as a C–H, C–O combination band.

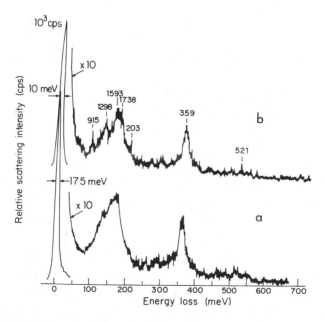

Figure 1.22. HREELS spectra of a stearic acid monolayer on Ag/AgO with a 6.5 eV primary beam energy: (a) "lower-resolution" survey scan, (b) "higher-resolution" scan. (From Wandass and Gardella [317], © 1985, Elsevier Science Publishers.)

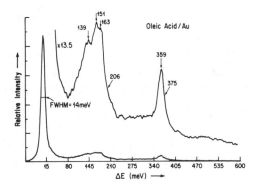

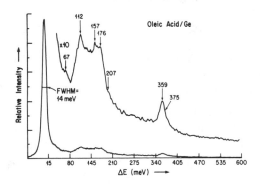

Figure 1.23. HREELS of a monolayer of oleic acid on Au (top), and the same monolayer on Ge (bottom). (From Wandass and Gardella [319], © 1985, Am. Chem. Soc.)

In a second paper, Wandass and Gardella reported on HREELS studies of LB monolayers and multilayers [318]. Here, they added some very important information to what already had been published [317]. First, they demonstrated that successful HREELS spectra can be obtained on multilayer samples (15 monolayers, film thickness of ~370 Å). Second, from differences in relative intensities of various vibrational loss features with respect to the layer thickness, they concluded that resonance interactions between vibration and electronic spectral features in the monolayers may exist. Third, they observed that methyl and methylene vibrations were intensified relative to others such as C–O, indicating that HREELS is extremely surface-sensitive, and the observed spectra was dominated by the topmost methyl layer. Finally, HREELS analysis of monolayers of *cis*-9-octadecenoic acid (oleic acid) on Au, Ag, Ge, and Al substrates demonstrated a complex relationship between the character of the observed spectra and the nature of the substrate (Figure 1.23) [319]. From these samples, they concluded that surface roughness factors have much to do with the intensities of both the elastic and loss peaks.

The HREELS results are interesting, and complement those from IR and Raman spectroscopies. The sensitivity that was demonstrated — especially in the spectra of 3:1 molar ratio of stearic/linoleic acid where the olefinic C–H could be detected clearly — shows that HREELS can be used to study systems with low surface coverage, and those with low concentration of host molecules.

1.2. Analysis of Surface Properties

A. Contact Angles and Surface Tension

The quality of stable monolayer and multilayer films can be estimated from wetting measurements. This is due to the fact that the *shape* of a liquid drop on a plane, homogeneous surface (which is the result of the free energy of this drop) is affected by the free energy of this surface. The force per unit length acting on a surface or interface is equal to the interfacial tension γ (mN m^{-1} or dyn cm^{-1}), and the work done on a system when increasing the area by dA at constant γ is γdA. The work of adhesion between a solid and a liquid is defined as

$$W_{SL} = \gamma_{SV} + \gamma_{LV} + \gamma_{SL} \, , \qquad 1.26$$

where W is the work, γ is the surface interfacial tension, and LV, SV, and SL refer to liquid–vapor, solid–vapor, and solid–liquid interfaces, respectively.

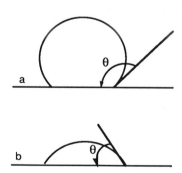

Figure 1.24. (a) A typical water contact angle on a methyl surface, where $\theta = 111°–115°$, and (b) a typical hexadecane contact angle on a methyl surface, where $\theta = 45°–46°$.

If γ_s^o is the surface energy in vacuum, then its energy in air or in a vapor will be made less by the spreading pressure of π, so that

$$\gamma_{SV} = \gamma_s^o - \pi \,. \qquad\qquad 1.27$$

When a liquid does not wet a surface completely, it forms an angle θ, the contact angle with the surface. Figure 1.24 presents typical water ($\theta = 111°–115°$), and hexadecane (HD, $\theta = 45°–46°$) contact angles on a surface comprised of methyl groups (e.g., OTS, on silicon).

Let us consider a virtual displacement of the contact region between the liquid and the surface (the triple line), so that the surface is wet by an additional area of 1 cm^2 (Figure 1.25). We have a:
• surface-free energy increase at the solid–liquid interface = γ_{SL} ,
• surface-free energy decrease at solid–vapor interface = γ_{SV},
• surface-free energy increase at the liquid–vapor interface = $\gamma_{LV}\cos\theta$.
By the principle of virtual work, we get

$$\gamma_{SL} + \gamma_{LV}\cos\theta = \gamma_{SV} = \gamma_s^o - \pi \,. \qquad\qquad 1.28$$

This equation describes the relationship between the free energy of the surface and a unique contact angle θ, *under ideal conditions*, for a system in equilibrium [320], and is known as Young's equation after Thomas Young [321], sometimes written as

$$\gamma_{LV}\cos\theta = \gamma_{SV} - \gamma_{SL} \,. \qquad\qquad 1.29$$

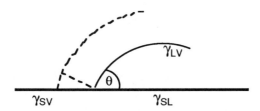

Figure 1.25. A virtual displacement of the contact region between the liquid and the surface.

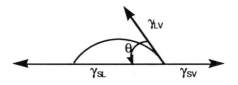

Figure 1.26. A liquid drop on a solid surface.

49

Figure 1.26 presents an illustration of the Young equation. The angle of a drop on a solid surface is the result of the balance between the cohesive forces in the liquid and the adhesive forces between the solid and the liquid. If there is no interaction between the solid and the liquid, the contact angle will be 180°. As the interaction between the solid and liquid increases, the liquid spreads until $\theta = 0$ and Equation 1.29 becomes

$$\gamma_{LV} = \gamma_{SV} - \gamma_{SL}.$$ 1.30

So far, we have discussed contact angles on ideal surfaces. We note that real surfaces rarely exhibit a true, unique, thermodynamic equilibrium contact angle, as defined by Equation 1.30 [322]. In this case, a different contact angle is measured when the drop has advanced (θ_a) or receded (θ_r) on the surface prior to measurement. In other words, the line defining the solid–liquid–gas interface, also known as the *three-phase line,* either has advanced or receded.

While this is the usual way of defining these angles, they are not the true maximum advancing and minimum receding angles for which we define the hysteresis as

$$\Delta\theta = \theta_{a,\max} - \theta_{r,\min}.$$ 1.31

Thus, if after the addition of liquid to the drop, or the withdrawal of liquid from the drop, the three-phase line moves, a new equilibrium is established between the drop and the surface, and the measured contact angles are not the *true* maximum or minimum ones. Therefore, we propose adopting a procedure of measuring the contact angle *prior* to any advancing or receding of the drop on the surface. Although quite challenging, this will give contact angles that may be as close as possible to the true maximum and true minimum; thus, a true hysteresis can be measured.

To achieve this measurement, a drop of a *fixed volume* should be used, and the addition and withdrawal of liquid from the drop should also be at fixed volumes. This way, the comparison between different surfaces will be reliable. It is thus recommended that one use an accurate syringe for this measurement, form the drop on the end of a flat hypodermic needle, lower the needle *slowly* toward the surface, and when the drop touches the surface, *keep the needle in the drop* (captive drop) and read the contact angle immediately. If the drop is allowed to fall from the needle to the surface, smaller contact angles usually are obtained because of mechanical vibrations. This is true especially for liquids, such as diiodomethane (CH_2I_2), that have high density and therefore are more affected by gravity. Also, if the needle is taken out of the drop, the reading may be lower than that of the true advancing contact angle, again because of vibrations. It also is recommended that different drop volumes be used for different liquids, and smaller drops for liquids with high density.

To establish a value for an advancing contact angle, a given small volume of liquid is added to the drop, and while the needle is in the drop (Figure 1.27), the contact angle is measured before the boundary of the drop has moved. It is suggested that contact angles be measured on both sides of the droplet, and that a number of readings be collected at different places on the surface and for different independent samples. With this procedure, the data set will have statistical meaning. The same procedure should be used for the receding contact angle measurements. Thus, while the needle is in the drop, a fixed volume of liquid is withdrawn, and the reading is done before the boundary of the drop has moved. Also, it may be convenient to add liquid to the drop, and subsequently withdraw the same volume from the drop.

Theoretically, repeating this procedure should give the same readings for the advancing angle every time a volume is added, and for the receding angle, every time a volume is withdrawn from the drop. With all this in mind, it should be established first that the liquid used does not have corrosive effects on the monolayer. It was found that, for example, methylene iodide is corrosive to monolayers of alkanethiols on gold [323].

The procedure alluded to in the preceding is somewhat similar to the tilted-drop method (Figure 1.28), where the advancing contact angle is measured at the bottom end, and the receding angle is at the top end of the drop [324]. Here, the contact angles are measured on a tilted sample where the drop almost moves, but before any change in the position of the drop on the surface has occurred. The water contact angle for a smooth solid surface exposing closely packed methyl groups (–CH$_3$) is 111°–115° [325–327], while the hexadecane contact angle for the same surface is 45°–46° [327].

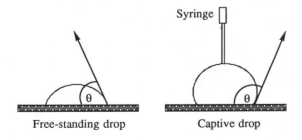

Figure 1.27. Schema of contact angle measurements of a free-standing (sessile) drop and of a captive drop.

51

Figure 1.28. A liquid drop on a tilted surface.

On the other hand, a solid exposing closely packed methylene groups ($-CH_2-$) is a smooth polyethylene surface. In this case, the water and hexadecane contact angles are $102°-103°$ [328–331] and $0°$ [329–331], respectively. Therefore, in disordered or liquid-like monolayers, as methylene groups are more exposed to the surface, the contact angles decrease.

Self-assembled monolayers of derivatized alkanethiols on gold surfaces provide a useful tool for surface engineering. A recent paper summarizes the studies of a large number of thiol monolayers adsorbed on gold [332]. In this paper, the reader may find a valuable collection of advancing contact angles (θ_a) for water and hexadecane. Table 1.3 summarizes some of the results.

The effect of the introduction of an ether linkage into the alkyl chain has been studied by Bain and Whitesides [333]. They prepared a series of thiols with an ether linkage [$CH_3-(CH_2)_n-O-(CH_2)_{16}-SH$] in different positions along the chain and studied the advancing contact angles (θ_a) with water, ethylene glycol, and hexadecane. They found that for sufficiently long alkyl groups (butyl group, $n = 3$), the measured contact angles approached those measured on a monolayer of docosanethiol ($C_{22}H_{45}SH$).

Furthermore, they discovered that the length of the alkyl group, for which the measured contact angle approached the value observed for the long-chain thiol, varied with the nature of the contacting liquid. Hexadecane was least sensitive to the position of the ether linkage, while water showed the highest sensitivity. Since it is not possible for hydrogen bonds to form with an oxygen atom buried under ~4 Å of alkyl chain, the suggested mechanism was long-range polar interaction, or penetration of the small water molecules into the monolayer. This monolayer is liquid-like about the soft oxygen linkage. Tillman *et al.* also found water to be most sensitive to the order of hydroxylated surfaces. The reader will find detailed discussion on this matter in the section on the building of SA multilayers (Section 3.2.F).

Table 1.3. Advancing contact angles on derivatized thiol monolayers on gold. (From Bain *et al.* [332], © 1988, Am. Chem. Soc.)

RSH	$\theta_a(H_2O)$	$\theta_a(HD)$
$HS(CH_2)_2(CF_2)_5CF_3$	118	71
$HS(CH_2)_{21}CH_3$	112	47
$HS(CH_2)_{17}CH=CH_2$	107	39
$HS(CH_2)_{11}Br$	83	0
$HS(CH_2)_{11}Cl$	83	0
$HS(CH_2)_{11}OCH_3$	74	35
$HS(CH_2)_{10}CO_2CH_3$	67	28
$HS(CH_2)_8CN$	64	0
$HS(CH_2)_{11}OH$	0	0
$HS(CH_2)_{15}CO_2H$	0	0

It has been found that hysteresis ($\Delta\theta$) is a function of the surface polarity [334], heterogeneity [335–339], and roughness [340–343], as well as the polarity, of the contacting liquid. Bain and Whitesides found that hysteresis is greater for contaminated surfaces and on monolayers in which a polar group is buried beneath the surface [333].

Surface roughness makes it more complicated to relate contact angles to the surface chemical composition. Moreover, it makes it difficult even to compare surfaces that have similar compositions but different degrees of roughness. Therefore, it was suggested that one define a true contact angle (θ_{true}), an observed contact angle (θ), and a roughness factor (r) — the latter of which is defined as the ratio between the true surface area and its geometrical area ($r \geq 1$) — and write the following relationship [344–347],

$$\cos\theta = r\cos\theta_{true} \, .\qquad\qquad 1.32$$

According to this equation, roughness should increase contact angles that are greater than 90°, and decrease those that are less than 90° [348, 349]. Indeed, it was found that the roughening of a smooth nonpolar surface resulted in an increase of the advancing contact angles, a decrease in the receding contact angles, and as a result, an increase in the hysteresis [340, 350, 351]. Recently, Bain *et al.* reported results of experiments carried out on gold substrates, with the differences in roughness on the scale of 100 Å [352]. They adsorbed a monolayer of octadecanethiol ($C_{18}H_{37}SH$) on these surfaces and found that the contact angles always lay within a three-degree range. They also evaporated gold onto the very rough, unpolished side of the silicon wafer (consisting of 10 μm pyramidal asperities) and found that both water and hexadecane advancing contact angles on these rough surfaces were not significantly different from those measured on the smooth surfaces, but the hysteresis was about twice as much.

These results suggest that while the absolute value of the advancing contact angle may be used for comparing very smooth surfaces with different chemical compositions, it is not sensitive enough to surface roughness. Therefore, a complete study of different surfaces should include careful measurements of both advancing and receding contact angles and a comparison of the hysteresis.

It is clear that one of the open questions in the area of wetting is that of length scale. The experiments described earlier suggest that there is no dramatic difference in wetting between surfaces that have roughness differences on the order of microns. It is important, therefore, to be able to measure contact angles on atomically smooth surfaces, and to introduce roughness in a controlled fashion. Although reported in scientific meetings, there has been no account in the literature for a good alkyltrichlorosilane monolayer on mica surfaces. Such surfaces represent the ideal substrates for the studies suggested here, since roughness resulting from the substrate has been completely eliminated, and effects of other parameters on contact angle hysteresis can be studied systematically. Another possibility is to use alkanethiol monolayers on atomically smooth gold surfaces. The problem in this case is that the length scale of atomically flat gold surfaces is only 1000–2000 Å, which results in unusually high hysteresis even on —CF$_3$ surfaces [353]. Of course, once large areas of atomically flat gold surfaces are available, this will be the ideal system for wetting studies.

So far, we have discussed surface roughness and its effect on hysteresis. There are, however, other parameters that affect hysteresis. These parameters include:

1. *Heterogeneity* of the solid surface — a surface composed of two types of molecules, provided that the domains are large enough (> 100 Å x 100 Å).

2. *Reorientation* of surface molecules on the surface — a reorganization of the polar groups (e.g., OH or COOH) on high free-energy surfaces, due to exposure to a low dielectric constant that reduces the surface free energy [354].

3. *Solubility* of a surface component.

4. *Interaction* between the surface and the liquid — an adsorption (desorption) of the liquid molecules that changes the free energy of the surface accordingly.

In the case of heterogeneous surfaces, the Cassie equation was suggested for the interpretation of wettability results [344, 345, 355–357],

$$\cos\theta = f_1 \cos\theta_1 + f_2 \cos\theta_2 , \qquad\qquad 1.33$$

where θ is the contact angle of a liquid on the heterogeneous surface composed of a fraction f_1 of one type of chemical groups and f_2 of a second type of chemical groups (where $f_1 + f_2 = 1$), and θ_1 and θ_2 are the contact angles of this liquid on the pure homogeneous surfaces of 1 and 2, respectively.

However, Israelachvili and Gee recently suggested a more general approach to the treatment of the wetting of heterogeneous surfaces [358]. They started with the Young-Dupré equation,

$$\gamma_L (1 + \cos\theta) = W , \qquad\qquad 1.34$$

where γ_L is the surface free energy of the liquid and W the work of cohesion of the liquid with the surface. If we now take two homogeneous surfaces with chemical groups 1 and 2, we get

$$\gamma_L (1 + \cos\theta_1) = W_1 , \qquad\qquad 1.35$$
$$\gamma_L (1 + \cos\theta_2) = W_2 , \qquad\qquad 1.36$$

Of course, for the heterogeneous surface, we should get

$$\gamma_L (1 + \cos\theta) = f_1 W_1 + f_2 W_2 . \qquad\qquad 1.37$$

Now, if the patches of 1 and 2 are very small (approaching molecular dimensions), then, according to Israelachvili and Gee, the polarizabilities or dipole moments of 1 and 2 have to be averaged, rather than the cohesion energy. If this is the case, then

$$\gamma_L (1 + \cos\theta_1) = W_1 \propto \{w_1 w_L\}^{1/2} , \qquad\qquad 1.38$$
$$\gamma_L (1 + \cos\theta_2) = W_2 \propto \{w_2 w_L\}^{1/2} , \qquad\qquad 1.39$$

and for the heterogeneous surface, we get

$$\gamma_L (1 + \cos\theta) = W \propto [(f_1 w_1 + f_2 w_2)w_L]^{1/2} . \qquad\qquad 1.40$$

By substituting Equations 1.38 and 1.39 into Equation 1.40, we get

$$\gamma_L (1 + \cos\theta)^2 = f_1 (1 + \cos\theta_1)^2 + f_2 (1 + \cos\theta_2)^2 , \qquad\qquad 1.41$$

which is suggested as a replacement for the Cassie equation.

The problem with the preceding approach is that it completely neglects the interactions among surface functional groups. It is reasonable to assume that the effective dipole moment of a surface OH group, for example, depends on the dipole moment of its neighbors, and that the cutoff distance for the long-range surface dipole–dipole interaction also should be a function of the surface dipoles (i.e., the cutoff distance for a C–F dipole will be shorter than that for a C≡N

55

one). However, it is not clear that, with the present precision of contact angle measurements, they can be the chosen observable to distinguish the suggested treatment of surface polarization, from the existing more simple approach.

When the advanced contact angles (θ_a) of homologous series of liquids are measured for a given surface, their cosine is usually a monotonic function of γ_L. This was first discovered by Zisman [359], who proposed the following function:

$$\cos\theta_{SLV} = a - b\gamma_L = 1 - \beta\,(\gamma_L - \gamma_C)\,. \qquad 1.42$$

The line extrapolates to zero θ at a certain value of γ_L, which Zisman called the critical surface tension γ_C. Zisman found that various series of solvents extrapolate to the same value; thus he proposed γ_C as a characteristic quantity of a given surface.

A collection of γ_C values for different surface compositions appears in Table 1.4. Figure 1.29 presents the Zisman plot for a monolayer of OTS on a silicon surface [11]. The critical surface tension γ_C is useful as a summary of wetting results. It is not unusual that two different research groups will report different contact angles on what otherwise is the same surface. These differences may result from very small differences in the liquid purity, as well as from statistical errors in the actual reading of contact angles by different people. Since γ_C allows a comparison of results from different groups, it is recommended that γ_C be calculated for every new surface system. Furthermore, since the value of β in Equation 1.47 is rather small (0.03 to 0.04), actual contact angles for various surfaces can be estimated from γ_L and γ_C.

Table 1.4. Values of γ_C for different surface compositions.

Surface Constitution	γ_C (dyn/cm, at 20°)
$-CF_3$	6
$-CF_2H$	15
$-CF_3$ and $-CF_2-$	17
$-CF_2$	18
$-CH_2-CF_3$	20
$-CH_3$ (crystal)	22
$-CH_3$ (OTS monolayer)	20
$-CH_3$ ($C_{22}H_{45}SH$ monolayer on Au)	19
$-CH_2$	31
$-CClH-CH_2-$	39
$-CCl_2-CH_2-$	40
$=CCl_2$	43

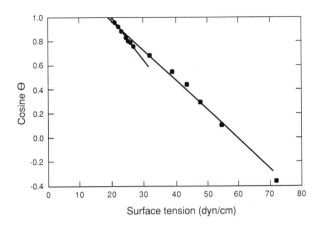

Figure 1.29. A Zisman plot of advancing contact angles on octadecyltrichlorosilane monolayer on silicon (OTS/Si). The short line represents extrapolation made using hydrocarbons, while the long line represents extrapolation made using polar solvents. Note that the two lines practically suggest the same value for γ_C. (From Tillman *et al.* [11], © 1988, Am. Chem. Soc.)

To summarize, wetting is the most sensitive measure of surface composition and structure, and shouldbe used, therefore, to study every new surface. Note that wetting properties of a surface relate to other properties, such as adhesion [342] and adsorption. The latter, for example, is an important consideration in the choice of polymeric materials for prosthetic parts, and for bioseparation membranes. The adsorption and, consequently, precipitation of proteins on the surface is detrimental in both cases. Similar surface-energy considerations are important in the design of biosensors (Section 5.8.C), where the adsorbed bioactive molecule should exhibit activity similar to that of such a molecule in solution. Thus, any structural changes that result from incompatibility of the biomolecule with the surface free energy will result in the apparent loss of bioactivity.

In our opinion, the current treatment of wettability of heterogeneous surfaces lacks a firm theoretical foundation. In the Cassie equation (1.33), a simple concentration-weighted averaging is performed on the contact angles of the two homogeneous surfaces. On the other hand, in the Israelachvili and Gee approach (1.41), such averaging is performed on the polarizabilities or dipole moments of the surface functionalities. Thus, both cases represent concentration-weighted interpolation between two extremes. Once those two extremes are fixed, environmental parameters, such as temperature and relative humidity, play no role in the interpolation. The definition of the contact angle is given by the Young equation (1.29). For homogeneous surfaces, evaluation of the interfacial free energies γ_{SV} and γ_{SL} involves only thermal averaging over the appropriate equilibrium ensemble. For heterogeneous surfaces, on the other hand, a proper statistical treatment should involve a thermal average of the related partition functions at each fixed configuration of surface functionalities, and not only at the two extreme cases corresponding to homogeneous surfaces. Subsequently, a quenched average of the resulting interfacial free energies, relative to a frozen probability distribution of those surface functionalities, should be performed. This can be demonstrated by simple lattice models, similar to the one recently discussed in our work on a concentration-driven surface transition in the wetting of mixed alkanethiol monolayers on gold [360].

We already have discussed experimental details of contact-angle measurements; we note that these measurements should be done in vibrationless conditions, preferably using a heavy table equipped with shock absorbers. Furthermore, for detailed, reproducible studies, one should have control of the temperature, relative humidity, and when volatile liquids are used, the vapor pressure of the liquid. This can be achieved by working in a closed, insulated compartment containing a small beaker with the liquid under study, and purged with water-saturated nitrogen.

A final note should be made concerning the purity of liquids used in contact-angle measurements. It is strongly recommended that one use caution when using commercial liquids, even when they are purchased as "analytical," "anhydrous," etc. We measured different HD contact angles for the same monolayer of OTS/Si when using different batches of the so-called anhydrous HD. Thus, all paraffin liquids should be percolated through basic alumina (activity one) until passing the *Zisman test* [361, 362]. In this test, a droplet of the liquid is placed on the surfaces of acidic and basic water and its spreading monitored with time. Ideally, the droplet should not spread at all, but it practically is required that the droplet does not spread for at least one hour.

B. Surface Second-Harmonic Generation Spectroscopy

Optical second-harmonic generation (SHG) is a highly directional, coherent process that arises from radiation of dipoles oscillating at 2ω in a medium nonlinearly induced by a laser field at frequency ω [363, 364], and is allowed only in systems without inversion symmetry under the electric dipole approximation. This generally is the case for surface modes because of the broken symmetry at the surface (e.g., water–air, solid–air, etc.), and therefore, SHG is a versatile probing method for the study of molecular orientation at surfaces. In other words, the electric quadratic contribution can dominate completely if sufficient field gradient exists at the interface. Thus, once the symmetry of the medium is correct, SHG is sensitive enough to detect even a submonolayer coverage. Equation 1.43 describes the surface nonlinear susceptibility $\chi_s^{(2)}$ [365–367],

$$\chi_{s,ijk}^{(2)} = N_s \langle T_{ijk}^{\lambda\mu\nu} \rangle \beta_{\lambda\mu\nu} , \qquad 1.43$$

where N_s is the surface density of molecules (molecules per unit area), β is the molecular nonlinear polarizability, T is the coordinate transformation connecting laboratory (x,y,z) and molecular (ξ,η,ζ) axes, and the angular brackets $\langle\rangle$ denote an average over the molecular orientations. In the SHG experiment, we measure $\chi^{(2)}$; however, it is the T that we need to know to find molecular orientation. This is rather difficult unless the molecules under study have a single β component along a specific molecular axid ξ, $\beta_{\xi\xi\xi}$, which is randomly oriented with respect to the surface normal. This usually is the case with dyes that exhibit large nonlinearities, where the largest β component is parallel to the direction of the molecular dipole. Thus, equations similar to those outlined here also are used in the study of molecular orientation in polled polymer films. In this case, we can write [365–367] the following:

$$\chi_{s,zzz}^{(2)} = N_s \langle \cos^3\theta \rangle \beta_{\xi\xi\xi} , \qquad 1.44$$

$$\chi_{s,zii}^{(2)} = \chi_{s,izi}^{(2)} = \chi_{s,iiz}^{(2)} = \frac{1}{2} N_s \langle \sin^2\theta \cos\theta \rangle \beta_{\xi\xi\xi}, \qquad 1.45$$

where $i = x,y$, and θ is the polar angle between the ξ axis and the surface normal that we chose as the z axis in the laboratory coordinates. Information about the molecular orientation now can be obtained from a simple ratio of the two independent components of $\chi^{(2)}$,

$$\frac{\chi_{s,zzz}^{(2)}}{\chi_{s,izi}^{(2)}} = 2\left[\frac{\langle\cos\theta\rangle}{\langle\cos\theta\sin^2\theta\rangle} - 1\right]. \qquad 1.46$$

If the orientational distribution is a sharply peaked function, Equation 1.46 is reduced to

$$\frac{\chi_{s,zzz}^{(2)}}{\chi_{s,izi}^{(2)}} = 2\left[\frac{1}{\sin^2\theta} - 1\right]. \qquad 1.47$$

Figure 1.30 shows a schematic description of an experimental setup used to observe SHG from monolayers at the water–air interface. The filter is placed to

59

separate the fundamental beam from the second-harmonic signal. The different components of $\chi^{(2)}$ can be measured selectively by the appropriate combination of the input (ω) and output (2ω) polarizers (Figure 1.30). This is achieved using the polarization null method in which a proper ω and 2ω polarization combination is used to obtain the vanishing of the effective surface $\chi^{(2)}$, which is given by

$$\chi_{s,\text{eff}}^{(2)} = a\chi_{s,zzz}^{(2)} + b\chi_{s,izi}^{(2)} = 0 , \qquad 1.48$$

where a and b are known coefficients.

Berkovic *et al.* used SHG to study the polymerization of a monolayer of vinyl stearate and octadecyl methacrylate at the water–air interface [368]. Kim and Liu measured molecular orientation of different chain-length alcohols at the water–air interface [369]. They reported that the orientation is strongly dependent on the surface concentration of the alcohol. Grubb *et al.* studied the orientation of monolayers at the water–air, and decane–water interfaces, and reported that the molecular orientation depends on the nature of the nonaqueous phase [370]. Mizrahi *et al.* used SHG to study anisotropic orientation distribution of polycrystalline LB monolayers [371]. They reported the first observation of a systematic influence on the properties of the prepared monolayer due to the trough used. Mizrahi *et al.* also used SHG to determine the microscopic symmetry of a *J*-aggregate-forming amphiphilic cyanine dye [372]. Berkovic *et al.* were the first to report that the third-harmonic generation (THG) signal (in reflection) from a monolayer of polydiacetylene is several times larger than that from the water subphase [372].

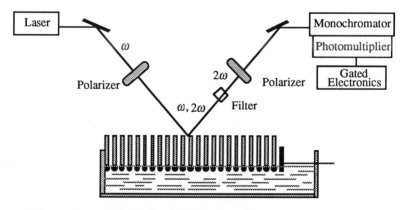

Figure 1.30. A schematic description of an experimental setup used to observe SHG from monolayers at the water–air interface. (From Shen [363], © 1989, Taylor & Francis.)

Sum frequency generation (SFG) is a nonlinear process similar to second-harmonic generation. Thus, while in SHG spectroscopy, only one frequency is used (i.e., $\omega_0 \rightarrow 2\omega_0$), in sum frequency generation spectroscopy, two different frequencies are used (i.e., ω_1, $\omega_2 \rightarrow \omega_1 + \omega_2$). As a second-order process, SFG cannot be used in a bulk with centrosymmetry, and therefore is highly surface-specific [374]. The basic idea of using SFG for vibrational spectroscopy is to convert the IR resonance signal to the visible so that it can be detected more easily.

Zhu *et al.* reported the first observation of a vibrational spectrum of a monolayer (coumarin 504 on fused silica) by IR-visible SFG [375]. Harris *et al.* used SFG to study monolayers adsorbed on metal and semiconductor surfaces [376]. They examined octadecanethiol on gold, cadmium stearate on silver, and stearic acid on germanium. Their results indicate that the shape and sign of the resonance signals is dependent on the relative magnitude and phase of the adsorbate and substrate nonlinear polarizations. Guyot-Sionnest *et al.* studied silane monolayers at the solid–air and solid–liquid interfaces [377]. Their results fundamentally are in agreement with IR and x-ray studies, indicating the the alkyl chains in the OTS monolayers essentially are oriented normal to the surface. They also found little difference between polar and nonpolar liquids (i.e., CCl_4–monolayer interface), indicating that the monolayer apparently is a closely packed assembly, with little or no incorporation of liquids at the interface, which could result in a randomization process. The authors could also monitor, using SFG, *in situ* adsorption of OTS on fused silica in real time.

In conclusion, SHG and sum frequency generation are powerful techniques for studies of interfaces. Recent contributions to scientific meetings indicate that it is possible to estimate *gauche*-bond concentration at chain termina using these techniques. This is very important when issues such as order–disorder transitions (Parts Two and Three) and surface reorganization are concerned. Note that these experiments are not as straightforward as, for example, FTIR, and usually are carried out by experienced spectroscopists.

C. Low-Energy Helium Diffraction

Low-energy atom diffraction is a recent addition to surface diffraction techniques [378–382]. This tool is most appropriate for probing the structure of the outermost atomic layer of any surface. (In monolayers of alkyl chains, it is a methyl layer.) The two main advantages of the technique are its extreme surface specificity and nondestructive nature. Therefore, a quantitative intensity analysis is possible, especially when the interaction potential between the probe atom (He) and the surface (CH_3) is known [383].

Chidsey *et al.* studied the surface of docosanethiol ($CH_3-(CH_2)_{21}-SH$) monolayers on a gold surface and cooled to below 100 K [384]. The substrate was a 1500 Å epitaxial film of Au(111) vapor deposited on mica at a surface temperature of ~225°C [385].

Figure 1.31(a) shows the diffraction at an azimuthal angle of 0°, where the set of peaks labeled $(-n,0)$ is most intense, and Figure 1.31(b) the diffraction at an azimuthal angle of 30°, where the peak labeled $(-1,-1)$ is most intense. Both spectra were taken at a surface temperature of 50 K. These spectra indicate that alkyl chains are packed hexagonally with a domain size of ~50 Å across (i.e., assemblies of ~7 x 7 molecules), and with a lattice distance of 5.01 ± 0.02 Å. This lattice distance is in complete agreement with the value of 4.97 ± 0.05 Å measured by Strong and Whitesides using transmission electron diffraction [308]. When surface temperature increases, the diffraction peaks gradually lose intensity, until finally disappearing at ~100 K. This loss of intensity means that there is temperature-induced vertical displacement of the CH_3 groups.

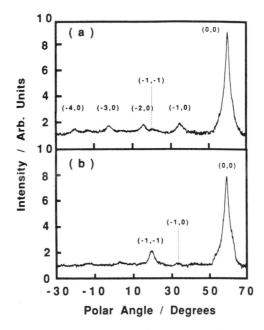

Figure 1.31. He diffraction of the surface of docosanethiol on Au(111), (a) at an azimuthal angle of 0°, and (b) at an azimuthal angle of 30°. (From Chidsey *et al.* [384], © 1989, Am. Inst. Phys.)

Preliminary results of diffraction from monolayers of $CH_3-(CH_2)_9-SH$ indicate that at 35 K, there is diffraction and the lattice constant is 5 Å [386].

Low-energy helium diffraction is an interesting new technique that gives information on the symmetry and spacing of chain terminal groups at the surface of a molecular assembly. Note that since samples are equilibrated at very low temperatures (e.g., 50 K), the resulting structural information may be different than the actual structure at 300 K. This by no means reduces the importance of this technique and its contribution to the studies of surface structure. An exciting possibility that low-energy helium diffraction may open is the study of molecular assemblies on structurally defined surfaces, i.e., epitaxy processes [386].

D. Optical Microscopy

Optical microscopy in the polarized and interference modes is very useful in the study of the size, shape, and orientation of domains, and to observe grain boundaries. The birefringence of a relatively thick LB film can be used to analyze its structure. Here, the polarizer and analyzer are set at 90° to one another, and molecular assemblies having optical axes in different directions are observed as contrasting regions. The use of monochromatic light helps to enhance these contrasts, since dark and light regions appear. Peterson [387] and then Peterson and Russell [388] used optical microscopy to study multilayers of ω-tricosenoic acid and to establish the epitaxy of its LB films. The addition of a Nomarski accessory to the optical microscope allows observation of transparent objects where refractive index or thickness differences are the only variations. The Nomarski microscope technique is basically a combination of working with polarized light and an interference method.

An interesting application of optical spectroscopy is in microdefect decoration and visualization. Lesieur *et al.* developed an electrodeposition procedure whereby the defects in the insulating organic film are filled with copper [389]. The film is deposited on a metal surface that serves as the electrode, which is dipped into an aqueous $CuSO_4$ solution. Copper electrodeposition occurs in pinholes and defects where an electrical contact between the electrode and solution exists. Peterson used this electrodeposition technique to study LB films of ω-tricosenoic acid [390]. Recently, Matsuda *et al.* used the technique to examine LB films of cadmium arachidate and a squarylium dye [391]. They could show that pinhole-free LB films can be constructed by utilizing amorphous films.

Braun *et al.* reported on a surface decoration by evaporation of 20 Å of Pt /C at an angle of 45° onto the LB film, and subsequently stabilizing the Pt/C layer with 200 Å of carbon [392]. They used the technique on different films deposited on different substrates.

E. Fluorescence Microscopy

The most important development in fluorescence microscopy of organic monolayers is the development of microscopy on the trough, allowing the study of monolayers at the water–air interface. Although McConnell *et al.* used fluorescence labeling in their studies of phospholipid monolayers [393–395], Lösche and Möhwald achieved the first direct visualization of the nucleation process in such monolayers [396]. Using a specially designed shallow trough, they inspected the underside of an illuminated (uv light) monolayer by having the objective lens of the fluoresecence microscope in the bottom of the trough. The resolution of the technique is ~2 μm over a field view of ~250 μm in diameter, and a time resolution of ~100 ms. A new design of a trough for an inversed fluorescence microscope was developed recently by Miyano and Mori [397].

Lösche *et al.* used this technique to study the phase transition of a phospholipid monolayer containing less than 1% of an amphiphilic fluorescent dye [398]. The critical property of this dye was its misibility in the liquid phase, and immisibility in the solid phase. Thus, once nucleation occurred, the dye was rejected from the solid domains, and a high contrast was achieved. Heckl *et al.* studied the first protein-containing phospholipid monolayer where the protein was LHCP, the light-harvesting chlorophyll protein B800-850 [399]. Duschl *et al.* used fluorescence microscopy to establish the two-dimensional crystal structure of a cyanine *J*-aggregate [400]. The first observations of LB monolayers of tetraphenylporphyrin (TPP) and tetrapyridylporphyrin (TPyP) derivatives were by Miller *et al.* [401]. In this case, the fluorescence spectra were very sensitive to molecular aggregation, thus providing enough contrast in the micrographs. Möhwald *et al.* continued the work on porphyrin aggregation, and studied both pure porphyrin monolayers and mixed monolayers of porphyrins with phospholipids [402].

Keller *et al.* [403], Moy *et al.* [404], Fischer *et al.* [405], and Lösche *et al.* [406] studied the the effect of electrical forces, arising from surface dipoles, on the shapes and arrangements of solid lipid domains at the water–air interface [407, 408]. Heckl *et al.* used an electric field to induce the movement of phospholipid domains at the water–air interface [409]. Möhwald *et al.* showed that due to long-range electrostatic forces, *J*-aggregates formed by adsorption from the subphase form periodic superstructures [410].

64

Suresh *et al.* observed the unstable growth of a liquid–condensed-liquid expanded interface in monolayers of myristic acid near its critical temperature [411]. The image reveals parallel rows of molecules with a side-by-side spacing of ~5 Å, in agreement with the expected alkyl–alkyl spacing in a closely packed, ordered monolayer.

Moore *et al.* used 1% of a stearic acid molecule labeled at the 12-position with methylamino NBD (*N*-methyl-7-nitro-2,1,3-benzoxadiazol-4-yl) to study the phase transition in LB monolayers of stearic acid [412]. They could observe clearly the evolution of film texture, measure its kinetics, and study the liquid–solid transition. Duschl *et al.* used fluorescence microscopy to examine cyanine dye–stearic acid mixed monolayers [413]. They reported that the dye-rich crystals had all the characteristic features of the "brick-stone" arrangement proposed from the packing in *J*-aggregates.

Riegler developed a fluorescence microscope for real-time studies of LB layer deposition [414], and Riegler and LeGrange used it in the first observation of monolayer phase transition on the meniscus in an LB transfer configuration [415]. This work is discussed in Section 2.1.E.1.

F. Electron Microscopy

Figure 1.32 presents a schematic diagram of a *scanning electron microscope* (SEM). Here, an electron beam from a tungsten filament is focused by magnetic lenses on an area of surface 50–150 Å in diameter in a high-vacuum chamber.

This is the same arrangement, in principle, as in Auger spectroscopy and low-energy electron diffraction (LEED). The differences between these experiments, however, are in the electron beam energies, the geometry of the bombardment, and the mode of observation. Since we are going to discuss different surface imaging techniques, it is important to describe the different interaction modes of an electron beam and atoms on a surface. The electron beam is composed of so-called primary electrons, and in the first case, this beam simply can be reflected from the surface (the top layer of atoms) without any energy interchange. Since the energy of primary electrons is high, much higher than the energy with which electrons are bound to nuclei, a dissipation of this energy in the sample will knock electrons out of the atoms. These are the secondary electrons, and their energies (lower than those of the primary electrons) can vary because, in most cases, a number of ionization processes may be induced by the primary electrons.

1. Scanning Electron Microscopy (SEM)

The SEM technique uses the intensity of the secondary electrons (usually a broad energy distribution) [417–419]. Some of these electrons recombine with ions at the surface, and as a result, photons are released. These photons are the basis for the SEM imaging capability, and the contrast in the image is a result of differences in scattering from different surface areas, as a result of geometry differences. SEM has been widely used for visualization of organic surfaces, especially in the study of surface morphology, domains, pinholes and defects, and patterns.

Fryer *et al.* reported on the imaging of a single phthalocyanine monolayer by SEM [304, 420]. Itoh *et al.* used SEM to study monolayers of zinc chlorophyll and stearic acids on SiO_2 substrates, reporting a phase separation whereby disk-like assemblies of stearic acid were formed, and seemed to be compressed especially along the direction of deposition [421].

Fromherz *et al.* used SEM for visualization of platinum–carbon replicas of cadmium arachidate multilayers [422]. (See earlier.) Wasserman *et al.* used SEM to investigate adhesion failure in silicon/11-trichlorosilylundecanethiol/gold samples [423]. Fujiki *et al.* reported on the direct patterning of phthalocyanine LB films [424], and Oguchi *et al.* used SEM to study etched patterns of LB films of acetylized poly(vinyl alcohol) [425].

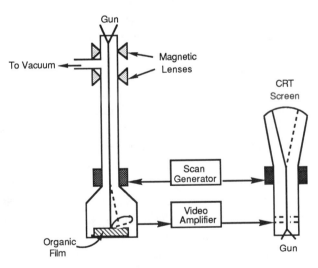

Figure 1.32. A schematic diagram of a scanning electron microscope. (Adapted from Hiemenez [416], © 1986, Marcel Dekker.)

66

2. Transmission Electron Microscopy (TEM)

This technique requires that the thickness of the film under study will be less than 300 Å; however, direct observation of one monolayer is not possible because of lack of contrast. The electron energy in TEM is very high (100 KeV) and the resolution ranges from 1000 Å in charge decoration to a few tens of Ångstroms in silver decoration. (See later.)

Barraud *et al.* developed an *in situ* silver decoration technique for the visualization of monolayer slow collapse [426]. By immersing behenic acid multilayers into an aqueous $AgNO_3$ solution, they introduced the silver heavy atoms into the polar planes of the multilayer, thus forming silver behenate. TEM micrographs revealed regions of parallel dark lines in 45–50 Å spacing, which is equal to the thickness of a bilayer, thus suggesting a folding mechanism for the slow-collapse process. Fischer and Sackmann used a charge decoration technique with TEM for studies of phospholipid monolayers in the coexistence regime [427]. Heckl *et al.* used the technique to study protein-containing phospholipid monolayers [399]. Barger *et al.* studied pure tetrakis-(cumylphenoxy)phthalocyanine monolayers and mixed monolayers with fatty acids, and discovered disk-like aggregates, more than monomolecular-layer-thick in all cases, indicating a significant degree of phase separation [428]. Braun *et al.* used TEM to study the surface characteristics of LB films of stearic, ω-tricosenoic, 8,12-diynoic acids and phthalocyaninatopolysiloxane [301]. Inoue *et al.* used both SEM and TEM to study the microstructure in LB films of the NLO-active 4-nitroaniline derivative, and observed three-dimensional microcrystallites even in mixed monolayers with icosanoic acid [429]. Rahman *et al.* used both SEM and TEM to study monolayers of polypyrrole on glass and copper substrates, respectively [307]. These micrographs reveal the smooth coverage of the polymer over large surface areas of both substrates.

In conclusion, SEM and TEM are powerful techniques for visualization of organic surfaces, especially in the study of surface morphology, domains, pinholes and defects, and patterns. Many reports lack information about monolayer quality, and disclose only the physical properties of interest. Thus, correlations between the structure of individual amphiphiles and the quality of films made thereof are not possible. Detailed information on surface morphology and its relation to film stability are vital to the developement of any monolayer-based technology. Therefore, attempts should be made to use SEM and TEM when new monolayers are made, so that such information can be provided.

G. Scanning Tunneling Microscopy (STM)

Scanning tunneling microscopy (STM) is quite a new tool for the investigation of very small areas of surfaces with extremely high level of precision [430,,431]. The original work in STM was on semiconductors, where their surfaces were imaged with unprecedented resolution [432–434]. Other exciting applications of STM were local work function imaging [435] and local electron spectroscopic information [434,,436–439], including, for example, the superconducting energy gap [440,,441]. The reader may find further information on STM in reviews [442–445] and in the proceedings of the International Conferences on Scanning Tunneling Microscopy [446–448].

In the preparation of a sample for STM studies, there are three main requirements: (*a*) that the sample will have some electrical conductivity; (*b*) that it will be flat, preferably atomically flat; and (*c*) that molecules and atoms on the surface will have limited mobility. Evaporated gold on mica satisfies the first two requirements [385, 430, 431, 449–453], platinum [454, 455], silver [454], and aluminum [456] also have been studied.

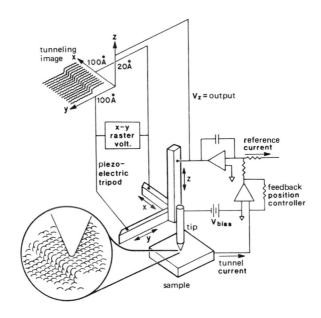

Figure 1.33. Schematic illustration of the scanning tunneling microscope. The tunnel current from tip to sample, induced by V_{bias}, is maintained constant by an electronic feedback system (right), which controls the tip position normal to the sample surface via the *z*-leg of the piezoelectric tripod. A record of the tripod *z*-leg feedback voltage, as the *x–y* tripod legs raster scan the tip laterally, constitutes a tunneling image, which is a form of replica of the sample surface (insert at bottom). (From Golovchenko [442], © 1986, Science.)

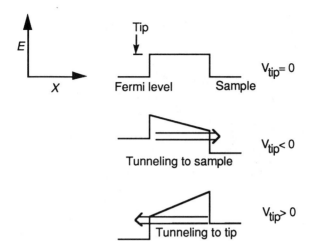

Figure 1.34. An energy diagram explaining tunneling in STM experiment.

In a typical STM experiment (Figure 1.33), an atomically sharp metallic tip (e.g., tungsten) is brought very close ($\leq$ 10 Å) to the surface. If a small potential difference (~1 V) is applied between the surface and the tip, a tunneling current will flow. This tunneling is due to the fact that wave functions extend into vacuum. Thus, as a result of the overlap between atoms on the tip and atoms on the surface, a current can flow between the tip and the surface. Since this current decreases about one order of magnitude per ~1 Å of the electrical gap width, accuracy of the order of 0.1 Å can be achieved. Now the tip is scanned on the surface while the current is monitored. Figure 1.34 presents an energy scheme for the tunneling.

There are two ways to carry out an STM experiment, the first being the constant current [430, 431], and the second the constant height approach [457]. In the first, the tip height above the sample is adjusted maintaining a constant current, and the image is obtained as a map of the tip height z, vs. the lateral coordinated x and y. In the latter and more limited approach, the height is kept constant and the current I is recorded as a function of the lateral coordinates.

The imaging of organic adsorbates by STM is challenging due to the low conductivity of most organic films, and there still are many controversial cases in the literature. In the case of π-systems, reports were published on imaging of layers of low-molecular-weight liquid crystalline materials on graphite [458–460], individual molecules of phthalocyanine on copper [461], and protoporphyrin IX dimethyl ester on gold (one monolayer) [462]. Studies of monolayers and multilayers of long alkyl chain materials were reported starting with the paper of Hallmark *et al.*, who compared STM and grazing-angle FTIR and also compared octadecyltrichlorosilane and Cd arachidate monolayers [58]. In a recent paper [463], they reported on the same study, but used evaporated gold on mica as substrates. The images of monolayers of OTS and CdA revealed a highly ordered, close-packed hexagonal structure that extended for hundreds of Ångstroms in the lateral direction. However, careful experiments with uncoated Au/mica gave the same results, indicating that they did not detect by STM the presence of these monolayers on the gold surfaces.

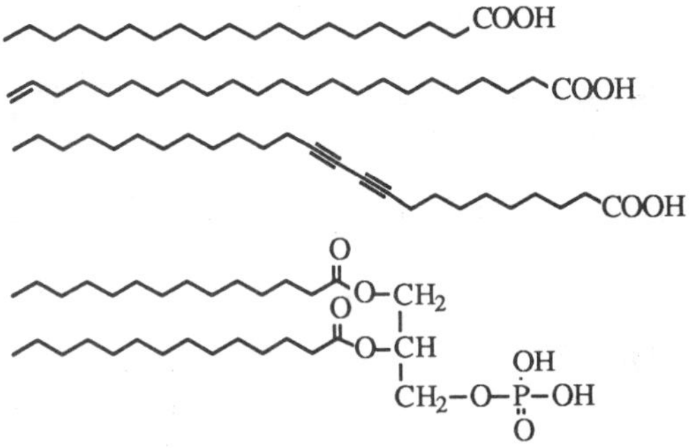

~~~~~~~~~~~~~~~COOH   arachidic acid (24 Å)

~~~~~~~~~~~~~~~COOH   ω-tricosenoic acid (27.5 Å)

~~~~~~~~~≡≡~~~~~~COOH   12,8-diynoic acid (15 Å and 10 Å)

$$O$$
~~~~~~~~~C(=O)O-CH$_2$
~~~~~~~~~C(=O)O-CH   dimyristoylphosphatidic acid (16.5 Å)
                    | OH
                    CH$_2$-O-P-OH
                             O

Braun *et al.* [392] and Fuchs [464] reported on the imaging of a monolayer of Cd arachidate, ω-tricosenoic acid, and 12,8-diynoic acid, and Hörber *et al.* [465] also reported on imaging Cd arachidate on graphite, and di-myristoyl-phosphatidic acid (DMPA) on both graphite and gold surfaces. These materials can be divided into two parts, those with an alkyl chain length longer than 20 Å, and those where the alkyl chain length is shorter than 20 Å. (The net alkyl chain lengths in the parentheses above were calculated based on the contribution of 1.25 Å and 1.5 Å per methylene and methyl groups, respectively, and for all-*trans* perpendicular orientation.)

For the first two, and in view of the careful study of Hallmark *et al.*, it seems that the alkyl chains are too long (>20 Å) to allow tunneling, and therefore, there may be some problems in the interpretation of the contrast in STM imaging of such organic materials. In the case of the 12,8-diynoic acid, there is a 15 Å alkyl chain above, and 10 Å alkyl chain below the diyne group. If this diyne π-system has conductivity much higher than that of a saturated alkyl chain, one could speculate that since the saturated chains above and below this conductor are short enough (<< 20 Å), electron tunneling may be possible. The idea of dividing an insulating alkyl chain by a conducting π-system needs further study, however, if correct, it may have an important implication for the imaging of organic monolayers by STM. In the DMPA monolayer, the alkyl chain length is 16.5 Å, short enough to allow tunneling. The fact that varying the support did not affect the measured lattice spacing is encouraging and may indicate that the images are indeed "real."

In an attempt to circumvent the conductivity problem, the imaging of single molecules adsorbed on the surface (e.g., phthalocyanine [461]) was studied. This idea was used successfully in recent papers, where the imaging of molecules adsorbed flat on conducting surfaces (e.g., graphite) is described. Spong *et al.* published the high-resolution STM image of a liquid–crystalline material, 5-nonyl-2-*n*-nonoxylphenylpyrimidine [466]. Another publication in this class of adsorbed liquid crystals on flat graphite surfaces is that of Foster and Frommer [467].

Rabe *et al.* published a paper in which they were able to get images of both octylcyanobiphenyl molecules and mixed cellulose ether (laurylmethyl) [468]. They found that localized etching of the graphite (at between –2.5 and –4V) results in a readily observed crystalline adsorbate phase of octylcyanobiphenyl. It seems that the change in surface structure and roughness promotes crystallization of a more stable phase that can be observed under experimental conditions. In the case of cellulose ether (a polymer), they studied Langmuir–Blodgett monolayers on graphite, this time with a positive bias of between +10 mV and +1 V, and were able to get a resolution better than 10 Å. Their pictures show beautifully how the polymer is extended over several hundreds of Ångstroms, and reveal characteristic kinks in the chain, resulting from the adsorption of a twisted ribbon.

In the area of biology, Beebe *et al.* [469] and Lee *et al.* [470] published on the imaging of DNA and double stranded RNA. In both cases, minor and major grooves in the double helixes of the adsorbed biomolecules could be resolved.

With all the preceding information in mind, there still is much controversy as to the actual meaning of the STM images. In many cases, images show scratches in a direction parallel to the tip scanning, suggesting that the tip penetrated into the organic film. In other cases, scanning in two different directions does not produce the same image. Moreover, the reproducibility in the case of LB and SA monolayers is rather poor, and the question of conductivity through the alkyl chains has not been resolved yet. There also is a question whether the tungsten tip is stable under the high electric field. Finally, the pressure applied on the organic surface in the STM experiment is $\approx 10^{-7}$ newtons, orders of magnitude larger than that applied in the atomic-force microscope (AFM) experiment (10^{-11}–10^{-10} newtons, see next section). Thus, I feel that the method of choice for organic thin films should be AFM.

71

H. Atomic Force Microscopy (AFM)

The *atomic force microscope* (AFM), in contrast to STM, can image both conductors and nonconductors with atomic resolution [471–476]. Here, the sample is scanned past a stationary tip mounted on a spring. The atoms in the tip interact with those on the surface, and the resulting attractive or repulsive forces deflect the spring. Most of the advances in the design of AFM have been done in Hansma's group.

In the present design, the size of the spring is <1 mm, and the forces used are $\leq 10^{-8}$ newtons, equivalent to a weight of one millionth of a gram. (Usually, the forces are even smaller, 10^{-10}–10^{-11} newtons). This is necessary so that the tip does not damage the surface.

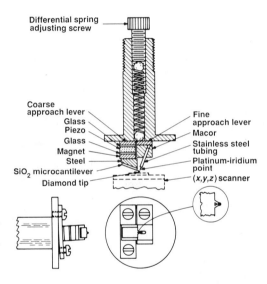

Figure 1.35. Setup of the AFM. The sample is shown by dashed lines. The sample is attached to the center of the (x,y,z) scanner and pressed against the tip of the force sensor. The central mounting minimizes the interference and distortions that a signal on one axis in an (x,y,z) scanner of the single-tube design imposes on the other axes. The silicon chip holding the silicon dioxide microcantilever is glued to a steel wedge that has a 20° angle. The steel wedge is held in place by a magnet on the force sensor, allowing easy positioning opposite to the platinum–iridium point. The approach of the point to the cantilever is achieved in three steps. The coarse approach lever with the cantilever (insert, lower left) is used to lower the gap between point and cantilever to ~10 μm. Further reduction is achieved with the differential spring mechanism that consists of the differential spring adjusting screw and its associated spring and ball and the fine approach lever holding the tip. The last stage of adjustment is the piezo located between the magnet and the coarse approach lever. To achieve a well-defined sharp tip, a piece of diamond dust 40 to 60 μm wide or a fragment of a shattered diamond is used (insert lower right). (From Marti *et al.* [475], © 1988, Science.)

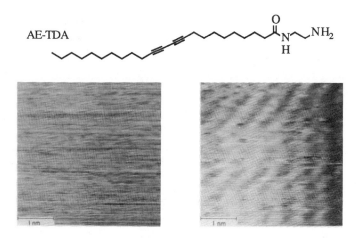

Figure 1.36. AFM images of the surface of a glass slide (left), and of a polymerized monolayer of AE-TDA on a glass slide (right). The rows of molecules are with a side-by-side spacing of 5 Å. (From Marti *et al.* [475], © 1988, Science.)

To monitor the deflection of the spring with a sub-Ångstrom resolution, Marti *et al.* used tunneling [477]. Figure 1.35 describes their basic AFM design [474, 477]. The sample is mounted on an *x,y,z* scanner [478] and pressed against a tip that is mounted on a spring. The deflection of the spring is monitored by sensing the tunneling current between the back of the spring and a sensing electrode.

Two recent papers from Hansma's group demonstrate the potential in the AFM technique. In the first [Gould, 479], they describe a molecular resolution image of crystals of DL-leucine. They were able to resolve arrays of molecules with lattice spacings that are consistent with x-ray diffraction data. In the second [Marti, 480], they image polymerized LB monolayers of *n*-(2-aminoethyl)-10,12-tricosdiynamide (AE-TDA, see chemical diagram above Figure 1.36). The image reveals parallel rows of molecules with a side-by-side spacing of ~5 Å, in agreement with the expected alkyl–alkyl spacing in a closely packed, ordered monolayer (Figure 1.36).

I. Scanning Ion Conductance Microscopy

The *scanning ion conductance microscope* (SICM) is a very recent addition to the family of scanning-probe microscopes [481]. In this method (Figure 1.37), a micropipette filled with electrolyte (0.1 M NaCl) is scanned laterally over the examined surface, while a feedback system lowers and raises it to keep the conductance constant [482]. The dimensions of the micropipette were ~0.1–0.2 μm and ~0.05–0.1 μm for the outer and inner diameters, respectively.

Since the tip follows the topography of the surface, the image can be obtained by measuring the voltage that the feedback system applied to the tip to keep conductance constant. Of course, when the tip approaches the surface, ion conductance decreases due to a decrease in the volume of electrolyte between the tip and the surface. The hardware and software was developed in Hansma's group, where the imaging of 0.8 μm diameter pores in a Nulepore membrane filter has been demonstrated. With their enhancement program, they could get information on the third dimension (depth) of the pores in the surface.

While the SICM at this stage can image with submicron resolution, it seems that future generations of the microscope, with pipettes that have much smaller tip openings, will make it possible to resolve images at the scale of 100 Å. When this is achieved, SICM will be an excellent method for the imaging of soft non-conducting monolayer and multilayer surfaces.

J. Surface Plasmon Microscopy

Plasmon surface polaritons (surface plasmons, or PSP) are surface electromagnetic modesthat travel along a metal-dielectric interface as bound nonradiative waves with their field amplitudes decaying exponentially perpendicular to the interface [483,484]. The dispersion relation for the case of an interface separating a metal and a dielectric is given by,

$$k_x^\infty = \frac{\omega}{c}\left[\frac{\varepsilon_m(\omega)\varepsilon_d(\omega)}{\varepsilon_m(\omega) + \varepsilon_d(\omega)}\right]^{1/2}, \qquad 1.49$$

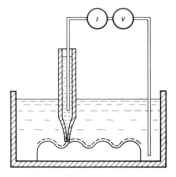

Figure 1.37. Schematic illustration of the scanning ion conductance microscope (SICM). (Hansma *et al.* [482], © 1989, Science.)

where k_x^∞ is the longitudinal component of the complex PSP wave vector, ω is the real frequency, c is the speed of light, $\omega_m(\omega) = \varepsilon_{mr}(\omega) + i\varepsilon_{mi}(\omega)$ is the complex frequency-dependent dielectric constant of the metal, and $\omega_d(\omega) = \varepsilon_{dr}(\omega) + i\varepsilon_{di}(\omega)$ is the complex frequency-dependent dielectric constant of the dielectric material (organic film) [485]. Since photons cannot be converted directly to PSP, a so-called plasmon coupler should be used. In the works described here, a prism in the Kretschmann configuration has been used [486]. Figure 1.38 presents a schematic setup of the SPM. The laser beam (ω_L)is coupled to the plasmon surface polariton (PSP) modes at the silver-coating (*ca.* 50 nm)–air interface via the 90° BK-7 glass prism in an ATR geometry. Resonance excitation of PSP in the laterally heterogeneous interface occurs at an angle θ_0 wherever the momentum matching condition $k^{||}_{photon} = k_{sp}$ between the parallel component of the photon wave vector $k^{||}_{photon}$ and the PSP wave vector k_{sp}, is fulfilled:

$$k^{||}_{photon} = \frac{\omega_L}{c} \varepsilon_{prism}^{1/2} \sin\theta_0 = k_x \ . \qquad 1.50$$

Any thin dielectric coating causes an angular shift of the resonance conditions, and therefore, it can be used to characterize optically LB monolayers and multilayers [487]. Note that the latter is sensitive to n_z (the perpendicular component of the refractive index of the dielectric). This is because the PSP field components are mostly E_z polarized. The resonance conditions also depend on the thickness of the coating that results in a thickness-dependent resonance for PSP excitation. In short, those areas that carry a PSP field couple out some of the intensity to the air side, due to the intrinsic roughness of the interface. This scattered light is focused by the lens of a camera on a screen, as in Figure 1.38.

Knoll *et al.* introduced *surface plasmon microscopy* (SPM) as a new imaging technique [19, 488–490]. This technique is useful especially for the imaging of low-contrast samples, such as heterogeneously organized lipid LB monolayers. In this case, there are two phases in coexistence, i.e., the expanded (fluid) and the condensed (crystalline). SPM allows the imaging of such systems without the addition of a fluorescent dye, which usually is used to enhance the contrast between the different domains. The contrast mechanism is illustrated in Figure 1.39.

Thus, when plane waves impinge upon the interface at an angle θ_0, they excite PSP only at the bare silver areas, while the covered silver areas remain dark. On the other hand, when an angle θ_1 is used, PSP is excited only at the covered silver areas, and the bare silver surface remains dark.

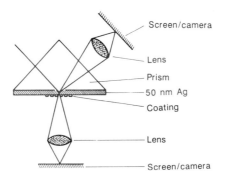

Figure 1.38. Schematic setup of SPM. (From Rothenhäusler *et al.* [488], © 1988, Nature.)

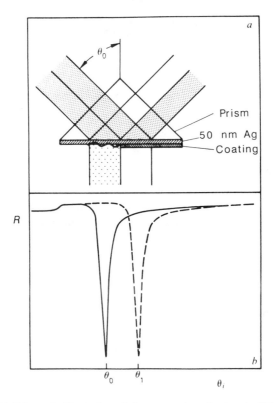

Figure 1.39. (*a*) Schematic illustration of the contrast mechanism in SPM. Half the silver film is covered by a thin coating, which shifts the resonance condition for the plasmon coupling. At θ_0, PSP are excited only in the bare silver areas (wavy arrow). A small fraction of this light is coupled out to the air side, while almost nothing is reradiated to the prism side. The coated regions are off resonance, so that all laser light is totally reflected. (*b*) Reflectivity R vs. angle of incidence θ_i for bare silver (full line), and for the coated regions (dashed line). (From Rothenhäusler *et al.* [488], © 1988, Nature.)

76

K. Auger Electron Spectroscopy

In *Auger electron spectroscopy* a 2–3 KeV electron beam interacts with the sample, causing ionization of an inner electron (e.g., a K electron) in a surface atom, thus creating a vacancy. To fill this vacancy, a less energetic electron from an outer shell (e.g., an L_i electron) drops into the K-shell, releasing the energy difference ($\Delta E = E_K - E_{L_i}$), which now can ionize an ever more outer electron (in general, an L_i electron). This outer electron is called the *Auger electron*. The energy difference $\Delta E - E_{L_i}$ is the kinetic energy of the Auger electron. Since these energy differences are specific to different elements, Auger spectroscopy allows the selective detection of different elements and thus the analysis of surface composition [491, 492]. Moreover, the use of SEM and Auger spectroscopy in conjunction with each other gives a map of the surface composition [419]. In general, a grazing-angle of incident electrons is needed to maximize the contribution of surface atoms to the Auger spectrum.

Barraud *et al.* used Auger spectroscopy to investigate the reaction of a first-coated monolayer of a fatty acid with the substrate [493]. Green and Faulkner investigated the electrochemical oxidation of an LB film of zinc phthalocyanine [494], where the incorporation of the PF_6^- anion in the oxidized films was observed in the Auger spectra. Laibinis *et al.* used SEM and Auger spectroscopies to create an element map in their studies of orthogonal self-assembled monolayers [495].

L. X-Ray Photoelectron Spectroscopy

In an *x-ray photoelectron spectroscopy* (XPS or ESCA) experiment, a sample is exposed to monochromatic x-radiation and the properties of inner-shell electrons are probed. If E_0 is the energy of the x-ray, and E_j is that of the electron in the atom (*s, p,* etc.), E_0-E_j will be the energy of the ejected electron. Since E_j is a function of the valence state of the atom, the electron spectrometer may give valuable structural information. For example, for 1s sulfur electrons, one observes an increase of over 5 V in the ionization energy as the valence state of the sulfur varies from -2 (R–S–R') to +6 (R–SO$_2$–R'). This effect is illustrated in Figure 1.40, where the oxidation of a thioether to a sulfoxide (S^{+4}) in a self-assembled monolayer was studied [496]. Because of the penetrating nature of x-rays, special techniques are used that can emphasize the contribution from atoms in different depth positions in a film. For example, the use of x-rays at a grazing angle to the surface emphasizes the contribution from surface atoms, and a study of photoelectron spectroscopy as a function of the takeoff angle offers an excellent way to study atom distribution in a film. This is illustrated in Figure 1.41, where the position of a sulfone (SO_2) group in a self-assembled monolayer was determined [497].

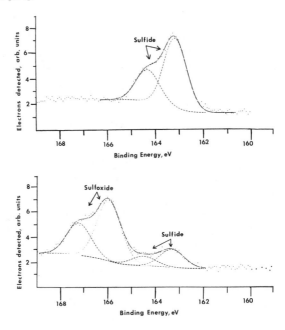

Figure 1.40. XPS spectra of monolayers. Top, high-resolution scan of a monolayer of CH_3–S–C_6H_4–O–$(CH_2)_{11}$–$SiCl_3$ on Si, and bottom, high-resolution scan of the same monolayer after oxidation with H_2O_2 /CH_3COOH at room temperature. (From Tillman *et al.* [496], © 1989, Am. Chem. Soc.)

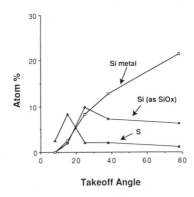

Figure 1.41. Atom percentage determined by XPS vs. electron takeoff angle for silicon and sulfur, in a monolayer of $C_{12}H_{25}$–SO_2–C_6H_4–O–$(CH_2)_{11}$–$SiCl_3$ / Si. (From Tillman *et al.* [497], © 1990, Am. Chem. Soc.)

Nuzzo *et al.* used XPS to study the adsorption of dialkyl disulfides and alkanethiols on gold surfaces [498, 499]. Wasserman *et al.* used the technique to establish radiation damage caused to a monolayer of tetradecyltrichlorosilane on silicon by exposure to x-rays from a synchrotron source [5]. Another important use of ESCA is in the analysis of surface functionality concentration in mixed monolayers. This approach has been used extensively where the molecular-level understanding of wetting is sought [332, 352, 360, 501–504].

Untereker *et al.* used ESCA to examine chemically modified SnO_2, TiO_2, and glassy carbon electrodes [505]. They found that bonding of organosilane layers to the electrode causes diminution of the native photoelectron intensity of the electrode elements, and suggested that this is due to electron interactions with the organosilane. Based on Carlson and McGuire [506], they suggested

$$I/I_{unreacted} = \exp(-d/\lambda) , \qquad\qquad 1.51$$

where I is the photoelectron current, d is the average organosilane layer thickness, and λ is the escape depth of the electrode element photoelectron through it. Using $\lambda = 11$ Å from a universal escape depth plot developed by Pamberg [507], and their value of $I/I_{unreacted} = 0.38$ for SnO_2 electrodes modified with a silane [505], they estimated $d \sim 11$ Å for the organosilane layer $(H_2N(CH_2)_2NH(CH_2)_3SiO—)$, which is in agreement with model-estimated 10 Å thickness of a monolayer. Similar calculations were done for other organosilanes on the different electrodes. Of course, the accuracy of thickness determination depends directly on the employed value of escape depth of photoelectrons through the organosilane; thus, one has to exercise caution. Nevertheless, ESCA can serve as an additional tool for film thickness estimation, and, therefore, estimation of coverage (mol cm^{-2}). For more information on escape depth, the reader may refer to References 508–512.

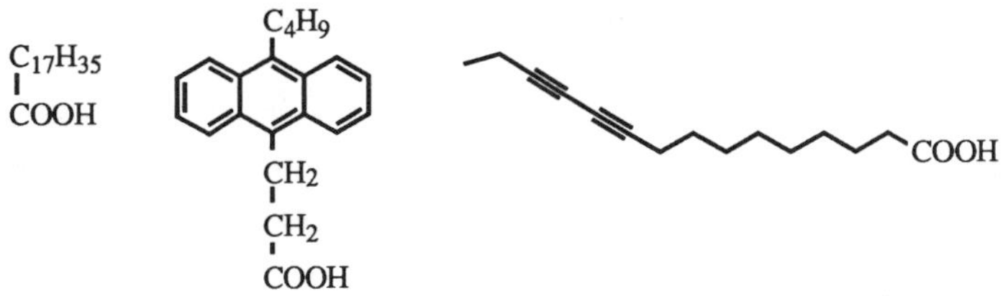

$C_{17}H_{35}$
|
COOH

C_4H_9

CH2
|
CH2
|
COOH

COOH

Brundle *et al.* reported that the escape depth of photoelectrons from metal surfaces through cadmium icosanoate (26.5 Å thick) was 36–41 Å [513]. Clark *et al.* used ESCA to investigate the effect of etching on samples of LB layers deposited in the (100) surface of *n*-type InP semiconductors [514]. In another paper, Clark *et al.* studied LB multilayers of cadmium stearate, 9-*n*-butyl-10-anthrylpropionic acid (and its cadmium salt), and a diacetylene polymer derived from pentacosa-10,12-diynoic acid (see chemical diagram) [515].

Their escape depth values for a fixed kinetic energy of 1170 eV are 45, 57, and 70 Å, for films of cadmium stearate, the diacetylene polymer, and the anthracene derivative, deposited at dipping pressures of 25, 15, and 15 dyn/cm, respectively. These numbers indicate the contribution of the more polarizable and ionizable π-system to the escape depth, and also can be related to differences in electron tunneling behavior in the different organic media.

Hazell *et al.* used take-off-angle ESCA experiments to study LB films of octadecylamine ($C_{18}H_{37}$-NH_2, ODA) on glass slides [516]. They looked at two molecular conformations in bilayers, first where the ODA films deposited on a hydrophilic substrate, and then where the deposition was on a hydrophobic substrate (Figure 1.42). Their conclusion was that discrepancies between model predictions and the variation in carbon composition with the angle suggest that LB films of ODA may not be as well-oriented or as stable as anticipated.

The suggested reason was that originally ordered films may not survive the drying out in the UHV. Kobayashi *et al.* investigated the composition of LB films of icosanoic acid deposited from aqueous subphase containing metal ions such as cadmium, calcium, barium, and lead in different pH regions [517]. They reported that the pH region for salt formation depended, as expected, on the nature of the metal ion.

Takahara *et al.* used XPS to study LB composite thin films of fluorocarbon amphiphiles with two fluorocarbon chains and a poly(vinyl alcohol)/(fluoro-carbon amphiphile) [518]. They reported that the photoelectron mean path for C_{1s} photoelectrons in the fluorocarbon film was *ca.* 40 Å. This value is similar (±10%) to the escape depth reported earlier for cadmium stearate [515]. However, the chain length in the case of the fluoroalkyl was C_{11} (CF_3–$(CF_2)_7$–CH_2–CH_2–) compared to the C_{17} chain length in the cadmium stearate film, therefore suggesting that the escape depth in the fluoroalkyl case is inherently smaller.

In conclusion, XPS is a useful technique for quantitative analysis of surface composition, but gives very little, or no, lateral resolution. Studies of monolayer surfaces are not straightforward. Thus, when mixed monolayers are under study, it is vital to study carefully monolayers of each of the mixture components, and hence to establish the "100%" surface concentration for each surface element. In the case of monolayers of alkanethiols on silver, it is very difficult to establish surface concentration of oxygen-containing groups (e.g., OH), since even under very careful evaporation, a silver surface will have between a third- and a half-monolayer of the oxide.

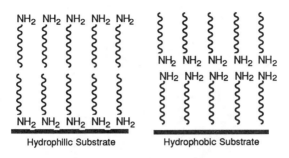

Hydrophilic Substrate   Hydrophobic Substrate

Figure 1.42. A tail-to-tail bilayer (left) and a head-to-head bilayer (right) of ODA.

Takeoff angle experiments are useful for studying the interior of thin organic films. These experiments may help establish the vertical structure, and thus the monolayer nature, of a thin organic film.

M. Surface Ionization Mass Spectroscopy

Secondary-ion mass spectroscopy (SIMS) is one of the most sensitive techniques available for studying surface composition. It started as a tool for the study of inorganic surfaces (metals and semiconductors) [519], and then became more applicable for the study of organic surfaces with the development of static or low-damage experimental conditions [520–522]. The technique allows investigation of surfaces with lateral resolution of ~1 μm, sensitivity in the 10^{-9}–10^{-12} g range [523], and depth resolution of ~10 Å.

The SIMS experiment involves two types of ions. The first is called a primary ion and it usually is Ar^+ or Xe^+ [524, 525], although other ions such as Cs^+ also were used [526]. To avoid beam damage, the primary-ion dose must be kept $\leq 10^{13}$ ions/cm^2 (static SIMS) [1]. The reader should remember that SIMS always is a destructive process. The difference between the dynamic and static SIMS processes is that in the latter, ions come out only from the outermost surface. The primary-ion beam impinges on the surface, ionizes molecules, and creates secondary ions that are detected in a mass spectrometer (Figure 1.43).

Once surface damage has been minimized, the secondary ions represent the molecular species in the film, similar in some respects to conventional mass spectrometry. Thus, protonated, deprotonated, metal-containing pseudo-molecular ions, as well as fragmented ions, were observed in high yields from organic surfaces [522].

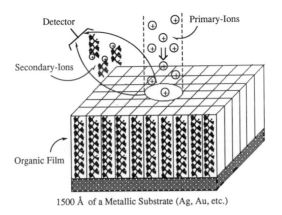

Detector   Primary-Ions

Secondary-Ions

Organic Film

1500 Å of a Metallic Substrate (Ag, Au, etc.)

Figure 1.43. A schematic description of the SIMS experiment (Adapted from Laxhuber *et al.* [524], © 1984, Elsevier Science Publishers.)

The desorption of these secondary ions, although induced by a collision cascade triggered by the ionization process, depends on molecular environment [527, 528], which makes it very attractive for the investigation of ordered organic systems, where substrate–organic as well as intermolecular interactions determine the packing and orientation of the molecules in the assemblies. SIMS is also a very sensitive tool for the study of surface reactions, and may be complimentary to XPS, Auger, and wettability experiments [529, 530].

Laxhuber *et al.* studied LB cadmium arachidate multilayers, and reported on the destruction rate in a dynamic (high ion current) SIMS experiment [531]. Toyokawa *et al.* studied cobalt stearate multilayers and monitored the Co^+ signal as a function of layers [532]. Laxhuber *et al.* used SIMS (again with dynamic conditions) to study double-layer samples of arachidic acid as a function of subphase Mg^{2+} concentration (10^{-4}–10^{-2} M) at a constant Cd^{2+} concentration of 3.6×10^{-4} M [524]. Their data revealed the cooperativity between weakly binding Mg^{2+} and strongly binding Cd^{2+} cations, where at small Mg^{2+} concentrations ($\leq 10^{-3}$ M) the head-group dissociation increased, thus increasing Cd^{2+} binding. However, for Mg^{2+} concentrations of $\geq 10^{-3}$ M, Cd^{2+} binding decreases; thus, a maximum of surface concentration of Cd^{2+} was obtained. This study demonstrates the utility of SIMS in the determination of binding mechanisms and their relationship with molecular organization as reflected in the pressure-area isotherm.

Gardella *et al.* empolyed static conditions in their work with organic surfaces [525, 533–538]. They observed molecular and quasi-molecular ions both in positive and negative ion spectra for fatty acids $(M-H)^-$ and $(M+H)^+$ [525]. However, one of the more interesting results in this study was the difference between saturated and unsaturated fatty acids. The presence of $(M+Ag)^+$ silver cationized molecules was observed only from monolayers of unsaturated acids. They suggested that the attachment of the silver ion is to the double bond in the chain, a process which apparently occurred during the bombardment process. They also reported in their reports that chemically oxidized silver substrate yielded higher intensities of substrate-cationized parent ions [525, 536]. The fact that experimental details, such as the oxidation procedure of the substrate, have such effects on the SIMS results emphasizes that much more systematic research has to be done before a detailed understanding and, therefore, control of all parameters in the SIMS experiments is at hand. Hook and Gardella studied LB films of polymerizable functionalized diacetylenes on silver, gold, and germanium substrates [537]. In this study, they reported that the chemical composition of the substrate has a strong influence on molecular ion emission. Moreover, they proposed that a molecular ion–substrate complex (formed only in the case of silver $(M+Ag)^+$), probably was a necessary step in the emission of the molecular ion from the monolayer film.

Wittmaack *et al.* further investigated the potential of SIMS for sputter depth profiling of organic systems [539]. They used both Ar^+ and O_2^+ beams at energies between 2 and 8 KeV. Their main conclusion is that sputter-depth profiles of organic layers on substrates of high atomic numbers are not easy to interpret. It seems that this observation should be adopted to other SIMS studies. It is not clear, for example, what the ionization mechanism is, when an $(M+Ag)^+$ ion is formed, and what the effect of molecular packing and orientation and its relation to the substrate is on the SIMS results. It thus is recommended that caution be exercised, before detailed and systematic studies are carried out and these questions are answered.

References

1. Blodgett, K. B. *J. Am. Chem. Soc.* **1935**, *57*, 1007.
2. Azzam, R. M. A.; Bashara, N. M. *Ellipsometry and Polarized Light;* North-Holland Publishing Company: Amsterdam, 1977.
3. McCrackin, F. L.; Passaglia, E.; Stromberg, R. R.; Steinberg, H. L. *J. Res. Natl. Bur. Stand, Sect A.* **1963**, *67*, 363.
4. Iwamoto, M.; Suzuki, M.; Hino, T. *Shin Ku* **1986**, *28*, 693.
5. Birzer, J. O.; Schulzer, H. J. *J. Colloid. Polym. Sci.* **1986**, *264*, 642.
6. Gaines, G. L., Jr. *Insoluble Monolayers at Liquid–Gas Interfaces*; Interscience: New York, 1966.
7. Allara, D. L.; Nuzzo, R. G. *Langmuir* **1985**, *1*, 45.

8. *Polymer Handbook*; Brandrup, J.; Immergut, E. H., Eds; John Wiley: New York, 1975, v13–v22.
9. Wasserman, S. R.; Whitesides, G. M.; Tidswell, I. M.; Ocko, B. M.; Pershan, P. S.; Axe, J. D. *J. Am. Chem. Soc.* **1989**, *111*, 5852.
10. Tillman, N.; Ulman, A.; Penner, T. L. *Langmuir* **1989**, *5*, 101.
11. Tillman, N.; Ulman, A.; Schildkraut, J. S.; Penner, T. L. *J. Am. Chem. Soc.* **1988**, *110*, 6136.
12. Porter, M. D.; Bright, T. B.; Allara, D. L.; Chidsey, C. E. D. *J. Am. Chem. Soc.* **1987**, *109*, 3559.
13. Rabe, J. P.; Novotny, V.; Swalen, J. D.; Rabolt, J. F. *Thin Solid Films* **1988**, *159*, 359.
14. *CRC Handbook of Chemistry and Physics*, 66th ed., 1985.
15. Rabe, J. P.; Knoll, W. K. *Opt. Commun.* **1986**, *57*, 189.
16. Francon, M. *Rev, Opt.* **1952**, 2.
17. Francon, M. *Le Microscope á Contraste de Phase et la Microscopie Interférentiel*, p. 121, Paris, 1953.
18. Nomarski, G. *Rev. Mét* **1955**, 2.
19. Rothenhäusler, B.; Duschl, C.; Knoll, W. *Thin Solid Films* **1988**, *159*, 323.
20. Segmuller, A. *Thin Solid Films* **1973**, *18*, 287.
21. Pomerantz, M.; Segmuller, A. *Thin Solid Films* **1980**, *68*, 33.
22. Nakagiri, T.; Sakai, K.; Lida, A.; Ishikawa, T.; Matushita, T. *Thin Solid Films* **1985**, *133*, 219.
23. Cuypers, P. A.; Hermens, W. T.; Hemker, H. C. *Anal. Biochem.* **1978**, *84*, 56.
24. Cuypers, P. A.; Corsel, J. W.; Janssen, M. P.; Kop, J. M. M.; Hemker, H. S.; *Surfactant Solution, [Proc. Int. Symp.], 4th Meeting*, Mittal, K. L.; Lindman, B., Eds., Plenum: New York, 1982, Volume 2, 1301–12.
25. Benferhat, R.; Drevillon, B.; Robin, P. *Thin Solid Films* **1988**, *156*, 295.
26. Ishino, Y.; Ishida, H. *Langmuir* **1988**, *4*, 1341.
27. Maoz, R; Sagiv, J. *J Colloid Interface Sci.* **1984**, *100*, 465.
28. Harrick, N. J. *Internal Reflection Spectroscopy*, Wiley-Interscience: New York, 1967.
29. Harrick, N. J.; du Pre, F. K. *Appl. Opt.* **1966**, *5*, 1739.
30. Haller, G. L.; Rice, R. W. *J. Phys. Chem.* **1970**, *74*, 4386.
31. Den Engelsen, D. *J. Opt. Soc. Am.* **1970**, *61*, 1460.
32. Blodgett, K. B.; Langmuir, I. *Phys. Rev.* **1937**, *51*, 964.
33. Tomar, M. S. *J. Phys. Chem.* **1974**, *78*, 947.
34. Zbinden, R. *Infrared Spectroscopy of High Polymers*, Academic Press: New York, 1964.
35. Fringeli, U. P.; Schadt M.; Rihak, P.; Gunthardt, Hs. H. *Z. Naturforsch. A* **1974**, *31*, 1098.
36. Mirabella, F. M., Jr. *Appl. Spectrosc. Rev.* **1985**, *21*, 45.
37. Kopp, F.; Fringeli, U. P.; Muehlethaler, K.; Guenthard, Hs. H. *Z. Naturfosch. C: Biosci.* **1975**, *30*, 711.
38. Ohnishi, T.; Ishitani, A.; Ishida, H.; Yamamoto, N.; Tsubomura, H. *J. Phys. Chem.* **1978**, *82*, 1989.
39. Hartstein, A.; Kirtley, J. R.; Tsang, J. C. *Phys. Rev. Lett.* **1980**, *45*, 201.
40. Okamura, E.; Umemura, J.; Takenaka, T. *Biochim. Biophys. Acta* **1985**, *812*, 139.
41. Kimura, F.; Umemura, J.; Takenaka, T. *Langmuir* **1986**, *2*, 96.
42. Maoz, R.; Sagiv, J. *Langmuir* **1987**, *3*, 1034.
43. Davies, G. H.; Yarwood, J. *Spectrochim. Acta Part A* **1987**, *43*, 1619.

44. Kamata, T.; Umemura, J.; Takenaka, T. *Bull. Inst. Chem. Res., Kyoto Univ.* **1987**, *65*, 170.
45. Kamata, T.; Umemura, J.; Takenaka, T. *Bull. Inst. Chem. Res., Kyoto Univ.* **1987**, *65*, 179.
46. Naselli, C.; Swalen, J. D.; Rabolt, J. F. *J. Chem. Phys.* **1989**, *90*, 3855.
47. Ishitani, A; Ishida, H.; Soeda, F.; Nagasawa, Y. *Anal. Chem.* **1982**, *54*, 682.
48. Blanke, J. F.; Vincent, S. E.; Overend, J. *Spectrochim. Acta, Part A* **1976**, *32*, 163.
49. Allara, D. L.; Swalen, J. D. *J. Phys. Chem.* **1982**, *86*, 2700.
50. Snyder, R. G.; Strauss, H.; Ellinger, C. A. *J. Phys. Che.* **1982**, *86*, 5145.
51. Snyder, R. G.; Maroncelli, M.; Strauss, H. L.; Hallmark, V. M. *J. Phys. Chem.* **1986**, *90*, 5623.
52. Hill, I. R.; Levin, I. W. *J. Chem. Phys.* **1979**, *70*, 842.
53. Allara, D. L.; Nuzzo, R. G. *Langmuir* **1985**, *1*, 52.
54. Golden, W. G.; Snyder, C. D.; Smith, B. *J. Phys. Chem.* **1982**, *86*, 4675.
55. Rabolt, J. F.; Burns, F. C.; Schlotter, N. E.; Swalen, J. D. *J. Electron. Spectrosc. Relat. Phenom.* **1983**, *30*, 29.
56. Bonnerot, A. Chollet, P. A.; Frisby, H.; Hoclet, M. *Chem. Phys.* **1985**, *97*, 365.
57. Naselli, C.; Rabe, J. P.; Rabolt, J. F.; Swalen, J. D. *Thin Solid Films* **1985**, *134*, 173.
58. Hallmark, V. M.; Leone, A.; Chiang, S. ; Swalene, J. D.; Rabolt, J. F. *Polym. Prepr. (Am. Chem. Soc. Div. Polym. Chrm.)* **1987**, *28*, 22.
59. Dote, J. L.; Mowery, R. L. *J. Phys. Chem.* **1988**, *92*, 1571.
60. Mowery, R. L.; Dote, J. L. *Mikrochim. Acta* **1988**, *2*, 69.
61. Cipriani, J.; Racins, S.; Dupeyrat, R.; Hasmonay, H.; Dupeyrat, M.; Levy, Y.; Ombert, C. *Opt. Commun.* **1974**, *11*, 70.
62. Lis, L. J.; Goheen, S. C.; Kauffman, J. W. *Biochim. Biophys. Res. Commun.* **1977**, *78*, 492.
63. Creighton, J. A. Blatchford, C. G.; Albrecht, M. G. *J. Phys. Chem.* **1980**, *84*, 2083.
64. Allen, C. S.; van Duyne, R. P. *Chem. Phys. Lett.* **1979**, *63*, 455.
65. Otto, A. *Surf. Sci.* **1978**, *75*, L392.
66. Sandroff, C. J.; Herschbach, D. R. *J. Phys. Chem.* **1982**, *86*, 3277.
67. Rabolt, J. F.; Santo, R.; Schlotter, N. E.; Swalen, J. D. *IBM J. Res. Dev.* **1982**, *26*, 209.
68. Barbaczy, E.; Dodge, F.; Rabolt, J. F. *Appl. Spectrosc.* **1987**, *41*, 176.
69. Sandroff, C. J.; Garoff, S.; Leung, K. P. *Chem. Phys. Lett.* **1983**, *96*, 547.
70. Joo, T. H.; Kim, K.; Kim, M.S. *J. Phys. Chem.* **1986**, *90*, 5816.
71. Kovacs, G. J.; Loutfy, R. O.; Vincett, P. S.; Jennings, C.; Aroca, R. *Langmuir* **1986**, *2*, 689.
72. Cotton, T. M.; Uphaus, R. A.; Möbius, D. *J. Phys. Chem.* **1986**, *90*, 6071.
73. Evans, S. D.; Urankar, E.; Ulman, A.; Ferris, N.. *J. Am. Chem. Soc.*, submitted.
74. Metiu, H.; Das, P. *Ann. Rev. Phys. Chem.* **1984**, *35*, 507.
75. Moskovits, M. *Rev. Mod. Phys.* **1985**, *57*, 783.
76. van Duyne, R. P. in *Surface Enhanced Raman Scattering* Chang, R. K.; Furtak, T. E., Eds., Plenum: New York, 1982.
77. Kotler, Z.; Nitzan, A. *J. Phys. Chem.* **1982**, *86*, 2011.
78. Zeman, E. J.; Carron, K. T.; Schatz, G. C.; van Duyne, R. P. *J. Chem. Phys.* **1987**, *87*, 4189.

79. Kim, J. -H.; Cotton, T. M.; Uphaus, R. A.; Möbius, D, *J. Phys. Chem.* **1989**, *93*, 3713.
80. Dierker, S. B.; Murray, C. A.; Legrange, J. D.; Schlotter, N. E. *Chem. Phys. Lett.* **1987**, *137*, 453.
81. Schlotter, N. E. *Proc. SPIE-Int. Soc. Opt. Eng.* **1987**, *822*, 110.
82. Schlotter, N. E.; Schaertel, S. A.; Kelty, S. P.; Howard, R. *Appl. Spectrosc.* **1988**, *42*, 746.
83. Zimba, C. G.; Hallmark, V. M.; Swalen, J. D.; Rabolt, J. F. *Mikrochim Acta* **1988**, *2*, 235.
84. Knoll, W.; Philpott, M. R.; Swalen, J. D.; Girlando, A. *J. Chem. Phys.* **1982**, *77*, 2254.
84. Rothenhäusler, B.; Knoll, W. *Surf. Sci.* **1987**, *191*, 585.
86. Rothenhäusler, B.; Knoll, W. *Opt. Commun.* **1987**, *63*, 301.
87. Rothenhäusler, B.; Knoll, W. *Appl. Phys. Lett.* **1987**, *51*, 783.
88. Rothenhäusler, B.; Knoll, W. *J. Opt. Soc. Am. B* **1988**, *5*, 1401.
89. Duschl, C.; Knoll, W. *J. Chem. Phys.* **1988**, *88*, 4062.
90. Orem, M. W.; Adamson, A. W. *J. Colloid Interface Sci.* **1970**, *34*, 473.
91. Kinloch, C. D.; McMullen, A. I. *J. Sci. Inst.* **1959**, *36*, 347.
92. Bergerson, J. A.; Gaines, G. L., Jr. *J. Colloid Interface Sci.* **1967**, *23*, 292.
93. Plaisance, M.; Ter-Minassian-Saraga, L. *CR* **1970**, *270*, 1269.
94. Shimojo, T.; Ohnishi, T. *J. Biochem. (Tokyo)* **1967**, *61*, 89.
95. Blank, M.; Essandoh, S. O. *Nature (London)* **1967**, *215*, 286.
96. Standish, M. M.; Pethica, B. A. *Biochem. Biophys. Acta* **1967**, *144*, 659.
97. Standish, M. M.; Pethica, B. A. *Trans. Faraday Soc.* **1968**, *64*, 1113.
98. Colacicco, G. *Nature (London)* **1971**, *233*, 202.
99. Paltauf, F.; Hauser, H.; Phillips, M. C. *Biochim. Biophys. Acta* **1971**, *249*, 539.
100. Babakov, A. V.; Myagkov, I. V.; Sotnikov, P. S.; Terekhov, O. P. *Biofizika* **1972**, *17*, 347.
101. Hill, M. W.; Lester, R. *Biochim. Biophys. Acta* **1972**, *282*, 18.
102. Colacicco, G.; Buckelew, A. R., Jr. *J. Colloid Interface Sci.* **1974**, *46*, 147.
103. Cleary, G. W.; Zatz, J. L. *J. Colloid Interface Sci.* **1973**, *45*, 507.
104. Oldani, D.; Hauser, H.; Nichols, B. W.; Phillips, M. C. *Biochim. Biophys. Acta* **1975**, *382*, 1.
105. Ksenzhek, O. S.; Gevod, V. S.; Koganov, M. M. *Dokl. Akad. Nauk SSSR* **1975**, *221*, 1211.
106. Ohki, S.; Sauve, R. *Biochim. Biophys. Acta* **1978**, *511*, 377.
107. Thomas, C.; Ter-Minassian-Saraga, L. *Bioelectrochem. Bioenerg.* **1978**, *5*, 357.
108. Thomas, C.; Ter-Minassian-Saraga, L. *Bioelectrochem. Bioenerg.* **1978**, *5*, 369.
109. Pawelek, J.; Gutkowski, P.; Glanowska, H.; Hanicka, M. *Bull. Acad. Pol. Sci. Ser. Sci. Chim.* **1978**, *26*, 571.
110. Rickard, W. F.; Sehgal, K. C.; Jackson, C. M. *Biochim. Biophys. Acta* **1979**, *552*, 1.
111. Rickard, W. F.; Sehgal, K. C.; Jackson, C. M. *Biochim. Biophys. Acta* **1979**, *552*, 11.
112. Maggio, B.; Lucy, J. A. *Adv. Exp. Med. Biol.* **1977**, *83*, 225.
113. Bangham, A. D.; Mason, W. *Br. J. Pharmacol.* **1979**, *66*, 259.
114. Okhi, S. *Physiol. Chem. Phys.* **1981**, *13*, 195.
115. Szundi, I. *Colloids Surf.* **1984**, *10*, 205.

116. Prats, M.; Teissie, J.; Tocanne, J. F. *Nature (London)* **1986**, *322*, 756.
117. Lamarche, F.; Techy, F.; Aghion, J.; Leblanc, R. M. *Colloid Surf.* **1988**, *30*, 209.
118. Van Mau, N.; Trudelle, Y.; Daumas, P.; Heinz, F. *Biophys. J.* **1988**, *54*, 563.
119. Vogel, V.; Möbius, D. *J. Colloid Interface. Sci.* **1988**, *126*, 408.
120. Maggio, B.; Ahkong, Q. F.; Lucy, J. A. *Biochem. J.* **1976**, *158*, 647.
121. Shah, D. O.; Schulman, J. H. *J. Lipid Res.* **1967**, *8*, 215.
122. Shah, D. O.; Schulman, J. H. *Lipids* **1967**, *2*, 21.
123. Colacicco, G. *Biochim. Biophys. Acta* **1972**, *266*, 313.
124. Colacicco, G. *Chem. Phys. Lipids* **1973**, *10*, 66.
125. Sacre, M. M.; El Mashak, E. M.; Tocanne, J. F. *Chem. Phys. Lipids* **1977**, 20, 305.
126. Okhi, S.; Kurland, R. *Biochim. Biophys. Acta* **1981**, *645*, 170.
127. Seimiya, T.; Heki, Y.; Oohinata, H. *Colloid Sirf.* **1981**, *3*, 37.
128. Lakhdar-Ghazal, F.; Tichadou, J. L.; Tocanne, J. F. *Eur. J. Biochem.* **1983**, *134*, 531.
129. Tocanne, J. F.; Tichadou, J. L.; Lakhdar-Ghazal, F. *Stud. Phys. Theor. Chem.* **1983**, *24*, 77.
130. Okhi, S.; Roy, S. Ohshima, H.; Leonards, K. *Biochemistry* **1984**, *23*, 6126.
131. DeSimone, J. A.; Heck, G. L.; DeSimone, S. K. in *Electr. Double Layers Biol.*, Blank, M., Ed., Plenum: New York, 1985, 17–29.
132. Seimiya, T.; Miyasaka, H. Kato, T.; Shirakawa, T.; Ohbu, K.; Iwahashi, M. *Chem. Phys. Lipids* **1987**, *43*, 161.
133. Cadenhead, D. A.; Demchak, R. J. *J. Colloid Interface Sci.* **1967**, *24*, 484.
134. Christodoulou, A. P.; Goldstein, A. B.; Rosano, H. L. J. Colloid Interface Sci. **1967**, *25*, 482.
135. Dreher, K. D.; Wilson, J. E. . *Colloid Interface Sci.* **1970**, *32*, 248.
136. Yamashita, T.; Nagai, T.; Hirota, K. *Bull. Chem. Soc. Jpn.* **1969**, *42*, 1145.
137. Goerke, J.; Harper, H. H.; Borowitz, M. *Advan. Exp. Med. Biol.* **1970**, *7*, 23.
138. Mingis, J.; Zobel, F. G. R.; Pethica, B. A.; Smart, C. *Proc. Roy. Soc. Ser. A* **1971**, *324*, 99.
139. Goddard, E. D.; Kung, H. C. *J. Colloid Interface Sci.* **1971**, *37*, 585.
140. Rosano, H. L.; Yin, D. -N.; Cante, C. J. *J. Colloid Interface Sci.* **1971**, *37*, 706.
141. Mingis, J.; Pethica, B. A. *J. Chem. Soc. Faraday Trans. 1* **1973**, *69*, 500.
142. Goddard, E. D. *Croat. Chem. Acta* **1973**, *45*, 13.
143. Kim, M. W.; Cannell, D. S. *Phys. Rev. A.* **1976**, *14*, 1299.
144. Volhardt, D.; Wuestneck, R.; Zastrow, L. *Colloid Polym. Sci.* **1978**, *256*, 983.
145. Volhardt, D.; Wuestneck, R. *Colloid Polym. Sci.* **1978**, *256*, 1095.
146. Tredgold, R. H.; Smith, G. W. *J. Phys. D.* **1981**, *14*, L193.
147. Casper, J.; Goormaghtigh, E.; Ferreira, J.; Brasseur, R.; Vandenbranden, M.; Ruysschaert, J. M. *J. Colloid Interface Sci.* **1983**, *91*, 546.
148. Tredgold, R. H.; Smith, G. W. *Thin Solid Films* **1983**, *99*, 215.
149. Bois, A. G.; Ivanova, M. G.; Panaitov, I. I. *Langmuir* **1987**, *3*, 215.
150. Georges, C.; Lewis, T. J.; Liewellyn, J. P.; Salvagno, S.; Taylor, D. M.; Stirling, C. J.; Vogel, V. *J. Chem. Soc. Faraday. Trans. 1* **1988**, *84*, 1531.
151. Bois, A. G.; Baret, J. F.; Kulkarni, V. S.; Panaiotov, I. Ivanova, M. *Langmuir* **1988**, *4*, 1358.
152. Willis, R. F. *J. Colloid Interface Sci.* **1971**, *35*, 1.
153. Ter-Minassian-Seraga, L.; Thomas, C. *J. Colloid Interface Sci.* **1974**, *48*, 42.

154. Plaisance, M.; Ter-Minassian-Seraga, L. *J. Colloid Interface Sci.* **1976**, *56*, 33.
155. Demchak, R. J.; Fort, T., Jr. *J. Colloid Interface Sci.* **1974**, *46*, 191.
156. Kuchhal, Y. K.; Katti, S. S.; Biswas, A. B. *J. Colloid Interface Sci.* **1974**, *49*, 48.
157. Mitchell, J. R.; Evans, M. T. A.; Adams, D. J. *J. Colloid Interface Sci.* **1971**, *36*, 406.
158. Malcolm, B. R. *J. Polym. Sci. Part C* **1971**, *34*, 87.
159. Thomas, C.; Ter-Minassian-Seraga, L. *J. Colloid Interface Sci.* **1976**, *56*, 412.
160. Phillips, M. C.; Jones, M. N.; Patrick, C. P.; Jones, N. B.; Rodgers, M. *J. Colloid Interface Sci.* **1979**, *72*, 98.
161. Ter-Minassian-Seraga, L. *Langmuir* **1986**, *2*, 24.
162. Jones, R.; Tredgold, R. H.; Hoorfar, A. *Thin Solid Films* **1985**, *123*, 307.
163. Seta, P.; Bienvenue, E.; D'Epenoux, B.; Tenebre, L.; Momenteau, M. *Photochem. Photobiol.* **1987**, *45*, 137.
164. Caspers, J.; Landuyt-Caufriez, M.; Dellers, M.; Ruysschaert, J. M. *Biochim. Biophys. Acta* **1979**, *554*, 23.
165. Van Mau, N.; Davion, T. L.; Gavach, C. *J. Electroanal. Chem. Interfacial Electrochem.* **1980**, *114*, 225.
166. Van Mau, N.; Davion, T. L.; Amblard, G. *J. Colloid Interface Sci.* **1983**, *91*, 138.
167. Matsumura, H.; Furusawa, K. Inokuma, S.; Kuwamura, T. *Chem. Lett.* **1986**, 453.
168. Reinach, P.; Brody, S. S. *Biochemistry* **1972**, *11*, 92.
169. Frommer, M. A.; Miller, I. R.; Khaiat, A. *Advan. Exp. Med. Biol.* **1970**, *7*, 119.
170. Sahay, A. K.; Mishra, B. K.; Behera, G. B.; Shah, D. O. *Indian J. Chem. Sect. A* **1988**, *27*, 561.
171. Ter-Minassian-Saraga, L.; Panaiotov, I.; Abitboul, J. S. *J. Colloid Interface Sci.* **1979**, *72*, 54.
172. Dragcevic, D.; Milunovic, M.; Pravdic, V. *Croat. Chem. Acta* **1986**, *59*, 397.
173. Oliveira, O. N., Jr.; Taylor, D. M.; Lewis, T. J. Salvagno, S.; Stirling, C. J. M. *J. Chem. Soc. Faraday Trans. 1* **1989**, *85*, 1009.
174. Evans, S. D.; Ulman, A. *Chem. Phys. Lett.* **1990**, *170*, 462.
175. Kuchhal, Y. K.; Katti, S. S.; Biswas, A. B. *J. Colloid Interface Sci.* **1974**, *49*, 48.
176. Harkins, W. D.; Fischer, E. K. *J. Chem. Phys.* **1933**, *1*, 852.
177. Vogel, V.; Mobius, D. *Thin Solid Films* **1988**, *159*, 73.
178. Nutting, G. C.; Harkins, W. D. *J. Am. Chem. Soc.* **1940**, *62*, 3155.
179. Ewers, W. E.; Sack, R. A. *Australian J. Chem.* **1954**, *7*, 40.
180. Hansen, R. S. *J. Phys. Chem.* **1959**, *63*, 637.
181. Myers, R. J.; Harkins, W. D. *J. Chem. Phys.* **1937**, *5*, 601.
182. Harkins, W. D.; Kirkwood, J. G. *J. Chem. Phys,* **1938**, *6*, 53.
183. Adamson, A. W. *Physical Chemistry of Surfaces* ; Wiley: New York, 1982.
184. Malcolm, B. R. *J. Colloid Interface Sci.* **1985**, *104*, 520.
185. Malcolm, B. R. *Thin Solid Films* **1985**, *134*, 201.
186. Daniel, M. F.; Hart, J. T. T. *J. Mol. Electron.* **1985**, *1*, 97.
187. Buhaenko, M. R.; Goodwin, J. W.; Richardson, R. M.; Daniel, M. F. *Thin Solid Films* **1985**, *134*, 217.
188. Halperin, K.; Ketterson, J. B.; Dutta, P. *Langmuir* **1989**, *5*, 161.
189. Messier, J.; Marc, G. *J. Phys. (Paris)* **1971**, *32*, 799.

190. Janzen, A.; Frederick, B. J. R.; Stillman, M. J. *J. Am. Chem. Soc.* **1979**, *101*, 6337.
191. Vandevyver, M.; Barraud, A., Raudel-Teixier, M.; Phillips, G. C. *J. Colloid Interface Sci.* **1982**, *85*, 571.
192. Cook, M. J.; Daniel, M. F.; Dunn, A. J.; Gold, A. A.; Thomson, A. J. *J. Chem. Soc. Chem. Commun.* **1986**, 863.
193. Pace, M. D.; Barger, W. R.; Snow, A. W. *J. Magn. Reson.* **1987**, *75*, 73.
194. Cook, M. J.; Dunn, A. J.; Gold, A. A.; Thomson, A. J.; Daniel, M. F. *J. Chem. Soc. Dalton Trans.* **1988**, 1583.
195. Palacin, S.; Ruaudel-Teixier, A.; Barraud, A. *J. Phys. Chem.* **1986**, *90*, 6237.
196. Palacin, S.; Barraud, A. *J. Chem. Soc. Chem. Commun.* **1989**, 45.
197. Barraud, A.; Lesieur, P.; Ruaudel-Teixier, A.; Vandevyver, M. *Thin Solid Films* **1982**, *134*, 195.
198. Barraud, A.; Lesieur, P.; Richard, J.; Ruaudel-Teixier, A.; Vandevyder, M. *Thin Solid Films* **1985**, *133*, 125.
199. Richard, J.; Vandevyver, M.; Lesieur, P.; Ruaudel-Teixier, A.; Barraud, A.; Bozio, R.; Pecile, C. *J. Chem. Phys.* **1987**, *86*, 2428.
200. Vandevyver, M.; Richard, J.; Barraud A.; Ruaudel-Teixier, A.; Lequan, M.; Lequan, R. M. *J. Chem. Phys.* **1987**, *86*, 6754.
201. Morand, J. P.; Lapouyade, R.; Delhaes, P.; Vandevyver, M.; Richard, J.; Barraud, A. *Synth. Met.* **1988**, *27*, B569.
202. Ikegami, K.; Kuroda, S.; Saito, M.; Saito, K.; Sugi, M.; Nakamura, T.; Matsumoyo, M.; Kawabata, Y. *Phys. Rev. B: Condens. Matter* **1987**, *35*, 3667.
203. Ikegami, K.; Kuroda, S.; Sugi, M.; Saito, M.; Iizima, S.; Nakamura, T.; Matsumoto, M.; Kawabata, Y.; Saito, G. *Synth. Met.* **1987**, *19*, 669.
204. Fugita, S.; Suga, K.; Fujihira, M.; Nakamura, T.; Kawabata, Y. *Chem. Lett.* **1987**, 1715.
205. Ikegami, K.; Kuroda, S.; Saito, M.; Saito, K.; Sugi, M.; Nakamura, T.; Matsumoto, M.; Kawabata, Y. *Thin Solid Films* **1988**, *160*, 139.
206. Ikegami, K.; Kuroda, S.; Saito, K.; Saito, M.; Sugi, M.; Nakamura, T.; Matsumoto, M.; Kawabata, Y.; Saito, G. *Synth. Met.* **1988**, *27*, B587.
207. Suga, K.; Fujita, S.; Fujihira, M. *J. Phys. Chem.* **1989**, *93*, 392.
208. Kuroda, S.; Sugi, M.; Iizima, S. *Thin Solid Films* **1983**, *99*, 21.
209. Iwasaki, T.; Wakabayashi, H.; Ishii, T.; Iriyaa, K. *Appl. Phys. Lett.* **1984**, *45*, 1089.
210. Iriyama, K.; Yoshiura, M.; Ozaki, Y.; Ishii, T.; Yasui, S. *Thin Solid Films* **1985**, *132*, 229.
211. Kuroda, S.; Ikegami, K.; Sugi, M.; Iizima, S.; *Solid State Commun.* **1986**, *58*, 493.
212. Kuroda, S.; Ikegami, K.; Saito, K.; Saito, M.; Sugi, M. *J. Phys. Chem. Jpn.* **1987**, *56*, 3319.
231. Kuroda, S.; Sugi, M.; Iizima, S. *Thin Solid Films* **1985**, *133*, 189.
214. Kuroda, S.; Ikegami, K.; Saito, K.; Saito, M.; Sugi, M. *Thin Solid Films* **1988**, *159*, 285.
215. McGregor, T. R.; Cryz, W.; Fenander, C. I.; Mann, J. A. Jr. *Adv. Chem.* **1975**, *144*, 308.
216. Cadenhead, D. A.; Mueller-Landau, F. *Adv. Chem.* **1975**, *144*, 294.
217. Genkin, M. V.; Davydov, R. M.; Shapiro, A. B.; Krylov, O. V. *Zh. Fiz. Khim.* **1977**, *51*, 1282.
218. Baglioni, P.; Carla, M.; Dei, L.; Martini, E. *J. Phys. Chem.* **1987**, *91*, 1460.

219. Bonosi, F.; Gabrielli, G.; Martini, G.; Ottaviani, M. F. *Langmuir* **1989**, *5*, 1037.
220. Kajzar, F.; Messier, J. *Chem. Phys.* **1981**, *63*, 123.
221. Nakahara, H.; Katoh, T.; Sato, M.; Fukuda, K. *Thin Solid Films* **1988**, *160*, 153.
222. Yoshino, K.; Manda, Y.; Sawada, K.; Morita, S.; Takahashi, H.; Sugimoto, R.; Onoda, M. *J. Phys. Soc. Jpn.* **1989**, *58*, 1320.
223. Adler, G. *Mol. Cryst. Liq. Cryst.* **1988**, *161(B)*, 309.
224. Rieutord, F.; Benattar, J. J.; Bosio, L. *J. Phys. (Les Ulis. Fr.)* **1986**, *47*, 1249; *ibid.* **1986**, *47*, 1849.
225. Holley, C.; Bernstein, S. *Phys, Rev.* **1936**, *49*, 403.
226. Holley, C.; Bernstein, S. *Phys, Rev.* **1937**, *52*, 525.
227. Pomerantz, M.; Segmüller, A. *Thin Solid Films* **1980**, *68*, 33.
228. Marra, W. C.; Eisenberger, P. E.; Cho, A. Y. *J. Appl. Phys.* **1979**, *50*, 6927.
229. Prakash, M.; Dutta, P.; Ketterson, J. B.; Abraham, B. M. *Chem. Phys. Lett.* **1984**, *111*, 395.
230. Prakash, M.; Dutta, P.; Ketterson, J. B. *Thin Solid Films* **1985**, *134*, 1.
231. Belbeoch, B.; Roulliay, M.; Tournarie, M. *J. Chim. Phys. Phys. -Chim. Biol.* **1985**, *82*, 701.
232. Allain, M.; Benattar, J. J.; Rieutord, F.; Robin, P. *Europhys. Lett.* **1987**, *3*, 309.
233. Rieutord, F.; Benattar, J. J.; Bosio, L.; Robin, P.; Blot, C.; De Kouchkovsky, R. *J. Phys. (Les Ulis. Fr.)* **1987**, *48*, 679.
234. Bosio, L.; Benattar, J. J.; Rieutord, F. *Rev. Phys. Appl.* **1987**, *22*, 775.
235. Outka, D. A.; Stöhr, J.; Rabe, J. P.; Swalen, J. D.; Rotermund, H. H. *Phys. Rev. Lett.* **1987**, *59*, 1321.
236. Outka, D. A.; Stöhr, J.; Rabe, J. P.; Swalen, J. S. *J. Chem. Phys.* **1988**, *88*, 4076.
237. Rabe, J.; Swalen, J. D.; Outka, D. A.; Stöhr, J. *Thin Solid Films* **1988**, *159*, 275.
238. Mizushima, K.; Nakayama, T.; Azuma, M. *Jpn. J. Appl. Phys.* **1987**, 26, 772.
239. Mizushima, K.; Egusa, S.; Azuma, M. *Jpn. J. Appl. Phys.* **1988**, *27*, 715.
240. Fromherz, P.; Oelschlägel, U.; Wilke, W. *Thin Solid Films* **1988**, *139*, 421.
241. Jark, W.; Comelli, G.; Russell, T. P.; Stöhr, J. *Thin Solid Films* **1989**, *170*, 309.
242. Kajiyama, T.; Hanada, I.; Shuto, K.; Oishi, Y. *Chem. Lett.* **1989**, 193.
243. Skita, V.; Filipkowski, M.; Garito, A.. F.; Blasic, J. K. *Phys. Rev. B* **1986**, *34*, 5826.
244. Skita, V.; Richardson, W.; Filipkowski, M.; Garito, A.. F.; Blasic, J. K. *J. Physique* **1986**, *47*, 1849.
245. Fischetti, R.; Skita, V.; Garito, A. F.; Blasie, J. K. *Phys. Rev. B* **1988**, *37*, 4877.
246. Dutta, P.; Peng, J. B.; Lin, B.; Ketterson, J B.; Prakash, M.; Georgopoulos, P.; Ehrlich, S. *Phys. Rev. Lett.* **1987**, *58*, 2228.
247. Lin, B.; Peng, J. B.; Ketterson, J. B.; Dutta, P. *Thin Solid Films* **1988**, *159*, 111.
248. Barton, S. W.; Thomas, B. N.; Flom, E. B.; Rice, S. A.; Lin, B.; Peng, J. B.; Ketterson, J. B.; Dutta, P. *J. Chem. Phys.* **1988**, *89*, 2257.
249. Lin, B.; Peng, J. B.; Ketterson, J. B.; Dutta, P.; Thomas, B. N.; Buontempo, J.; Rice, S. A. *J. Chem. Phys.* **1988**, *90*, 2393.

250. Grundy, M. J.; Richardson, R. M.; Roser, S. J.; Penfold, J.; Ward, R. C. *Thin Solid Films* **1988**, *159*, 43.
251. Buhaenko, M. R.; Grundy, M. J.; Richardson, R. M.; Roser, S. J. *Thin Solid Films* **1988**, *159*, 253.
252. Kjaer, K.; Als-Nielser, J.; Helm, C. A.; Tippmann-Krayer, P.; Möhwald, H. *Thin Solid Films* **1988**, *159*, 17.
253. Möhwald, H. *Thin Solid Films* **1988**, *159*, 1.
254. Richardson, W.; Blasie, J. K. *Phys. Rev. B* **1989**, *39*, 12165.
255. L'vov, Yu. M.; Svergun, D.; Geigin, L. A.; Pearson, C.; Petty, M. C. *Phil. Mag. Lett.* **1989**, *59*, 317.
256. Pomerantz, M.; Segmüller, A.; Netzer, L.; Sagiv, J. *Thin Solid Films* **1985**, *132*, 153.
257. Wasserman, S. R.; Whitesides, G. M.; Tidswell, I. M.; Ocko, B. M.; Pershan, P. S.; Axe, J. D. *J. Am. Chem. Soc.* **1989**, *111*, 5852.
258. Grayer Wolf, S.; Landau, E. M.; Lahav, M.; Leiserowitz, L.; Deutsch, M.; Kjaer, K.; Als-Nielsen, J. *Thin Solid Films* **1988**, *159*, 29.
259. Grayer Wolf, S.; Deutsch, M.; Landau, E. M.; Lahav, M.; Leiserowitz, L.; Kjaer, K.; Als-Nielsen, J. *Science* **1988**, *243*, 1286.
260. Landau, E. M.; Grayer Wolf, S.; Sagiv, J.; Deutsch, M.; Kjaer, K.; Als-Nielsen, J.; Leiserowitz, L.; Lahav, M. *Pure Appl. Chem.* **1989**, *61*, 673.
261. Day, D.; Lando, J. B. *Macromol.* **1980**, *13*, 1478.
262. Tredgold, R. H.; Vickers, A. J.; Hoorfar, A.; Hodge, P.; Khoshdel, E. *J. Phys. D* **1985**, *18*, 1139.
263. Tredgold, R. H.; Evans, S. D.; Hodge, P.; Hoorfar, A. *Thin Solid Films* **1988**, *160*, 99.
264. Oyanagi, H.; Yoneyama, M.; Ikegami, K.; Sugi, M.; Kuroda, S. -I.; Ishiguro, T.; Matsushita, T. *Thin Slid Films* **1988**, *159*, 435.
265. Fujiki, M.; Tabei, H. *Langmuir* **1988**, *4*, 320.
266. Dent, N.; Grundy,.M. J.; Richardson, R. M.; Roser, S. J.; McKeown, N. B.; Cook, M. J. *J. de Chim. Phys.* **1988**, *85*, 1003.
267. Oyanagi, H.; Sugi, M.; Kuroda, S. -I.; Iizima, S.; Ishiguro, T. *Thin Solid Films* **1985**, *133*, 181.
268. Duschl, C.; Frey, W.; Helm, C.; Als-Nielsen, J.; Möhwald, Knoll, W. *Thin Solid Films* **1988**, *159*, 379.
269. Kim, S.; Furuki, M.; Pu, L. S.; Nakahara, H.; Fukuda, K. *Thin Solid Films* **1988**, *159*, 227.
270. Nakano, A.; Shimizu, S.; Takahashi, T.; Nakahara, H.; Fukuda, K. *Thin Solid Films* **1988**, *159*, 303.
271. Imazeki, S.; Takeda, M.; Tomioka, T.; Kakuta, A.; Mukoh, A.; Nakahara, T. *Thin Solid Films* **1985**, *134*, 27.
272. Jones, R.; Tredgold, R. H.; Hoorfar, A.; Allen, R. A.; Hodge, P. *Thin Solid Films* **1985**, *134*, 57.
273. Tredgold, R. H.; Allen, R. A.; Hodge, P. *Thin Solid Films* **1987**, *155*, 343.
274. Ruaudel-Teixier, A.; Barraud, A.; Vandevyver, M.; Beobeoch, B.; Roulliay, M. *J. Chim. Phys. Phys. -Chim. Biol.* **1985**, *82*, 711.
275. Otsuka, A.; Saito, G. *Synth. Met.* **1988**, *27*, B575.
276. Hirota, S.; Itoh, U.; Sugi, M. *Thin Solid Films* **1985**, *134*, 67.
277. Banerjie, A.; Lando, J. B. *Thin Solid Films* **1980**, *68*, 67.
278. Nakahara, H.; Miyata, S.; Wang, T. T.; Tasaka, S. *Thin Solid Films* **1986**, *141*, 165.

279. Wunderlich, W., *Macromolecular Physics*, Vol. 1, Academic Press: New York, 1973, p. 97.
280. Yang, X. Q.; Chen, J.; Hale, P. D.; Inagaki, T.; Skotheim, T. A.; Okamoto, Y.; Samuelson, L.; Tripathy, S.; Hong, K.; Rubner, M. F.; denBoer, M. L. *Synth. Met.* **1989**, *28*, C251.
281. Watanabe, I.; Hong, K.; Rubner, M. F. *Synth. Met.* **1989**, *28*, C473.
282. Havinga, E.; de Wael, J. *Rec. Trav. Chim.* **1937**, *56*, 375.
283. Germer, L. H.; Storks, K. H. *J. Chem. Phys.* **1938**, *6*, 280.
284. Russell, G. J.; Petty, M. C.; Peterson, I. R.; Roberts, G. G.; Lloyd, J. P.; Kan, K. K. *J. Mater. Sci.* **1984**, *3*, 25.
285. Bonnerot, A.; Chollet, P. A.; Frisby, H.; Hoclet, M. *Chem. Phys.* **1985**, *97*, 365.
286. Garoff, S.; Deckman, H. W.; Dunsmuir, J. H.; Alvarez, M. S.; Bloch, J. M. *J. Physique* **1986**, *47*, 710.
287. Vogel, V.; Wöll, C. *J. Chem. Phys.* **1986**, *84*, 5200.
288. Vogel, V.; Wöll, C. *Thin Solid Films* **1988**, *159*, 429.
289. Riegler, J. E. *J. Phys. Chem.* **1989**, *93*, 6475.
290. Peterson, I. R.; Russell, G. J. *Philos. Mag. A.* **1984**, *49*, 463.
291. Peterson, I. R.; Russell, G. J. *Thin Solid Films* **1985**, *134*, 143.
292. Jones, C. A.; Russell, G. J.; Petty, M. C.; Roberts, G. G. *Philos. Mag. B.* **1986**, *54*, L89.
293. Peterson, I. R.; Russell, G. J.; Earls, J. D.; Girling, I. R. *Thin Solid Films* **1988**, *161*, 325.
294. Lieser, G.; Tieke, B.; Wegner, G. *Thin Solid Films* **1980**, *68*, 77.
295. Sotnikov, P. S.; Bannikov, V. S.; Troitskiy, V. I.; Ilchenko, A. Ya. *J. Mol. Electron.* **1989**, *5*, 155.
296. Naegele, D.; Lando, J. B.; Ringsdorf, H. *Macromol.* **1977**, *10*, 1339.
297. Duda, G.; Schouten, A. J.; Arndt, T.; Lieser, G.; Schmidt, G. F.; Bubeck, C.;Wegner, G. *Thin Solid Films* **1988**, *159*, 221.
298. Neal, D. B.; Petty, M. C.; Roberts, G. G.; Ahmad, M. M.; Feast, W. J.; Girling, I. R.; Cade, N. A.; Kolinsky, P. V.; Peterson, I. R. *Electron. Lett.* **1986**, *22*, 460.
299. Heard, D.; Roberts, G. G.; Holcroft, B.; Goringe, M. J. *Thin Solid Films* **1988**, *160*, 491.
300. Earls, J. D.; Peterson, I. R.; Russell, G. J.; Girling, I. R.; Cade, N. A. *J. Mol. Electron.* **1986**, *2*, 85.
301. Czikkely, V.; Försterling, H. D.; Kuhn, H. *Chem. Phys. Lett.* **1970**, *6*, 11.
302. Duschl, C.; Frey, W.; Knoll, W. *Thin Solid Films* **1988**, *160*, 251.
303. Luk, S. Y.; Mayers, F. R.; Williams, J. O. *Thin Solid Films* **1988**, *157*, 69.
304. Fryer, J. R.; Hann, R. A.; Eyres, B. L. *Nature* **1985**, *313*, 382.
305. Troitskiy, V. I.; Bannikov, V. S.; Berzina, T. S. *J. Mol. ELectron.* **1989**, *5*, 147.
306. Frisby, H.; Bonnerot, A. *Chem. Phys.* **1989**, *130*, 353.
307. Rahman, A. K. M.; Samuelson, L.; Minehan, D.; Clough, S.; Tripathy, S.; Inagaki, T.; Yang, X. Q.; Skotheim, T. A.; Okamoto, Y. *Synth. Met.* **1989**, *28*, C237.
308. Strong, L.; Whitesides, G. M. *Langmuir* **1988**, *4*, 546.
309. Nicklow, R. M.; Pomerantz, M.; Segmüller, A. *Phys. Rev. B.* **1981**, *23*, 1081.
310. Hayter, J. R.; Highfield, R. R.; Pullman, B. J.; Thomas, R. K.; McMullen, A. I.; Penfold, J. *J Chem. Soc. Faraday. Trans. I* **1981**, *77*, 1437.

311. Highfield, R. R.; Thomas, R. K.; Cummings, P.; Gregory, D.; Mingis, J. *Thin Solid Films* **1983**, *99*, 165.
312. Penfold, J.; Ward, R. C.; Williams, W. G. *J. Phys. E: Sci. Instrum.* **1987**, *20*, 1411.
313. Stamm, M.; Reiter, G.; Hüttenbach, S. *Physica B* **1989**, *156–157*, 564.
314. Fisher, G. B.; Sexton, B. A. *Phys. Rev. Lett.* **1980**, *44*, 683.
315. Ibach, H.; Mills, D. L., *Electron Energy Loss Spectroscopy and Surface Vibrations*, Academic Press: New York, 1982.
316. Ibach, H. *Surf. Sci.* **1977**, *66*, 56.
317. Wandass, J. H.; Gardella, J. A., Jr. *Surf. Sci.* **1985**, *150*, L107.
318. Wandass, J. H.; Gardella, J. A., Jr. *Langmuir* **1986**, *2*, 543.
319. Wandass, J. H.; Gardella, J. A., Jr. *Langmuir* **1987**, *3*, 183.
320. Adamson, A. W.; Ling, I. *Adv. Chem. Ser.* **1964**, *43*, 57.
321. Young, T. *Miscellaneous Works;* Peacock, G., Ed.; Murray: London, 1855, Vol. 1, p. 418.
322. Young, T. *Philos. Trans. R. Soc. London* **1805**, *95*, 65.
323. Whitesides, G. M., private communication.
324. For a review on various techniques for measuring contact angles, see Neumann, A. W.; Good, R. J. in *Surface and Colloid Science*, Good, R. J., Stromberg, R. R., Eds., Plenum: New York, 1979.
325. Ray, B. R.; Bartell, F. E. *Colloid Sci.* **1953**, *8*, 214.
326. Adam, N. K.; Elliott, G. E. P. *J. Chem. Soc.* **1962**, 2206.
327. Fox, H. W.; Zisman, W. A. *J. Colloid Sci.* **1952**, *7*, 428.
328. Schonhorn, H. *Nature* **1966**, *210*, 896.
329. Holms-Farley, S. R.; Reamey, R. H.; McCathy, T. J.; Deutch, J.; Whitesides, G. M. *Langmuir* **1985**, *1*, 725.
330. Holms-Farley, S. R.; Whitesides, G. M. *Langmuir* **1986**, *2*, 266.
331. Holms-Farley, S. R.; Whitesides, G. M. *Langmuir* **1987**, *3*, 62.
332. Bain, C, D.; Evall, J.; Whitesides, G. M. *J. Am. Chem. Soc.* **1989**, *111*, 7155.
333. Bain, C. D.; Whitesides, G. M. *J. Am. Chem. Soc.* **1988**, *110*, 5897.
334. Holmes-Farley, S. R. Ph.D. Thesis, Harvard, 1986.
335. Penn, L. S.; Miller, S. J. *J. Colloid Interface Sci.* **1980**, *78*, 238.
336. Dettre, R. H.; Johnson, R. E. *J. Phes. Chem.* **1965**, *69*, 1507.
337. Neumann, A. W.; Good, R. J. *J. Colloid Interface Sci.* **1972**, *38*, 341.
338. Schwartz, L. W.; Garoff, S. *Langmuir* **1985**, *1*, 219.
339. Schwartz, L. W.; Garoff, S. *J. Colloid Interface Sci.* **1985**, *106*, 422.
340. Dettre, R. H.; Johnson, R. E. *Adv. Chem. Ser.* **1964**, *43*, 136.
341. Joanny, J. F.; de Gennes, P. G. *J. Chem. Phys.* **1984**, *81*, 552.
342. de Gennes, P. G. *Rev. Mod. Phys.* **1985**, *57*, 827.
343. Ray, B. R.; Bartell, F. E. *J. Colloid Sci.* **1953**, *8*, 214.
344. Wenzel, R. N. *Ind. Eng. Chem.* **1936**, *28*, 988.
345. Wenzel, R. N. *J. Phys. Colloid Chem.* **1949**, *53*, 1466.
346. Shuttleworth, R.; Bailey, G. L. J. *Discuss. Faraday Soc.* **1948**, *3*, 162.
347. Cassie, A. B. D. *Discuss. Faraday Soc.* **1948**, *3*, 11.
348. Good, R. J., *J. Am. Chem. Soc.* **1952**, *74*, 5041.
349. Eick, J. D.; Good, R. J.; Neumann, A. W. *J. Colloid Interface Sci.* **1975**, *53*, 235.
350. Bartell, F. E.; Shepard, J. W. *J. Phys. Chem.* **1953**, *57*, 211.
351. Shepard, J. W.; Bartell, F. E. *J. Phys. Chem.* **1953**, *57*, 458.

352. Bain, C. D.; Troughton, E. B.; Tau, Y-T.; Evall. J.; Whitesides, G. M.; Nuzzo, R. G. *J. Am. Chem. Soc.* **1989**, *111*, 321.
353. Evans, S. D.; Ulman, A., unpublished results.
354. See Ref. 10, where the initial advancing contact angle of a hydroxylated surface (~30°) increases to ~ 60° after exposure to boiling chloroform for *ca.* 10 min.
355. Baxter, S.; Cassie, A. B. D. *Text. Ind.* **1945**, *36*, T57.
356. Cassie, A. B. D.; Baxter, S. *Trans Faraday Soc.* **1944**, *40*, 546.
357. Dettre, R. H.; Johnson, R. H., Jr. *Symp. Contact Angle Bristol,* **1966.**
358. Israelachvili, J. N.; Gee, M. L. *Langmuir* **1989**, *5*, 288.
359 Zisman, W. A. *Adv. Chem. Ser.* **1964**, *43*, 1.
360. Ulman, A.; Evans, S. D.; Shnidman, Y.; Sharma, Y.; Eilers, J. E.; Chang, J. C. *J. Am. Chem. Soc.* **1991**, *113*, 0000.
361. Zisman, W. A. *J. Chem. Phys.* **1941**, *9*, 534.
362. Zisman, W. A. *J. Chem. Phys.* **1941**, *9*, 729.
363. Shen, Y. R. *Liquid Cryst.* **1989**, *5*, 635, and references therein.
364. Prasad, P. N.; Williams, D. J. *An Introduction to Nonlinear Optics in Organic Systems*, Wiley: New York, 1990.
365. Heinz, T. F.; Tom, H. W. K.; Shen, Y. R. *Phys. Rev. A* **1983**, *28*, 1883.
366. Rasing, Th.; Shen, Y. R.; Kim, M. W.; Valint, P., Jr.; Bock, J. *Phys. Rev. A* **1985**, *31*, 537.
367. Rasing, Th.; Berkovic, G.; Shen, Y. R.; Grubb, S. G.; Kim, M. W. *Chem. Phys. Lett.* **1986**, *130*, 1.
368. Berkovic, G.; Rasing, Th.; Shen, Y. R. *J. Chem. Phys.* **1986**, *85*, 7374.
369. Kim, M. W.; Liu, S. -N. *Am. Inst. Phys. Conf. Proc.* **1988**, *172*, 477.
370. Grubb, S. G.; Kim, M. W.; Rasing, Th.; Shen, Y. R. *Langmuir* **1988**, *4*, 452.
371. Mizrahi, V.; Stegeman, G. I.; Knoll, W. *Chem. Phys. Lett.* **1989**, *156*, 392.
372. Mizrahi, V.; Stegeman, G. I.; Knoll, W. *Phys. Rev. A* **1989**, *39*, 3555.
373. Berkovic, G.; Superfine, R.; Guyot-Sionnest, P.; Shen, Y. R.; Prasad, P. N. *J. Opt. Soc. Am. B* **1988**, *5*, 668.
374. Shen, Y. R. *The Principles of Nonlinear Optics* , Wiley: New York, 1984, Chap. 35.
375. Zhu, X. D.; Suhr, H.; Shen, Y. R. *Phys. Rev. B* **1987**, *35*, 3047.
376. Harris, A. L.; Chidsey, C. E. D.; Levinos, N. J.; Loiacono, D. N. *Chem. Phys. Lett.* **1987**, *141*, 350.
377. Guyot-Sionnest, P.; Superfine, R.; Hunt, J H.; Shen, Y. R. *Chem. Phys. Lett.*
378. Toennies, J. P. *J. Vac. Sci. Technol.* **1987**, A5, 440.
379. Kern, K.; Comsa, G. in *Adv. Chem. Phys.*, Lawley, K. P., Ed.; Wiley: London, 1989.
380. Barker, J. A.; Auerbach, D. L. *Surf. Sci. Rep.* **1983**, *4*, 59.
381. Danielson, L.; Ruiz, J. C.; Schwartz, C.; Scoles, G.; Huston, J. M. *Faraday Discuss. Chem. Phys.* **1985**, *80*, 1.
382. Ruiz-Suarez, J. C.; Klein, M. L.; Moller, M. A. ; Rowtree, P. A.; Scoles, G.; Xu, J. *Phys, Rev. Lett.* **1988**, *61*, 710.
383. Kohl, K. H.; Kohlhase, A.; Faubel, M.; Staemmler, V. *Mol. Phys.* **1985**, *55*, 1255.
384. Chidsey, C. E. D.; Liu, G. -Y.; Rowntree, P.; Scoles, G. *J. Chem. Phys.* **1989**, *91*, 4421.
385. Chidsey, C. E. D.; Loiacono, D. N.; Sleator, T.; Nakahara, S. *Surf. Sci.* **1988**, *200*, 45.
386. Scoles, G., private communication.
387. Peterson, I. R. *Thin Solid Films* **1984**, *116*, 357.

388. Peterson, I. R.; Russell, G. J. *Thin Solid Films* **1985**, *134*, 143.
389. Lesieur, P.; Barraud, A.; Vandevyver, M. *Thin Solid Films* **1987**, *152*, 155.
390. Peterson, I. R. *J. Mol. Electron.* **1986**, *2*, 1.
391. Matsuda, H.; Sasaki, K.; Kawada, H.; Eguchi, K.; Nakagiri, T. *J. Mol. Electron.* **1989**, *5*, 107.
392. Braun, H. G.; Fuchs, H.; Schrepp, W. *Thin Solid Films* **1988**, *159*, 301.
393. von Tscharner, V.; McConnell, H. M. *Biophys. J.* **1981**, *36*, 409.
394. McConnell, H. M.; Tamm, L. K.; Weis, R. M. *Proc. Natl. Acad. Sci. USA* **1984**, *81*, 3249.
395. Weis, R. M.; McConnell, H. M. *Nature (London)* **1984**, *310*, 47.
396. Lösche, M.; Möhwald, H. *Rev. Sci. Instrum.* **1984**, *55*, 1968.
397. Miyano, K.; Mori, A. *Jpn. J. Appl. Phys.* **1989**, *28*, 252.
398. Lösche, M.; Helm, C.; Mattes, H. D.; Möhwald, H. *Thin Solid Films* **1985**, *133*, 51.
399. Heckl, W. M.; Lösche, M. Möhwald, H. *Thin Solid Films* **1985**, *133*, 73.
400. Duschl, C.; Lösche, M. Miller, A.; Fischer, A.; Möhwald, H.; Knoll, W. *Thin Solid Films* **1985**, *133*, 65.
401. Miller, A.; Knoll, W.; Möhwald, H.; Ruaudel-Teixier, A. *Thin Solid Films* **1985**, *133*, 83.
402. Möhwald, H.; Miller, A..; Stich, W.; Knoll. W.; Ruaudel-Teixier, A.; Lehmann, T.; Fuhrhop, J. -H. *Thin Solid Films* **1986**, *141*, 261.
403. Keller, D. J.; McConnell, H. M.; Moy, V. T. *J. Phys. Chem.* **1986**, *90*, 2311.
404. Moy, V. T.; Keller, D. J.; Gaub, H.; McConnell, H. M. *J. Phys. Chem.* **1986**, *90*, 3198.
405. Fischer, A.; Losche, M.; Möhwald, H.; Sackmann, E. *J. Phys. Lett.* **1984**, *45*, L785.
406. Lösche, M.; Möhwald, H. *Eur. Biophys.* **1986**, *11*, 35.
407. McConnell, H. M.; Keller, D. J. *Proc. Natl. Acad. Sci. USA* **1987**, *84*, 4706.
408. Möhwald, H. *J. Mol. Electron.* **1988**, *4*, 47.
409. Heckl, W. M.; Miller, A.; Möhwald, H. *Thin Solid Films* **1988**, *159*, 125.
410. Möhwald, H.; Kirstein, S.; Haas, H.; Flörsheimer, M. *J. de Chim. Phys.* **1988**, *85*, 1010.
411. Suresh, K. A.; Nittmann, J.; Rondelez, F. *Europhys. Lett.* **1988**, *6*, 437.
412. Moore, B.; Knobler, C. M.; Broseta, D.; Rondelez, F. *J. Chem. Soc., Faraday Trans. 2* **1986**, *82*, 1753.
413. Duschl, C.; Kemper, D.; Frey, W.; Meller, P.; Ringsdorf, H.; Knoll, W. *J. Phys. Chem.* **1989**, *93*, 4587.
414. Riegler, J. E. *Rev. Sci. Instrum.* **1988**, *59*, 2220.
415. Riegler, J. E.; LeGrange, J. D. *Phys. Rev. Lett.* **1988**, *61*, 2492.
416. Hiemenz, P. C. *Principles of Colloid and Surface Chemistry;* Marcel Dekker: New York, 1986).
417. Janssen, A. P.; Akhter, P.; Harland, C. J.; Venables, J. A. *Surf. Sci.* **1980**, *93*, 453.
418. Vandevyver, M.; Legressus, C. *Proc. 6th Int. Conf. on X-R ay Optics, Tokyo* , University of Tokyo Press, 1972, p. 467.
419. Barraud, A.; Rosilio, P. H.; Legressus, C..; Okuzumi, H.; Mogami, A. in *Proc. 8th Int. Conf. on Electron Microscopy, Canberra* , Sanders, J. V.; Goodchild, D. J., Eds., The Australian Academy of Science, Canberra, 1974, Part 1, p. 682.
420. Fryer, J. R. *Mol. Cryst. Liq. Cryst.* **1986**, *137*, 49.

421. Itoh, K.; Yokota, T.; Fujushima, A.; Honda, K. *Thin Solid Films* **1986**, *144*, L115.
422. Fromherz, P.; Kemper, C.; Maass, E. *Thin Solid Films* **1988**, *159*, 405.
423. Wasserman, S. R.; Biebuyck, H.; Whitesides, G. M. *J. Mater. Res.* **1989**, *4*, 886.
424. Fujiki, M.; Tabei, H.; Imamura, S. *Jpn. J. Appl. Phys.* **1987**, *26*, 1224.
425. Oguchi, K.; Yoden, T.; Kosaka, Y.; Watanabe, M.; Sanui, K.; Ogata, N. *Thin Solid Films* **1988**, *161*, 305.
426. Barraud, A.; Leloup, J.; Maire, P.; Ruaudel-Teixier, A. *Thin Solid Films* **1985**, *133*, 133.
427. Fischer, A.; Sackmann, E. *Nature* **1985**, *313*, 299
428. Barger, W.; Dote, J.; Klusty, M.; Mowery, R.; Price, R.; Snow, A. *Thin Solid Films* **1988**, *159*, 369.
429. Inoue, T.; Tase, K.; Okada, M.; Okada, S.; Matsuda, H.; Nakanishi, H.; Kato, M. *Jpn. J. Appl. Phys.* **1988**, *27*, 1635.
430. Binnig, G.; Rohre, H. *Helv. Phys. Acta* **1982**, *55*, 726.
431. Binnig, G.; Rohre, H.; Gerber, C.; Weibel, E. *Phys. Rev. Lett.* **1982**, *49*, 57.
432. Binnig, G.; Rohre, H.; Gerber, C.; Weibel, E. *Phys. Rev. Lett.* **1983**, *50*, 120.
433. Becker, R. S.; Golovchenko, J. A.; McRae, E. G.; Swartzentruber,.B. S. *Phys. Rev. Lett.* **1985**, *55*, 2028.
434. Tromp, R. M.; Hamers, Demuth, J. E. *Phys. Rev.* **1986**, *34B*, 1388.
435. Binnig, G.; Rohrer, H. *Surf. Sci.* **1983**, *126*, 236.
436. Binnig, G.; Frank, K. H.; Fuchs, H.; Garcia, N.; Reihl, B.; Rohrer, H.; Salvan, F.; Williams, J. A. *Phys. Rev. Lett.* **1985**, *55*, 991.
437. Becker, R. S.; Golovchenko, J. A.; Harman, D. H..; Swartzentruber,.B. S. *Phys. Rev. Lett.* **1985**, *55*, 2032.
438. Becker, R. S.; Golovchenko, J. A.; Swartzentruber,.B. S. *Phys. Rev. Lett.* **1985**, *55*, 987.
439. Hamers, R. J.; Tromp, R. M.; Demuth, J. E. *Phys. Rev. Lett.* **1986**, *56*, 1972.
440. Elrod, S. E.; de Lozanne, A. L.; Quate, C. F. *Appl. Phys. Lett.* **1984**, *45*, 1240.
441. Lozanne, A. L.; Elrod, S. A.; Quate, C. F. *Phys. Rev. Lett.* **1985**, *54*, 2433.
442. Golovchenko, J. A. *Science (Washington)* **1986**, *232*, 48.
443. Binnig, G.; Rohrer, H. *Angew. Chem.* **1987**, *99*, 622; *Angew. Chem. Int. Ed. Engl.* **1987**, *26*, 606.
444. Hansma, P. K.; Tersoff, J. *J. Appl. Phys.* **1987**, *61*, R1.
445. Rabe, J. P. *Angew. Chem.* **1989**, *101*, 117.
446. Proc. 1st Int. Conf. Scanning Tunneling Microscopy, *Surf. Sci.* **1987**, *181*.
447. Proc. 2nd Int. Conf. Scanning Tunneling Microscopy, *J. Vac. Sci. Technol.* **1988**, *A6*, 275.
448. Proc. 3rd Int. Conf. Scanning Tunneling Microscopy, *J. Microsc.* (Oxford), in press.
449. Binnig, G.; Rohre, H.; Gerber, C.; Weibel, E. *Surf, Sci.* **1983**, *131*, L379.
450. Binnig, G.; Rohre, H.; Gerber, C.; Stroll, E. *Surf. Sic.* **1984**, *144*, 321.
451. Kaiser, W. J.; Jaklevic, R. C. *Surf. Sci.* **1987**, *181*, 55.
452. Kaiser, W. J.; Jaklevic, R. C. *Surf. Sci.* **1987**, *182*, L227.
453. Hallmark, V. M.; Chiang, S.; Rabolt, J. F.; Swalen, R. J. *Phys. Rev. Lett.* **1987**, *59*, 2879.

454. Behm, R. J.; Hoesler, W.; Ritter, E.; Binnig, G. *Phys. Rev. Lett.* **1986**, *56*, 228.
455. Ritter, E.; Behm, R. J.; Potsche, G.; Wintterlin, J. *Surf. Sci.* **1987**, *181*, 403.
456. Walmsley, D. G.; Turner, R. J.; Hagan, H. P.; Garcia, N.; Baró, A.; Vazquez, L.; Paganini, R. *J. Vac. Sci. Technol.* **1988**, *A6*, 404.
457. Bryant, A.; Smith, D. P. E.; Quate, C. F. *Appl. Phys. Lett.* **1986**, *48*, 832.
458. Rabe, J. P.; Sano, M.; Batchelder, D.; Kalatchev, A. A. *J. Microsc. (Oxford)* **1988**, *152*, 573.
459. Foster, J. S.; Frommer, J. E. *Nature* **1988**, *333*, 542.
460. Spong, J. K.; Mizes, H. A.; LaComb, L. J. Jr.; Dovek, M. M.; Frommer, J. E.; Foster, J. S. *Nature* **1989**, *338*, 137.
461. Lipple, P. H.; Wilson, R. J.; Miller, R. D.; Wöll, Ch.; Chiang, S. *Phys. Rev. Lett.* **1989**, *62*, 171.
462. Coombs, J. H.; Pathica, J. B.; Welland, M. E. *Thin Solid Films,* **1988**, *159*, 293.
463. Hallmark, V. M.; Leone, A.; Chiang, S.; Swalen, J. D.; Rabolt, J. F. *Mikrochim. Acta* **1989**, *11*, 39.
464. Fuchs, H. *Physica Scripta* **1988**, *38*, 264.
465. Hörber, J. K. H.; Hänsch, T. W.; Heckl, W. M.; Möhwald, H. *Chem. Phys. Lett.* **1988**, *145*, 151.
466. Spong, J. K.; Mizes, H. A.; LaComb, L. J. Jr.; Dovek, M. M.; Frommer, J. F.; Forster, J. S. *Nature* **1989**, *338*, 137.
467. Foster, J. S.; Frommer, J. E. *Nature* **1988**, *333*, 542.
468. Rabe, J. P.; Buchholz, S.; Ritcey, A. M. *J. Vac. Sci. Technol.* **1990**,******, 0000.
469. Beebe, T. P.; Wilson, T. E.; Ogletree, D. F.; Katz, J. E.; Balhorn, R.; Salmeron, M.; Siekhaus, W. J. *Science* **1989**, *243*, 370.
470. Lee, G.; Arscott, P. G.; Bloomfield, V. A.; Evans, D. F. *Science* **1989**, *244*, 475.
471. Binnig, G.; Quate, C. F.; Gerber, Ch. *Phys. Rev. Lett.* **1986**, *56*, 930.
472. Binnig, G.; Gerber, Ch.; Stoll, E.; Albrecht, T. R.; Quate, C. F. *Europhys. Lett.* **1987**, *3*, 1281.
473. McClelland, G. M.; Erlandsson, R.; Chiang, S. *Review of Progress in Quantitative Nondestructive Evaluation 6* , Plenum: New York, 1987.
474. Marti, O.; Drake, B.; Hansma, P. K. *Appl. Phys. Lett.* **1987**, *51*, 484.
475. Marti, O.; Ribi, H. O.; Drake, B.; Albrecht, T. R.; Quate, C. F.; Hansma, P. K. *Science* **1988**, *239*, 50.
476. Marti, O.; Drake, B.; Gould, S.; Hansma, P. K. *Proc. SPIE* **1988**, *897*, 22.
477. Sonnenfield, R.; Schneir, J.; Drake, B.; Hansma, P. K.; Aspnes, D. E. *Appl. Phys. Lett.* **1987**, *50*, 1742.
478. Binnig, G.; Smith, D. P. E. *Rev. Sci. Instrum.* **1986**, *57*, 1688.
479. Gould, S.; Marti, O.; Drake, B.; Hellemans, L.; Bracker, C. E.; Hansma, P. K.; Keder. N. L.; Eddy, M. M.; Stucky, G. D. *Nature* **1988**, *332*, 332.
480. Marti, O.; Ribi, H. O.; Drake, B.; Albrecht, T. R.; Quate, C. F.; Hansma, P. K. *Science* **1988**, *239*, 50.
481. Muralt, P; Pohl, D. W. *Appl. Phys. Lett.* **1986**, *48*, 514.
482. Hansma, P. K.; Drake, B.; Marti, O.; Gould, S. A. C.; Prater, C. B. *Science* **1989**, *243*, 641.
483. Raether, H. In *Physics of Thin Films;* Hass, G.; Francombe, M. H.; Hoffmann, R. W., Eds.; Academic Press: New York, 1977; Vol. 9, pp.. 145–261.

484. Burstein, E.; Chen, W. P.; Hartstein, A. *J. Vac. Sci. Technol.* **1974,** *11,* 1004.
485. Hickel, W.; Duda, G.; Jurich, M.; Kröhl, T.; Rochford, K.; Stegeman, G. I.; Swalen, J. D.; Wegner, G.; Knoll, W. *Langmuir* **1990,** *6,* 1403.
486. Kretschmann, E. *Opt. Commun.* **1972,** *6,* 185.
487. Gordon, J. G., II; Swalen, J. D. *Opt. Commun.* **1977,** *22,* 374.
488. Rothenhäusler, B.; Duschl, C.; Knoll, W. *Nature (London)* **1988,** *332,* 615.
489. Hickel, W.; Knoll, W. *Acta Metall.* **1989,** *37,* 2144.
490. Hickel, W.; Kamp, D.; Knoll, W. *Nature (London)* **1989,** *339,* 186.
491. Somorjai, G. A.; Szalkowski, F. J. *Adv. High Temp. Chem.* **1971,** *4,* 137.
492. Smith, T. *J. Appl. Phys.* **1972,** *43,* 2964.
493. Barraud, A.; Rosilio, C.; Ruaudel-Teixier, A. *Thin Solid Films* **1980,** *68,* 7.
494. Green, J. M.; Faulkner, L. R. *J. Am. Chem. Soc.* **1983,** *105,* 2950.
495. Laibinis, P. E.; Hickman, J. J.; Wrighton, M. S.; Whitesides, G. M. *Science* **1989,** *245,* 835.
496. Tillman, N.; Ulman, A.; Elman, J. *Langmuir* **1989,** *5,* 1020.
497. Tillman, N.; Ulman, A.; Elman, *Langmuir* **1990,** *6,* 1512.
498. Nuzzo, R. G.; Zegarski, B. R.; Dubois, L. H. *J. Am. Chem. Soc.* **1987,** *109,* 733.
499. Nuzzo, R. G.; Fusco, F. A.; Allara, D. L. *J. Am. Chem. Soc.* **1987,** *109,* 2358.
500. Wasserman, S. R.; Whitesides, G. M.; Tidswell, I. M.; Ocko, B. M.; Pershan, P. S.; Axe, J. D. *J. Am. Chem. Soc.* **1989,** *111,* 5852.
501. Ulman, A.; Evans, S. D.; Shnidman, Y.; Sharma, R.; Eilers, J. E. *Conference Proceedings, Contact Angles and Wetting Phenomena,* submitted.
502. Bain, C. D.; Whitesides, G. M. *Science* **1988,** *240,* 62.
503. Bain, C. D.; Whitesides, G. M. *J. Am. Chem. Soc.* **1988,** *110,* 3665.
504. Bain, C. D.; Whitesides, G. M. *J. Am. Chem. Soc.* **1989,** *111,* 7164.
505. Untereker, D. F.; Lennox, J. C.; Wier, L. M.; Moses, P. R.; Murray, R. W. *J. Electroanal. Chem.* **1977,** *8,* 309.
506. Carlson, T. A.; McGuire, G. E. *J. Electron Spectrosc.* **1972/73,** *1,* 161.
507. Pamberg, P. W. *Anal. Chem.* **1973,** *45,* 549A.
508. Penn, D. R. *J. Vac. Sci. Technol.* **1976,** *13,* 221.
509. Penn, D. R. *J. Electron Spectrosc. Relat. Phenom.* **1976,** *9,* 29.
510. Flitch, R.; Raider, S. I. *J. Vac. Sci. Technol.* **1975,** *12,* 305.
511. Powell, C. F. *Surf. Sci.* **1974,** *44,* 29.
512. Ebel, M. F. *Surf. Interface Anal.* **1981,** *3,* 173.
513. Brundle, C. R.; Hosper, H.; Swalen, J. D. *J. Chem. Phys.* **1979,** *70,* 5190.
514. Clark, D. T.; Fox, T.; Roberts, G. G.; Sykes, R. W. *Thin Solid Films* **1980,** *70,* 261.
515. Clark, D. T.; Fox, Y. C. T.; Roberts, G. G. *J. Electron. Spectr. Related Phenom.* **1981,** *22,* 173.
516. Hazell, L. B.; Rizvi, A. A.; Brown, A. A.; Ainsworth S. *Spectrochim. Acta* **1985,** *40B,* 739.
517. Kobayashi, K.; Takaoka, K.; Ochiai, S. *Thin Solid Films* **1988,** *159,* 267.
518. Takahara, A.; Morotomi, N.; Hiraoka, S.; Higashi, N.; Kunitake, T.; Kajiyama, T. *Macromol.* **1989,** *22,* 617.
519. Benninghoven, A. *Surf. Sci.* **1975,** *53,* 596.
520. Benninghoven, A. *Int. J. Mass Spectrom. Ion Phys.* **1983,** *53,* 85.
521. Day, R. J.; Unger, S. E.; Cooks, R. G. *Anal. Chem.* **1980,** *52,* 555A.

522. Benninghoven, A., in *Ion Formation from Organic Solids* , Benninghoven, A., Ed., Springer: Berlin, 1983, p. 64.
523. Benninghoven, A.; Niehuis, E.; Friese, T.; Greifendorf, D.; Steffens, P. *Org. Mass Spectrom.* **1984**, *19*, 346.
524. Laxhuber, L.; Möhwald, H.; Hashmi, M. *Colloids and Interfaces* **1984**, *10*, 225.
525. Wandass, J. H.; Gardella, J. A., Jr. *J. Am. Chem. Soc.* **1985**, *107*, 6192.
526. Bolbach, G.; Blais, J. -C.; Hebert, N. *Mol. Cryst. Liq. Cryst.* **1988**, *156*, 361.
527. Hébert, N.; Blais, J. C.; Bolbach, G.; Inchaouh, J. *Int. J. Mass Spectrom. Ion Processes* **1986**, *70*, 45.
528. Cooks, R. G.; Busch, K. L. *J. Int. J. Mass Spectrom. Ion Processes* **1983**, *53*, 111.
529. Unger, S. E.; Cooks, R. G. *Surf. Sci. Lett.* **1982**, *116*, 211.
530. Chait, B. I.; Field, F. H. *J. Am. Chem. Soc.* **1985**, *107*, 6743.
531. Laxhuber, L.; Möhwald, H.; Hashnii, M. *Int. J. Mass Spectrom. Ion Phys.* **1983**, *51*, 93.
532. Toyakawa, F.; Abe, H.; Furoya, K.; Kikuchi, T. *Surf. Sci.* **1983**, *133*, L429.
533. Gardella, J. A., Jr.; Hercules, D. M. *Anal. Chem.* **1980**, *52*, 226.
534. Gardella, J. A., Jr.; Hercules, D. M. *Anal. Chem.* **1981**, *53*, 1879
535. Gardella, J. A., Jr.; Novak, F. A.; Hercules, D. M. *Anal. Chem.* **1984**, *56*, 1371.
536. Gardella, J. A., Jr.; Wandass, J. H.; Cornellio, P. A.; Schmitt, R. L., in *Springer Series in Chemical Physics Vol. 44: Secondary Ion Mass Spectrometry SIMS V* , Benninghoven, A.; Colton, R. J.; Simons, D. S.; Werner, H. W., Eds., Springer: Berlin, 1986, p. 534.
537. Hook, K. J.; Gardella, J. A., Jr. *J. Vac. Sci. Tech. A* **1989**, *7*, 1795.
538. Wandas, J. H.; Schmitt, R. L.; Gardella, J. A., Jr. *Appl. Surf. Sci.* **1989**, *40*, 85.
539. Wittmaack, K.; Laxhuber, L.; Möhwald, H. *Nucl. Instr. Meth. Phys. Res.* **1987**, *B18*, 639.

PART TWO

LANGMUIR–BLODGETT FILMS

In this part, we use the term *Langmuir–Blodgett films* (LB films) to denote monolayers and multilayers transferred from the water–air interface (or liquid–gas interface in general) onto a solid substrate. The molecular film at the water–air interface is denoted as *Langmuir film*.

The origin of the *oil-on-water* subject began quite early. Some will go back as far as to the Babylonians, who spread oil droplets on a water surface and used the behavior of the film to foretell the future [1], or to Aristotle, who spoke on the subject of "spreading oil on troubled water" [2]. However, it seems that Benjamin Franklin carried out the first preparation of a Langmuir film when he showed that a layer of oil (which was, apparently, a few nanometers thick) had a calming influence on the water in the Clapham ponds [3]. It was in the 19th century that Agnes Pockels prepared the first monolayers at the water–air interface [4-7]. Those who read her reports will appreciate the ingenious work, especially in view of the computer-controlled, expensive instrumentation in today's modern laboratories.

It was about 25 years later — and after reports by Rayleigh on the nature of these layers [8] and by Devaux [9] and Hardy [10], who reported on the amphiphilic structure of molecules that form Langmuir films — that Langmuir carried out the first systematic study of monolayers of amphiphilic molecules at the water–air interface [11]. The first study on a deposition of multilayers of long-chain carboxylic acid onto a solid substrate was carried out by Blodgett [12, 13]. The reader may find an account for most of this early work in a book by Gaines [14].

2.1. Preparation Methods

Langmuir–Blodgett (LB) was, indeed, the first technique to provide the chemist with the practical (and still the most extensively studied) capability to construct ordered molecular assemblies [15–23]. Let us consider first the preparation of monolayers at the water–air interface, and start with a discussion on troughs.

A. The Trough

It was Agnes Pockels who first designed a trough with barriers for manipulation of the film at the water–air interface. In this design, the barriers were placed across the edges of the trough [4]. Since then, the trough has been dramatically changed and developed [24, 25]. Today, the trough is fully computerized with state-of-the-art electronics. Figure 2.1 presents a schematic description of a one-bath trough.

In this scheme, *a* is a bath, usually made out of Teflon; *b,* a moving barrier that allows control of the pressure applied on the monolayer; *c,* a motor that moves the barrier; *d,* a control device that gets information from a pressure sensor on the water surface and controls the pressure; *e,* a balance that measures the surface pressure; *f,* a motor with a gearbox that lowers and raises the substrate; and *g,* a solid substrate.

Fox and Zisman were the first to report on a trough milled from a solid piece of Teflon [26], and today, Teflon is widely used for the water bath(s) and barrier(s). This is mainly because of cleanliness advantages, since it can be treated with strong oxidizing agents such as sulfochromic acid, or a mixture of nitric and hydrochloric acid to ensure a complete removal of organic contaminants. The balance most commonly used to measure the surface pressure is the Wilhelmy plate [27]. The reader will find a discussion on surface pressure, and on the Wilhelmy [28–52], Langmuir [11, 53–92], and other methods [11, 63–67] of surface pressure measurements, in Gaines' monograph [14] and Adamson's book [68].

The development of fully automated troughs for building monomolecular layers and multilayers started in the mid-1970s [69]. Since then there has been a continuous effort to master the speed and quality of deposition, and thus, vibration-free and stall-free troughs using servomechanism for both the barrier and substrate lift were introduced [70]. The interest of materials scientists in the construction of organic superlattices from large molecules and polymers that form very viscous films has brought about the need for temperature-controlled troughs.

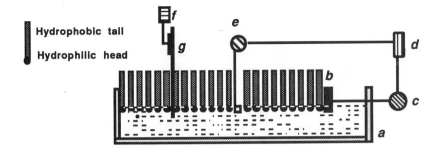

Figure 2.1. A trough for deposition of monolayers on solid substrates.

Such a trough, in which film deposition can be studied over the entire temperature range from near freezing to near boiling, was developed by Dutta *et al.* [71]. It allows a deeper understanding of phases and transition in the LB film, and thus provides a more complete pressure-area isotherm. It also makes possible a successful deposition of compounds that do not form LB films at room temperature. Furthermore, the trough also allows the study of air-sensitive molecules, since it provides a controlled environment.

The need for a continuous fabrication of thick, noncentrosymmetric organic films for nonlinear optical (NLO) applications was the driving force for the development of troughs with two baths in which a film of molecule A is formed on one bath, and a film of molecule B on the other. Different transfer geometries and mechanisms were suggested in the literature [72–77]. A rotating arm first was suggested by Bikerman [72], and more recently was adopted by different groups [73–76]. Barraud *et al.* [73] and Holcroft *et al.* [75] developed troughs with spherically shaped substrate holders. In the Barraud trough, there are programmable dipping arms that are stationary; i.e., they move up and down, but not from bath A to bath B, and an auxiliary rotating arm takes the substrate from one arm to the other. In the Holcroft design, on the other hand, the substrate is on the cylinder that can oscillate or rotate about a central axis and thus allow the substrate to move through the floating monolayers. The major advantage in the Barraud design is that the system can work symmetrically and exchange two samples simultaneously, one from A to B and one from B to A. In the Holcroft design, the advantage is that the trough retains the benefits of the constant-perimeter principle. The major question that can be raised, when a *rotating drum* with a large surface area constantly rotates about the central axis, is how much material is being transferred from one monolayer to the other. This may be a serious problem when molecule A and molecule B have dipole moments in opposite directions relative to the monolayer surface, or when A is an electron donor and B an electron acceptor. The incorporation of A molecules in film B in the first case may affect the NLO properties of the ABABAB superlattice formed.

Different designs have been suggested for a double-bath trough with arms that can move vertically and also transfer the substrate from bath A to bath B [76, 77]. In the Daniel *et al.* design, there is only one arm, and the substrate is transferred from bath A to bath B under the subphase through a special *gate* using a Teflon leaf [76]. Here, again, there may be some concern as to how much exchange of molecules between the A and B regions is possible. Lin *et al.* addressed this concern by transferring the substrate above and under the subphase from one arm to the other [77]. Another double-bath trough that offers a similar solution to the possible mixing problem is the KSV 5000 system (Figure 2.2). The difference from the Lin system is that here, there is one movable arm above the subphase and another moveable arm in the subphase. That is, both the upper and lower arms operate in bath A and in bath B, which allows for only one substrate to be used at a time.

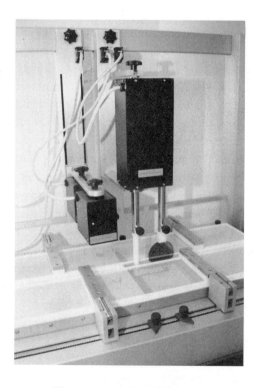

Figure 2.2. The KSV 5000 trough.

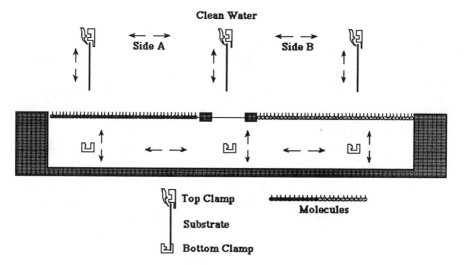

Clean Water

Side A   Side B

Top Clamp   Molecules

Substrate

Bottom Clamp

Figure 2.3. A cross-section view of LB film deposition in the KSV 5000 trough.

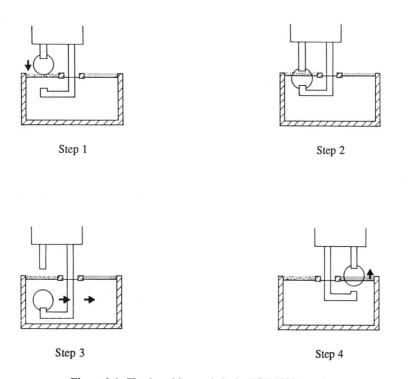

Step 1   Step 2

Step 3   Step 4

Figure 2.4. The deposition cycle in the KSV 5000 trough.

The KSV design also provides a clean subphase area, so that the coating process can start with a retracting mode, without having the substrate dipped through one of the floating films. This design seems the best so far in terms of cleanliness and flexibility.

Let us now consider the deposition cycle in the KSV 5000 system (Figure 2.3). In this process, there are four steps (Figure 2.4): In Step 1, a hydrophobic substrate is brought down through either of the films. Alternatively, a hydrophilic substrate can be brought down through the clean phase in the middle, and the starting point is in the subphase. In Step 2, the substrate is transferred from the upper arm to the lower arm, which is in the subphase. In Step 3, the substrate is brought down to the bottom of the *well* and is transferred to the center of the second bath; in the final step, the lower arm brings the substrate up through the film, and when the upper arm takes hold of it, it is transferred to the position in Step 1, above the film in the first bath.

Miyano and Maeda introduced the first trough with four independent, moveable barriers [78]. This design provides a uniform film compression and thus allows the study of stress inhomogeneity in very viscous, e.g., polymeric, films. In this case, it is sometimes useful to have a floating film with a width similar to that of the substrate to ensure a smooth transfer. Kumehara *et al.* also report on construction of a similar trough, with the same width as the solid substrate [79]. The authors reported that good LB films successfully could be prepared of trivalent metal (La^{3+}, Al^{3+}) stearate. A four-barrier trough may also help in the compression of a film to give a more uniform, stress-free monolayer.

Other troughs dedicated to different experiments also have been reported. Yun and Bloch designed a trough dedicated to x-ray study of monolayers at the water–air interface [80]. Their design addresses four important issues: (*a*) maintaining the subphase level above the rim of the trough, which is very important, since x-rays are incident upon the interface at very shallow angles of the order of milliradians; (*b*) maintaining the monolayers stablity over the several hours when synchrotron radiation is used; (*c*) making the trough fit an x-ray diffractometer; and (*d*) maintaining the subphase surface at a fixed level. This trough was used successfully by the authors to perform the first surface extended x-ray fine structure (SEXAFS) experiment on liquid surfaces, when they studied monolayers of Mn-stearate. Lösche and Möhwald [81], Seul and McConnell [82], and then Miyano and Mori [82] reported on the design of troughs for studies of fluorescence spectroscopy of monolayers at the water–air interface.

B. Experimental Considerations in the Preparation of LB Films

The fabrication of multilayers by the LB method is not a straightforward process. This is true especially when electro-optic and electronic materials are concerned. Here, reproducibility of results is crucial, and extreme care thus should be taken to control the fine details of the fabrication. The reader may find a discussion on this issue in a review by Tredgold [23], in a very detailed and technical paper by Mingins and Owens [84], and in a paper by Peterson [85]. In this section, we address briefly some of the technical issues concerning the environment, the subphase, the substrate, the amphiphiles, and cleaning between experiments.

1. The Environment

While it is true that extreme care should be taken in the fabrication of LB films, it is not practical — but rather very expensive — to have, for example, complete atmosphere, temperature, and humidity control in the laboratory. It thus is recommended that the LB trough be installed in a laminar hood to ensure a dust-free environment. We usually put the bath on a heavy piece of marble, which is placed on shock absorbers to minimize vibrations. The cleanliness of the air also should be ensured by removing any organic solvents and material. Such contaminants usually are concentrated on the water surface, and thus their damaging effect may be orders of magnitude higher than what one could expect from their low concentration in the air.

2. The Subphase

The most commonly used subphase is water, although mercury and other materials [14], as well as glycerol [73, 86], also have been used. The water used is deionized through a Millipore system that also contains a final filter to remove bacteria. As the feed, it is recommended that "house" deionized water be used (provided that it is of high enough quality). Finally, the water is passed through a Millipore filter to remove residual dust immediately before use. In most cases, the pH of the subphase has to be adjusted and controlled, as well as its ionic content. Different amphiphiles show different sensitivity to these parameters, and it usually takes some time before the final subphase pH and ionic content is established [87]. It sometimes is impossible to remove organic comtaminants completely just by filtration through an active carbon filter. Again, these organic contaminants could concentrate at the water–air interface and their effect would be detrimental. In certain systems, breakdown products from the ion-exchange resins affect the surface properties of the water as well [88, 89]. Traditionally, one oxidative stage, using basic permanganate, has been sufficient to ensure a complete removal of these organic contaminants, although it is normal to add a second distillation step without any additives.

To summarize, we recommend using a standard value of the surface tension of water of 72.0 mN m^{-1} at 25°C, established by Harkins and Alexander [90, 91]. Pallas and Pethica have published also on a variety of methods by which they obtained this surface tension value [92], while a higher value of 73.0 ± 0.04 mN m^{-1} at 20 ± 0.1°C also has been mentioned [84].

3. The Substrate

The LB technique is unique, since monolayers can be transferred to many different substrates. Most LB depositions have involved hydrophilic substrates where the monolayers are transferred in the retraction mode [93]. Transparent hydrophilic substrates, such as glass [94, 95] and quartz [96], allow the study of spectra in the transmission mode. Other hydrophilic substrates are aluminum [97–100], chromium [12, 74], and tin [101], all in their oxide forms (e.g., Al_2O_3/Al). One of the most commonly used substrates today is the silicon wafer. A high-quality wafer can be purchased, polished on one side, but should carefully be cleaned prior to any use.

In a typical procedure, the silicon wafers are cleaned by heating at 90°C in a mixture of 30% H_2O_2 and concentrated H_2SO_4 (30:70 v/v) for 30 min [102]. *Caution! A solution of H_2O_2 in H_2SO_4 ("piranha" solution) is a very strong oxidant and reacts violently with many organic materials. It should be handled with extreme care.* Wasserman *et al.* found that cleaned silicon substrates can be stored under clean water for 48 h and still retain high hydrophilicity [103]. According to Carim *et al.*, the treatment with piranha solution yields substrates with a surface of silicon dioxide (SiO_2) [104], with Si–OH group concentration of ~5 x 10^{14} per cm^2 [105, 106]. This concentration of OH groups is approximately equivalent to the number of alkyl chains that can be packed in a cm^2 (assuming ~20 Å^2 per alkyl chain, discussed later).

Gold is an oxide-free substrate, and thus is a preferred substrate for reflection spectroscopy studies. However, since the surface energy of clean gold is of the order of 1000 mJ/m^2 [107], it is contaminated easily, which results in an uneven quality of LB films [16]. Evaporated gold substrates (e.g., Au/Cr/Si) can be cleaned by a brief ($\leq$ 5 min) exposure to piranha solution; however, long exposures increase the gold surface roughness. For example, we have found differences between the contact angles on $CF_3-(CF_2)_9-CH_2-CH_2-SH$ monolayers on gold, depending on the gold treatment [108]. Thus, contact angles on monolayers adsorbed on the gold immediately after removal from the evaporator always were larger than those on monolayers adsorbed on gold after treatment with the piranha solution. However, the advantage of this cleaning is that it results in a very hydrophobic surface that is completely wetted by water [109, 110]. Such a gold surface is excellent as a reference in FTIR spectroscopy. We have found that if such hydrophilic gold is used immediately after cleaning, its IR spectrum does not show absorption due to hydrocarbons.

Gold substrates also can be cleaned by soxhlet in CCl_3CH_3 for 30 min [93], or by argon plasma (high energy, 3–5 minutes). However, both cleaning procedures, while apparently removing organic contaminants, do not yield a hydrophilic surface. Note that gold comes negatively charged from the argon plasma and cannot be used as a reference in surface potential measurements. For that, we recommend using the hydrophilic gold described earlier. Once the gold has been cleaned, it yields uniform oleophobic film deposition [93].

Silver (Ag/Si) should be used directly upon removal from the evaporator. Otherwise, it can be cleaned either with CCl_3CH_3 [93] or argon plasma.

Gallium arsenide (GaAs) wafers can be made hydrophilic either by sonication for five minutes in 5% HF solution, or treated with bromine-methanol or piranha solution [93]. Tredgold and El-Badawy showed that hydrophilic GaAs undergoes deposition in retraction [111]. Cadmium telluride (CdTe) also can be treated with bromine-methanol, and was reported to yield adequate deposition [111].

Hydrophobic silicon substrates are prepared by cleaning the surface first as described, and subsequently reacting with a silinizing reagent (trimethylchloro-silane [$(CH_3)_3SiCl$], dimethyldichlorosilane [$(CH_3)_2SiCl_2$], or, preferably, OTS) [112–114]. We note, however, that the nature of the silanized surface is quiite different, depending whether $(CH_3)_3SiCl$ or $(CH_3)_2SiCl_2$ is used.

The cleaning procedure for glass substrates that yields highly active surfaces for SA monolayers includes a treatment in H_2SO_4/H_2O_2, 4:1 mixture at 120°C (see previous caution remark on piranha solution) for 20 min, followed by water, ethanol, and acetone rinses, and drying in the nonoxidizing portion of a Bunsen burner flame for *ca.* 10 sec [115].

Freshly cleaved mica is an atomically flat substrate and has been widely used both in LB experiments [116] and as a substrate for preparation of very flat gold substrates [117].

109

4. Amphiphiles and Solvents

The purity of the organic amphiphiles under study is of great importance. Any contaminant in the amphiphile will be incorporated in the monolayer, even if it is soluble in the subphase. Usually, neither elemental analysis nor spectroscopic methods such as proton NMR are sufficient to ensure purity. The only way we know of ensuring purification to the level required for the preparation of LB monolayers is preparative gas chromatography (GC) for volatile molecules, and preparative high-pressure liquid chromatography (HPLC) for complex molecules and polymers. It is also recommended that these techniques be used as the standard analytical method for the determination of the purity of amphiphiles. This is true especially when prepolymerized amphiphiles are used, where contamination monomers and oligomers should be strictly avoided. Moreover, the HPLC procedure allows better control of molecular weight distribution and the study of monolayer properties as a function of this parameter. Also, it is possible to collect fractions with different average molecular weight and with narrow molecular weight distribution. These fractions, which differ only in the number of repeat units, may yield monolayers with different stability, rheology, uniformity, and transferability properties. A study of the different monolayer properties as a function of the molecular weight of the polymer will establish the exact material properties needed for the successful construction of stable multilayer films.

The same purity criteria should be implemented for the organic solvents used to prepare the spreading solution, and the inorganic acids, bases, and salts added to the water subphase. Different batches, even from the same supplier, may have differences sufficient to cause irreproducibility in results. This is true especially for organic solvents. Here, we do not recommend distillation, since the result may be even worse for the small quantities needed in the laboratory. The best way to achieve high-quality solvents is to pass the solvent through a dry column of basic alumina. Sometimes, one treatment is not enough, and the semipure solvent has to be percolated through a fresh basic alumina column again. Pure chloroform has to be stored in a dark bottle, under nitrogen, in the freezer.

5. Between Experiments

Another important issue is the treatment of the trough between experiments; in other words, the cleaning procedures. *It should be emphasized that any surface in the trough should be free of surface-active materials.* This means that no organic detergent is allowed, and only inorganic cleaning solutions should be used. If the trough is made out of glass, it is recommended that one use, as a final stage, a rapid (not more than 1 min) treatment with a mixture of 50 vol % water and 50 vol % nitric acid to which is added ~4% of 40% hydrofluoric acid. If the trough is made out of Teflon, it is recommended that sulfochromic acid be used. In both cases, the cleaning stages must be followed by rinsing with large volumes of pure water. The same cleaning procedure is recommended for all the glassware used in the experiment.

C. What Is the Amphiphile?

All of us know from our chemistry studies that there are materials that are soluble in water and others that are not. For example, inorganic salts or salts of organic acids are soluble in water, while lipids are not soluble in water, but do dissolve in nonpolar solvents, such as CCl_4, pontane, etc. The reason for this difference in solubility is a result of the different possible interactions that exist between the solute and the solvent in the two cases. Thus, in water (dielectric constant $\varepsilon = 81$), salts dissociate and the ions are stabilized by the electrostatic (Coulomb) interactions with the water and with the H_3O^+ and OH^- ions. This stabilization compensates for the energy loss due to the breaking of hydrogen bonds between water molecules. However, when lipids dissolve in water, there is a breaking of hydrogen bonds with no interaction that compensates for it. Materials that are soluble in water are *hydrophilic*, which means they like water, while those that dissolve in nonpolar solvents are called *hydrophobic*, which means they are "afraid of" water. The *amphiphile* is a molecule that is insoluble in water, with one end that is hydrophilic, and, therefore, is preferentially immersed in the water, and the other that is hydrophobic, and preferentially resides in the air (or in the nonpolar solvent). A classical example of an amphiphile is stearic acid ($C_{17}H_{35}CO_2H$), where the long hydrocarbon "tail" ($C_{17}H_{35}$—) is hydrophobic, and the carboxylic acid head group (—CO_2H) is hydrophilic. This carboxylic group can dissociate in the water and become negatively charged, and it usually is assumed that the first one or two methylene groups in a monolayer of stearic acid also are in the water (Figure 2.5). This can be justified easily based on water solubilities of carboxylic acids; for example, propanoic acid (CH_3CH_2COOH) is soluble in water. Since the amphiphiles have one end that is hydrophilic and the other that is hydrophobic, they like to locate in interfaces such as between air and water, or between oil and water. This is the reason they also are called *surfactants* (surface active).

111

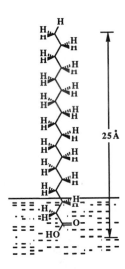

Figure 2.5. A stearic acid molecule on the water–air interface.

Table 2.1. The effect of different functional groups on film formation of C_{16}-compounds (Adapted from Adam [118].)

Very Weak (*no film*)	Weak (*unstable film*)	Strong (*stable film with* C_{16} *chain*)	Very Strong (C_{16} *compounds* *dissolve*)
Hydrocarbon	$-CH_2OCH_3$	$-CH_2OH$	$-SO_3^-$
$-CH_2I$	$-C_6H_4OCH_3$	^-COOH	$-OSO_3^-$
$-CH_2Br$	$-COOCH_3$	$-CN$	$-C_6H_4SO_4^-$
$-CH_2Cl$		$-CONH_2$	$-NR_4^+$
$-NO_2$		$-CH=NOH$	
		$-C_6H_4OH$	
		$-CH_2COCH_3$	
		$-NHCONH_2$	
		$-NHCOCH_3$	

It should be noted, that the solubility of an amphiphilic molecule in water depends on the balance between the alkyl chain length and the strength of its hydrophilic head. For example, while hexadecanoic acid [$CH_3-(CH_2)_{14}-CO_2H$] forms stable Langmuir films, hexadecyltrimethylammonium chloride [$CH_3-(CH_2)_{15}-N(CH_3)_3^+Cl^-$] is too soluble in water to allow the formation of a monolayer. In this case, however, above their critical micellar concentration, such molecules may form micelles or vesicles (close structures whose walls are made of bilayers of amphiphilic molecules) in water. The reader may find a discussion on the properties of monolayer films, their formation and stability, in Gaines' book [14]. It is useful, however, to present a table here from Adam's book that summarizes the properties of different head groups (Table 2.1) [118].

Amphiphilic molecules are not special at all. All of us use them in our everyday life (soaps and detergents), and even more important, cell membranes in our body are built from amphiphiles (phospholipids). Many scientists use organic synthesis to construct surfactants for different purposes, such as stabilization emulsions. Such molecular engineering is very important, especially in the design of amphiphiles with bulky groups, such as chromophores, donors, acceptors, etc. The reader may find interesting information on this subject in papers by Kuhn, Möbius, and Bucher [15], and by Blinov [119].

D. The Pressure-Area Isotherm

In a typical experiment, a drop of a dilute solution (~1%) of an amphiphilic molecule in a volatile solvent (e.g., $CHCl_3$) is spread on the water–air interface of a trough. The solvent that evaporates leaves a monolayer of the molecules in what is called a *two-dimensional gas* due to the relatively large distances between the molecules (Figure 2.6). Of course, the amount of the amphiphile should be calculated so that the resulting film would be a monomolecular layer. At this stage, the barrier moves and compresses the molecules on the water–air interface, while the pressure and area per molcule are recorded. The surface pressure, hence the force per unit length of the barrier, is $\pi = \sigma_0 - \sigma$ (in newtons/meter, or dyn/cm). Thus, π is the difference between the surface tension of pure water (σ_0), and that of the water covered with a monolayer (σ). Since we know the total number of molecules and the total area that the monolayer occupies, we can calculate the area per molecules (Å^2) and construct a π-A isotherm that describes the surface pressure as a function of the area per molecule. Figure 2.7 presents a schematic isotherm of stearic acid on 0.01 M HCl. The reader also is referred to the discussion on π-A isotherms in Gaines' book [14].

The pressure-area isotherm is rich in information on stability of the monolayer at the water–air interface, the reorientation of molecules in the two-dimensional system, phase transitions, and conformational transformations. We shall expand on some of these issues when we discuss the structure of LB monolayers. (See Section 2.3.A.)

113

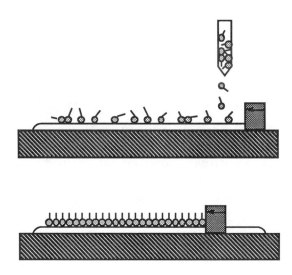

Figure 2.6. A monolayer of molecules in a two-dimensional gas (top), and in the compressed form (bottom).

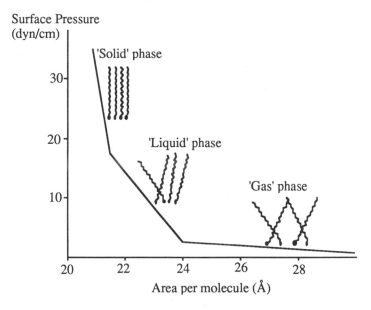

Figure 2.7. A scheme of surface pressure (π) vs. area per molecule (A) for stearic acid on 0.01M HCl.

The area per molecule in the "gas" state is large, and ideally there should be no interaction (lateral adhesion) between the molecules, and therefore, the surface pressure is low. (For example, at ~50 Å^2, the surface pressure is extremely low, ~0.1 – 0.5 dyn/cm.) However, even when a surface pressure approaches zero, surfactant molecules have a natural tendency to aggregate. (Examples are diacetylene amphiphiles.) This aggregation is discussed later. For an ideal two-dimensional gaseous phase, the molecules must have negligible size compared to the interface area, and thus will obey the equation $\pi A = kT$, which is a two-dimensional ideal gas equation, where A is the area per molecule, π is the surface pressure, and T is the absolute temperature. Notice that A is equivalent to V, and π is equivalent to P in the ideal gas law.

As the barrier moves, the molecules are compressed, the intermolecular distance decreases, the surface pressure increases, and in the case of stearic acid, a phase transition can be observed in the isotherm. (Note that there are many cases where no such phase transition can be observed.) This first phase transition is assigned to a transition from the "gas" to the "liquid" state (also known as liquid-expanded, LE, state), which occurs at ~24 Å^2 for a monolayer of stearic acid. In the liquid phase, the monolayer is coherent, except that the molecules occupy a larger area than in the condensed phase. The molecules have more degrees of freedom and *gauche* conformations can be found in the alkyl chains. When the barrier compresses the film further, a second phase transition can be observed, from the "liquid" to the "solid" state (also known as the liquid-condensed, LC, state). At this stage, the area per molecule is ~20 Å^2 (for a single alkyl chain, e.g., stearic acid). In this condensed phase the molecules are closely packed and uniformly oriented.

If further pressure is applied on the monolayer, it collapses due to mechanical instability [120], and a sharp decrease in the pressure is observed. This collapse pressure is a function of the temperature, the pH of the subphase, and the speed by which the barrier is moved. There are materials that exhibit two instances of collapse pressure. These are, for example, the polypeptides that first collapse into bilayers, which then collapse into microcrystallites [121].

E. Film Transfer

1. The Meniscus

The subject of the *meniscus* and its shape is related closely to contact angle and surface tension (Section 1.2.A). A meniscus is formed in the junction of three phases (triple line), and Figure 2.8 represents a thin vertical solid substrate

115

suspended at a liquid from the arm of a trough. The result of the surface tension and contact angle is the entrainment of a meniscus around the perimeter of the plate.

To understand what happens when the liquid is covered with a monolayer, let us consider first a clean liquid surface, and an experiment similar to the Wilhelmy plate measurement of surface tension γ. In this experiment, we measure the weight of the meniscus (w), which is equal to the upward force created by the interaction of the liquid with the substrate surface $(P\gamma\cos\theta)$, where θ is the static contact angle as shown in Figure 2.8, and P is the perimeter of the plate $(P = 2(l + t))$, where l is the length and t the thickness of the plate, so

$$w = P\gamma\cos\theta. \qquad\qquad 2.1$$

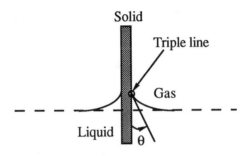

Figure 2.8. A meniscus around a vertical plate. θ is the static contact angle.

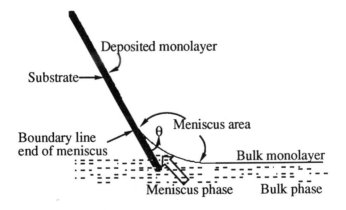

Figure 2.9. Different regions in the deposition process of an LB film. (Adapted from Riegler and LeGrange [3], © 1988, Am. Phys. Soc.)

116

It is clear that the shape of the meniscus is a function of the surface tension, which in turn is a function of the nature of the surface. Once the surface tension is reduced by the existence of an organic layer, it cannot support a large w, and the contact angle will be larger. The reader is referred to Reference 122 for a more detailed discussion on the meniscus.

Recently, Riegler and LeGrange showed that "liquid-to-solid" phase transition can occur in the meniscus region, at surface pressures that are lower than the transition pressure indicated by the π-A isotherm [123, 124]. Figure 2.9 presents the different regions in the monolayer and subphase. Using the fluorescence microscope configuration, they could detect the growth of solid domains in the liquid monolayer at the different stages of the transfer process. It could thus be suggested that although no pressure or temperature gradients existed between the bulk subphase and the meniscus phase, the state of the monolayer was different in these two regions. Therefore, it cannot be assumed that there is a one-to-one transfer between the water and the substrate.

2. Vertical Deposition

Let us now discuss the two methods of transfer of monolayers from the water–air interface onto a solid substrate. The first, and more conventional, method is the *vertical deposition,* which was demonstrated first by Blodgett and Langmuir [125]. They showed that a monolayer of amphiphiles at the water–air interface can be deposited by the displacement of a vertical plate (Figure 2.10).

While this process still is not understood completely, we know that there is a certain critical velocity U_m above which the transfer process does not work. This velocity is of the order of a few cm/sec in the best cases, and usually is much smaller for viscous layers (e.g., polymeric amphiphiles).

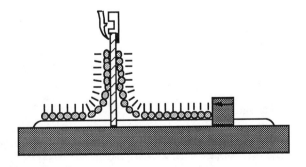

Figure 2.10. Deposition of a monolayer from the water–air interface to a vertical plate.

Gaines suggests using the term *reactive interface* to describe the cases where deposition rates are relatively high [39], and de Gennes, looking at the meniscus in terms of interfacial energies, calculated the maximum velocity of deposition U_m [126]. He looked at a liquid, covered with a surfactant film, being pulled out along a vertical plate. As the plate is being displaced, the triple line does not move, the liquid drains downwards, and a new, smaller, dynamic contact angle is formed (θ_d). Using the values of $\theta = 0.2°$, $\bar{\gamma} = 50$ dyn/cm (where $\bar{\gamma} = \gamma - \pi$, π being the surface pressure in the trough), and $\eta = 10^{-2}$ poises, he could calculate $U_m \sim 60$ mm min^{-1}, which is very similar to observed deposition rates [127]. We note, however, that a very recent paper by Grundy *et al.* indicates that high-quality LB films can be obtained at high deposition rates (120 ± 5 mm min^{-1}) [127a]. This result has important implications for device fabrication.

Buhaenko and Richardson measured recently the forces of emersion (upstroke), and immersion (downstroke) during LB deposition [128]. They defined a system at equilibrium as being where the forces applied to the barrier and the substrate are insufficient to cause deposition. Figure 2.11 presents a schematic diagram of their system. The change in free energy can be written as

$$\gamma_{DF} - \gamma_{SF} + \gamma_0 = \pi - \left(\frac{F}{P}\right)_e , \qquad 2.2$$

where γ_{DF} is the surface free energy of the deposited film, γ_{SL} is the surface free energy of the substrate–liquid interface, γ_{SF} is the surface free energy of the spread film, γ_0 is the surface free energy of the clean subphase, π is the surface pressure of the spread monolayer (maintained by the barrier), F is the force exerted by the substrate on its holder, and P is the perimeter of the triple line. The subscripts e and i stand for emersion and immersion, respectively.

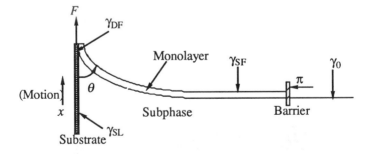

Figure 2.11. A schematic model for a cross-section at the triple line during an upstroke deposition of a monolayer. (Adapted from Buhaenko and Richardson [128], © 1988, Elsevier Science Publishers.)

According to the Young equation (see Section 1.2.A), the following relationship can be written for the triple line,

$$\gamma_{DF} = \gamma_{SL} + \gamma_{SF}\cos\theta \,, \qquad 2.3$$

where θ is the contact angle at the triple line. By combining Equations 2.2 and 2.3, we get for the emersion,

$$-\left(\frac{F}{P}\right)_e = \gamma_{FS}\cos\theta \,. \qquad 2.4$$

Similarly, for the immersion the same formula can be developed,

$$-\left(\frac{F}{P}\right)_i = \gamma_{FS}\cos\theta \,. \qquad 2.5$$

Note that although Equations 2.4 and 2.5 are similar, the values of the contact angle θ in these cases are completely different.

Blake and Haynes proposed an interesting jump model to explain dipping speed [129]. In this model, they assumed an activation energy barrier to the motion of the triple line contact angle from one site to the other, and used the following equation,

$$U = K\delta\exp\left\{\frac{\gamma}{nkT}\ (\cos\theta - \cos\theta_{static})\right\} \,, \qquad 2.6$$

where n is the number of adsorption sites per unit area, U is the speed of the meniscus, δ is the distance between two adsorption sites, K is the number of jumps per second, and γ is the surface tension of the liquid. Buhaneko and Richardson verified experimentally the model suggested by Petrov, which is based on the jump model [130], and suggested the existence of a critical deposition velocity U_m, above which the meniscus advances faster than a precursor film of molecules from the monolayer can adsorb on the substrate. In molecular terms, it means that the interactions of the molecules in the monolayer frontier (which is, in fact, the triple line) with the substrate determines the success of the deposition at a given rate. Thus a *reactive* deposition will occur when molecules adsorb spontaneously onto the substrate at the same speed that a new clean area becomes available as a result of substrate withdrawal. If this is the case, the adsorption is accompanied by draining of the subphase and no entrainment of subphase occurs. On the other hand, if the speed by which a new clean substrate area is created is higher than the monolayer adsorption (*non-reactive* deposition), the monolayer is practically forced onto the substrate, entrainment of subphase will occur, and a distortion of the meniscus will be observed [130]. It also is reasonable to assume

119

that a forced-deposited film will be easy to remove due to lack of adsorption, thus reducing the deposition ratio. Note that reactive deposition is akin to self-assembly. (See Part Three.) Accordingly, surfaces that exhibit reactive deposition usually are good substrates for self-assembly. For example, a monolayer of arachidic acid will manifest a reactive transfer to highly hydrophilic, clean glass. Using a suitable solvent, arachidic acid will self-assemble on the same substrate.

Buhaenko et al. studied the influence of shear viscosity (η^s) on maximum possible deposition rate (U_m) as a function of chain length in fatty acids (n), and in two isomers of tricosenoic acid [131]. Figure 2.12 presents U_m as a function of n. It is interesting that in the case of $n < 21$, U_m is hardly affected by n; the authors suggested that this may indicate that below a certain value of η^s, U_m is determined by another property, and they proposed that this property is the rate of expulsion of the aqueous subphase during the deposition process. Although the speed of dipping was found to depend on the entrainment characteristics of the film [127, 132], it is hard to accept without further evidence that such a difference exists between C_{18} and C_{22} carboxylic acids. A different explanation that can be proposed is that at this pressure, the monolayers are highly defective solids, with high concentration of *gauche* conformations. This model may be supported by the work of Barton et al., who studied monolayers of eicosanol [133].

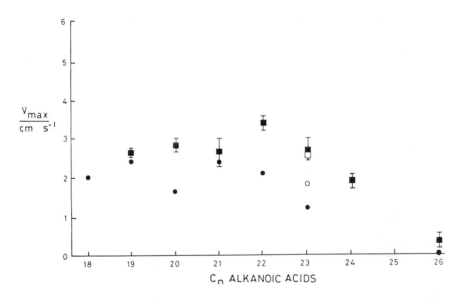

Figure 2.12. Maximum deposition speed U_m vs. number of carbon atoms in chain (n) for varous fatty acids. (Adapted from Buhaenko et al. [131], © 1985, Elsevier Science Publishers.)

120

Langmuir and Blodgett showed over 50 years ago that when a substrate is moved through the water–air interface, a monolayer is transferred either in the dipping or the retracting of the substrate through the water surface [12, 125, 134]. However, in spite of the long history of this classical technique, the details of monolayer deposition still are not understood completely, and very subtle changes in the experimental conditions may result in dramatic changes in the deposition process. The first and obviously important factor is the stability and homogeneity of the monolayer at the water–air interface. We already have discussed the strict experimental considerations required in these experiments; it is important, however, to add that at the water–air interface, the kinetics of a chemical reaction may be competely different from that in the bulk. For example, oxidation of cholesterol in a two-dimensional monolayer assembly is quite fast (~45 min) [135]. This also may be an important consideration for photoactive dyes, where *cis-trans* isomerization or oxidation can change the bulk physical properties of the assembly under study.

One of the most detailed engineering-oriented studies is that of Biddle *et al.* [136]. They monitor the barrier movement as a function of time (creep test) to evaluate the stability of brasidic acid film at the water–air interface. (Brassidic acid is *trans*-13-docosenoic acid, $CH_3-(CH_2)_7-CH=CH-(CH_2)_{11}-COOH$). Figure 2.13 presents the effect of $CdCl_2$ concentration in the subphase on creep tests at 30 mN m^{-1}. Higher Cd^{2+} concentrations lead to the greater stability of the monolayer.

These stability tests confirm the observation of many reseachers that the addition of divalent ions to the water subphase, and the apparent formation of the acid salt, enhances the stability and deposition of fatty acid monolayers compared to those of the free acid layers [63, 87, 129, 137–149].

The effect of the water subphase age on monolayer stability was studied, and it was found that the best results could be obtained on a two-day-old subphase, although change in the subphase could not be detected. The effects of temperature and pH on monolayer stability were also studied. Finally, a processing window that allows reproducible production of homogeneous films was constructed (Figure 2.14). Thus, when an appropriate engineering approach is used, the problem of making reproducible, high-quality, thick LB fims may be solved. It is clear, however, that the process described in this work, although long and tedious, should be adopted when the preparation of device-quality films are considered. Once the stability of the monolayer at the water–air interface has been studied fully, the transfer process can be addressed. Let us turn now to the different types of LB films.

121

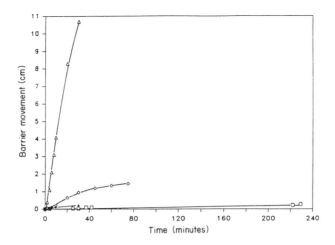

Figure 2.13. The effect of CdCl$_2$ concentration on creep tests at 30 mN m^{-1} (CdCl$_2$ concentration: Δ, 10^{-5} M; ◇ ,10^{-4} M; +, 10^{-3} M; □ , 5 x 10^{-3} M). (From Biddle *et al.* [136], © 1985, Elsevier Science Publishers.)

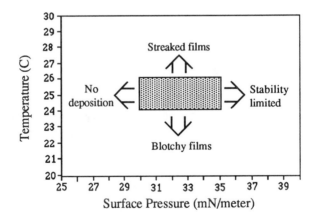

Figure 2.14. Processing window for brassidic acid monolayers. The parameters used were "aged" water or low concentration of CdCl$_2$, 0.5–1 mg/mL brassidic acid concentration, spreading area of 60 Å^2/molecule, dwell period of 10 mim, compression rate of 2 cm/min and dipping velocity of 1–3 mm/min. (Adopted from Biddle *et al.* [136], © 1985, Elsevier Science Publishers.)

When a substrate is moved through the monolayer at the water–air interface, the monolayer can be transferred during emersion (retraction or upstroke) or immersion (dipping or downstroke). A monolayer usually will be transferred

122

during retraction when the substrate surface is hydrophilic, and the hydrophilic head groups interact with the surface. On the other hand, if the substrate surface is hydrophobic, the monolayer will be transferred in the immersion, and the hydrophobic alkyl chains interact with the surface. If the deposition process starts with a hydrophilic substrate, it becomes hydrophobic after the first monolayer transfer, and thus the second monolayer will be transferred in the immersion. This is the most usual mode of multilayer formation for amphiphilic molecules in which the head group is very hydrophilic (—COOH, —PO$_3$H$_2$, etc.), and the tail is an alkyl chain. This mode is called the Y-type deposition. For very hydrophilic head groups, this is the most stable deposition mode, since the interactions between adjacent monolayers are either hydrophobic–hydrophobic, or hydrophilic–hydrophilic (Figure 2.15).

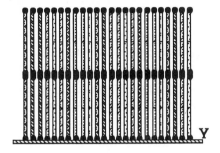

Figure 2.15. A Y-type multilayer film (centrosymmetric).

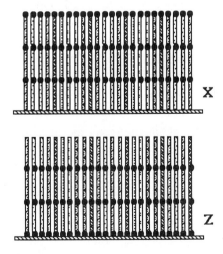

Figure 2.16. X- and Z-types of deposition of LB multilayers (noncentrosymmetric).

123

It was reported, however, in the early papers of Langmuir and Blodgett that films also can be formed only in downstroke (X-type), and that deposition speed may affect the deposition mode [125, 134, 135]. A third possible deposition is when films are formed in upstroke only (Z-type, Figure 2.16). These are cases where the head group is not so hydrophilic (e.g., —COOMe [150]), or where the alkyl chain is terminated by a weak polar group (e.g., —NO$_2$) [151]. In both cases the interactions between adjacent monolayers are hydrophilic–hydrophobic, and therefore, these multilayers are less stable than the Y-type systems. We note, however, that X deposition of relatively nonpolar amphiphiles such as esters give ordered films, whereas Y deposition is "pathological." Both X- and Z-type films are noncentrosymmetric (Figure 2.16), and, therefore, may have importance in NLO applicaions (see Part Five). Finally, it is worthwhile to emphasize that X, Y, and Z depositions do not necessarily lead to X-, Y-, and Z-type films [152].

At this point, it is essential to introduce the concept of transfer ratio. It already was observed by Blodgett that the amount of amphiphile that could be deposited on a glass slide depended on several factors [12]. The deposition ratio is defined as A_l/A_s, where A_s is the area of the substrate coated with a monolayer, and A_l is the decrease of area occupied by that monolayer at the water–air interface (at constant pressure). Honig *et al.* defined an *ideal* Y-type film as a multilayer system with a constant transfer ratio of one for both upstroke and downstroke [113]. An ideal X-type film can be defined accordingly as a layer system where the transfer ratio is always one for the downstroke and zero for the upstroke. In practice, there are deviations from these ideal definitions. Thus, there are cases where the transfer ratio is ≤ 1, but constant for succeeding dippings. In other cases, the transfer ratio is not equal for upstroke and downstroke depositions in the case of Y-type films, or is not zero in upstroke depositions for X-type films. This case is, in fact, a mixed X–Y-type film, and it was observed early on that X and Y films appear to be identical in structure [153–157]. This is a clear manifestation of the inherent instability of X films, since it suggests that the molecules in the X film "flip over." The mechanism of this process is not clear at all although several mechanisms have been suggested [159–161]. For such X–Y films, a deposition can be defined by the ratio θ [113],

$$\theta = \frac{A_l^u}{A_l^d},$$

2.7

where A_l^u is the transfer ratio for the upstroke deposition, and A_l^d for the downstroke deposition. Of course, a Y film should yield $\theta = 1$, while for an X film, $\theta = 0$, and for an ideal Z film, $\theta = \infty$. Once the decrease of area of the monolayer at the water–air interface is plotted as a function of time during the deposition process, the deposition ratio for successive layers can be measured, and information on the deposition nature can be obtained.

In trying to understand the deposition process better, and to elucidate the rearrangement mechanism, precision contact angle measurements during deposition should be carried out. It is clear that since the nature of the substrate — that is, whether it is hydrophilic or hydrophobic — affects the nature of deposition, a careful characterization of the substrate during deposition may help in understanding this process. Gaines measured the contact angle during deposition of calcium arachidate (using the capillary-rise method [162]) [163]. Neuman alone, and with Swanson, studied contact angles during deposition of calcium stearate, but only on the upstroke [164, 165]. Clint and Walker measured contact angles on immersion by the capillary-rise method [166]. Neumann developed a direct weighing method for determining the contact angle of a liquid on a solid substrate [167, 168], and Neumann and Good replaced the capillary with a Wilhelmy plate [169]. Peng *et al.* measured the static and dynamic contact angles between a monolayer-covered water surface and a mica substrate [170]. Their results indicate that the dynamic contact angle on the upstroke is $\sim 18°$ smaller for the first layer than for subsequent layers, whereas the angle for the downstroke is established with the first layer. This means that one monolayer is sufficient to mask completely the substrate, and it is in agreement with the observation that the nature of the substrate influences the deposition of only the first layer. The static contact angle on the upstroke (polar deposition of lead stearate on lead stearate) averages $6°$ greater than the dynamic angle. (See earlier discussion on the meniscus.) We note that a recent paper by Robinson *et al.* on variations in the surface free energy of fatty acid LB multilayers present results that do not agree with Peng *et al.* [170a]. Thus, it indicates that the difference between the immersion and withdrawal angles decreased with increasing film thickness for the first 18 layers.

Peng *et al.* monitored both the contact angle and the transfer ratio during deposition of lead stearate and cadmium stearate [116]. Their observations are important, since they demonstrate the problems associated with the use of LB films in molecular devices. It clearly could be concluded that the deposition process is far from being ideal, and that X-type deposition began at different times and in different parts of the film. It thus was suggested that X-type deposition should be considered pathological, since its signature means a highly defected film. They also suggested that incomplete conversion of the fatty acids to their salt may be responsible for the behavior [171], since if the free acid molecules peel during the upstroke while salt molecules remain, a Y-type film with partial monolayers will be formed. Indeed, SEM could detect 100 Å-scale defects clearly in the films.

Fukuda and Shiozawa studied a family of long-chain esters and established conditions for the formation of X-type and Y-type films [150]. They were able to show that a Y-type deposition is favored by high surface pressure, low temperature, and rapid withdrawal. They also noted that although some X films can retain the structure characterized by the deposition process, it may be transient and may change into the stable form of the bulk crystal (which is similar to a Y-film).

Hasmonay *et al.* studied the composition and transfer mechanism of LB multilayers of stearates [148]. They examined a long series of experiments and arrived at the conclusion that Y-layers are observed only if stearic acid is present. This probably is due to the stabilization that the polar interaction of the carboxylic groups have on the bilayer structure.

One of the most detailed studies of the deposition mechanism was carried out by Peterson *et al.* in their work with ω-tricosenoic acid [132, 172, 173]. (ω-tricosenoic acid is $(CH_2=CH-(CH_2)_{20}-COOH.)$ They suggested a reorganization model for monolayer drainage in this case [132], and found an odd-even effect for the maximum drainage speed of *n*-alkanoic acids [173], which is very similar to the behavior of their melting points [174]. They also reported on epitaxy in multilayers of ω-tricosenoic acid [173], and on heteroepitaxy between an octadecanoic acid bilayer and subsequently deposited ω-tricosenoic acid [172]. An interesting result in this work was that tilted modifications of both octadecanoic and ω-tricosenoic acids on an HF-etched silicon substrate could be prepared by deposition at high speed from an acidic substrate [172]. This means that the free volume in the deposited monolayer may be a function of the deposition rate, and at high rates this volume increases, probably due to the shear forces in the meniscus.

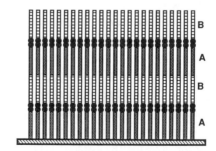

Figure 2.17. A Y-type ABAB LB multilayer film.

126

We have discussed in this section the noncentrosymmetry arrangement of molecules in the X- and Z-type multilayers, and explained the source of their instability. This stability problem can be eliminated by the use of two different amphiphiles. In this case, one monolayer is transferred in immersion (A) and the other in the emersion mode (B). Thus, a stable ABAB, Y-type film is obtained, but with no plane of symmetry (noncentrosymmetry), since each bilayer consists of two different types of molecules (Figure 2.17). Such films are useful for different applications (e.g., second-order NLO, pyroelectric, piezoelectric, etc.). Blinov *et al.* [175, 176] and Daniel *et al.* [177, 178] were the pioneers in this area, and since their original contributions there have been many papers dealing with the formation and structure of ABAB films. [179–187] We shall discuss these studies in detail in Section 5.1.A.1.

We note that the ABABAB is only one possible configuration. In principle, it is possible to prepare other combinations such as AABBAA, ABBABB, etc., and, therefore, to tailor organic films with interesting physical properties.

3. Schaefer's Method for the Preparation of LB Films

Another way of building LB multilayer structures is the *horizontal lifting* method (*Schaefer's method*), which was introduced by Langmuir and Schaefer in 1938 [188]. This method, however, was invented about 1000 years ago by the Japanese, who call it *sumi nagashi* [189]. In this technique, the artist spreads Chinese ink, which is a suspension of carbon particles in a protein solution, on a water surface. The protein monolayer thus formed is picked up by paper, using horizontal lifting.

Schaefer's method is useful for the deposition of very rigid films, which are at the two-dimensional-solid region in the π-A diagram (Figure 2.7). In this method, a compressed monolayer first is formed at the water–air interface (1, Figure 2.18), then a flat substrate is placed horizontally on the monolayer film (2, 3, Figure 2.18). When this substrate is lifted and separated from the water surface, the monolayer is transferred onto the substrate (4, Figure 2.18), keeping, in theory, the same molecular direction (X-type, Figure 2.16).

Kato reported on deposition of high-quality X-type LB films of ethyl stearate and octadecyl acrylate using a fully automated horizontal lifting [190, 191]. However, cadmium and lanthanum arachidate give LB films with x-ray diffraction corresponding to Y-type deposition. This overturn of molecules during horizontal lifting also was reported by Kato for cadmium pentacosa-10,12-diynoate [192].

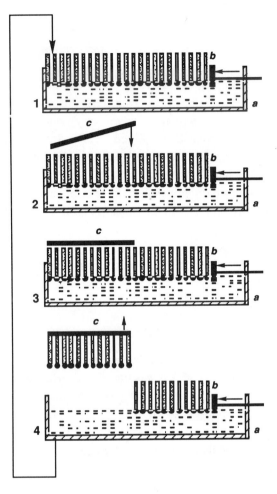

Figure 2.18. The Schaefer method of transfer: *a.* trough, *b.* barrier, *c.* substrate.

The overturn probably occurs because once the substrate is just lifted, a meniscus is formed around it with a triple line, as described previously. Further lifting disrupts this meniscus and as a result, the monolayer from the water–air interface in the periphery is "sucked in," thus forming a bilayer. In fact, in many cases, heterogeneous films with different number of layers in different places were formed during the lifting process. There were different oral reports on attempts to solve the overturn problem, where a collar was placed around the substrate after it made contact with the organic film, thus separating it from the other parts of the film. No details have been published yet on any success. It can be expected that monolayers of polymeric amphiphiles may be good candidates for this horizontal lifting because of their high viscosity.

Future fully automated lifting probably will be done in a long trough where the film will be compressed to have a width similar to that of the substrate. This compressed film then will be divided into equal sections, and an automatic arm holding the substrate will pick up the sections, one after the other (Figure 2.19). Such a fully automated trough can be extended to a production line easily with a series of single-bath troughs, each with a different film, and robots with substrates moving around and picking different film sections according to the material engineering requirements.

Once the practical problems are solved, the Schaefer technique will be widely used because of its great advantages. The first advantage is that horizontal deposition rate is not reduced with the increased film viscosity, and therefore, polymeric films that give thermally stable monolayers can be used. The second advantage is the formation of noncentrosymmetric X-type multilayers that can be used for different applications. The third, and yet more generally important advantage of this horizontal lifting technique is the ability to construct *organic superlattices*. By superlattice, we mean a close-packed, ordered, three-dimensional molecular organizate, which exhibits new physical characteristics and is created by the repeated overlay of monomolecular layers of different types of organic molecules. This kind of material engineering at the molecular level has interesting possibilities, where superlattices with different functionalities can be prepared. Such superlattices may enable the construction of molecular integrated devices, since the different layers may have different functions, such as amplification, optical processing, electronic transmitting, etc.

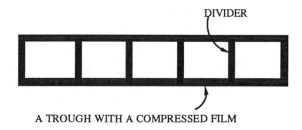

DIVIDER

A TROUGH WITH A COMPRESSED FILM

Figure 2.19. A trough with dividers for horizontal lifting of rigid LB films.

129

F. Monolayers at the Water–Oil Interface

Most of the early work in the area of monolayers at the water–oil interface considered potential differences due to spread monolayers at the interface [193–206]. Robb and Alexander compared monolayers of dodecyl triethylammonium bromide on the water–air and water–hexane interfaces [207]. They reported that π-A approaches 2 kT at both interfaces, but was reduced to 1 kT when enough salt was added to the aqueous subphase. Taylor et $al.$ studied the state of aggregation in pure and mixed monolayers of phospholipids at the water–heptane interface (structures shown) as models for biomembranes [208].

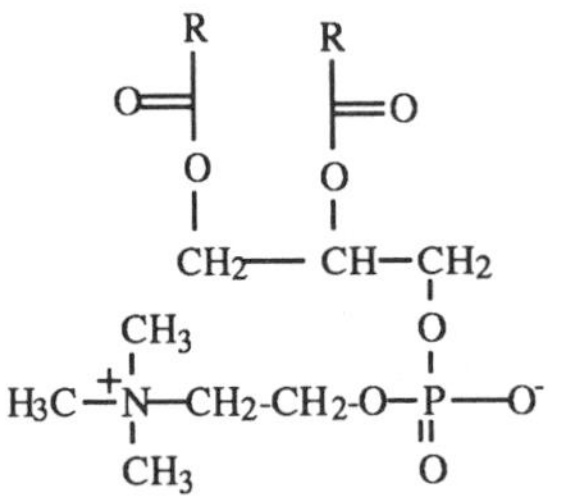

Phosphatidylcholin 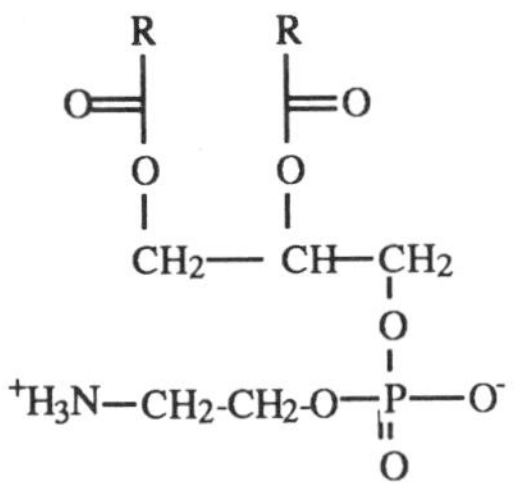   Phosphatidylethanolamine

They reported that the longer the chain (C_{22} vs. C_{16}) and the lower the temperature, the more the changes occur upon compression approximate to a classical first-order phase transition, implying the formation of highly aggregated clusters in the condensation process.

The striking phenomenon in a mixed film of C_{16} derivatives of the two phospholipids is that each component condenses out of the expanded state separately. The implication of this behavior is that in the fully condensed, close-packed state of the mixed phospholipids, the monolayer is probably mosaic with two kinds of small clusters, each rich in one or the other of the two components. In years following Robb and Alexander studies, much work had been devoted to the examination of phase properties of phospholipid monolayers as models for natural membranes. Llerenas and Mingis studied phase transition in 1,2- and 1,3-distearoyl-sn-glycerophospho-choline and found evidence of a change in the ordering parameter of these isomers in the solid-condensed phase [209]. Bell et $al.$ developed a statistical mechanical model in which the monolayer molecules occupy sites on a two-dimensional lattice and adopt one of two possible orientational states (lattice gas model) [210]. By appropriate choice of the interaction energies of the molecules in these states (attractive or repulsive), and the use of order–disorder statistics, they could calculate the feature characteristics of the second-order phase transition observed in these systems. Using monolayers at the water–oil interface, Ohki and Ohki measured surface pressure for phosphatidylserine monolayers at the water–hexadecane interface at 70 $Å^2$/molecule and reported $\pi > 40$ mN m^{-1} [211]. They deduced that the surface pressure of a lipid bilayer also should be quite high and suggested the value of ~45 mN m^{-1} in the liquid–crystalline state. This study indicates that monolayers at the water–oil interface apparently are good models for the study of bilayer membranes, since they represent half of this biological system. Bell and Wilson used a monomer–dimer lattice gas model for the calculation of patterns of isotherms corresponding to four quite different types of monolayer behavior [212]. These calculations, although quite simple compared to today's molecular dynamics, provide a great deal of help in understanding of packing and order at the water–oil interface.

Yue *et al.* [213] and Taylor *et al.* [214] published detailed studies on phospholipid monolayers at the water–heptane interface. They studied the π–A isotherm as a function of temperature (5°–40°C) and showed that the phase transition is very temperature-sensitive. Using the Clausius–Clapeyron equation, they calculated heat of the phase change and reported that it changes with temperature but not salt concentration or pH.

Hayashi *et al.* studied surface viscosities of dipalmitoylphosphatidyl-ethanolamine monolayers at the water–hexane interface as a function of bulk pH [215]. They found that monolayers expanded when the pH was shifted from neutral to to alkaline (probably due to the repulsion between the anionic phosphate head groups), and that led to an increase in surface pressure at constant area. This increase was accompanied by the reduction of surface viscosity that was attributed to disintegration of the zwitterionic structure in the condensed monolayer. In another study from the same Japanese group, Hayashi *et al.* report on the interaction between L-diacylglycerophoaphoethanolamine and *n*-alkanes in monolayers and bilayers [216]. They studied the pressure-area isotherms at the water–oil interface for alkanes ranging from *n*-hexane to *n*-hexadecane, and found a strong dependency of the phase transition from the expanded liquid to the condensed phase on the chain length of the alkane. Using differential scanning calorimetry, they studied the phase transition temperature and its entropy for the same lipid in the different alkanes. The temperatures of gel-to-liquid crystalline phase transition for these bilayer systems were changed in the same way as the monolayers, thus suggesting that the incorporated *n*-alkanes have an entirely parallel effects on the phase transition in these two condensed films.

131

Mingis *et al.* reported on an extensive study of the effect of chain length (C_{14} to C_{22}) on phase transition in phospholipid monolayers at the aqueous sodium chloride–*n*-heptane interface [217]. In the lecithin series (1,2-di-acyl glycero-phosphocholines), they found second-order phase transitions that are very sensitive to the alkyl chain length and, at a given temperature, move to a higher surface pressure and lower area/molecule as the chain length is decreased. Finally, for the di-C_{12} derivative, no phase transition could be distinguished. They calculated the entropies of compression and found that they vary linearly with the chain length, with a slope of Rln3 per methylene unit (where R is the gas constant). The entropy change for each alkyl chain is $(n–1)$Rln3, where n is the chain length. These results — together with the comparison of the heat of the monolayer phase transitions, the calorimetric heat of melting of the alkyl chains, the heat of fusion, and heat of solution of the corresponding alkanes — brought the authors to suggest that the chains in the phospholipid monolayers are fully flexible (liquid-like) at low densities, but very restricted (crystalline-like) at the condensed phase. This is an important conclusion that has implications beyond the immediate subject of phospholipid systems. The question of packing, orientation, and free rotation of the alkyl chains in a two-dimensional assembly is discussed in Part Four.

Desai and Stroeve used electrodes to study transport across lipid monolayers at the water–oil interface [218]. These new measurements can help in the understanding of ion transport across natural biological membranes. Davies and Jones studied the interaction of liposomes containing main intrinsic membrane proteins of the human erythrocyte membrane with phospholipid monolayers at the water–air and water–decane interfaces [219]. They found that when these proteins are present in proteolipid vesicles, their transfer into the monolayer phase is facilitated greatly.

This research is an example of a study that can be defined under the general subject of recognition. In this new exciting area of research, monolayers are used as model systems for biological membranes, and experiments are designed to try and mimic processes that occur on the cell membrane. With the vision of cellular diagnostics and cellular therapy, there will be a need to distinguish among cells (select cancer cells in the presence of healthy ones), and to be able to target a drug to a specific cell. This will require an understanding of the changes that occur in the cell membrane as a result of a disease. It can thus be expected that the research area of recognition will expand and that lipid monolayers both at the water–air and water–oil interfaces will become more complex trying to mimic the cell membrane and its functionalities.

2.2. Thermal Stability of LB Films

A. Order–Disorder Transitions

One of the early works on the thermal stability of LB films is that of Menter and Tabor, who used relaxation electron diffraction to study the effect of heating on LB monolayers of fatty acids $(CH_3-(CH_2)_n-CO_2H$, where $n = 7$ to $45)$ [220]. They concluded that the short range order regularity in these LB films is only stable well below the melting point of the bulk material. They suggested that as the temperature rises, the randomness in the position (i.e., the place of one amphiphile relative to its neighbors) and then in the orientation (i.e., the tilt angle) of the alkyl chains increases. In an earlier work, the same authors suggested that some orientation is maintained up to the melting point. We note that these results are not in agreement with recent studies (see later in this section) that conclude that the randomization of the orientation is the first step, and only at much higher temperatures is the head group affected.

The first study on order–disorder transitions in LB multilayers was carried out by Naselli *et al.* using FTIR [221, 222]. Using heating and cooling rates of 20 and 8°C min^{-1}, respectively, they reported a two-step melting process in an LB monolayer of cadmium arachidate (CA) on silver. Figure 2.20 presents a series of spectra of different temperatures. The band intensity in the C–H stretching region began to show relative intensity variations above 65°C, while the head group vibrations at 1545 and 1432 cm^{-1} showed no such changes. At 90°C, a considerable change in the intensities of both $v_a(CH_2)$ and $v_s(CH_2)$ at 2918 and 2850 cm^{-1}, respectively, was observed, again, with no substantial change in the head group stretching region. At 125°C, however, more dramatic changes were detected both in the C–H and head group stretching regions. These results suggest that a progressive disordering in the alkyl chains started with a pre-melting stage, which is triggered in the 45°–60°C region, and slowly developed up to the melting point (110°C, confirmed by DSC studies). The second stage is above melting, the monolayer is liquid-like (as both $v_a(CH_2)$ and $v_s(CH_2)$ are shifted to higher frequencies), and a complete randomization of the alkyl chains occurs. The fact that an irreversible process occurs once the alkyl chains are heated to a temperature >100°C may be, at first glance, at odds with the fact that liquid paraffin crystallizes in a structure with all-*trans* hydrocarbon chains. Thus, the reader could ask: Why should monolayers be trapped in a supercooled state when the free energy strongly favors an all-*trans* conformation? The answer to this question will be divided into two parts. First, in monolayers, when head groups still are pinned to the surface, chain motion is restricted.

133

Thus, releasing the entangled alkyl chains may take a very long time, similar to time scales of similar processes in glassy polymers. Second, in multilayer systems, cross-layer mobility at the molten state can make chain entanglement even more problematic. We conclude, therefore, that monolayers are trapped in a supercooled state because of kinetic rather than thermodynamic reasons. One important issue that was not addressed in this work is monolayer desorption and material loss. Note that the loss of a few percent of material should create free volume and considerable disorder.

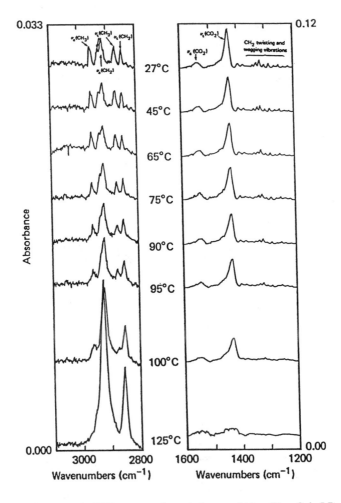

Figure 2.20. Grazing angle FTIR spectra of a cadmium arachidate film of six LB monolayers on silver at various temperatures. (From Nasselli *et al.* [221], © 1985, Am. Inst. Phys.)

Table 2.2. Advancing contact angles (deg) measured before and after heating and cooling to ambient temperature. (Adapted from Cohen *et al.* [223], © 1986, Am. Chem. Soc.)

Film Type	Before			After		
	H_2O	HD[a]	BCH[a]	H_2O	HD	BCH
CA/CA/OTS/Al	115 ± 2	47 ± 1	53 ± 1	109 ± 2	42 ± 1	48 ± 1
CA (LB)	108 ± 2	43 ± 1	49 ± 1	100 ± 2	32 ± 2	41 ± 2
AA (SA)	105 ± 2	45 ± 1	51 ± 1	97 ± 2	30 ± 2	39 ± 2

[a]HD — hexadecane, BCH — bicyclohexyl.

Cohen *et al.* studied the thermal stability of CA monolayers on aluminum oxide and of CA bilayers on an OTS monolayer [223]. For comparison, they studied a self-assembled monolayer of arachidic acid (AA) on aluminum oxide. In all cases, they used both contact angles and FTIR to examine the monolayers prior to and after the heating process. The part of their results concerning LB films is presented in Table 2.2. The grazing-angle FTIR spectra of CA/CA/OTS/Al at different temperatures revealed that this system represents the weakest film-to-substrate bonding, i.e., with only van der Waals interaction. Indeed, the film shows the most severe disorder upon heating (starting between 83 and 103°C), both in the hydrocarbon (C–H) and head (CO_2^-) stretching regions. The ratio of the $v_a(CH_3)$ (2962 cm^{-1}) and $v_a(CH_2)$ (2918 cm^{-1}) vibrations, which is an indication of the orientation of the alkyl chain relative to the surface, changed dramatically. Also, a shift toward higher wave numbers was observed. This shift is an indication of the liquid-like nature of the monolayer after the heating process. These results essentially are in agreement with Nasselli *et al.* [221, 222]. We note, however, that since CA was deposited on OTS/Al, thermally induced changes in the OTS layer could affect the behavior of the CA bilayer. This has not been discussed in the paper, and we feel that the order in the CA bilayer cannot be separated from that of the OTS monolayer.

While the FTIR spectrum of AA on aluminum (which represents the strongest film-to-surface bonding of an LB film) is not presented in this paper, the spectra of CA/Al at different temperatures revealed no change in the COO$^-$ stretching region up to 125°C. On the other hand, in the C–H stretching region, there were changes (starting between 76 and 100°C) less dramatic than in the CA/Ag and CA/CA/OTS/Al cases. The reasons for this order enhancement presumably are the ionic interactions that exist between the CA molecules and the oxide-covered metallic surface [224]. The interaction between CA and the native aluminum oxide is stronger than between CA and the native silver oxide, as is apparent from a comparison between these two studies.

Barbaczy *et al.* [225] and Rabe *et al.* [226, 227] reported on variable-temperature Raman studies of CA LB multilayers using waveguide Raman spectroscopy. This work is important because of (*a*) the wide temperature range, $-90°-100°C$, and (*b*) the use of Raman spectroscopy. The C–H stretching region is broadened upon heating from $-90°$ to $16°C$, indicating less molecular motion in the low temperature range. The Raman results tend to support the two-stage disordering mechanism, since no significant *gauche* contribution could be detected in the Raman spectra below $90°C$. Scanning microellipsometry and interference contrast microscopy results indicated lateral diffusion ($D \leq 10^{-10}$ $cm^2\ sec^{-1}$), which is typical of a solid phase at a temperature as high as $100°C$ [227]. Thus, it was suggested that under $100°C$, the order–disorder transition is characterized by a strong increase in rotation (libration) about the chain axis, analogous to bulk hydrocarbon materials [228, 229].

Rothberg *et al.* looked at monolayers and multilayers of cadmium stearate (CS) on sapphire using pulsed opto-acoustic spectroscopy [230]. In this experiment, the configuration is similar to that in the ATR experiment; hence, the CH_2 symmetric and antisymmetric stretching vibrations, which have transition dipole moments in the plane of the substrate, will adsorb the incident polarization strongly. When a film of five layers of CS on sapphire was heated to $65°C$, there was a very slight broadening of the peaks and a blue shift in the frequencies. However, a major decrease in the integrated intensity was detected. This is an indication of a decrease in the original ordering of the alkyl chains, which apparently could not be restored even after cooling to $20°C$. Further heating (to $100°C$) reduces the integrated intensity to ~1/3 of its original value, while cooling to $20°C$ restores the spectrum obtained at $65°C$. The same behavior was observed in a monolayer of CS on sapphire; however, a careful examination of peak positions suggests that the multilayer is slightly more ordered than the monolayer. Rothberg *et al.* also observed an inhomogeneous broadening of the peaks, indicating a probable difference in the order parameter of methylene units toward chain ends and in the middle, with the latter being more ordered. This is in agreement with what was found in molecular dynamics simulations. (See Section 4.3.A.2.)

Riegler used electron diffraction to study the thermal behavior of LB monolayers [231]. This is one of the more important studies because of the electron diffraction sensitivity to positional order and to the packing symmetry. The important result in this paper is that a pre-melting disordering could be observed clearly. A sharp decrease in the diffraction peak intensities was detected at ~35°C for cadmium stearate (C_{18}), ~55°C for cadmium arachidate (C_{20}), and ~75°C for cadmium behenate (C_{22}), with no apparent change in the hexagonal geometry of the two-dimensional assembly. It was suggested, therefore, that the intensity decrease was caused by thermally random tilt orientational disorder, or bending (i.e., *trans-gauche* isomerization) of the chains.

Richardson and Blasie studied the thermal melting of arachidic acid (AA) monolayers, deposited between two bilayers of polymerized diacetylene (D), by high-resolution x-ray diffraction over the temperature range of 30°–80°C [232]. We note that the authors used 1 mM $CdCl_2$ as a subphase in their experiments; thus, although using the abbreviation AA, they really may have investigated the cadmium salt and not the free acid. They looked at the structure of one [DDAADD], three [DD(AA)$_3$DD], and 10 bilayers [DD(AA)$_{10}$DD] of arachidic acid. Heating of the multilayer systems resulted, according to their model, in a two-stage process, where the first stage was a "continuous" untilting of the all-*trans* chains (increased thickness), followed by a second chain-melting stage, where kinks and jogs were being developed. The most important result in this study is that these are simultaneous processes; i.e., the chain untilting stage is not completed before the formation of kinks and jogs starts. Furthermore, the magnitude of these thermally induced changes decreases with the increasing film thickness — $(AA)_{10} < AA)_3 < AA$ — which indicates that the average stability of an individual monolayer in a multilayer film increases with the increasing number of layers.

Order–disorder transitions in alkanethiol monolayers on gold fit into the discussion here. We feel that the structural information in both cases should be considered together, although the systems are definitely different. This is because there are common rules in the packing of alkyl chains (Part Four), notwithstanding the nature of the specific surface physical or chemical forces involved in the formation of the assembly.

Chidsey *et al.* studied the surface of docosanethiol (CH_3–$(CH_2)_{21}$–SH) monolayers on gold(111) surfaces using helium diffraction [233]. There, as surface temperature increased from 50 K, the diffraction peaks (from the surface CH_3 layer) gradually lose intensity, until finally disappearing at ~100 K. It is reasonable to assume that the terminal CH_3 groups will start to rotate first, simply because van der Waals stabilizations at the interface are weaker than in the bulk. Thus, although Barbaczy *et al.* started their experiments at a temperature higher than that reported by Chidsey *et al.* for a complete randomization of the surface CH_3 groups, these studies are nevertheless complementary. Nuzzo *et al.* studied temperature-dependent phase behavior in long-chain docosanethiol monolayers on gold(111) surfaces [234]. In this study, they cooled the monolayers to 80 K, and observed significant changes in the IR spectrum. It seems that the conclusion most significant to the present discussion is the evidence of *gauche* conformations concentrated at the chain termina. It was suggested further that a population of *gauche* bonds exists at temperatures of the order of 200 K.

To summarize, the structure of monolayers at room temperature still is not understood fully. Whether monolayers are paracrystalline, polycrystalline, hexatic or liquid crystalline has not been fully established, and we believe it varies from system to system. The extent of motion and thermal disorder, even at room temperature, is clear from molecular dynamics simulations [235, 236]. Dealing with monolayer systems at elevated temperatures, is even more complicated. The nature of the pre-melting transition, e.g. — randomization of chain tilt, *trans-gauche* isomerization, chain untilting, onset of rotator phases, etc. — has not been firmly established in any monolayer system. Of course, it is probable that more than one mechanism may be at work simultaneously.

We note that while FTIR may not be sensitive enough to detect the early thermally induced changes in LB films, Raman spectroscopy has been shown to be well suited to detect *gauche* conformations. However, the reader should bear in mind that the number of surface *gauche* bonds is small compared to the bulk multilayer sample, and that chain termina within the film may have considerably fewer *gauche* bonds than those in the outermost layer. Therefore, the Raman data discussed previously do not exclude surface *gauche* conformations. We note, however, that substrate effects have not been mentioned in the preceding discussions, simply because this issue never has been studied in relation to order–disorder transitions. It also is essential to stress that small differences in preparation procedures may have considerable effects on the structure, and hence on the stability of the molecular assembly. Thus, a comparison between different samples, even of the same amphiphile, is not straightforward.

It is apparent from electron diffraction (although taken from monolayers), and especially from x-ray diffraction data, that thermal processes begin at relatively low temperatures, well below the melting point. We note that the difference in IR spectra at 65°C and 75°C in Figure 2.20, i.e., the increase in the $v_a(CH_2)$ intensity, is at a temperature only slightly above the one deduced from electron diffraction from a single monolayer (55°C) [231]. Therefore, these results are not that different.

Taking all the preceding qualifying statements into consideration, one would conclude that suggesting a comprehensive picture for order–disorder transitions in monolayer and multilayer systems is practically impossible. Nevertheless, we propose that there are general properties that are common to all systems, regardless of the amphiphile nature, the mode of adsorption, and the substrate. Hence, we present a possible scenario here for order–disorder transitions, but strongly advise the reader to recall that this is only a general picture.

• *At 50 K* (–223°C): We suggest that any molecular assembly (LB of SA, monolayer or multilayer) is completely frozen. Helium diffraction could be detected from the surface methyl layer of docosanethiol on gold(111) (Figure 2.21a) [233]. Recent detailed molecular dynamics simulations of monolayers of docosanethiol on (111) gold at 50 K offer support for this picture [235, 236]. We propose that this should be the case for other LB and SA monolayers as well. We note, however, that while methyl-terminated monolayers apparently reorganize to the all-*trans* conformation when placed at 50 K, this may not be the case for hydroxy-terminated monolayers [237]. There, molecular dynamics simulations suggest that surface reorganization leads to H-bonded hydroxyl groups, and the reorganization back to the ordered assembly is not favorable, kinetically or thermodynamically [238].

• *At 100 K* (–173°C): The terminal methyl groups in a monolayer of docosanethiol on gold(111) rotate freely [233]. We believe that the surface methyl groups will be the first to disorder in other molecular assembles, too, although temperatures may be somewhat different.

• *At 183 K* (–90°C): There should be *gauche* bonds in the monolayer of docosanethiol on gold(111). Raman spectroscopy could not detect their existence in the LB multilayers of CA, probably because of their low total concentration [225].

• *At 200 K* (–73°C): FTIR studies suggest that there are *gauche* bonds concentrated at the chain termina (Figure 2.21b) [234]. Molecular dynamics simulations at 300 K suggest the same conclusion [235, 236]. It may be reasonable to propose that disorder starts from chain termina and penetrates slowly into the bulk of the assembly.

• *At 289 K* (16°C): No disorder could be detected in LB multilayers of CA, either by IR [221, 222] or Raman spectroscopy [6].

• *Between 297 and 353 K* (room temperature and 80°C): The assemblies continuously change. A slow increase in *gauche* bonds concentration competing with chain untilting was found for AA monolayers deposited between two bilayers of polymerized diacetylene by high-resolution x-ray diffraction [232]. We suggest that a similar process may occur in other assemblies, depending upon chain length and the initial free volume in the system.

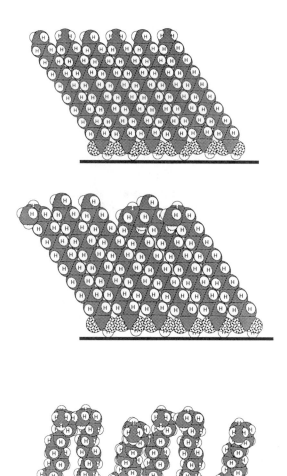

Figure 2.21. An assembly of stearic acid: (a) all-*trans* (50–200 K); (b) *gauche* bonds at the chain termina (starting at ~200 K); (c) irreversibly disordered (> 400 K).

• *Above 353–363 K* (80°–90°C): Melting was apparent in AA monolayers deposited between two bilayers of polymerized diacetylene; kinks and jogs probably are distributed along the chains. On the other hand, both IR studies of CA LB films on silver [221, 222], and Raman studies of CA LB films [226], do not suggest that such a severe disorder occurs, but indicate rather that a strong increase in rotation about the chain axis may be the disorder process. These results imply that thermal stability of molecular assemblies is not only a strong function of chain length (van der Waals forces) [231], but also of the nature of head groups and the interactions between them (electrostatic forces, H-bonds, etc.).

• *Around 402 K* (125°C): Randomization of the carboxylic head group occurred in LB films of CA on silver [221, 222], and in CA monolayers on Al_2O_3 [223] (Figure 2.21c).

With this collection of data in mind, we are still left with the issue of reversibility reported in some cases. We suggest that since *gauche* bonds concentrate at the monolayer–air interface (chains termina), reversibility is possible in methyl-terminated chains at relatively low temperatures. However, once conformational disorder (kinks and jogs) penetrate deeper into the bulk and chain entanglement occurs, the disorder becomes irreversible. This explanation may get some support from the work of Evans *et al.* on surface reorganization and wetting. They observed that the high free-energy surface of $HO–(CH_2)_{11}–SH$ on gold reorganized with time, showing an increase in the advancing water contact angle from ~20 to 60° in 30 min [237]. While this change was irreversible, it could be prevented by keeping the surface under a high dielectric constant liquid, such as water or glacial acetic acid. On the other hand, the hydroxylated surface resulting from the $LiAlH_4$ reduction of methyl trichlorosilyltricosanoate (C_{23}) showed an increase in the advancing water contact angle from ~30 to ~60° upon exposure to boiling CCl_4 for 10 min, which then could partially be decrease to ~45° upon 30 h immersion in clean water [239]. (See Section 3.2.F for further discussion.) These results, and the recent study in our laboratory on the reorganization of $HO–(CH_2)_{21}–SH$ monolayers on gold (showing an increase in the advancing water contact angle from ~25 to ~60° in 350 min, at a temperature of 50°C) [237], suggest that disorder, indeed, may start from the monolayer surface, where *gauche* bonds already exist, and penetrate slowly down into the bulk, with kinetics that are a strong function of chain length (monolayer viscosity).

Thermal stability is an important issue for any practical application. The preceding data and discussion point out the importance of order–disorder studies for new molecular assemblies. Caution is called for when considering literature data, including the aforementioned discussion. We still are far from being able to suggest detailed molecular mechanisms for the events occurring at different temperatures. However, we believe that this section may serve as a guideline when considering possible explanations for observations made in the laboratory.

B. Monolayer Desorption

Claesson and Berg studied the room temperature stability of cadmium arachidate monolayers by the surface force technique [240, 241]. They found that the stability of arachidic acid monolayers deposited on a hydrophobic substrate is much higher in a buffered 3×10^{-4} M $CdCl_2$ solution than in pure water. Furthermore, they reported that the adhesion force between the two head groups is much larger in the $CdCl_2$ (i.e., for the cadmium salt, $(-COO^-)_2Cd^{2+}$), than in water (i.e., for the acid, $-COOH$), and that no electrostatic repulsion existed between the layers in the Cd^{2+} solution at pH 6.6. This study explains what has been observed experimentally for many years, that fatty acid multilayer films are easy to assemble from a subphase containing heavy atoms, especially Cd^{2+}.

Laxhuber et al. studied the thermal desorption of arachidic acid from gold and reported a first-order process with an activation energy of 0.84 eV [242]. Jones and Webby used a quartz–crystal microbalance to study the thermal stability of LB films under vacuum [243]. Films (1–12 monolayers) of stearic acid and dipalmitoyl phosphatidyl choline were studied, and while loss of water was detected in both cases, LB mass loss revealed second-order kinetics for both films with very different activation energies (e.g., low activation energy 0.02 eV, and high activation energy 0.20 eV, for a six-monolayer film of stearic acid). In a detailed paper, Laxhuber and Möhwald presented a mathematical model and analyzed the thermodesorption of LB multilayers [244]. They suggested that for a film of discrete multilayers, if a molecule from layer n desorbs only after the molecule above it did, then first-order reaction in coverage by an individual monolayer should be assumed. Tippmann-Krayer et al. reported on the thermostability and photodesorption of LB films [245]. They studied multilayers of cadmium arachidate and suggested that the action of light is both to create defects and to provide energy to overcome the binding energy.

The clear second-order kinetics reported by Jones and Webby is a unique example in these thermodesorption studies. It seems that to explain this result, one needs to adopt a suggestion that they rejected in their paper, i.e., molecular association prior to desorption, meaning that in their case, dimers rather than monomers desorb. To justify this assumption, let us consider the intermolecular interactions in the LB film. These interactions can be divided into the van der Waals forces between the hydrocarbon chains and the electrostatic interactions between the head groups. Since the relative contributions of these two energies depends on the alkyl chain length, it is reasonable to assume that in the arachidic acid films (C_{20}), van der Waals interactions are stronger than in the stearic acid systems (C_{18}). Therefore, if the temperature required to break the van der Waals forces is high enough to break the electrostatic interactions (C_{20}), molecules will desorb one by one, and first-order kinetics will be observed. However, if van der Waals interactions are disrupted in a temperature in which electrostatic interactions still hold (C_{18}), dimers will be desorb and second-order kinetics will be observed.

2.3. The Structure of Hydrocarbon LB Films

Since the begining of this field of research, scientists have devoted most of their work to the study of simple amphiphiles, mainly straight-chain hydrocarbon compounds, with no — or very simple — functional groups. Thus, the majority of papers in the literature are still on LB films of fatty acids, perhaps because of the ready availability of these materials in pure form. Although very limited from the material point of view, these studies have established the crucial understanding of packing and order at the molecular level. In this section, we discuss structural issues in Langmuir monolayers at the water–air interface, and in monolayers and multilayers deposited on a solid substrate.

During the writing of this manuscript, it has been brought to our attention that a new book is being prepared, primarily on the structure of thin organic films, by R. H. Tredgold. We decided, therefore, to limit the discussion to the structure of monolayers at the water–air interface, and LB films of simple alkyl chain molecules.

A. Structure at the Water–Air Interface

In our earlier discussion on the pressure-area isotherm (Figure 2.7), we used a simplified picture, where the isotherm was divided into three main parts: the gas-, liquid-, and solid-like phases. In reality, the picture is more complicated; however, there is a consensus that the low-density limit is similar to a two-dimensional gas, where molecules are separated (although aggregation may occur), and that the highest-density limit is a two-dimensional solid. Between these extremes, there may be a variety of phases, such as orientationally disordered (liquid expended, LE), orientationally ordered (liquid condensed, LC), and striped [246–248]. It was only in recent years that Pathica *et al.* carried out experiments on extremely pure materials, and could suggest thermodynamic fixed points for monolayers. These three points are the surface pressure for equilibrium spreading [249], the liquid–gas [250–254], and the LE-to-LC phase transitions [253, 254]. Measurements of area per molecule as a function of surface pressure showed that monolayers undergo a two-dimensional gas–liquid phase transition at surface pressures lower than 0.1 dyn/cm [250–254]. At higher surface pressures, a second transition appears, in most cases as a kink in the isotherm, and was interpreted as a second-order LE-to-LC phase transition, or a first-order liquid–solid transition of a contaminated system [256, 257]. However, after the careful work of Pathica *et al.*, this kink became a nice and sharp first-order LE-to-LC phase transition [253, 255]. It is important to remark that measurements of monolayers at the water–air interface are very sensitive to minute temperature changes, and therefore, they should involve good temperature control.

The detailed study of monolayer structure at the water–air interface is of great importance. First, it helps us understand how molecules organize in two dimensions; however, this is not unique, since it also can be deduced from the structure of a deposited monolayer. (See later in this section.) Second, unfolding the structures of different phases and structural changes during phase transitions reveals the dynamics of molecular organization, which is related to processes that occur during the self-assembly of monolayers. There, molecules spontaneously organize without any outside force (Part Three), and in this process they get into unit cells, thus forming the two-dimensional solid. It is this understanding of molecular organization dynamics that directs the design of amphiphiles and allows the construction of thin films with useful properties.

Dutta *et al.* used synchrotron radiation to study monolayers of lead stearate $[(C_{17}H_{35}COO)_2Pb]$, and lignoceric acid $(C_{23}H_{47}COOH)$ at the water–air interface [258]. For lead stearate monolayers at high surface pressures of 18–25 dyn/cm, they determined an area per unit cell (area per molecule) of 17.8 Å^2, with correlation length of ~250 Å (> 50 molecules). Reduction of the pressure to below the "knee" in the π–A isotherm (increasing area per molecule) was accompanied by a slow decrease in peak height, indicating a first-order melting transition, although no such transition was indicated in the isotherm.

In an interesting study, Barton *et al.* studied LB monolayers of heneicosanol $(C_{21}H_{43}OH)$ at the water–air interface [133]. We note that in this monolayer, the size of the head group (OH) is smaller than the cross-section area of the alkyl chain, and therefore, the symmetry of the assembly is governed by the hydrocarbons. The authors reported that at a constant pressure ($\pi = 30$ dyn/cm), there was a phase transition in the temperature range, $16.3 < T < 21.3°$. They interpreted this result as a hexagonal-to-pseudohexagonal phase transition (analogous to rotator I-to-rotator II in lamellar crystalline *n*-paraffins Figure 2.22) [259]. The reader should not consider rotator II as a phase where the alkyl chains are free to rotate about their molecular axes.

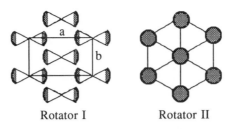

Rotator I   Rotator II

Figure 2.22. Representation of rotator I and rotator II phases. (After Unger [259], © 1983, Am. Chem. Soc.)

The difference between the two rotator phases is marked by the overall statistics of orientation of molecules relative to each other; hence, the long-range order in the layer. In rotator II, the CCC planes of the molecules are randomly oriented about their molecular axes; on the other hand, in the rotator I phase — that is, a uniaxially disordered structure — the CCC planes are oriented. As a result, intermolecular interaction and the effect of racheting on the packing is maximized in the latter. (See also Section 4.2.A.)

In a subsequent paper, Lin *et al.* studied the kinetics of phase transition in monolayers of heneicosanol at the water–air interface [77]. They found that at $T = 5°C$, they observed only the pseudohexagonal phase (rotator I, Figure 2.22), while at $T > 20°C$, only the hexagonal phase (rotator II, Figure 2.22) could be observed. The transition from pseudohexagonal to hexagonal occurred at $10 \leq T \leq 20°C$, and its rate was strongly temperature-dependent.

Recently, Lin *et al.* reported the phase diagram of heneicosanoic acid ($C_{20}H_{41}COOH$) at the water–air interface, in the range 1–8°C [260]. They reported (*a*) two distinct distorted-hexagonal (DH) solid structures at high pressures, both with chains that essentially are perpendicular to the surface; (*b*) two distinct tilted-DH structures at intermediate pressures; and (*c*) an expanded solid at low pressures. They concluded that all plateaus seen in the isotherm mark in-plane structural transitions, while all kinks mark tilting transitions.

Parallel to the work done in the U.S. by Dutta *et al.*, there was work in the area done in collaboration between Als-Nielsen's group in Denmark and Möhwald's group in Germany. In the latter, they occasionally used a combination of synchrotron x-ray diffraction and fluorescence microscopy to study monolayers at the water–air interface. Kjaer *et al.* studied phospholipid monolayers and reported that the transition to an orientationally ordered phase occurred at a pressure that was more than 15 mN m^{-1} below the pressure where

145

the transition to a positionally ordered phase occurred [261, 262]. This means that an intermediate phase exists with long-range orientational and short-range positional order. In a more recent paper, Kjaer *et al.* reported an x-ray reflection studies of arachidic acid at the water–air interface [263]. Their results demonstrated that above lateral pressures greater than 1 mN/m^{-1}, there exists a two-dimensional crystalline order at the water–air interface, with alkyl chains that are uniformly tilted, and with correlation length $\xi = 150$ Å. As the applied pressure increased, the tilt decreased continuously from 33° to 0°. The molecules in the tilted assemblies are in a pseudohexagonal lattice, and tilted toward a nearest-neighbor chain.

The question at this point is how these results relate to deposited LB films of fatty acids, or to self-assembled monolayers of alkanethiols. Now, since the OH head group in heneicosanol is smaller than the cross-section area of the alkyl chain, the system at the water–air interface can be compared to lamellar crystalline hydrocarbons $C_{23}H_{48}$ and $C_{25}H_{52}$, where the two rotator phases have almost the same volume per molecule and energy ($\Delta H_{trans} \sim 50$ cal/mol) [259, 264–266]. This may not be the case in LB films of fatty acids, where the carboxylic head group that is larger introduces free volume into the assembly, and thus imposes a tilt on the alkyl chains. Furthermore, once the head group is anchored to the surface, even if rotation is possible, it may result in the formation of a *gauche* bond with an energy penalty of ~500–700 cal/mol, much larger than the energy difference cited earlier. Also, our calculations show that the activation energy for rotation about the molecular axis in an assembly of dodecanethiol molecules on gold is quite high, and that the most stable arrangement is that of perpendicular alkyl chains in a pseudohexagonal symmetry [267]. (See Section 4.2.A.)

B. LB Films on Solid Substrates

The structure of LB monolayers and multilayers has been studied intensively in the last 10 years. Various experimental techniques have been employed for determining the packing and molecular orientation in these assemblies, and in Part One, we discussed them in great detail. The reader will find that one of the most studied systems is that of cadmium arachidate (CA), and although one of the first x-ray studies was published in the 1970s, it still is being studied today. This is because very high quality monolayers can be prepared, and therefore, are used to examine improvements in analytical tools. In this section, we are concerned mainly with the structure of saturated fatty acid monolayers and multilayers.

Takenaka *et al.* used ATR to study a 33-layer samples of stearic acid (SA) deposited on a Ge ATR plate [268, 269]. They reported that the tilt angle between the alkyl chain and the surface normal was between 25° and 35°. Ohnishi *et al.* used x-ray photoelectron and ATR FTIR spectroscopy to study monolayers and multilayers of CA [270]. They reported that the Cd^{2+} interacted with the glass substrate and that the alkyl chains are mainly perpendicular to the surface. Chollet used FTIR to study monolayers of behenic acid and calcium behenate on CaF_2 substrates [271]. They reported that while the alkyl chains were tilted 25 ± 4° in the acid monolayers, those in the calcium salt monolayers were tilted only 8 ± 5° from the surface normal. In another study, Chollet and Messier reported a tilt angle of 23 ± 2° in monolayers of behenic acid on CaF_2 [272, 273]. Allara and Swalen studied monolayers of CA on evaporated silver using grazing-angle FTIR [274]. They suggested a small tilt angle of the alkyl chain (although no specific value was indicated) and some twisting of the first few methylene groups adjacent to the carboxylic group.

Rabolt *et al.* repeated these experiments and reported the same results [275]. In another work, Rabolt *et al.* also studied the perdeuterated CA and found, again, that within few degrees, the chains are perpendicular to the surface [276]. Prakash *et al.* used x-ray diffraction to study LB films of lead stearate [277, 278]. They concluded that the LB films were very similar to epitaxially grown bulk crystals, with a very basic difference; i.e., in the LB assemblies there is order in the planes, but little correlation between different planes, similar to the smectic-B liquid crystal phase [279]. Bonnerot *et al.* used FTIR and electron diffraction to study monolayers and multilayers of behenic acid on carbon and aluminum substrates [280], and reported a tilt angle similar to that reported before for these monolayers on CaF_2 (23°). However, the most important result in this study is the structural transition observed when the thickness increased from one to seven monolayers.

Vogel and Wöll studied mixed monolayers of arachidic acid and methyl stearate on Cu, Ag, and Au single crystals (111) [281]. They reported that these monolayers form two-dimensional crystals with positional coherence that extended over the whole monolayer-coated metal substrate. Apparently, when single crystals are used as substrates, there is a strong epitaxy effect, and the monolayer is aligned with respect to the metal substrate. Kimura *et al.* studied LB films of stearic acid on a Ge plate, with 1–9 monolayers, using ATR FTIR [282]. They examined the CH_2 scissoring band, and suggested that there is a major difference between the stearic acid molecule in the first monolayer, and those in an LB film thicker than two monolayers. Thus, while in the thick film the molecules crystallize in a monoclinic form, and are tilted ~30° from the

147

surface normal, in the first monolayer they are in a hexagonal or pseudohexagonal subcell packing, with the alkyl chains perpendicular to the surface and free to rotate about their molecular axes [267]. Garoff *et al.* studied monolayers of cadmium stearate on SiO/C/Ni substrates [283]. They reported that the molecules are in a hexagonal symmetry, and are tilted from the surface normal. Their conclusion was that the monolayer is liquid-like, exhibiting a correlation length of 40 Å, and that the chains have significant rotation motion about their long axes. They compared the structure to that of a stacked hexatic phase [284].

We note that the idea of free rotation in the LB film is wrong. (See also the preceding discussion on the two rotator phases.) Once the free volume needed for such free rotation to occur and the resulting loss in van der Waals attraction energy are considered, it becomes apparent that free rotation is not possible at room temperature. The reason for this confusion, however, is the uniaxial distribution of molecular tilt angles, which makes it look like there is free rotation, while the fact is that close-packed assemblies of molecules, in hexagonal symmetry, are tilted in different directions. (See Reference 16 for more discussion.)

Skita *et al.* studied x-ray diffraction from LB films and could show that the surface of the outermost monolayer (at the monolayer–air interface) is disordered and that deposition of an additional monolayer induced ordering of this surface [285]. Outka *et al.* [286, 287] and Rabe *et al.* [288] used near-edge x-ray absorption fine structure (NEXAFS) to study molecular orientation in CA and calcium arachidate (CaA) monolayers on Si (111). They found that the alkyl chains in monolayers of CA were practically perpendicular to the surface, while in CaA the chains were tilted by $33 \pm 5°$ (similar to the tilt in alkanethiol monolayers on gold, discussed in Sections 3.3.A.4. and 3.3.A.5.) They suggested the formula $\tan\phi = nR/D$, where ϕ is the tilt angle from the surface normal, and $R = 2.52$ Å is the distance between second-nearest-neighbor carbon atoms in the alkyl chain, and D is the minimum van der Waals separation between the chains (which we calculated to be 4.24 Å [267]), and $n = 0, 1, 2$ to describe these distinct tilt angles (Figure 2.23). According to this formula, the tilt angles should be only $0°$, $30.7°$, $49.9°$, etc. Of course, since the tilt can be coupled with a rotation around the molecular axis, the tilt angle can fluctuate somewhat around this expected value. We shall discuss the issue of distance tilt angles in Section 4.2.A, where we describe our work on molecular mechanics simulations of SA monolayers of alkanethiols on gold [267].

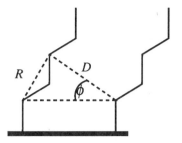

Figure 2.23. Efficient packing of alkyl chains.

There are two structural reports that are related to the aforementioned study and its suggested picture of efficient packing of alkyl chains. The first is that of Guyot-Sionnest *et al.,* who used sum-frequency vibrational spectroscopy to study Langmuir monolayers of pentadecanoic acid [CH_3–$(CH_2)_{13}$–COOH] [289]. They reported a tilt angle of $35°$ from the normal for the terminal methyl group, which suggests that the alkyl chains were nearly perpendicular to the water surface. The second is that of Dote and Mowery, who used grazing-angle FTIR spectroscopy to study LB monolayers of SA on aluminum surfaces (Al_2O_3/Al) [290, 291]. They calculated a tilt angle of $\phi = 12°$ for the fresh monolayers; however, after 168 h, they found a significant diminution in the methylene absorption ($v_a CH_2$). This change means that the alkyl chains became more perpendicular to the surface, which is in agreement with the idea of distinct tilt angle. There are two possible mechanisms that can be proposed to explain the change in tilt angle: (*a*) The monolayer is crystalline, and the observed $12°$ tilt angle is a weighted average of two kinds of domains, i.e., one having an average tilt angle of $\phi \sim 0°$ and the other $\phi \sim 30°$. The $\phi \sim 0°$ domains grow at the expense of the $\phi \sim 30°$ domains, which means that the monolayer should shrink due to the decrease in free volume. (*b*) The monolayer is amorphous, and the $12°$ tilt is a weighted average of different molecular tilts. In this case, the process reported was a slow crystallization process, where crystalline domains were formed, with an average tilt angle of $\phi \sim 0°$.

Mizushima *et al.* compared the structure of cadmium stearate LB films on glass, aluminum, and SiO_2 substrates, and reported that only in the latter was the monolayer ordered [292, 293]. Fischetti *et al.* studied the structure of multilayers of AA deposited on hydrophobic surfaces (OTS/Si) [294]. Using high-resolution x-ray diffraction, they could distinguish differences in the electron density of the polymethylene groups near the carboxyl head group and the terminal methyl group. They concluded that lattice disorder propagated away from the OTS–substrate surface, and that the introduction of bilayers of diacetylene derivatives into the AA multilayer films destroyed the long-range order.

2.4. LB Films of Fluorocarbon Amphiphiles

One of the ways to increase van der Waals interaction, and therefore, order and stability in molecular assemblies, is to replace the hydrocarbon by a rod-like fluorocarbon chain [295]. Langmuir and Schaefer reported on difficulties depositing perfluorinated carboxylic acids when the subphase contained regular metal ions [63]. However, Nakahara et $al.$ discovered that a normal π–A isotherm could be obtained with Al^{3+} in the subphase, and they reported on the first preparation of relatively thick LB film of henecosafluoroundecanoic acid $(C_{10}F_{21}COOH)$ [296]. The area per molecule in this study was 33.2 $Å^2$, which, by comparison to the cross-section area of the perfluoroalkyl (~25 $Å^2$), suggests that the chains are tilted relative to the surface normal. Figure 2.24 shows the π–A isotherm for the monolayers on barium or aluminum subphases.

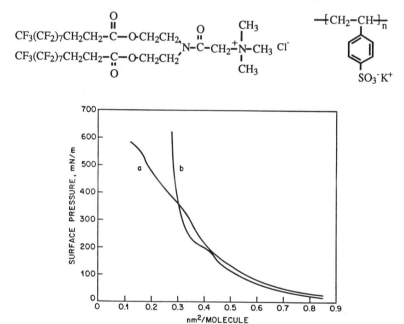

Figure 2.24. Surface-area isotherms at 20°C for henecosafluoroundecanoic acid monolayer on a Ba^{2+}-containing subphase (a), and on Al^{3+}-containing subphase (b). (From Nakahara et $al.$ [296], © 1986, Elsevier Science Publishers.)

150

Takahara *et al.* reported on the successful deposition of LB multilayer films of a two-chain fluorocarbon amphiphile [297]. Using potassium poly(styrene-sulfonate) in the subphase, they successfully deposited monolayers and multilayers. The area per molecule was 50 Å^2, suggesting that in this case, the molecules are quite normal to the surface, which is in agreement with the 8° tilt calculated from x-ray diffraction data. Their transfer ratio was very close to unity, indicating that an almost Y-type film was obtained.

Nasselli *et. al.* studied the order–disorder transition in LB films of semiperfluorinated cadmium nonadecanoate $[CF_3–(CF_2)_7–(CH_2)_{10}–COO^-]_2Cd^{2+}$ [298]. These are molecules with two parts that have very different properties. Thus, while the hydrocarbon part (the smaller in cross-section area) is known to undergo *trans-gauche* isomerization at elevated temperatures, the perfluoroalkyl part is known to remain rod-like even in the melt phase [299]. This difference clearly is being manifested in the structure, and in the behavior of these monolayers at elevated temperatures. The IR spectra in this case indicates the introduction of *gauche* bonds above 75°C. The difference between this case and those indicated previously is in the excess free volume imposed on the hydrocarbon parts by the perfluoroalkyl groups. This disordering is accompanied by further tilting of the fluorocarbon part of the molecules. These observations are reversible up to 125°C, apparently due to the rigid nature of the fluorocarbon chains that, therefore, are able to repack in the original arrangement upon reduction of temperature. This repacking imposes order on the hydrocarbon part of the molecules, thus forcing it to reorient.

2.5. LB Films Containing Ionophores

The transport of ions through biological membranes is a very important subject. Some antibiotic drugs (e.g., A23187) have been shown to bind and transport cations across natural and artificial membranes [300–303]. Such binding and transport of Ca^{2+} ions, for example, may have a control on cellular functions such as thyroid secretion [304], insulin release [305], cell fusion [306], and mitogenesis [307].

There are three kinds of experiments in this area, the first being the use of the pure ionophore on the water–air interface; the second involves its incorporation in a monolayer of a fatty acid; and the third includes its functionalization with a long alkyl chain, and using it as an amphiphile.

Caspers *et al.* studied the role of membrane charge density on the ionophore–ion complexation process at the water–lipid interface [308]. They incorporated the antibiotic into palmitic acid monolayers, and by changing the composition of the subphase (pH, nature of cation) and of the mixed film, studied the complexation process. They clearly established that the surface charge density has a role in the ionophore–ion complexation process:

$$\text{Antibiotic} \quad + \quad M^+ \quad \rightleftharpoons \quad \text{Antibiotic-}M^+$$
$$\text{\textit{monolayer}} \qquad \text{\textit{at the interface}} \qquad \text{\textit{monolayer}}$$

Their measurements showed that for a high negative surface charge density (basic pH, high value of molar function of charged lipid), the antibiotic exists in its complex form at the interface.

Ferreira *et al.* studied the antibiotic A23187 at the water–air interface [309]. Using radioactivity measurements with $^{45}Ca^{2+}$, they could show that at pH $\geq$ 5, two negatively charged ionophore molecules and one Ca^{2+} ion form a neutral complex, and that the activity sequence for bivalent ions is $Ca^{2+} > Mg^{2+} \gg Ba^{2+}$.

Winter and Manecke reported on the only experiment with polymerizable monolayers derived from crown ether surfactants [310]. They prepared dibenzo-18-crown-6, 1,10-diaza-18-crown-6, and 1,7-diaza-15-crown-5 derivatives and found that all formed stable monolayers at the water–air interface.

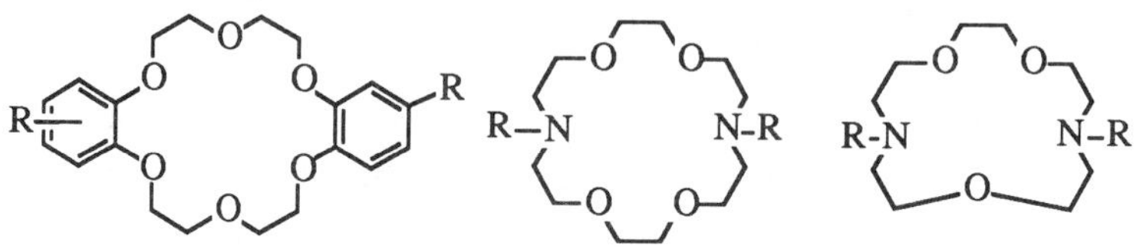

dibenzo-18-crown-6   1,10-diaza-18-crown-6   1,7-diaza-15-crown-5

Their π–A isotherms were featureless and, at 20°C, the monolayers were stable with a collapse area of, for example, ~60 Å^2/molecule, for the dibenzo-18-crown-6 derivatives. The polymerization of the monolayers was done at 10 mN/m^{-1}, where the area per molecule was 160 Å^2. During the uv irradiated polymerization, a decrease in the area was observed, which indicates that in the polymer monolayer the molecules are more densely packed.

Matsumura *et al.* published two papers on monolayers of crown ether compounds [311, 312]. In both, they used the same octadecyloxymethylene-18-crown-6 as their model compound. (See structure in next page.) In the first investigation [311], they studied monolayers of the ionophore derivative on different aqueous salt solutions. Stable monolayers could be formed with *ca.* 100 Å^2/molecule, and metal ion interaction with the membrane was established using surface potential measurements. They reported linear relationships between the surface potential and ion concentration in solution, and affinity sequences of $K^+ > Cs^+ > Na^+$ and $Ba^{2+} > Ca^{2+}$. In the second study [312], they indicated that an electrical double layer is generated by the complex formation at the membrane–solution interface. This double layer could be well described by the Stern model for the electrical double layer; by using this model, the binding constants of metal ions with the crown ethers at the interface can be evaluated.

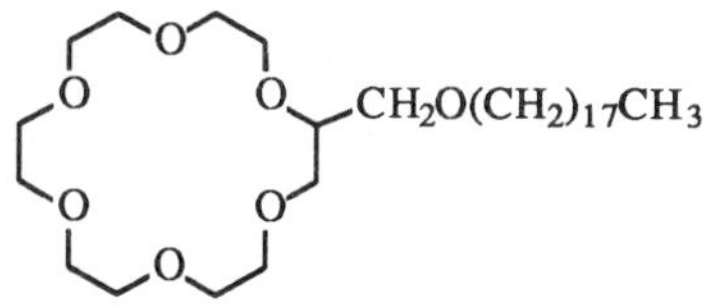

Gold *et al.* published the only study of macromolecules with pendant crown ether groups [313]. Using 4'-vinyl-benzo-18-crown-6 as a precursor, they prepared the polymer by standard free-radical polymerization.

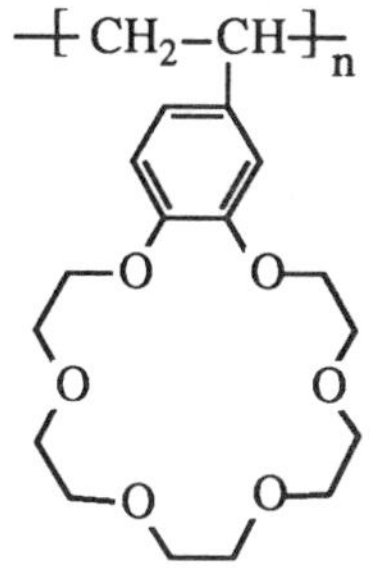

The π–A isotherm on pure water subphase showed a phase transition at ~ 40, with collapse at ~20 Å^2/molecule. Phase transition was even more pronounced when the subphase contained 0.1 M of an alkali metal salt. The authors suggested that in the region of the isotherm where the surface pressure was rising nearly linearly (70–40 Å^2/molecule), the pressure is due to repulsion between macromolecules. Their explanation for the inflection in the isotherm is quite reasonable. They suggest that these phase transitions reflect conformational transitions wherein the highly expanded macromolecule becomes more compact, as pendant crown moieties (possibly the benzo-part) are forced off the water. They also found that the order of collapse pressure of the film ($Li^+ >> Na^+ \sim Cs^+ > Rb^+ > K^+$) is reminiscent of the sequence observed in the binding of alkali metal ions to the crown ether in solution. Another important result in this study was that there is a significant difference between the binding behavior of the polymer in solution and at the water–air interface. Thus, while the polymer in CH_2Cl_2 is able to extract Rb^+ and Cs^+ picrates from aqueous solution more efficiently than

K^+ picrate, at the water–air interface the membrane appears to bind K^+ most strongly. This presumably is because at the water–air interface, the conformation of the polymer is highly extended and crown moities are likely to be oriented parallel to the water surface, thus not favoring efficient binding of Rb^+ and Cs^+, which is in a 2:1 ligand–ion ratio. This means that it should be possible to change binding characteristics of crown ethers at the water–air interface by appropriately designing the macromolecule. For example, changing the length and nature of the spacer (i.e., hydrophilic vs. hydrophobic), or a substitution on the polymer backbone, may change the affinity of the pendant crown ether groups to different metal ions.

2.6. LB Films of Liquid–Crystal Compounds

So far, we have discussed molecular assemblies, where interfaces were the ordering environment, and no inherent order existed in the system. Liquid–crystal phases (LC), on the other hand, are materials that have inherently ordered-layer structures, formed by self-organization of mesogenic compounds [314]. Therefore, by having a liquid–crystalline group in an amphiphile, one could get enhanced order, thermal stability, and interesting physical properties. Furthermore, the study of liquid–crystals at the water–air interface in a systematic way should add to our understanding of the two-dimensional organization, and the effect of the director on the relative orientation of molecules in the layers of a multilayer film.

The incorporation of mesogenic groups into amphiphilic structures, and the study of the properties of monolayers thereof, started in the early 1970s [315] and dramatically increased during the 1980s. Thus, a combination of amphiphilic and mesogenic properties of thermotropic LCs enabled detection of the formation of successive, discrete, and liquid–crystal-like multilayers at the water–air interface.

The structures of 4-heptylphenyl 4-hexanolyoxobenzoate (I) and 2-methybutyl 4-(4-hexyloxybenzoyloxy)benzoate (II) are shown in the next diagram.

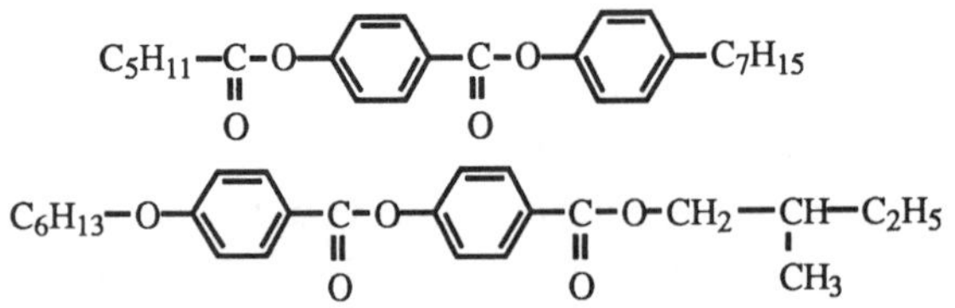

I - T_{C-N} = 38°C; T_{N-I} = 54°C

II - T_{C-N} = 43°C; T_{N-I} = 54°C

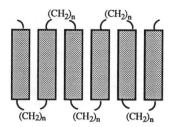

Figure 2.25. Suggested smectic-like arrangement of azo-, and azoxybenzene-containing main-chain polymeric liquid crystals at the water–air interface.

T_{C-N} is crystalline–nematic transition temperature, and T_{N-I} nematic–isotropic transition temperature [315–320]. In this case of calamitic or rod-shaped molecules, monolayers were found to collapse by a rolling-over mechanism, forming multilayers that were two, three, and four molecules thick [317].

The incorporation of other rod-shaped mesogenic units — such as biphenyl [319, 321], stilbene [322, 323], azobenzene [175, 324–326], benzylideneaniline [327], or pipyridine [328, 329] — into amphiphilic-type molecules was accomplished long before their potential as LC materials was studied or even recognized. We mention these papers here both because of the synthetic interest and the characterization of their monolayers at the water–air interface. Two examples of main-chain LC monolayer-forming polymers containing azo- and azoxybenzene units are presented in the next diagram. T_{C-N} again is the crystalline–nematic transition temperature, T_{N-I} is the nematic–isotropic transition temperature.

$$A_0 = 15 \text{ Å}^2$$
$$T_{C-I} = 226°C$$

$$A_0 = 25 \text{ Å}^2$$
$$T_{C-N} = 198°C; \quad T_{N-I} = 201°C$$

Suresh *et al.* incorporated azo-, and azoxybenzene derivatives into main-chain polymers (i.e., where the mesogen is a part of the chain, see structure above) and studied their properties at the water–air interface [320]. Surface pressure data reveals that upon compression, these mesogenic groups tilt from planar to vertical orientation (edge-on), thus occupying a very small area per molecule (Figure 2.25). This behavior is indicative of the dominant role these aromatic ring systems have on the two-dimensional order. Note that with this smectic-like configuration, the alkyl chains are outside the aromatic core, thus

155

facing both the aqueous subphase and air. No further evidence was provided for this molecular arrangement. This two-dimensional lamellar assembly of a rigid aromatic core with hydrocarbon chains acting as a diluent opens exciting possibilities in the design of materials with interesting physical properties and was further developed and demonstrated by Orthmann and Wegner [330].

Vickers *et al.* studied LC side-chain (biphenyl) polymeric LB films as optical wave guides [331]. Albrecht *et al.* studied monolayers of rod-shaped and disc-shaped LC compounds at the water–air interface [332]. While the rod-shaped amphiphiles form monolayers similar in behavior to conventional amphiphiles, the discotic LC behaved in a way suggesting that the molecules are edge-on at the interface, thus maximizing the π–π interactions. This is another example, therefore, of molecular arrangement whereby a rigid π-core is embedded in a liquid-like hydrocarbon medium.

The reader is referred to a recent review by Laschewsky on monolayers and LB films of discotic LCs [333]. It is beyond the scope of this book to discuss the different possible mesophases and their structures. Examples of a calamitic LC, a monomer of calamitic polymeric LC, and a discotic LC (from Reference 332) are shown in the diagram.

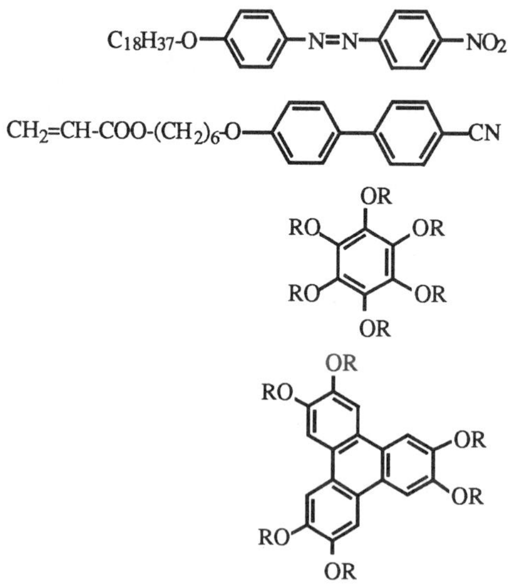

$C_{18}H_{37}$-O—⬡—N=N—⬡—NO_2 $T_{C-S} = 92^\circ C$; $T_{S-I} = 94^\circ C$

CH_2=CH-COO-$(CH_2)_6$-O—⬡—⬡—CN $T_{C-I} = 70^\circ C$

R = -CO-C_7H_{15}

$T_{C-Drd} = 81^\circ C$; $T_{Drd-I} = 87^\circ C$

R = -C_5H_{11}

$T_{C-Dho} = 68^\circ C$; $T_{Dho-I} = 123^\circ C$

The reader may find the definition of different discotic LC phases in Reference 333. Note these structures are molecules where the substituents are an electron donor and electron acceptor groups, thus forming a large molecular dipole moment. Such molecules are important for nonlinear optical (NLO) applications, which are discussed in Section 5.2. We note here that the inherent layer structure and order of LC make such LB films both interesting and important for material engineering. Indeed, the appreciation of these properties has triggered the synthesis and detailed studies of many interesting materials.

Richardson *et al.* studied the deposition and characterization of LB films of organometallic LC [186, 334], and Carpenter *et al.* studied LB films of side-chain polymers containing liquid–crystalline NLO chromophores [335]. (See Section 5.1.B.1 for further discussions.) Vandevyver *et al.* prepared and studied a family of new 3,5-diaryl-1,2-dithiolium LC salts (next diagram) associated with either diamagnetic (BF_4^-) or paramagnetic ($TCNQ^-$) anions [336].

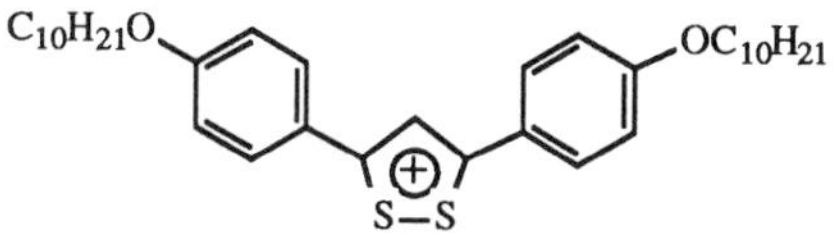

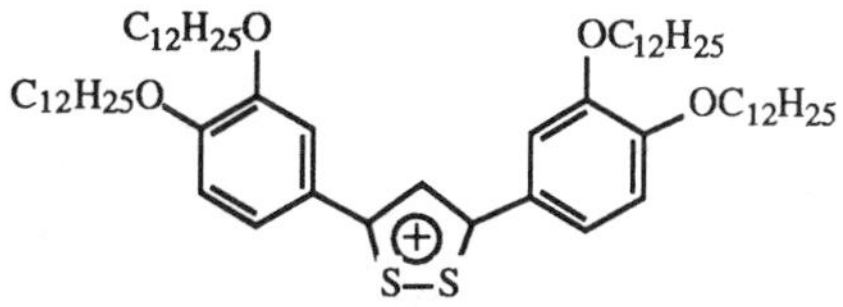

The authors report that the materials form very good, optically clear LB films with no light scattering. These are materials with interesting NLO and magnetic properties, although the authors mention only the lack of electrical conductivity in their report. The disubstituted salts (one substituent per phenyl) were found to belong to the smectic-S_A class, while tetrasubstituted salts formed the discotic LC phase (D_h). It was found that the tetrasubstituted salts (BF_4^- lead to films of better quality than the corresponding disubstituted analogues. This is probably because of (*a*) the better match between the cross-section areas of the heteroaromatic and aliphatic parts, (*b*) the better balance between the hydrophilic and hydrophobic parts, and (*c*) the better charge delocalization due to four aloxyn groups.

Sakuhara *et al.* investigated a terphenyl LC compound (next diagram), and reported that the hydrophilicity of the substrate and the ratio of the LC and CA determine the mode of transfer [337].

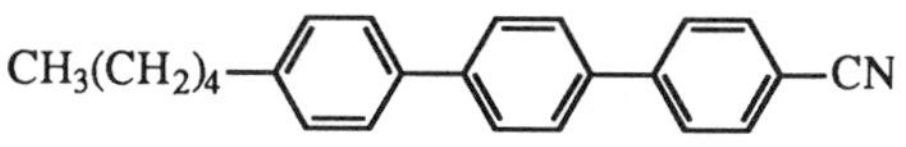

Thus, the pure LC gives Z-type deposition on hydrophilic substrates, which is not surprising, since the head group is not highly hydrophilic. On the other hand, on hydrophobic surfaces, the deposition starts as Y-type, but gradually changes to Z-type. This material is interesting because of (*a*) its short alkyl chain (C_5) and (*b*) its weak hydrophilic headgroup. This work is an important lesson in material engineering because it demonstrates that the inherent order of the LC material may be high enough so that a very short alkyl chain can be used. Let us consider that we want to design a film for an electro-optic application. The ratio between the electro-optic chromophore (in this case, the mesogenic terphenyl) and the alkyl chain determined how much useful material can be packed into this film, and hence, determined its performance. Furthermore, it proves that with the order given by the LC unit, there is no need for a strong hydrophilic head group, and, therefore, Z-type deposition becomes more stable, and even preferred.

Matsumoto *et al.* studied monolayers and LB films of amphiphilic dyes with mesogenic groups in the hydrophobic part (PhCy, see next diagram) [338]. While the C_{18} derivatives showed the expected amphiphilic behavior, the mesogenic derivatives had an unusual behavior at the water–air interface, indicating the formation of crystallites at the interface. Although interesting, the PhCy derivatives do not seem to have any advantage, and seem to indicate that once the aromatic–aliphatic ratio is too high, molecules tend to crystallize and not form an ordered film on water.

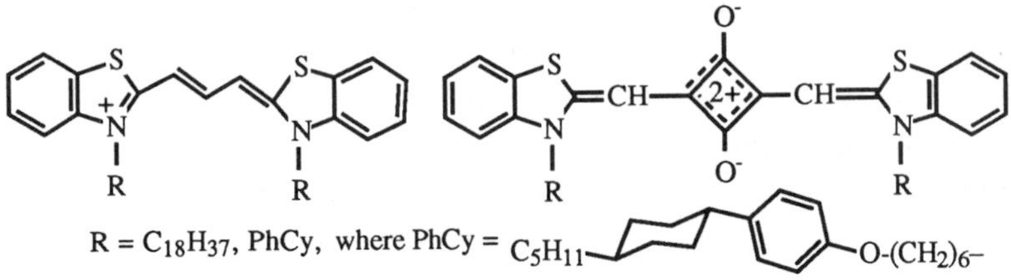

$R = C_{18}H_{37}$, PhCy, where PhCy = C_5H_{11}—⬡—⬡—O-$(CH_2)_6$—

Cook *et al.* studied LB films of 1,4,8,11,15,18-hexaoctyl-22,25-bis-(carboxypropeyl)phthalocyanine (next diagram), an asymmetrically substituted phthalocyanine, and reported that these materials form stable monolayers at the water–air interface that can be transferred onto hydrophilic silica substrates [339–341].

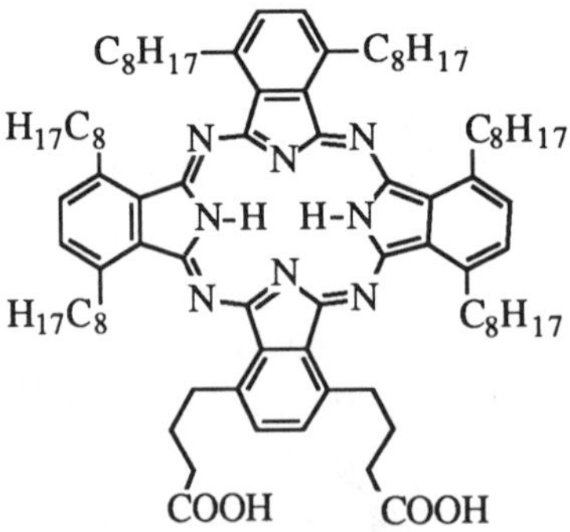

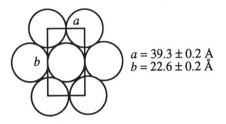

$$a = 39.3 \pm 0.2 \text{ Å}$$
$$b = 22.6 \pm 0.2 \text{ Å}$$

Figure 2.26. A schematic representation of the hexagonal mesophase, D_1.

When a monolayer film of the phthalocyanine derivative was heated, there was a remarkable change in the optical spectrum. This, by comparison to the spectrum of the bulk material, indicated a phase transition from the low-temperature herring-bone packing, to a high-temperature hexagonal packing (D_1, Figure 2.26). The authors did not indicate if this is a reversible transition.

Again, from the point of view of material engineering, the reader is reminded that phthalocyanines are very stable, highly aromatic systems, and that copper phthalocyanine is the major component of the photoconductor in every copy machine. A hexagonal packing as described, once extended in the c-direction (e.g., in polymeric phthalocyanines as described in Reference 330), may prove to be an important arangement for large π-systems.

Before closing this discussion, we note that to the best of our knowledge, none of the systems cited as having bulk LC transitions have actually been shown to undergo LC transitions *in the LB film*. It is apparent, however, that only the surface of the field of LB films of LCs has been scratched. The potential in these fascinating, ordered systems has not been fully comprehended, and much more fundamental science is needed before the scope and limitations of these materials are established.

2.7. LB Films of Porphyrins and Phthalocyanines

The porphyrin is one of the most important among biomolecules. It is involved in the fundamental processes of life, such as oxygen transfer and storage, electron transfer, and synthesis of amino acids. The porphine is the parent molecule in the family (Figure 2.27), and while all natural porphyrins are substituted at positions 2,3,7,8,12,13,17,18, the most stable synthetic porphyrin is 5,10,15,20-tetraphenylporphyrin (TPP). The phthalocyanine (PC, Figure 2.27) is a very stable, planar, synthetic aromatic macrocycle.

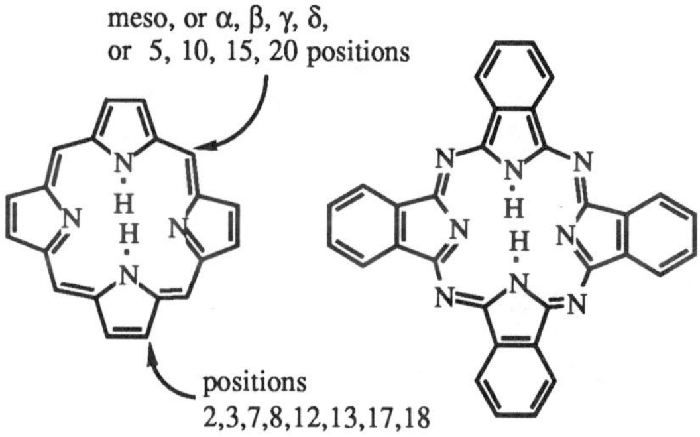

meso, or α, β, γ, δ,
or 5, 10, 15, 20 positions

positions
2,3,7,8,12,13,17,18

Figure 2.27. Porphine (left) and phthalocyanine (right) structures.

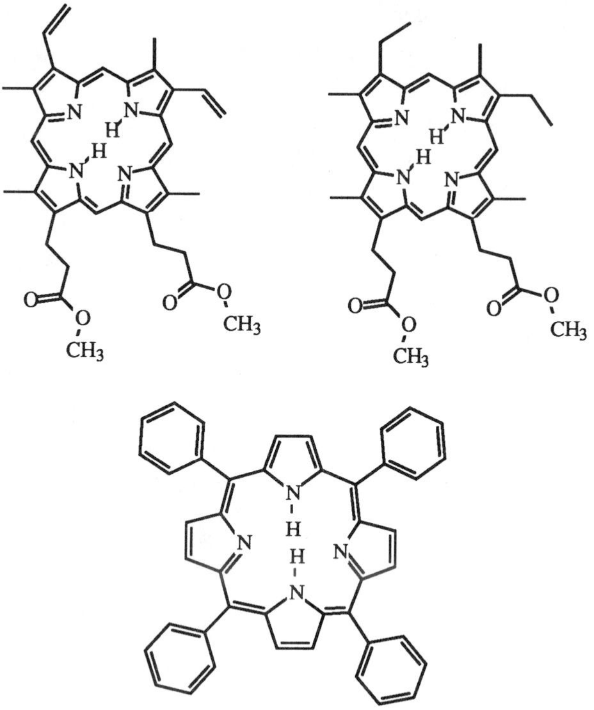

Figure 2.28. Protoporphyrin IX dimethyl ester (left), mesoporphyrin IX dimethyl ester (right), and TPP (bottom).

The syntheses of TPP and PC and their metallocomplexes are straightforward; however, complex molecules with more interesting substitutions may require lengthy synthetic routes. This is why, in many cases, either natural porphyrins or TPP derivatives are used. Two of the natural occurring porphyrins and TPP are presented in Fig. 2.28.

The emphasis in this part of the book is more on LB films of pure porphyrin and phthalocyanine derivatives, and less on their monolayers with other amphiphiles. The reader may find an account of some of these mixed monolayers in other parts, e.g., LB monolayers of phospholipids, or energy transfer in LB films.

A. LB Films of Porphyrins

The research on porphyrin monolayers was started by Alexander over 50 years ago [342]. Of particular interest is the use of porphyrins and metallo-porphyrins for the modeling of biological membranes [343, 344]. Some of the pioneering investigations on the effects of environment on reaction of porphyrins, such as μ-oxo dimer formation from dioctadecyl ester of mesoporphyrin iron (III) complex [345], photooxidation of protoporphyrin dimethylester [346], or photoreduction of TPP derivatives, was carried out in the 1970s [347]. Whitten *et al.* used mesoporphyrin IX dioctadecyl ester, meso-tetra(4-carboxyphenyl) porphyrin tetradihydro-cholestryl ester, and meso-tetra(2-hexanoylaminophenyl) porphyrin to study metallation and dimerization reactions at the water–air interface [348–350]. Stable films could be made and transferred only as mixtures with arachidic acid, but no reference to possible phase separation was made. In view of the work by Vandervyver *et al.* (see later in this section) it is likely that in these cases there was some phase separation also. We note that TPP derivatives are not planar due to the dihedral angle between the phenyl rings and the porphyrin. On the other hand, molecules such as the mesoporphyrin derivative are flat, and therefore, may form assemblies with the macrocyclic ring perpendicular to the surface (like a deck of cards).

Vandevyver *et al.* studied the structure of LB films of two metalloporphyrin derivatives, one a TPP substituted by one 5-(4-alkoxyphenyl)-10, 15, 20-tritolyl-porphyrin (M = Cu, Co), and the other by four alkoxy groups (5,10,15,20-tetra-(4-alkoxyphenyl)porphyrin (M = Cu, Co) [351]. None of the derivatives could form a stable film for transfer; therefore, a porphyrin–behenic acid mixture (1:4, respectively) was used. Utilizing UV-visible linear dichroism, IR linear dichroism, and EPR, they concluded that the metalloporphyrins were partially excluded and organize beside, or completely excluded from, the films. It seems that porphyrin derivatives that are too hydrophobic do not form stable, pure films, and once their alkyl chain substituents are not compatible in length with those of the fatty acids, they are completely excluded.

Jones *et al.* reported that the non amphiphilic protoporphyrin IX dimethyl ester, mesoporphyrin IX dimethyl ester, and meso-tetra(4-carbomethoxyphenyl)-porphyrin formed stable films that could be transferred as Z-type films to solid substrates, all with no long alkyl chains [352]. The area per molecule was ~55 $Å^2$ for the first two porphyrins, suggesting an approximate vertical packing of the rings. However, the area per molecule for the TPP derivative (~65 $Å^2$) and the high irreversibility of its π–A isotherm suggest packing of flat laying trimers (the calculated area per molecule being ~220 $Å^2$) or tilted dimers. This is in agreement with the early observation that protoporphyrin IX dimethyl ester forms three-dimensional clusters on the water surface [353].

Ruaudel-Teixier *et al.* reported on the first synthesis of amphiphilic porphyrin molecules (next diagram) [354]. In the first study, they replaced the phenyl rings in TPP by pyridine rings, quaternized with $C_{20}H_{41}Br$.

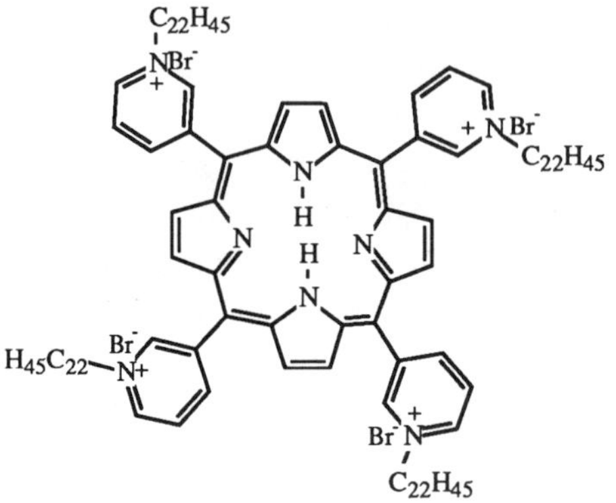

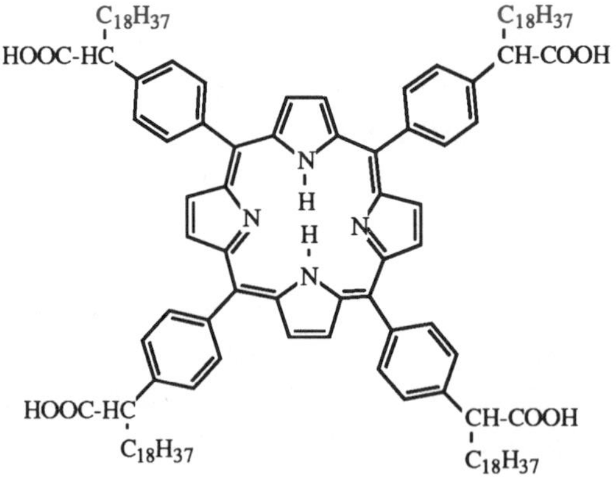

162

The pyridinium nitrogen is highly hydrophilic, while the long C_{20} hydrocarbon served as the hydrophobic part, tetra(3-eicosylpyridinium)porphyrin bromide (shown in diagram, top). The $\pi-A$ isotherm on pure water showed an area per molecule of ~160 Å^2. (The calculated molecular area is 220 Å^2.) X-ray diffraction revealed a d-spacing of 41 ± 1 Å, suggesting tilting, interpenetrating, and folding of the alkyl chains in a Y-type film. This is a reasonable explanation, considering that the four alkyl chains together occupy only 100 Å^2. Therefore, by tilting the macrocycle, which reduces its cross-section area, and the alkyl chains, which increases their cross-section area, more close-packing and stability are achieved. Polarized resonance Raman spectroscopy suggests that the macrocycles lie flat on the substrate with a tilt angle smaller than 15°.

In the second study, Lesieur *et al.* used TPP, substituted in the *meta* or *para* positions with an α-branched docosanoic acid [tetra[4-oxy(2-docosanoic acid)]phenyl-porphyrin (shown in previous diagram, bottom) [355]. The $\pi-A$ isotherm gave only 100 Å^2 as the area per molecule, and IR spectra show that protonation of the porphyrin occurred, thus providing extra stability to the film through electrostatic interaction.

Bardwell and Bolton studied LB films of 5-(4-carboxyphenyl)-10,15,20-tritolylporphyrin [356]. This TPP derivative with no long-chain alkyl groups formed well-behaved monolayers at the water–air interface, with an area per molecule suggesting that the macrocycle is oriented perpendicular to the surface. Spectroscopic studies of a transferred monolayer revealed the existence of three distinct species. At low pH, a monomer and aggregate coexist, while at pH > 7.3, a new aggregated species was identified. The structures of the two aggregates were not determined; however, from fluorescence spectroscopy, it was clear that these are not face-to-face dimers.

Tredgold *et al.* studied x-ray diffraction from LB films of Ag(I) and Au(III) mesoporphyrin IX dimethyl ester complexes [357]. Their results confirm the close-packing of the porphyrins on their edge in a face-to-face manner; however, their studies suggest that these complexes form microcrystallites shortly after film production.

McArdle and Ruaudel-Teixier reported the synthesis and properties of copper cyanoporphyrins in monolayer and multilayer films [358]. We note that cyano-substitution alters the reduction potential of the macrocycle, and the tetracyano derivative of TPP is reduced in ~1 V more positive potential than the parent molecule [359], and hence, suggests a direction in molecular design of strong electron acceptor π-systems for thin organic films. Diquaternized isomers (*syn* and *anti* $C_{18}H_{37}$— derivatives) were studied; however, both isomers probably contained small amounts of mono-, tris-, and tetraquaternized molecules. The dichained porphyrins yielded unstable films that could not be transferred onto a solid substrate. The area per molecule of 110 Å^2 for the dichained tetracyanoporphyrin suggests that the macrocycles are mostly vertically oriented. (The calculated area per molecule for a vertical molecule is 90 Å^2.) On the other hand, the tetraquaternized porphyrin was a good film former. 3,7,12,17-Tetracyano-5,10,15,20-tetra(3-pyridil) porphyrin is shown in diagram.

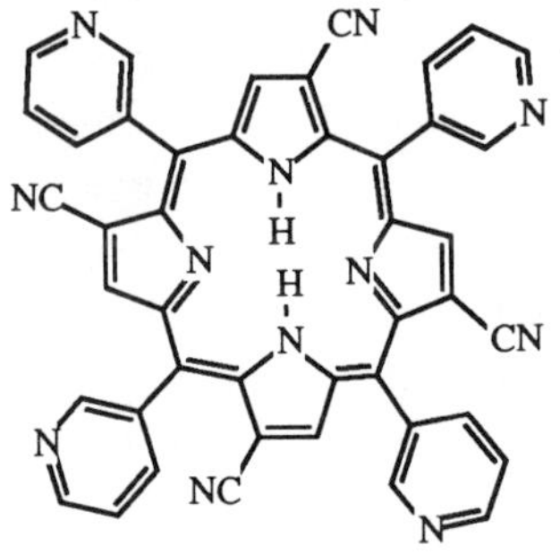

Miller *et al.* studied LB films of a 5,10-di[4-oxy(2-docosanoic acid)]-15,20-diphenylporphyrin [360]. They reported that stable monolayers could be formed (90 Å^2 on pure water, and 118 Å^2 area per molecule on 10^{-3} M CdCl$_2$, at 25 dyn/cm), and that the absorption spectra of the monolayer on the surface was strongly influenced by the pressure and the presence of Cd^{2+} ions in the subphase. Fluorescence microscopy of the monolayer revealed homogeneous molecular distribution for pressures above 1 dyn/cm.

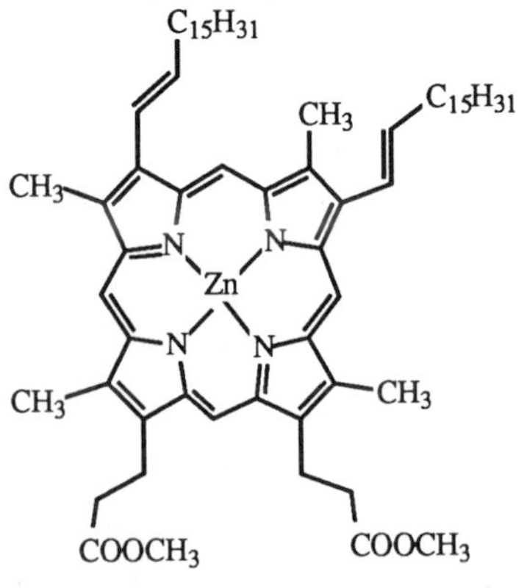

Möhwald *et al.* studied intermolecular interactions in monolayers of porphyrins [361]. They investigated different metalloporphyrins (some mentioned earlier); however, zinc 3,8-bis(1'-heptadecenyl)deuteroporphyrin dimethyl ester (shown in diagram before this paragraph) represents the kind of molecular design mentioned earlier. Here, a planar, amphiphilic porphyrin was designed with the long hydrocarbon chains substituted at the 3,8 positions. The π–A isotherm for this molecule showed a surface pressure of 30 dyn/cm at 66 $Å^2$ area per molecule. Optical spectroscopy indicates that the molecules aggregate, and from dichroic ratios an angle of 35° with respect to the surface normal was calculated. Fluorescence spectroscopy reveals that the film is amorphous with no single-crystal-type pattern.

Ringuet *et al.* reported a new amphiphilic porphyrin, again with two hydrocarbon chains substituted on the macrocycle (7,17-dioctadecyl-3,8,13,18-tetramethylporphine-2, 12-dipropanoic acid) [362]. Although the authors reported a reproducible and stable film formation, it seems that a molecule with the two alkyl chains substituted on adjacent pyrrole rings should be preferred.

Tredgold *et al.* studied alternating layers of porphyrins and fatty acids [363]. The most significant result in this study is the superiority of the mesoporphyrin IX diol (copper mesoporphyrin IX diol, shown in next diagram) as film-former material. The introduction of stronger hydrophilic OH groups gave this molecule an amphiphilic character, thus changing dramatically its behavior. X-ray diffraction results indicate lattice repeat distance identical to the expected value, suggesting that these mixed monolayers are true superlattices. SEM results further support this suggestion, showing a featureless surface.

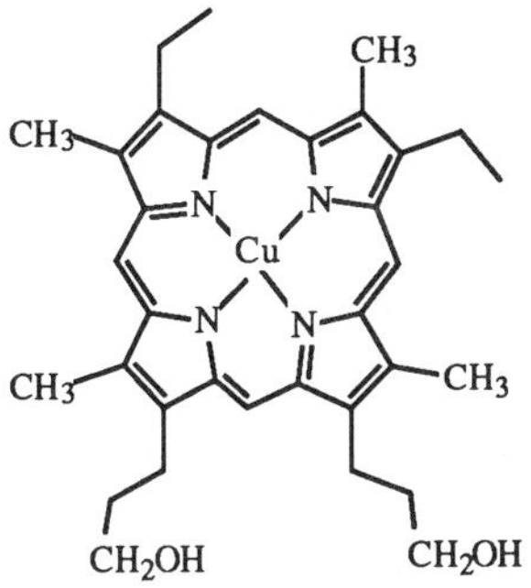

Luk and Williams reported on LB films of porphyrin dimers (Figure 2.29) [364]. The π–A isotherm for the monomeric porphyrin was featureless [365], and gave an area per molecule of 60 $Å^2$, indicating stacking of the porphyrins perpendicular to the surface. Irradiation of a 60-layer film yielded the dimer, confirmed both by the elimination of the CO stretching in the IR spectra and the changes in the optical spectra. The interest in this work stems from the electrical and optical properties of porphyrin and phthalocyanine polymeric LB films.

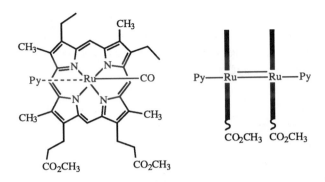

Figure 2.29. Ruthenium carbonyl mesoporphyrin IX pyridate (left), and a side-view scheme of its dimer (right).

Schick *et al.* reported a detailed characterization of monolayers of 5,10, 15,20-tetrakis(4-octyloxyphenyl)porphyrin (shown in next diagram) and the corresponding Cu(II), Zn(II), and Co(II) complexes [366]. Optical spectra of the assemblies indicate that these porphyrins exist in domains of at least seven molecules, independent of the metal ion. This result is different from that reported by Vandevyver *et al.*, where the alkyl chains were C_{12} and longer [351]. We note, however, that the dichroic ratio reported by schick *et al.* actually indicates that the orientation axis is ~55° (i.e., the magic angle), indistinguishable from a random distribution.

Apparently, in the present case, the ratio between the flexible alkoxy chains and the rigid macrocyclic π-system is just right, and the molecules are organized at the water–air interface like discotic liquid crystals, i.e., like an array of cards. A similar liquid–crystalline behavior was reported for octaoctyl-substituted phthalocyanine [341]. It was found that the members of the aggregates are

166

separated by ~4–5 Å, slipped by ~47° with respect to one another, and tilted (in a domain-fixed axis system) by ~40°.

B. LB Films of Phthalocyanines

Baker et al. were the first to report the successful preparation of LB films of metal-free PC [367]. The π–A isotherm for the unsubstituted PC, spread using dilithium salt (Li$_2$PC), yielded nontransferrable monolayers with an area per molecule of ~100 Å^2, which is an intermediate value between the area of the edge (~40 Å^2) and that of face (~160 Å^2) of the PC molecule. The addition of mesitilene (1,3,5-trimethylbenzene) to the spreading mixture gave transferable monolayers with an area per molecule of ~40 Å^2, indicating molecules that are stacked perpendicular to the surface like a deck of cards. Kovacs et al. reported the surface pressure induced dimorphic conversion in LB films of zinc PC [368].

Tetra-*tert*-butylphthalocyanine (TBP, shown in diagram) is one of the most studied PC derivatives. LB films of TBP were first reported by Baker et al.,who found that the monolayers were transferable, with an area per molecule of ~220 Å^2 [367].

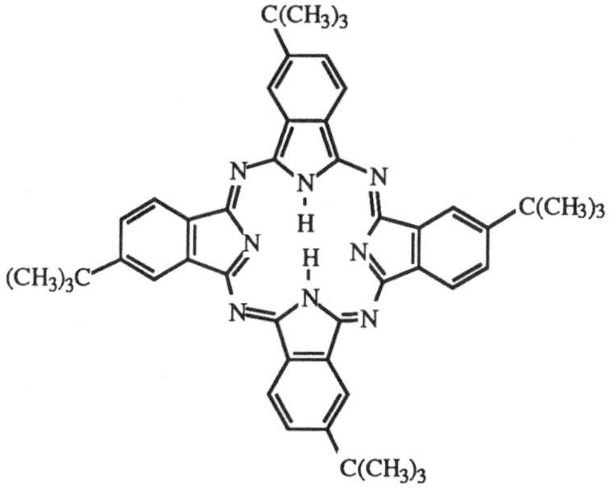

Kovacs et al. reported that monolayers of TBP are tough, adherent, and ultrathin (~17 Å), and can be heated to 150°C for 1 h with no apparent loss of optical density or change in spectrum shape [368]. Hann et al. studied monolayers of copper and zinc TBP [369]. TEM of one monolayer of Cu-TBP on amorphous carbon shows small ordered domains in an amorphous matrix. Aroca et al. [370] and Kovacs et al. [371] studied SERS of LB films of TBP on silver and indium islands. The reader may find valuable SERS data in this paper; however, as mentioned in Part One, no orientational information can be deduced from an

167

SERS experiment. Hua *et al.* reported the first monolayers of silicon TBP (TBPSiCl$_2$) [372]. The π–A isotherm is very steep, and the suggested area per molecule is 62 Å^2, which is smaller than 68 Å^2 calculated for the area of a molecule vertical to the surface (assuming a TBP ring edge of 20 Å and a ring spacing of 3.4 Å). Such molecular spacing is smaller than the Cl–Si–Cl bond length total of 4.04 Å, and suggests, therefore, that the macrocycles aggregate in a staggered arrangement, and somewhat slipped, as shown schematically in Fig. 2.30. Such a slipped stacking structure, forced by the Si–Cl bonds, results in the lack of blue shifted Q-bands seen in other systems [373].

Fujiki and Tabei studied LB films of TBP and of its Ni(II), Cu(II), and Pb(II) complexes, prepared by horizontal lifting [373]. At 5°C, the area per molecule was 32–43 Å^2 for the TBP and its Ni(II) and Cu(II) complexes. These values are different from those reported before (as high as 92 Å^2 per molecule for TBPZn) [367-369, 374], and strongly suggest that differences in laboratory practices may account for large differences in the π–A isotherm, and hence, the stability of the monolayer at the water–air interface. (See our discussion earlier on experimental considerations, Section 2.1.B.) X-ray diffraction data gave a *d*-spacing of 3.3–3.4 Å, suggesting a strong intermolecular π–π interaction. The large blue shift in the optical spectra of the LB films, compared to solution spectra, indicates that in this case, there is an effective π–π interaction in a "true" staggered stacking with no slippage.

Hirano *et al.* reported that the ATR FTIR spectra of Cu(II)TBP deposited on polyethylene was enhanced by evaporated Ni film (27 Å) [375]. The peaks that were enhanced were 1292, 1259, 1140, and 1078 cm^{-1}. Aroca *et al.* reported on fluorescence enhancement from LB monolayers of TBP on silver islands [376]. Cook *et al.* used EPR at low temperatures (20 or 25 K) to study the association and orientation of Cu(II)TBP in LB films [377]. They reported that Cu(II)TBP aggregated in CH$_2$Cl$_2$, but not in toluene (both at 25 K), and that in their LB films, the macrocycles are stacked like a deck of cards, with the face of the ring at $80 \pm 10°$ to the substrate surface.

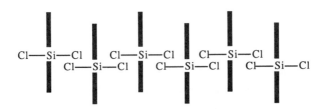

Figure 2.30. A schematic view of TBPSiCl$_2$ aggregation at the water–air interface.

Roberts *et al.* studied LB films of copper phthalocyanine tris($CH_2NHC_3H_7$-*iso*) [374]. (See next diagram.) The area per molecule was 57 Å^2, and the authors suggest that the molecules are tilted edge-on in the water with the short chains pointing upward and overlapping with the neighboring molecules. The tilted orientation was confirmed by Yoneyama *et al.*, who calculated $<\cos^2\theta> = 0.64$ from polarized optical spectra of LB films of the copper PC derivative, suggesting an average tilt angle of 37° from the surface normal [378]. Near-edge x-ray absorption fine structure (NEXAFS) studies of similar LB films further confirmed the suggested structure [379].

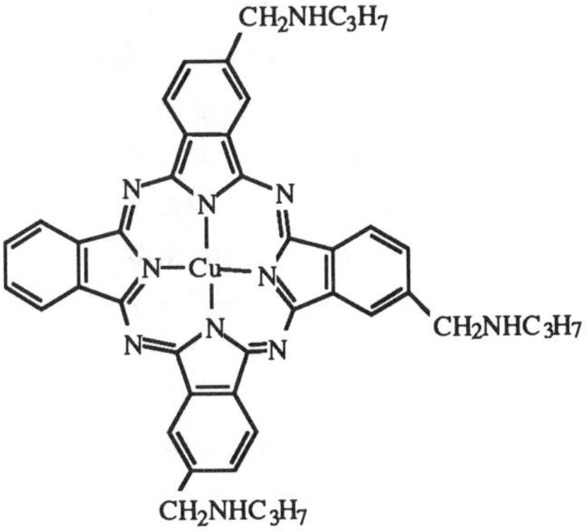

$CH_2NHC_3H_7$

$CH_2NHC_3H_7$

$CH_2NHC_3H_7$

Kalina and Crane reported preparation and characterization of LB films of copper 2,3,9,10,16,17,23,24-octa(dodecyloxymethyl)phthalocyanine (shown in next diagram) [380].

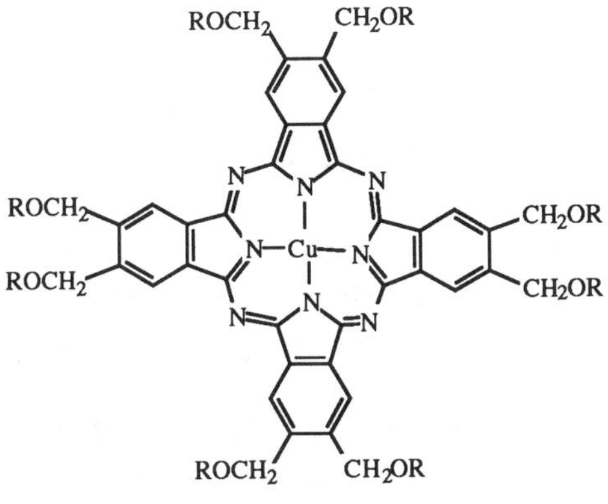

$ROCH_2$ CH_2OR

$ROCH_2$ CH_2OR

$ROCH_2$ CH_2OR

$ROCH_2$ CH_2OR

The π–A isotherm indicates an area per molecule of ~180 Å in agreement with molecules oriented with their planar rings parallel to the water–air interface. This

169

is not unexpected, since the area per alkyl chain is ~20 Å, which means that even without tinting, the eight alkyl chains occupy an area of ~160 Å. It generally can be expected that PC molecules with eight alkyl substituents will prefer this orientation, since one can consider this molecule a macroamphiphile, where the PC macrocycle is the hydrophilic head group, and the eight chains are the hydrophobic part. Films as thick as 151 monolayers could be built, but the exact structure of the films has not been determined. Optical spectra indicate some aggregation, which supports a Y-type structure where the PC molecules in two bilayers are face-to-face.

Octaalkoxy- or octaalkylphthalocyanines (copper 1,4,8,11,15,18,22,25-octa -substitutedphthalocyanine, R = C_nH_{2n+1}, OC_nH_{2n+1}, shown in next diagram) represent just one example of a growing family of discotic liquid crystals [381]. However, we already have discussed asymmetrically substituted PC derivatives as discotic liquid–crystalline amphiphiles. Although the asymmetrically substituted PCs show better monolayer behavior at the water–air interface than the octasubstituted PCs, it may be useful just from the molecular design approach to discuss these materials briefly.

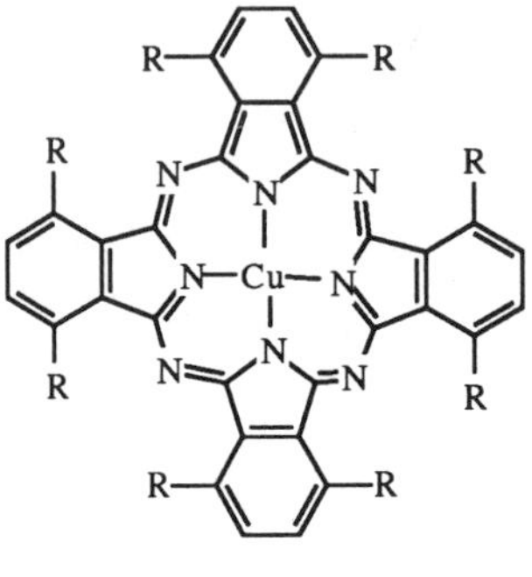

Cook *et al.* studied a family of octaalkoxyphthalocyanine derivatives [382]. (See previous diagram, where R = OC_nH_{2n+1}, $n = 3–5$.) They found that, in general, the metallophthalocyanines (Cu^{2+}, Zn^{2+}) yielded monolayers with greater stability, and values for area per molecule significantly larger than those of the metal-free compounds. In general, these materials did not produce good deposited films.

McKeown *et al.* studied a series of octaalkyl PCs (as in aforementioned diagram, where R = C_nH_{2n+1}, $n = 4–8$), and reported that all derivatives yielded monolayers with π–A isotherms with values of area per molecule lower than the minimum estimated, and that films were too rigid to be deposited [340]. These studies, together with the one discussed earlier [380], suggest that the PC

molecule is not hydrophilic enough to serve as a head group for a monolayer with
n-alkyl chains perpendicular to the surface, and that oxygen atoms at the PC
periphery are needed to give it the hydrophilic nature. Once alkoxy groups are
used, the chains should be long enough (e.g., $C_{12}H_{25}$) to compete with
crystallization and provide monolayer stability.

Ogawa *et al.* reported the preparation of a number of PC derivatives that
show highly anisotropic spectra [383]. (See next diagram.) They reported that
good monolayers at the water–air interface could be obtained with areas per
molecule of between 47 Å^2 (for II) to 90 Å^2 (for I). However, only monolayers
of III(a) (mixture of isomers) exhibited negligible hysteresis during compression
and expansion, and dipping transfer ratios of 0.9–1.0 for both down and up
strokes in a Y-type deposition. The other compounds could be deposited by
horizontal lifting, but no details are given. UV–Vis, FTIR, and x-ray diffraction
studies all suggest that the molecular planes are nearly perpendicular to the
substrate surface, and almost parallel to the dipping direction.

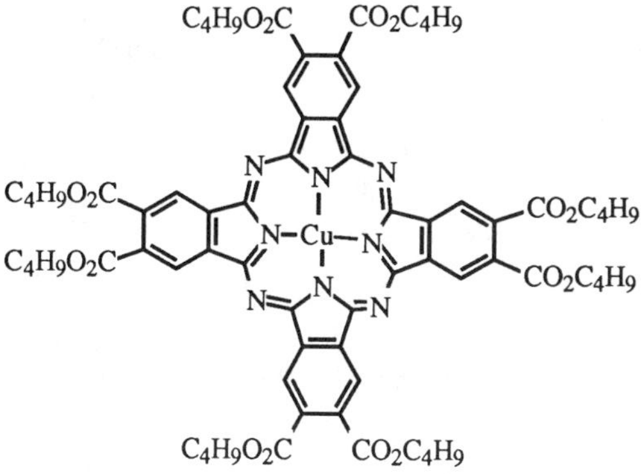

$C_4H_9O_2C$ $CO_2C_4H_9$

$C_4H_9O_2C$

$C_4H_9O_2C$

$CO_2C_4H_9$

$CO_2C_4H_9$

$C_4H_9O_2C$ $CO_2C_4H_9$

I

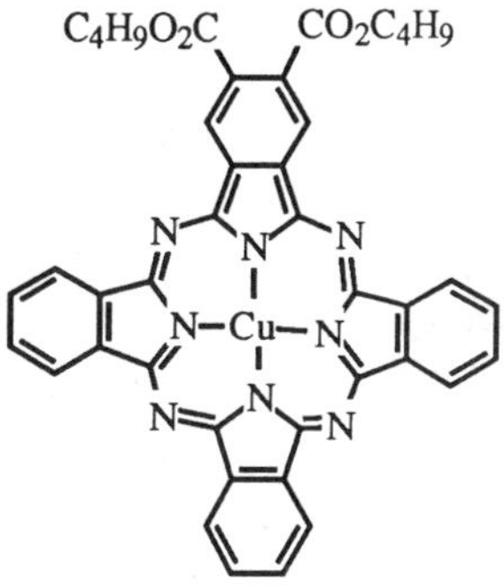

$C_4H_9O_2C$ $CO_2C_4H_9$

II

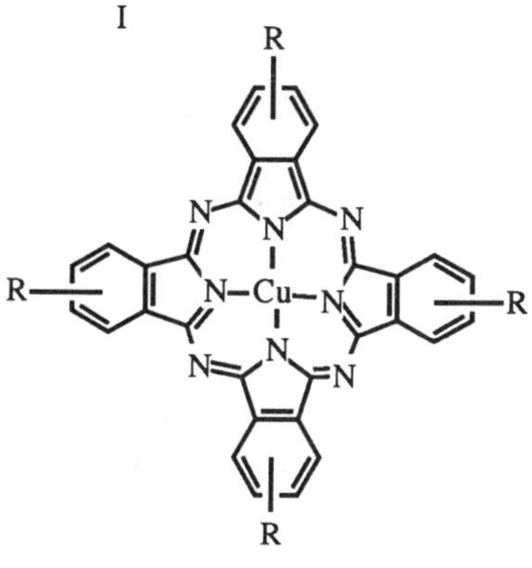

R

R

R

R

III

(a) - R = $-CO_2$-C_4H_9

(b) - R = $-CO_2$

(c) - R = $-CO_2$

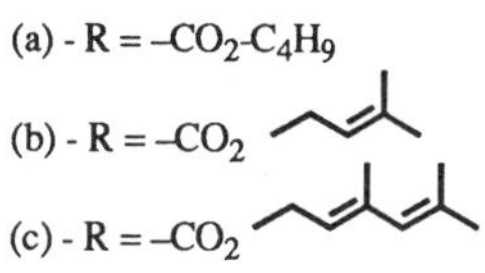

Fujiki *et al.* prepared LB films of two isomeric tetraoctadecyl amide derivatives by both vertical dipping and horizontal lifting techniques [384–386]. The difference between the two isomers is in the amide group connection to the macrocycle. Thus, while one is a derivative of octadecylamine (a), the other is a derivative of stearic acid ((b), see next diagram).

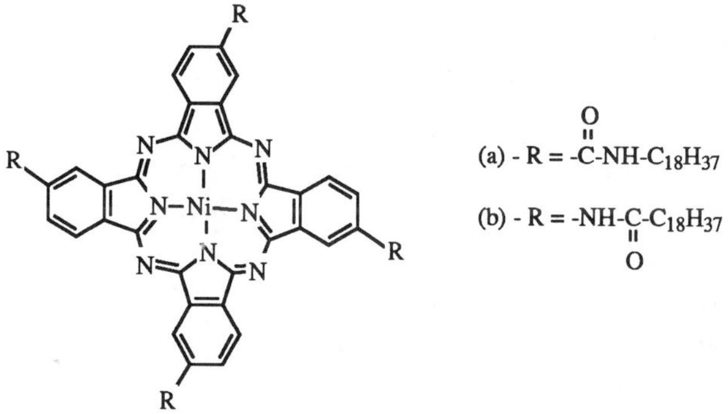

$$
\text{(a) - R} = -\overset{\displaystyle O}{\overset{\|}{C}}-NH-C_{18}H_{37}
$$

$$
\text{(b) - R} = -NH-\underset{\displaystyle O}{\overset{\|}{C}}-C_{18}H_{37}
$$

This difference defines the packing of these macromolecules, since in (a), the sp^2 carbonyl is connected to the PC, which means that the amide group is at the PC molecular plane, and only the heptadecyl group can be perpendicular to the surface. On the other hand, in (b), the nitrogen is connected to the PC ring, resulting in the stearoyl group being perpendicular to the surface. This, of course, means that the area per molecule in a close-packed assembly should be smaller in the case of (b). Indeed, the reported area per molecule was 120 and 112 Å^2 for (a), and (b), respectively. It is beyond the scope of this manuscript to discuss in detail the structure of the different LB films; however, from UV–Vis polarized spectra, it became clear that the number of molecules in the different assemblies are 5–14, and five, for (a), and (b), respectively.

Troitskiy *et al.* published a detailed study on the deposition and electron diffraction of alternating LB films of a tetra-3-octadecylsulphamoylphthalocyanine (shown in diagram next page) with barium behenate [387]. The motivation for this work was to introduce ultrathin insulating films between molecular assemblies that have electronic properties, and therefore, to develop the control on the structure vertical to the surface, at the molecular level. They prepared films containing bilayers of the two components altering in seven different ways, and suggested that the VOPC layers are amorphous, while those of the CuPC probably are more crystalline, although their data is not good enough to suggest structural details. They also suggested tilt angles of 13° and 11° for the VO and CuPC rings, respectively. For the alkyl chains in the bilayers, they suggested a tilt angle of ~40°, which is in agreement with the tilt angle in multilayers of barium behenate. The interest in the paper is more in the material design, and less in the structural data and its interpretation.

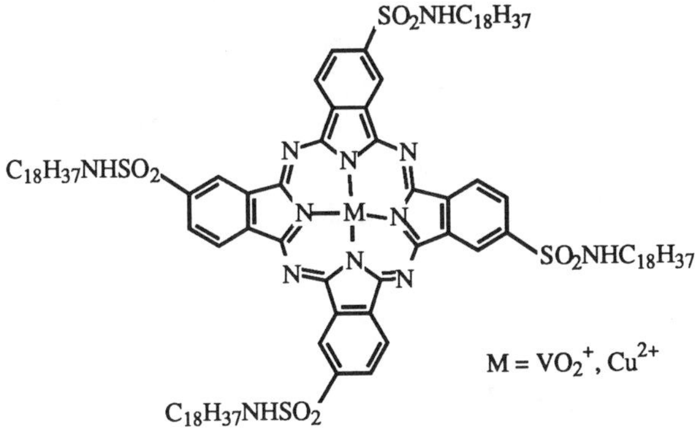

$SO_2NHC_{18}H_{37}$

$C_{18}H_{37}NHSO_2$

M

$N-M-N$

$SO_2NHC_{18}H_{37}$

$M = VO_2^+, Cu^{2+}$

$C_{18}H_{37}NHSO_2$

Snow *et al.* prepared and studied tetrakis(cumylphenoxy)phthalocyanine, an interesting PC derivative, with liquid–crystalline-like substituents (structure shown in diagram) [388–393].

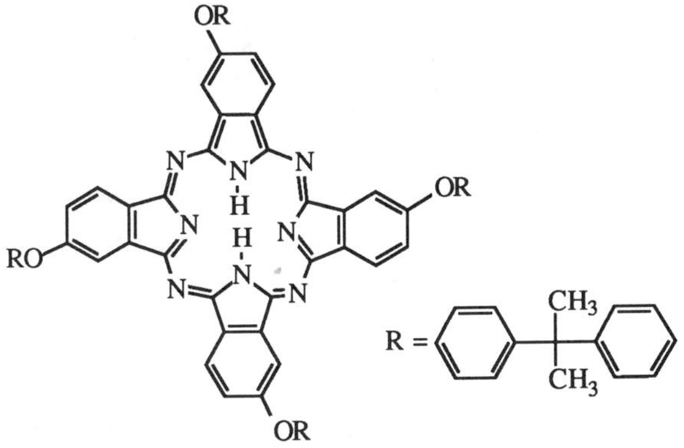

OR

OR

RO

$R = $

CH_3

CH_3

OR

This molecule is of interest because the cross-section area of the substituents is much larger than that of a normal alkyl chain, and therefore, the requirement of minimized free volume in the assembly may be easier to accomplish. It was found that molecular association in solution and in the LB films was related. The area per molecule that was affected by the complexing metal also was closely related to the degree of association. The π–A isotherms, electronic spectra, and x-ray diffraction indicate that the PC rings are perpendicular to the surface, and cofacially oriented within the aggregate. Resonance Raman scattering studies support these structures [390]. TEM studies showed that spread films are thicker than a monomolecular layer [393].

Palacin *et al.* prepared and studied vanadyl tetraoctadecyl tetrapyridino[3,4-b:3',4'-g:3",4"-l:3''',4'''-q] porphyrazinium bromide (shown in next diagram) [394]. This molecule is another example of molecular design, where the hydrophilic pyridinium is a part of the macrocycle. An interesting result of this study is that in mixed monolayers of the macrocycle with stearic acid, the carboxylate ion replaces the bromide, as evident from IR studies, thus providing an effective contribution of eight alkyl chains per macrocycle. This arrangement has excellent stability, which is expressed in the high collapse pressures as high as 50 mN m^{-1}.

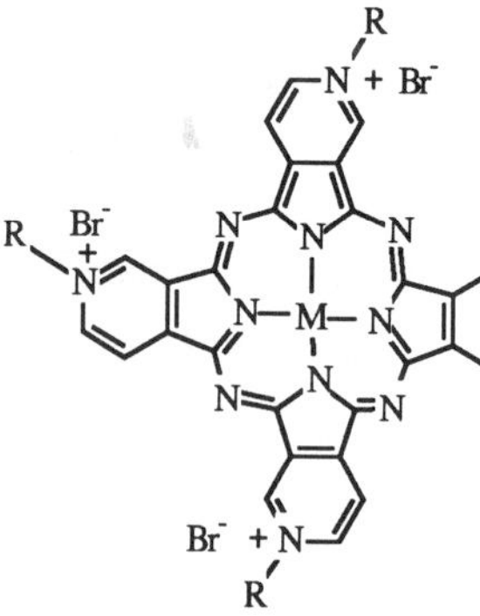

$R = C_{18}H_{37}, CH_3$
$M = VO_2^+, Co^{2+}$

In a recent paper, Palacin and Barraud took this principle further, when they used the cobalt tetramethyl derivative as the macro-head group, while introducing the alkyl chain through the replacement of the bromine anion by ω-tricosenoate, and reported that well-ordered films could be produced by vertical deposition [395].

Shutt *et al.* prepared an LB film from a new amphiphilic, two-ring phthalocyanine, (HO)GePc-O-SiPc(OSi(n-C$_6$H$_{13}$)$_3$ [396]. (See next diagram.)

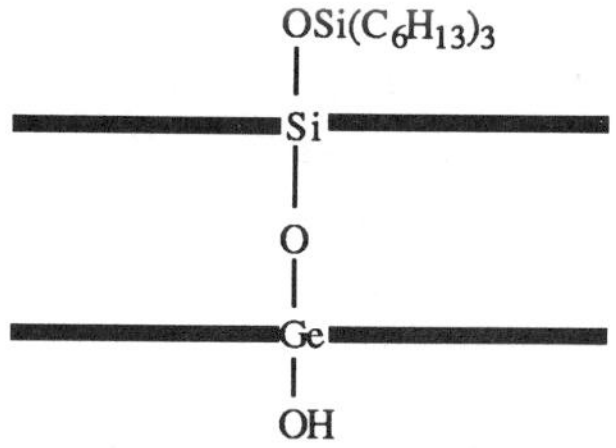

This PC-dimer forms good monolayers at the water–air interface, with area per molecule of 170 Å^2 (20°C), and with the rings parallel to the water surface. The hydrophilic OH "head" and the hydrophobic Si(C$_6$H$_{13}$)$_3$ "tail" help to ensure the desired molecular orientation, as well as provide high solubility. The monolayers thus formed are stable and robust, and can be deposited to form good multilayer films. The main interest in these monolayers is their apparent high anisotropy.

It should be possible, in principle, to extend the synthesis to trimers and tetramers of PC units. Otherwise, one could use the top monolayer, chemically modify it (i.e., by removing the trialkylsilane) to form a hydroxylated surface, and further adsorb other monolayers on it to form a superlattice having designed physical properties (e.g., semiconducting properties). Ko *et al.* [397] and Wang *et al.* [398] reported on microsensors for halogen gases using LB films of these PCs.

Orthmann and Wegner reported the synthesis of a PC polymer and the formation of monolayers and multilayers thereof [330]. This is an excellent example of molecular engineering where concepts known from the area of discotic liquid crystals are used. Thus, this polymer conceptually is a column of discs, where the alkyl substituents, OR_1 and OR_2, are of different chain lengths, and the polymer is prepared from a mixture of isomers to avoid possible crystallization. (See next diagram.)

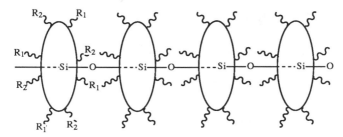

The polymer forms good monolayers with an area per molecule of 67 Å^2, in agreement with the cross-section area of a repeat unit determined from x-ray diffraction. Monolayers on the water surface had no preferential orientation; however, deposited monolayers were found to be highly anisotropic (Figure 2.31).

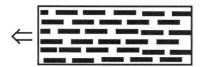 

Figure 2.31. Schematic of a monolayer of the PC polymer at the water–air interface (right) and glass substrate (left). The arrow denotes the dipping direction (order exaggerated).

This is an important phenomenon, since the only explanation for the anisotropy is that order could be introduced during deposition, probably at the meniscus, where shear forces are maximized. From the material engineering point of view, this concept of a rigid polymeric π-system, embedded in a liquid paraffin and thus forming stable monolayers and LB films, in principle, can be generalized to other systems. Hence, the order and orientation of other rigid rod-like polymers also should be enhanced during deposition, thus providing a highly anisotropic layer. Such high anisotropy is very important for applications such as third-order nonlinear optics.

In closing the discussion on LB films of PCs, we emphasize that, in general, these materials are notorious for forming microcrystals, and, for example, large band shifts in uv spectra of their LB films may be the result of such a behavior. Therefore, unless otherwise demonstrated, caution is called for when reading the reports cited.

2.8. Polymeric LB Films

Since the beginning of research in the area of LB films, it was clear that the low chemical, mechanical, and thermal stability of most LB films are major obstacles for their use in a wide range of technologies. Two basic approaches to the stability problem have been employed. In the first, polymerizable monomeric amphiphiles are deposited at the water–air interface, and polymerization (either photo-stimulated or condensation) is induced. Here, again, there are, in principle, two possibilities; i.e., either the polymer is formed at the water–air interface and then is transferred onto the substrate, or the monomeric film is transferred, and polymerization takes place in the film form. In the second approach, preformed polymers bearing both hyrophilic and hydrophobic groups are used. Although the first approach requires a longer process, the latter has the drawback that the polymer forms much more viscous films at the water–air interface, and thus makes it very difficult to transfer onto a substrate at a reasonable rate. On the other hand, polymerization of LB films is accompanied by shrinkage, since van der Waals interactions are replaced by chemical bonds, which makes such a process inadequate for any optical applications. In this part, we discuss polymerizable LB films, and comment on the three general families of monomeric amphiphiles — those with one or more double bonds, those with a diacetylene group, and finally, those with an oxirane group. We then turn to a discussion on prepolymerized amphiphiles.

A. Polymerizable LB Films

The first reports on polymerization in monolayers appeared in the literature in 1935 [399, 400], and then very few reports appeared again until the beginning of the 1960s [401–406]. This early work is accounted for in Gaines' book [14].

1. Amphiphiles Containing One or More Double Bonds

The aforementioned early works gave very little structural information on the polymerization process at the water–air interface. In fact it, was known only that this polymerization is different from that in the bulk [403–406]. It was the pioneering, detailed, and systematic work of Ringsdorf [407] and Lando [408] that helped establish our current understanding of these processes.

Ackerman et al. studied the polymerization of octadecyl methacrylate and N-octadecyl acrylamide at the water–air interface [407]. They used both irradiation and oxygen-initiated polymerization, and found that the latter yielded peroxides from a copolymerization of oxygen with the monomers.

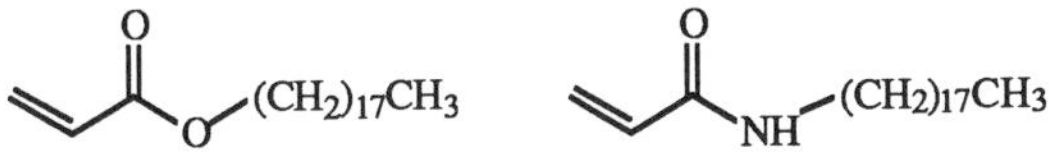

The first attempt to prepare a two-dimensional crystalline polymer was made by Cemel et al. [408]. They used ^{60}Co γ-radiation to initiate polymerization in monolayers of vinyl stearate (next diagram).

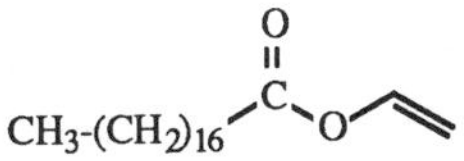

Wetting studies indicated that both the monomer and polymer monolayers exhibited closely packed methyl groups at their surfaces, and x-ray diffraction revealed that both have some single-crystal character. The polymerization was monitored by IR-ATR spectroscopy, and was found to depend upon temperature, irradiation doses, and oxygen.

Puterman et al. continued the research on the polymerization of vinyl stearate in monolayers and extended it to mixed monolayers of the monomer with ethyl stearate [409]. They found that in both monolayers and multilayers there was a solid solution of the two components, and that when the concentration of ethyl stearate exeeded 50%, the extent of polymerization was lowered.

177

The first photopolymerization of maleic (*cis*), and fumaric (*trans*) ester and amide derivatives was reported by Ackerman *et al.* [410].

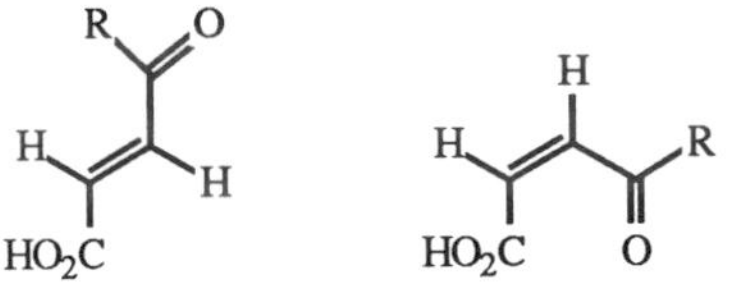

R = O(CH$_2$)$_{17}$CH$_3$
R = NH(CH$_2$)$_{17}$CH$_3$

Fumaric   Maleic

Dubault *et al.* studied the photopolymerization of mono- and diacrylic esters (shown in next diagram) at the water–air interface [411, 412]. In this diacrylic monomer, the polymers can cross-link, thus giving rise to severe changes in surface viscosity. This process was interpreted as a *gel point* where the number of cross-links is sufficient to form an infinite network.

An interesting study of polymerization of vinyl stearate, octadecyl methacrylate, and *N*-octadecyl acrylamide was reported by Lando and Fort [413].

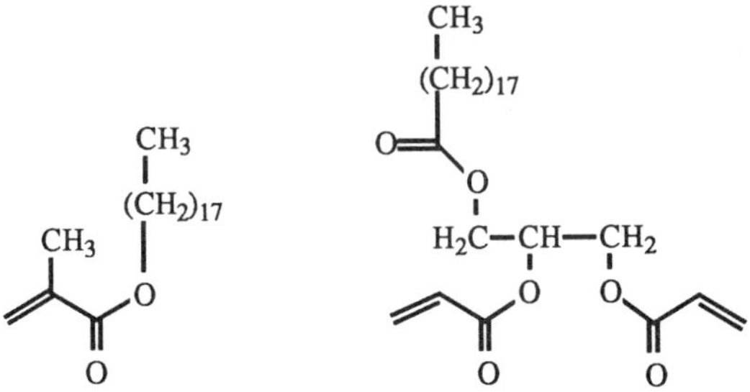

Octadecyl methacrylate   1-octadecyloxy-2,3-diacryloxypropane

They found that the shape of the kinetic curves was not affected by variation of surface pressure, but the rate of reaction was faster at lower surface pressures. They rationalized this behavior by suggesting that islands of ordered vinyl stearate are reacting at low surface pressures, and since polymerization is accompanied by expansion, it can be accommodated easily when these islands are not closely packed, thus resulting in a rate enhancement.

Naegele *et al.* studied in detail the photopolymerization of cadmium octadecyl fumarate in monolayers [414]. They found hexagonal packing in approximately the first 10 layers with the alkyl chain perpendicular to the substrate surface, while in thicker fims some rearrangement to the monoclinic structure was observed. By polymerization of alternating cadmium octadecyl fumarate–cadmium stearate multilayers it could be shown that the reaction occurred independently from the nature of the neighboring layer in the separate plane.

Naegele *at al.* also studied the effect of orientation of the monomers on kinetics and polymer characteristics in the photopolymerization of octadecyl methacrylate and octadecyl acrylate [414]. They used both temperature and surface pressure to change the orientation of alkyl chains in the film, and reported that the polymerization rate passed through a maximum with respect to surface pressure. This is in agreement with the idea that polymerization occurs in ordered islands, and accompanied by expansion.

Enkelmann and Lando studied the polymerization of ordered tail-to-tail bilayers of vinyl stearate under the water surface [152].

Barraud *et al.* studied the polymerization in ω-docosenyl 2,4-pentadienoate [415]. This molecule can undergo selective polymerization that depends upon the initiation.

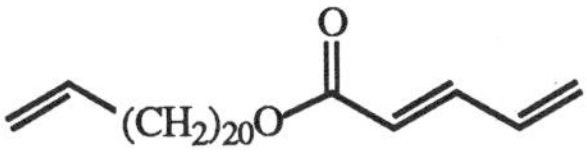

Banerjie and Lando studied γ-radiation-induced (^{60}Co) polymerization of *N*-octadecyl acrylamide in oriented ultrathin films [416], and Fariss *et al.* used electron beams to polymerize α-octadecylacrylic acid (shown in next diagram) [417]. In the latter case, the studies were made in an attempt to develop electron-beam-resist materials, and we shall discuss this issue in more detail in Section 5.7.

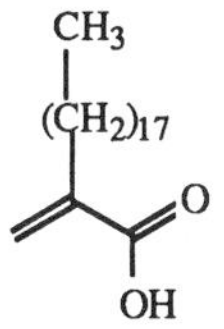

Rabe *et al.* studied the photopolymerization of octadecyl maleic acid and octadecyl fumaric acid at the water–air interface [418]. It seems that the main result in these studies is that while octadecyl fumaric acid expanded during polymerization, octadecyl maleic acid did not. A 1:3 mixture of the fumarate and maleate, respectively, preserved the film's area on polymerization. We note that a possible explanation to the observations cited here is that the two materials form separate phases, and that the expansion of one is compensated by the contraction of the other, with this contraction being caused by the expansion of the other phase. Thus, although the critical feature in this work is that it can lead to a crack-free resist, strain can still occur within each phase.

179

Berkovic *et al.* used nonlinear optics to study the polymerization of vinyl stearate and octadecyl methacrylate at the water–air interface [419]. By monitoring the second-harmonic signal, they could determine the reaction kinetics. Rapid polymerization occurred when the monolayers were exposed to uv light, while slower polymerization occurred when initiated through the water subphase.

Miyashita *et al.* studied the photopolymerization of *N*-octadecyl acrylamide as potential deep uv photoresist materials [420]. They deposited monolayers in the solid analogue phase, and subsequently irradiated the multilayer film. A large difference in solubility was found between the monomeric and polymeric LB films.

In a more recent report, Uchida *et al.* studied photopolymerization of a number of polymerizable amphiphiles [421]. One of the more interesting systems was a zwitterionic tetraalkylammonium amphiphile that formed stable monolayers at the water–air interface. (See next diagram for one example from this work.) These monolayers could be stabilized by polymerization, and the hydrophilic sulfonate surface was stable even in air. This surface showed an excellent blood compatibility and repelled blood platelets.

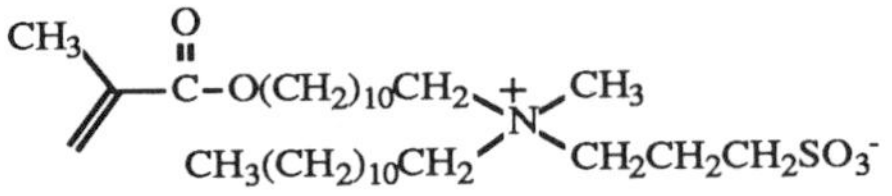

$$CH_3 \diagdown \quad \overset{O}{\overset{\|}{C}}-O(CH_2)_{10}CH_2 \diagdown \overset{+}{\underset{CH_3(CH_2)_{10}CH_2 \diagup}{N}} \diagdown \overset{CH_3}{CH_2CH_2CH_2SO_3^-}$$

In 1977, Barraud and collaborators used ω-tricosenoic acid (TCA, next diagram) to achieve cross-linking by a solid-state electron-induced polymerization [422–425].

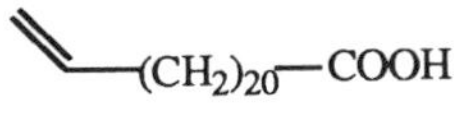

$$\diagdown \diagup \diagdown_{(CH_2)_{20}}—COOH$$

In this amphiphile, the double bond at the hydrophobic end of the molecule, and thus polymerization, has a very small effect on the bonding at the hydrophilic end of the molecule. Ogawa *et al.* studied x-ray [426] and electron-beam [427] polymerization in LB films prepared from the calcium salt of TCA.

Tanaka *et al.* studied the photoreaction of *trans*-4-octadecanoxy-cinnamic acid in monolayers [428]. Although this material formed stable monolayers at the water–air interface, it was impossible to build multilayer films on a substrate. The photoreaction in the monolayer (on glass) yielded the cyclodimer, 4,4'-dioctadecan-oxy-β-truxinic acid, from a *syn* head-to-head alignment [429]. (See next diagram for structures of the cinnamic acid and its cyclodimer.) Thus, it can be concluded that the molecules in the monolayer have a crystalline order, as required for such a lattice-controlled reaction [430].

180

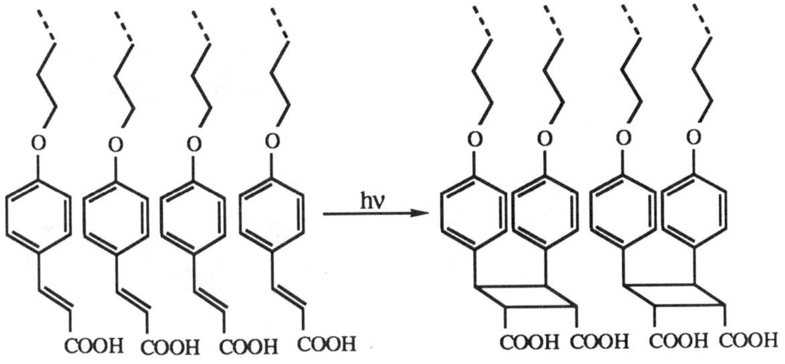

Koch *et al.* studied the photoreactions in monolayers of the same cinnamic acid derivative, and in monolayers of amphiphiles containing two cinnamic acid groups [431].

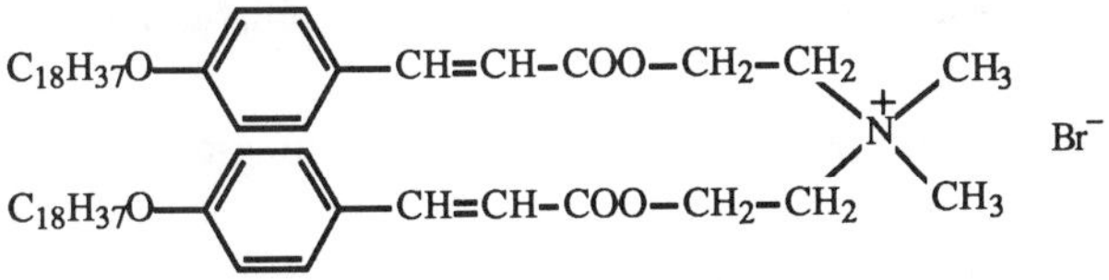

They found that in the double-chain amphiphiles, there are two possible photoreactions, intramolecular and intermolecular. In the first, the products are dimers similar to that found for the single-chain amphiphile, while in the latter, oligomeric and polymeric photoproducts are formed. We note, however, that in the experiments of Kock *et al.*, products were identified in vesicles. Although they demonstrated that monolayers of the two-chain cinnamates photoreact through bleaching of the absorption, they did not actually do product identification from the monolayers. It cannot be assumed with confidence, that the monolayers and the more fluid bilayer liposomes give the same reactions.

Hirota *et al.* studied the polymerization and optical properties of mixed LB films of β-parinaric acid (PA, shown in the next diagram) and stearic acid (SA) [432]. They found that the magnitude of the fraction of the polymerization region in the mixed LB films depended on the concentration of the PA, and as this concentration increased the layer structure of the film disappeared. This probably is due to the fact that the photoreactions cause severe changes in molecular orientation. Such changes cause changes in the thickness of the individual monolayers, thus disrupting the vertical order in the film.

all-*trans* $C_2H_5\text{---}(CH\text{=}CH)_4\text{---}(CH_2)_7\text{---}COOH$

Laschewsky and Ringsdorf studied the polymerization of amphiphilic dienes in LB films [433]. (See next diagram for structures.) They found that while all amphiphiles formed monolayers at the water–air interface, only the acids and alcohol formed multilayers.

$$CH_3\text{-}(CH_2)_n\text{---}CH\!\!=\!\!CH\text{---}CH\!\!=\!\!CH\text{---}CH_2OH$$

$$CH_3\text{-}(CH_2)_n\text{---}CH\!\!=\!\!CH\text{---}CH\!\!=\!\!CH\text{---}CHO$$

$$CH_3\text{-}(CH_2)_n\text{---}CH\!\!=\!\!CH\text{---}CH\!\!=\!\!CH\text{---}COOH$$

$$n = 12, 16$$

Photopolymerization was fast and occurred in two steps. In the first, a soluble polymer was the product (probably a cyclobutane derivative), while in the second (irradiation at $\lambda < 200$ nm), an insoluble polymer was formed. The polymerization produced changes in the layer spacing that were manifested in defects observed by SEM. However, the longer C_{22} amphiphiles yielded defect-free films.

Fukuda *et al.* investigated the polymerization of long-chain derivatives of α-amino acids, and diynoic, trienoic, and acrylic acids [434]. (Structures appear in next diagram.) They reported that these amino acid derivatives polycondensate in the Y-type fims, and that the amine group reacts with the ester to give an amide and a free alcohol, which could be removed by extraction.

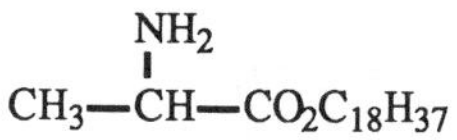

$$CH_3\text{---}\underset{\overset{|}{NH_2}}{CH}\text{---}CO_2C_{18}H_{37}$$

Octadecyl L-alanine

$$\underset{\overset{|}{CH_2\text{---}CO_2C_{18}H_{37}}}{H_2N\text{---}CH\text{---}CO_2C_{18}H_{37}}$$

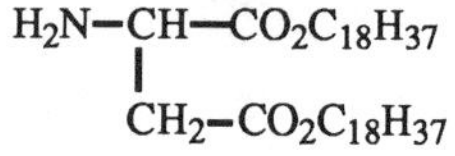

Dioctadecyl L-glutamate

2. Diacetylenic Amphiphiles

The utilization of polydiacetylenes for nonlinear guided wave devices has been proposed in numerous publications [435–439]. Theoretical calculations showing that one-dimensional systems of conjugated π-electrons should have large nonlinear susceptibilities, and that the magnitude of this nonlinearity is dependent on the conjugation length, were published by Rustagi and Ducuing [440], Cojan *et al.* [441], and Agrawal *et al.* [442]. It is clear, however, that the extended π-conjugation is not enough, and that high order, preferably that of a single crystal, is required.

Nevertheless, large single crystals of polydiacetylenes are very hard to grow; therefore, ordered two-dimensional assemblies that exhibit an inherent high-order parameter have been considered and studied as a possible alternative. The polymerization of diacetylenes in the solid state was discovered by Wegner [443]. The mechanism of the polymerization has been well studied and is rationalized to proceed as radical stepwise 1,4 addition to the conjugate triple bond [443, 444]. It is a topochemical reaction, i.e., requiring that the molecules are regularly packed in a very specific arrangement [444, 445]. Figure 2.32 shows that topological polymerization can occur only when (a) the difference between the translational period of the monomer (d_1) and the polymer (d_2) approaches zero (in other words, when the van der Waals distance is close to the chemical bond length; typically, d_1 for reactive diacetylenes is between 4.7 Å and 5.2 Å, and d_2 is almost constantly ~4.93 Å); (b) s_1 is smaller than 4.0 Å, with the lowest limit of 3.4 Å; and (c) the angle between the diacetylene rod and the translational vector α is close to 45°. Note that d_1, s_1, and α are related by $s_1 = d_1 \cdot \sin\alpha$. The polymerization reaction yields highly conjugated triple bonds with the two resonance forms,

Acetylenic (A)   Butatrienic (B) ·

The diacetylenes are, perhaps, one of the most systematically studied groups of materials. They have been used not only in LB films, but also as lipid membranes in liposomes, as models for biomembranes, and in tubules. The reader may find a description of this area in two reviews from Ringsdorf's group [446, 447]. The early systematic studies on diacetylenic amphiphiles were done by Tieke et al. [448], Tieke and Wegner [449], Day and Ringsdorf [450, 451], and Day et al. [452]. It was shown very early that these ultrathin polyacetylenic films of

$$CH_3\text{-}(CH_2)_m\text{—}C\equiv C\text{—}C\equiv C\text{—}(CH_2)_8\text{-}COOH \ ,$$

where $m \geq 8$, are very stable and have interesting physical properties, e.g., photoconductivity [453].

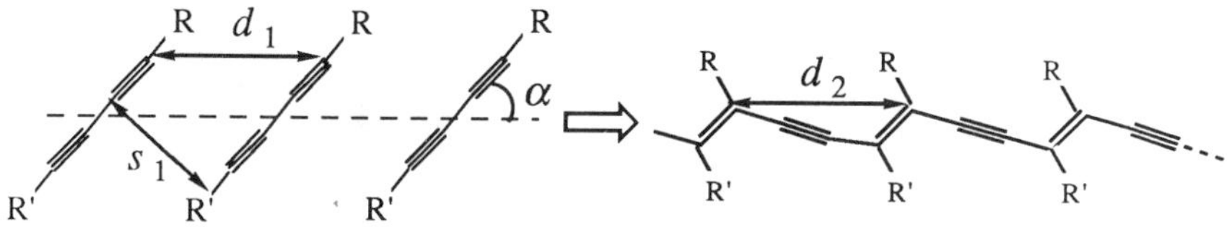

Figure 2.32. The topological principle in diacetylene polymerization.

183

The polymerization of diacetylenes is accompanied by a phase change due to reorientation of the side chains, and therefore, a change of the entire unit cell. The structural effects of the polymerization are not fully understood. Polymerization in the bulk gives blue crystals, and the same is true for polymerized LB films (that are colorless initially). However, the blue LB polydiacetylene films, in which register does not exist between successive layers, change their color to red upon annealing at or above 59°C, or on long exposure to uv light [454].

Raman spectroscopy suggested that changes in the spectrum during polymerization were a result of changes in the chromophore environment and structure [455, 456]. Tieke and Bloor showed, using Raman spectroscopy, that the phase change occurring during the photopolymerization of LB films of 10,12-tricosadiynoic acid ($m = 9$, previous chemical notation) involved an intermediate phase [457].

Day and Lando studied the morphology of crystalline diacetylene monolayers ($m = 15$, previous chemical notation) polymerized at the water–nitrogen interface [458]. They transferred the polymerized monolayers to porous substrates (e.g., an electron microscope grid) and reported that bilayer membranes were strong enough to span macroscopic holes. The polydiacetylene monolayers exhibited high birefringence, indicating a high degree of backbone orientation over macroscopic distances. When the polymerization was carried out in an ambient saturated with chloroform (and chloroform also was used as a spreading solvent), it was slowed down, and consequently larger monolayer crystallites were formed.

Lieser et al. used electron diffraction and small-angle x-ray technique to study the structure, phase transition, and polymerizability of cadmium salts of some diacetylene monocarboxylic acids ($m = 8$, 9, 11, and 13, previous chemical notation) [459]. They reported that photopolymerization did not destruct the LB assembly, and that crystallinity in both the monomer and the polymer was restricted to the $(CH_2)_n$ segments.

Tieke and Lieser studied the influence of the structure of long-chain diynoic acids on their polymerization properties in LB films [460]. They studied acids with the general formula:

$$CH_3-(CH_2)_{m-1}-C\equiv C-C\equiv C-(CH_2)_n\text{-COOH} ,$$

where $m = 10$, $n = 8$; $m = 16$, $n = 2$; and $m = 16$, $n = 0$. It was found that the acids with the diacetylene near the carboxylic head group formed the most stable monolayers. To understand the reason for this result, the reader should remember that short alkyl chains do not form solid-like assemblies because the van der Waals attraction energy is not enough to compensate for the entropic

184

effects. Thus, once the diacetylene is in the middle of a given alkyl chain, it divides it into two relatively short, liquid-like parts, each of which does not contribute large stabilization energy to the assembly. On the other hand, when the diacetylene is near the head group, thus leaving a C_{16}-long alkyl chain, a stable, solid-like assembly can be formed. The formation of such a stable assembly, however, does not guarantee that the diacetylene groups will be organized in the topology required for a good photopolymerization reaction. Indeed, it was found that the LB films with diacetylene units in the middle had a packing more favorable for polymerization. This is because the two alkyl chains on both sides of the polymerizing diacetylene provided the flexibility needed for the changing of the crystal structure during the polymerization reaction.

Bubeck *et al.* reported that the photopolymerization of diacetylene LB films can be sensitized by surface-active cyanine, acridinium, and anthraquinone dyes embedded in the layers [461]. The fact that both electron donor cyanine dyes and electron acceptor acridinium and anthraquinone dyes yield the same polymers suggests that electron transfer both to and from the diacetylene may initiate the polymerization reaction. In other words, both cationic and anionic polymerization mechanisms are possible, and the structure of the polydiacetylene thus formed is not a function of the mechanism.

Kajzar and Messier studied the solid-state polymerization and optical properties of diacetylene LB films [462]. They studied the polymerization as a function of the monovalent and divalent ion in the systems

$$CH_3\!-\!(CH_2)_{17}\!-\!C\!\equiv\!C\!-\!C\!\equiv\!C\!-\!CO_2^-\,X^+ \qquad\qquad X \equiv H, NH_4, Ag, Na$$
$$\text{and}\quad [CH_3\!-\!(CH_2)_{17}\!-\!C\!\equiv\!C\!-\!C\!\equiv\!C\!-\!CO_2^-]_2M^{2+} \qquad M \equiv Cd, Cu, Hg, Mn \ .$$

It was found that while the free acid and monovalent salts (except Na^+) polymerized in the solid state, all divalent salts did not. This is primarily because of the topology imposed by the divalent ions on the system, where the two diacetylene units do not have any possible motion in the direction normal to the layer surface; thus, no position adjustment during polymerization is possible. The polymers were not stable for more than a few weeks.

Tieke and Weiss studied the effects of preparation condition and of additives on the morphology of LB films of amphiphilic diacetylenes [463]. Yoshioka *et al.* studied the photopolymerization of LB films of amphiphiles containing phenyl and diacetylene groups [464].

$$\text{(phenyl)}\!-\!(CH_2)_m\!-\!C\!\equiv\!C\!-\!C\!\equiv\!C\!-\!(CH_2)_8\!-\!COOH$$

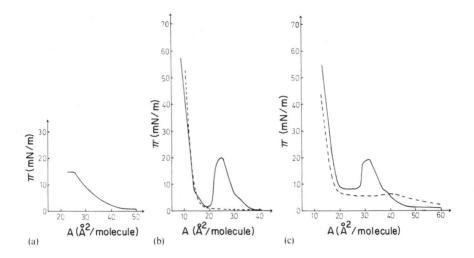

Figure 2.33. Pressure-area isotherms of (a) $C_6H_5-C\equiv C-C\equiv C-(CH_2)_8-COOH$, (b) $C_6H_5-(CH_2)_2-C\equiv C-C\equiv C-(CH_2)_8-COOH$, and (c) $C_6H_5-(CH_2)_3-C\equiv C-C\equiv C-(CH_2)_8-COOH$, at 5°C. The full lines represent the water subphase containing cadmium ions, and the broken lines refer to the pure water subphase. (From Yoshioka *et al.* [464], © 1985, Elsevier Science Publishers.)

This work is of special interest, since it contains both the diacetylene and another π-system that can serve as a model for more complex systems. Hence, it should point out a direction where two different functionalities are engineered in the same layer, and the intergroup interaction can be changed systematically, thus giving rise to new physical phenomena. Figure 2.33 presents the π–A isotherm for the three materials on pure water and on 3 x 10^{-4} M CdCl$_2$. In the present study, the separation of the phenyl and diacetylene groups was controlled by changing m (0, 2, 3). It is clear that the number of methylene units separating the two π-systems has a profound effect on the monolayer at the water–air interface. Thus, while in the case where $m = 0$ (ϕDA 0-8) no surface pressure could be observed at all areas, on pure water, for $m = 2, 3$ (ϕDA 2-8, and ϕDA 3-8, respectively), there was a steep rise at molecular areas smaller than the expected 25–30 Å^2. The addition of Cd^{2+} ions to the subphase results in some stabilization of the monolayer at the expected area per molecule, but is followed by a collapse at ~15 mN m^{-1}, which then is followed by a steep rise similar to the previous one. This behavior can be rationalized if a double layer was formed without film collapse in the second condensed region. LB films of φDA 0-8 did not polymerize, indicating the lack of correct topology or of the flexibility needed to allow changes in the unit cell in the polymerization process. Polymerization of φDA 3-8 yielded a red film.

Chen *et al.* studied LB polydiacetylene LB films of

$$CH_3-(CH_2)_{15}-C{\equiv}C-C{\equiv}C-(CH_2)_8-COOH\ ,$$

using surface enhanced Raman scattering (SERS) [465]. They reported large shifts ($\sim$50 cm^{-1}) to lower energies in both the C=C and C≡C stretching modes as the films' thickness changed from a monolayer or a bilayer to two or more bilayers, attributing them to delocalization of the electronic states in the polymer backbone. However, it is more reasonable to assume that electronic interactions do not occur between two π-systems separated by distances as in these films, and that the observed shifts are a result of structural changes in the film as it becomes thicker. Indeed, in a later paper, Chen *et al.* suggested that a disorder-to-order transformation occurs as the film becomes thicker [466].

Tieke and Weiss studied a series of amphiphilic diacetylenes with pyridine and 2,2'-bipyridine head groups [467]. (See next diagram.) They found that diacetylenic esters of isonicotinic acid form good monolayers and exhibit polymerization properties in the crystalline state that are similar to those of corresponding fatty acids. This indicates that the bulky headgroup does not prevent in this case good overlap of the diacetylenic groups. Since these esters form polymers that are soluble in chloroform, they could compare LB films of prepolymerized amphiphiles with those polymerized after deposition. The two types of polymeric films exhibited different morphologies, while the prepolymerized one formed inhomogeneous films that were not monomolecular.

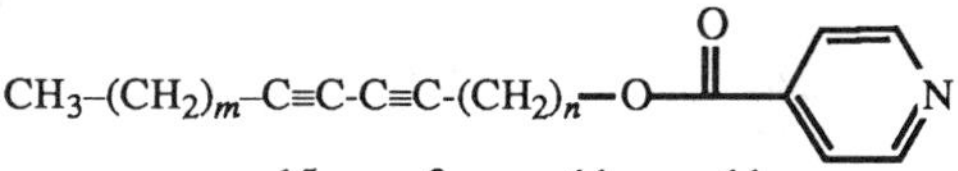

$$CH_3-(CH_2)_m-C{\equiv}C-C{\equiv}C-(CH_2)_n-O-\overset{\overset{\displaystyle O}{\|}}{C}-\langle\!\!\!\bigcirc\!\!\!\rangle N$$

$$m = 15, n = 3; m = 11, n = 11$$

Laxhuber *et al.* [468] used reflection spectroscopy (He-Ne laser at 632.8 nm) to study photothermal reaction kinetics and thermal desorption in diacetylene LB films of

$$CH_3-(CH_2)_{11}-C{\equiv}C-C{\equiv}C-(CH_2)_8-COOH\ .$$

They found that formation of the polymer results in an increase of film thickness by 1.9 Å, and that the kinetics of polymerization can be described in two consecutive light-induced reactions. In the first fast one, there is light absorption but no structural change, while in the second, the absorption is coupled with an immediate structural change.

Burzynski *et al.* studied monolayer and multilayer LB films of

$$CH_3-(CH_2)_{11}-C\equiv C-C\equiv C-(CH_2)_8-COOH$$

formed on a pure water subphase, using the resonance Raman optical-waveguide method [469]. The molecular organizations in both a monolayer and three-layer sample were the same, which indicates that the structural change reported by Chen *et al.* occurs only in films thicker than three layers. The blue form of the polydiacetylene was investigated and the blue-to-red conversion was found to involve a two-phase heterogeneous process.

Shutt and Rickert studied polydiacetylene salts of

$$CH_3-(CH_2)_{15}-C\equiv C-C\equiv C-(CH_2)_8-COOH$$

as thin film dielectrics in metal–LB film–semiconductor devices [470]. (See also Part Five.) They found that a transition took place when the concentration of $CdCl_2$ in the subphase increased from 10^{-5} to 10^{-3} M at 20°C, and that Cd^{2+} formed the best monolayers ($Cd^{2+} > Ba^{2+} > Mg^{2+}$). Using x-ray diffraction, they calculated a spacing of 27 Å for the Mg^{2+} salt, in agreement with the value determined for the free acid form [471]. The spacing calculated for the Cd^{2+} salt was 34 Å, implying that the polymerized monolayers are tilted at ~30°, in agreement with Day and Lando, who reported the tilt angle of 33° for the same monomer [458]. It can be concluded, therefore, that since that same tilt angle was found for the monomer and polymer states, the observed tilt is not caused by the polymerization process. Note that these tilt angles are an average between the tilt of the chains above and those below the π-system, which may not be the same.

Nakahara *et al.* used uv photoelectron spectroscopy to study the photopolymerization of

$$CH_3-(CH_2)_9-C\equiv C-C\equiv C-(CH_2)_8-COOH$$

in LB films [472]. They reported that the ionization threshold energy of the polymer (5.1 eV) was 1.6 eV smaller than that of the monomer (6.7 eV), and that no change of the threshold was found between the blue and red forms of the polymer. They suggested that in both forms, the dominant resonance structure probably is the acetylenic one; however, other spectroscopic studies suggest a difference in structure between the blue and red forms. Thus, although there is a large difference in the ionization threshold energy between the C≡C and C=C, the energy difference between the two conjugated π-systems is negligible, which is exhibited in their photoelectron spectra.

Kaneko *et al.* reported the first reversible color change for a short annealing in LB polydiacetylene films of [473]

$$[CH_3\text{-}(CH_2)_n\text{-}C\equiv C\text{-}C\equiv C\text{-}(CH_2)_8\text{-}COO^-]_2Cd^{2+}.$$

X-ray diffraction revealed that the reversible property is closely related to the layer structure. Raman spectroscopy indicated that the $C\equiv C$ to $C=C$ ratio of Raman peaks decreased with the increasing annealing temperature, indicating that the structures of the red and blue forms are completely different.

Göbel and Möhwald studied the topochemical polymerization of a diacetylene lipid with two chains (next diagram) as a function of the structure at the water–air interface [474]. They reported that there are two different lipid states formed at high surface pressure at different temperatures, and that there is no transition between them. Figure 2.34 presents the π–A isotherms for the molecule at temperatures of 24° and 10°C (pH 5.5). The difference in the crystalline nature of the two states is reflected in their morphology, as is apparent from fluorescence spectroscopy data. The polymerization process of the two states is very sensitive to their structural detail, and was suggested to give either a linear polymer in the case where the monomer structure is ordered (high temperature) or a polymer network where it is disordered (low temperature).

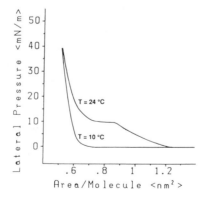

Figure 2.34. Pressure-area isotherms at temperatures 24° and 10°C (pH 5.5). (From Göbel and Möhwald [474], © 1988, Elsevier Science Publishers.)

189

Sotnikov *et al.* studied a series of tetraynoic fatty acids having the following general formula [475]:

$$C_nH_{2n+1}-C\equiv C-C\equiv C-(CH_2)_4-C\equiv C-C\equiv C-(CH_2)_8-COOH .$$

The films are different in molecular packing with respect to diacetylene films, probably due to the introduction of a second diacetylene group into the chain. As a result, no evidence for polymerization could be provided. It seems that such materials may be of interest and that decoupling of the two diacetylene groups may be required to achieve polymerization.

Ozaki *et al.* used diacetylene-terminated fatty acids

$$H-C\equiv C-C\equiv C-(CH_2)_n-COOH, \quad n = 8.\ 14.\ 20 ,$$

to prepare polyacene in an LB film [476]. In the first step, photopolymerization yielded polyacetylene $(CH=CH)_n$, which then was heat-treated ($200°–300°C$) to give the polyacene. (See reaction diagram.) An I_2-treated polyacene monolayer showed a conductivity of 10^{-3} S/cm. While not large, this conductivity indicates that with such innovative chemistry, one can design novel materials with interesting physical properties.

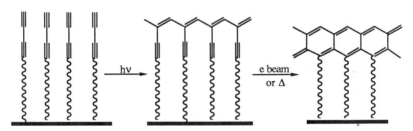

One general comment on diacetylene amphiphiles is that they tend to form films having a large number of defects, cracks and grain boundaries, when polymerized in the transferred film rather than at the water–air interface. This problem limits the utilization of such systems in applications such as NLO.

3. Amphiphiles Containing an Oxirane Group

Delaney *et al.* studied the preparation of monolayers and LB films from long-chain carboxylic acids having oxirane groups at various positions along the chain [477, 478]. (See reaction diagram.) They found that only the oxirane-terminated long-chain acids formed good monolayers, and that all other derivatives were too soluble. This C_{21} molecule had an area per molecule of 27 Å^2, which is larger that the 21 Å^2 that is the regular value for simple fatty acids. This increase is probably due to the bulky oxirane group at the terminus. Nevertheless, the π–A isotherm for this compound shows a very high collapse pressure, higher than 50 mN m^{-1}.

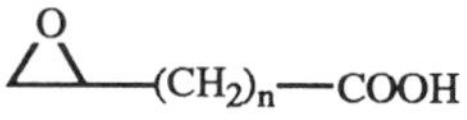

n = 8, 10, 14, 18

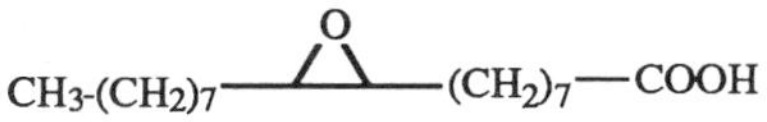

B. LB Films of Preformed Polymers

The interest in polymeric LB films clearly is driven by their potential applications as boundary lubricants, electrical insulators, NLO layers in waveguides, electrochromic and photochromic materials, etc. In this chapter, we discuss different families of polymeric amphiphiles and their LB films.

A *polymer* is a macromolecule that can be designed, like any other molecule, to have alternating hydrophilic and hydrophobic properties. Such properties can be built into the different parts of the polymer [447]. Thus, it is possible to design the backbone to be water-soluble (e.g., polyethylene oxide, polyvinyl alcohol, etc.), to introduce hydrophilic spacer groups into it (e.g., few ethylene oxide units), and to attach hydrophobic groups (or spacer groups) as side-chains. This is especially useful when the polymer side-chains are large, planar, conjugated chromophores or mesogenic groups. In this case, the hydrophilic alkyl chains are attached to the pendant groups. Less common are polymers with hydrophobic backbones and hydrophilic side-chains. This synthetic flexibility offers a control on the packing and orientation of polymers in two-dimensional assemblies, while, at the same time, providing the modification capability of polymer mechanical and stability properties.

Once an amphiphilic polymer has been spread at the water–air interface, it can reorganize and adopt a configuration where the hydrophilic parts are in the water subphase and the hydrophobic parts are in the air, thus forming a monolayer. The transfer of such a monolayer onto a substrate is not straightforward and is a function of the properties of the polymer under study. Since the viscosity of the monolayer increases with the increasing molecular weight of the polymer, the difficulty of monolayer transfer also is a function of molecular weight. On the other hand, the thermal and mechanical stabilities of polymeric monolayers also are improved with the increasing molecular weight. The design of an amphiphilic polymer, therefore, is not an easy one, and, at the time of this writing, this area is purely empirical.

With all the preceding notes on molecular engineering of polymers, the factors that determine whether or not a simple polymer is suitable to form a monolayer at the water–air interface are different from those discussed previously for small molecules [479]. The polymers do not have to be highly insoluble, but rather have high affinity to the interface (free energy of adsorption). Crisp suggested dividing simple polymers into two classes, where the first are formers of condensed films, and the second are formers of expanded films at the water–air interface [480]. For example, poly(ethyl methacrylate) belongs to the first class, while poly(ethyl acrylate) belongs to the second. The reader may find accounts for the early work on polymeric LB films in Gaines' book [14] and in a review by Goddard [481].

Puggelli and Gabrielli studied bidimensional mixtures in monolayers of poly(vinyl acetate) (PVAc), poly(methyl methacrylate) (PMMA), and poly(vinyl butyrate) (PVB) at the water–air interface [482]. The polymer–interface interactions are the apparent determining factor in the behavior of these mixtures. In another study, Gabrielli and Guarini studied the collapse mechanism of PMMA monolayers [483]. Gabrielli and Maddii investigated mixed monolayers of PMMA and stearic acid, oleic acid, and myristic acid [484]. They found that PMMA was compatible only with oleic acid, and that both are oriented horizontally to the water–air interface.

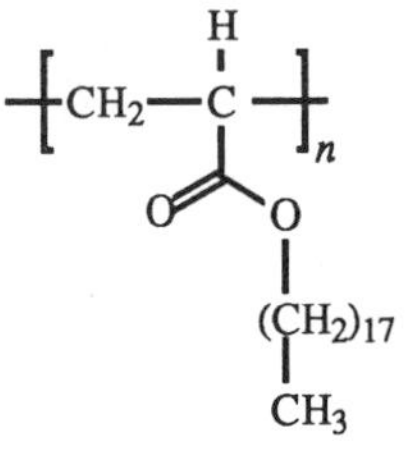

Poly(octadecyl acrylate) 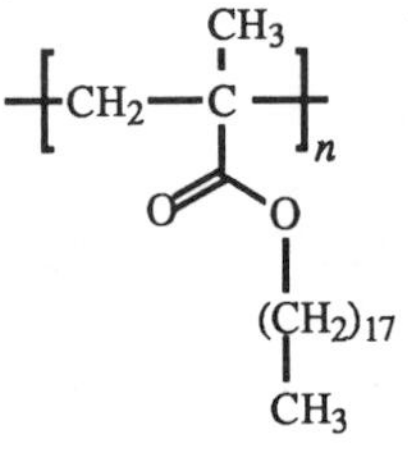   Poly(octadecyl methacrylate)

Mumby *et al.* studied monolayers and multilayers of poly(octadecyl acrylate) (PODA) having a weight-average molecular weight $M_N = 13,000$, and of poly(octadecyl methacrylate) (PODMA), with $M_N = 97,200$ [485, 486]. The monolayers were stable, and expansion–compression cycles reveal complete reproducibility. The minimum stable area per molecule was ~21 $Å^2$ and ~22 $Å^2$ for PODA and PODMA, respectively. These results indicate very close packing, since in the case of PODA, the area is only ~5% larger than the 20 $Å^2$ commonly used as a cross-section area in close-packed LB films of saturated alkyl chains [487, 488]. Attempts to form multilayers resulted in partial success, since only six monolayers could be deposited, apparently due to the lack of a hydrophilic group independent of the hydrophobic side chains.

Grazing-angle FTIR studies of monolayers transferred onto aluminized surfaces suggest that the polymer backbone is at the surface, with the alkyl chains primarily perpendicular to the surface (based on the relatively low intensity of the CH_2 vibrations).

Duda *et al.* studied monolayers of atactic and isotactic PODMA (different average molecular weights), and of copolymers with dodecyl methacrylate [489]. It was found that the atactic polymer forms more stable monolayers at the water–air interface than their isotactic analogue. Also, a liquid-like state in the π–A isotherm of PODMA could be induced by copolymerization with the shorter C_{12} alkyl chain methacrylate ester, and that at this state, monolayer transfer was especially successful.

Recently, Naito reported the study of poly(isobuly methacrylate) (PIBM, M_N = 2,3 x 10^5) LB films [490]. This is an interesting work, since the branched alkyl chain (*iso*-butyl) has a cross-section area larger than that of a normal chain, but it is a small group and can reorganize to accommodate a smaller area. Indeed, the experimental area per molecule was ~19 $Å^2$. This fit between the head group and the tail suggest that close-packed monolayers of PIBM should be hydrophobic on both sides, since the oxygen atoms are "buried" in the monolayer bulk. Experimental wetting measurements confirm this model, and water contact angles on the fifth and sixth monolayer in a Y-type multilayer film are almost the same ($73°$–$74°$). Another important parameter in polymeric LB films is their viscosity. In this case, PIBM monolayers at the water–air interface are not so viscous (2.5 x 10^{-2} g/s, at 10 dyn/cm). The monolayer deposited well on hydrophobic surfaces as amorphous assemblies, and proved to have very few pineholes.

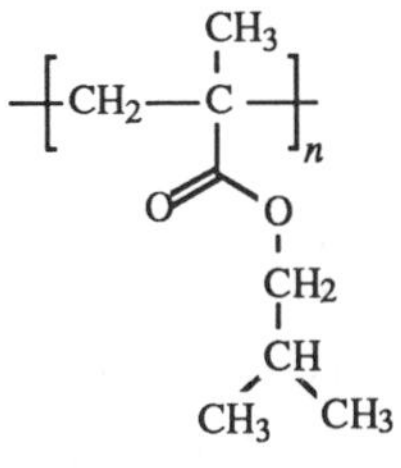

Poly(isobutyl methacrylate)

In 1982, Tredgold and Winter published the first [491] in a long series of papers on LB films of preformed polymers [357, 492–499]. They studied monolayers of copolymers of maleic anhydride and different vinyl derivatives (*n*-octadecyl vinylether, 1-octadecen, and styrene). (See next diagram.)

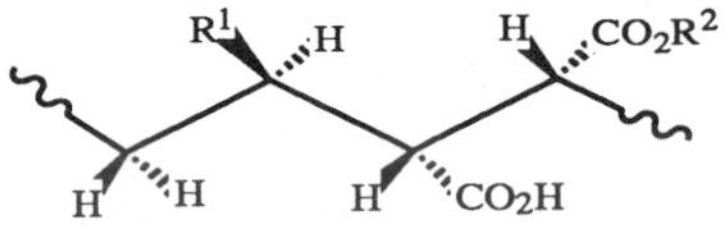

Initially, they used the anhydride and depended on its hydrolysis at the water–air interface; however, they reacted the copolymers later with different alcohols, thus introducing another R group to the polymer, while retaining one carboxylic group. Note that the two substituents R^1 and R^2 offer a large variety of materials, and therefore, a control on the T_g of the polymer, its solubility and spreading properties, the packing density in the monolayer, and its monolayer physical properties. The major difference between R^1 and R^2 is that when R^1 is a long chain and R^2 a methyl group, the polymer chains pack tightly at the interface, yielding stable, hard monolayers with melting points above 300°C. On the other hand, when R^2 is the long chain, it yields low packing density and a melting point of ~120°C, apparently due to the fact that both carboxylic groups are in the water, thus forcing the alkyl chain to bend around the polymer backbone to minimize its interactions with the water. The x-ray structure of a family of polymer monolayers indicated that all contain a regular layer structure [357]. An interesting collection of materials containing aromatic, vinyl, acetylene, and diacetylene groups was reviewed for electron-beam applications [497].

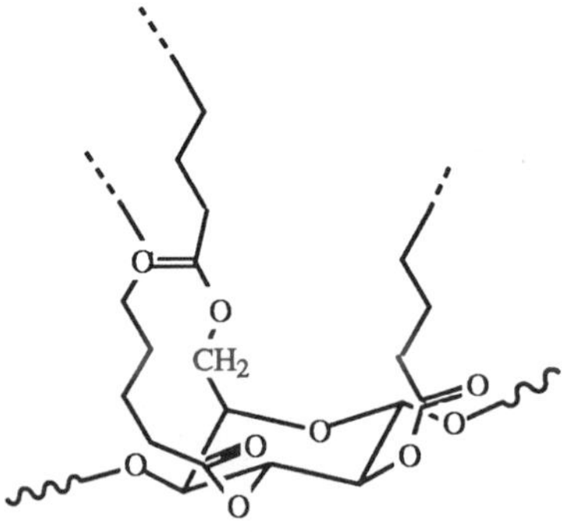

Kawaguchi *et al.* studied monolayers and multilayers of various cellulose esters [500]. (See structure in diagram before this paragraph.) The materials formed stable monolayers and exhibited a phase transition from the expanded to the condensed phase. The monolayer could be transferred by horizontal lifting to yield X-type films. These films were examined by polarized IR spectroscopy and were found to be well ordered, with the alkyl chains parallel to one another and perpendicular to the substrate.

194

Lovelock *et al.* studied monolayers of poly(octadecyl vinylether) on silicon dioxide [501].

$$\sim\!\!\left[CH_2\!-\!\underset{\underset{O(CH_2)_{17}CH_3}{|}}{CH}\right]_n\!\!\sim$$

One of the interesting ideas in the design of polymers for LB films was developed by Ringsdorf *et al.* [502]. The concept to use a spacer group (mostly hydrophilic) to decouple the motion of the polymer from that of the lipid membrane originally was demonstrated by Elbert *et al.* [503]. They prepared monolayers from a polymer with hydrophilic phosphate groups using a tetraethylene oxide spacer to link a glycerol diether to the polymer chain. (See next diagram.)

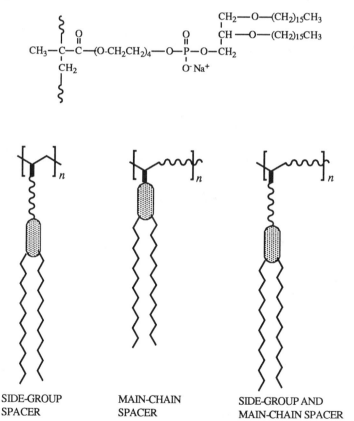

SIDE-GROUP   MAIN-CHAIN   SIDE-GROUP AND
SPACER   SPACER   MAIN-CHAIN SPACER

Figure 2.35. Decoupling of different motions in polymeric amphiphiles by the introduction of hydrophilic spacer groups. (Adapted from Elbert *et al.* [503], © 1985, Am. Chem. Soc.)

The different possibilities of introducing spacer groups into a polymer are presented in Figure 2.35. Ringsdorf *et al.* prepared three series of amphiphilic copolymers with hydrophilic spacers and with different values of monomer feed ($n = 0, 1, 5,$ or 10) [502]. The first polymer they studied was one with only a side-group spacer:

$$
\begin{array}{l}
CH_3\text{-}(CH_2)_{14}\text{-}CH_2\text{-}O\text{---}CH_2 \\
CH_3\text{-}(CH_2)_{14}\text{-}CH_2\text{-}O\text{---}CH \\
\qquad\qquad\qquad\quad CH_2\cdot COO\text{---}CH_2CH_2\text{---}COO\text{---}CH_2CH_2\text{---}OOC\text{---}C\text{---}CH_3
\end{array}
\qquad
\begin{array}{l}
CH_2 \\
\end{array}
$$

Here, the π–A isotherm exhibits a solid analogue phase with a collapse area of ~40 Å^2 per repeat unit. This area, which is comparable with the packing of the monomer, indicates that the alkyl chains in the assembly are tightly packed, and that a short side-group spacer was enough to provide flexibility, and thus to promote efficient organization of the macromolecule at the water–air interface.

$$
\begin{array}{l}
CH_3\text{-}(CH_2)_{14}\text{-}CH_2\text{-}O\text{-}\text{---}CH_2 \\
CH_3\text{-}(CH_2)_{14}\text{-}CH_2\text{-}O\text{-}\text{---}CH \\
\qquad\qquad\qquad\quad CH_2COO\text{---}CH_2CH_2\text{---}COO\text{---}CH_2CH_2\text{---}OOC\text{---}C\text{---}CH_3 \\
\qquad\qquad\qquad\qquad\qquad\qquad\qquad\qquad\qquad\qquad\qquad\quad CH_2 \\
\qquad\qquad\qquad\qquad\qquad\qquad\qquad HO\text{---}CH_2CH_2\text{---}OOC\text{---}CH
\end{array}
$$

$$
\begin{array}{l}
CH_3\text{-}(CH_2)_{16}\text{-}CH_2\text{-}O\text{-}\text{---}CH_2 \\
CH_3\text{-}(CH_2)_{16}\text{-}CH_2\text{-}O\text{-}\text{---}CH \\
\qquad\qquad\qquad\quad CH_2\text{-}OOC\text{---}C\text{---}CH_3 \\
\qquad\qquad\qquad\qquad\qquad\quad CH_2 \\
\qquad HO\text{---}CH_2CH_2\text{---}OOC\text{---}CH
\end{array}
$$

$$
\begin{array}{l}
CH_3\text{-}(CH_2)_{16}\text{-}CH_2 \\
\qquad\qquad\qquad\qquad\quad NH\text{---}CO\text{---}CH_2CH_2\text{---}COO\text{---}CH_2CH_2\text{---}OOC\text{---}C\text{---}CH_3 \\
CH_3\text{-}(CH_2)_{16}\text{-}CH_2 \\
\qquad\qquad\qquad\qquad\qquad\qquad\qquad\qquad\qquad\qquad\qquad\quad CH_2 \\
\qquad\qquad\qquad\qquad\qquad\quad HO\text{---}CH_2CH_2\text{---}OOC\text{---}CH
\end{array}
$$

To further achieve decoupling of the polymer backbone and hydrophobic side-chains, flexible main-chain hydrophilic spacer groups were incorporated. The effect of lipid-to-comonomer ratio on the packing of the monolayers also was studied. For the copolymer with a ratio of 1:1 ($n = 1$, see aforementioned diagram), a solid analogue phase with tightly packed alkyl chains and high collapse pressure was observed. On the other hand, in the case of the copolymer with a ratio of 1:5 ($n = 5$), a liquid analogue phase was favored, followed by a solid analogue phase. These results reflect the increased mobility of the monolayer due to the increased spacer length.

The concept just described has been developed further in the Ringsdorf group and was adopted by many other researchers. It is beyond the scope of this review to describe all the systems in detail; however, we shall describe some significant trends. Biddle *et al.* prepared and studied a large number of copolymers with acrylic acid as the comonomer (next diagram) [504]. They changed the side-chain-to-main-chain spacer ratio (m) and studied the behavior of these copolymers at the water–air interface. Figure 2.36 presents the onset of the fluid analogue region as a function of m. Thus, as the amount of the hydrophilic main-chain spacer increased, the onset of the fluid analogue phase occurred at higher and higher areas per repeat unit.

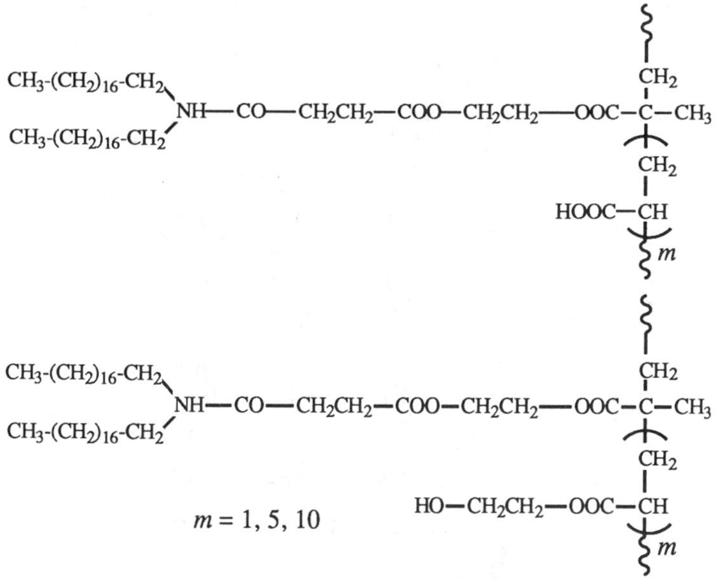

$$CH_3\text{-}(CH_2)_{16}\text{-}CH_2 \diagdown$$
$$\qquad\qquad\qquad NH\text{—}CO\text{—}CH_2CH_2\text{—}COO\text{—}CH_2CH_2\text{—}OOC\text{-}\underset{\underset{\textstyle HOOC\text{—}CH}{|}}{\overset{\overset{\textstyle CH_2}{|}}{\underset{\textstyle CH_2}{C}}}\text{-}CH_3 \quad \}m$$
$$CH_3\text{-}(CH_2)_{16}\text{-}CH_2 \diagup$$

$$CH_3\text{-}(CH_2)_{16}\text{-}CH_2 \diagdown$$
$$\qquad\qquad\qquad NH\text{—}CO\text{—}CH_2CH_2\text{—}COO\text{—}CH_2CH_2\text{—}OOC\text{-}\underset{\underset{\textstyle HO\text{—}CH_2CH_2\text{—}OOC\text{—}CH}{|}}{\overset{\overset{\textstyle CH_2}{|}}{\underset{\textstyle CH_2}{C}}}\text{-}CH_3 \quad \}m$$
$$CH_3\text{-}(CH_2)_{16}\text{-}CH_2 \diagup$$

$$m = 1, 5, 10$$

Note that all the systems had a solid analogue phase at ~40 Å^2 per repeat unit. We conclude, therefore, that the introduction of spacer groups into a polymeric amphiphile is an important tool in the design of LB films with a high degree of order and stability. Figure 2.37 presents the variable temperature x-ray study of a 30-layer sample of the copolymer with $m = 10$.

197

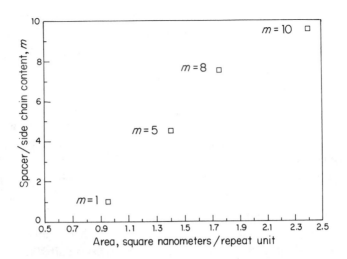

Figure 2.36. Onset of the fluid analogue region vs. copolymer composition. (From Biddle *et al.* [504], © 1988, Elsevier Science Publishers.)

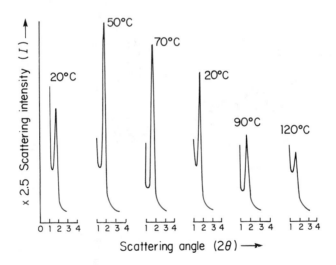

Figure 2.37. Variable temperature x-ray study of the copolymer with $m = 10$. (From Biddle *et al.* [504], © 1988, Elsevier Science Publishers.)

Schneider *et al.* used polarized IR, in both grazing-angle and transmission modes, to study a series of copolymers containing varying concentrations of hydrophilic main-chain spacers [505]. The most important result in this study is that the introduction of high concentrations of side-chain spacer groups allows

198

more flexibility of the polymer backbone, which results in a higher degree of side-chain orientation.

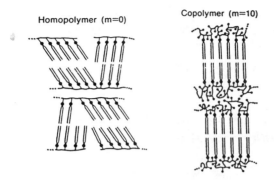

Figure 2.38. Model of the orientation of alkyl side chains in bilayers of homopolymer ($m = 0$) and copolymer ($m = 10$). (From Schneider *et al.* [505], © 1989, Am. Chem. Soc.)

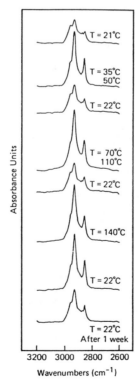

Figure 2.39. IR spectra ($E \perp$) of CH_2 stretching region of the copolymer with $m = 10$ as a function of temperature (From Schneider *et al.* [505], © 1989, Am. Chem. Soc.)

199

Figure 2.38 presents a model of the orientation of alkyl side chains in the bilayer of the homopolymer ($m = 0$), and a copolymer with $m = 10$. The thermal stability of the copolymer ($m = 10$, melting transition 33.5°C) was exceptionally good. It was found that even heating to a temperature as high as 140°C did not destroy the order completely; in fact, a high degree of orientation returned after remaining at 2°C for up to one week. Figure 2.39 presents the IR spectra of the CH_2 stretching region of the copolymer with $m = 10$ as a function of temperature.

One of the first polymers to be investigated extensively as a monolayer at the water–air interface was polysiloxane [506–515]. Bornett and Zisman studied a family of poly[methyl(n-alkylsiloxane], and poly[methyl(3,3,3-trifluoropropyl)-siloxane] $\{(CH_3)_3Si-[OSi-(CH_3)(R)]_n-OSi(CH_3)_3$, where R = CH_3, C_2H_5, C_4H_9, and where R = $CH_2CH_2CF_3$, respectively, $n = 12$ or $14\}$ [516]. They investigated the relationships of film pressure vs. area, surface potentials, and the vertical component of surface dipole moment as a function of the number of repeat units and of the alkyl chain length. In this as well as all early studies, there are suggestions for molecular orientation at the water–air interface. This, however, does not have any support from spectra (since it was not available at this time), and therefore, we do not discuss it here in detail.

Close in structure to the polysiloxanes are the polyethers. Watanabe *et al.* studied monolayers of a family of poly(vinyl alkylals) at the water–air interface, and their transfer to solid substrates (next diagram) [517–520]. These copolymers formed stable films at the water–air interface with an area per molecule of ~30 Å^2, irrespective of the side-chain length and degree of polymerization.

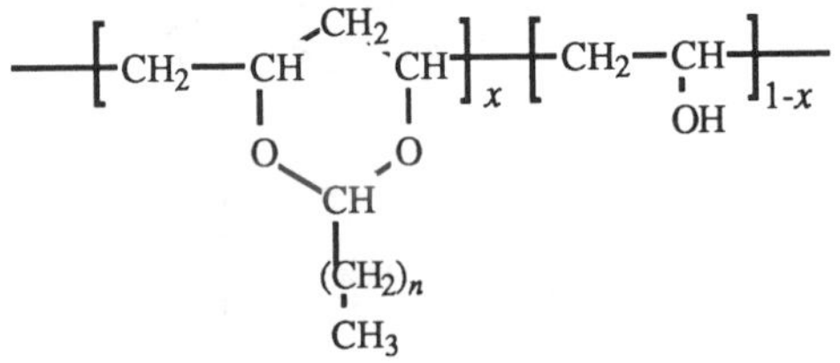

$$
-\left[CH_2-CH \begin{array}{c} CH_2 \\ CH \end{array}\right]_x \left[CH_2-CH \right]_{1-x}
$$

The monolayer transferred exclusively in the Y mode, with a transfer ratio of unity to form excellent multilayer films (100 layers). The odd–even effect in the wettability of the built-up films confirmed their Y-type structure. These very closely packed polymer films may be interesting as nonactive intermediate layers in organic superlattices.

Another group of polymers is made up of those where the main chain contains a chromophore. Suresh *et al.* studied LB films of liquid–crystalline polymers containing azobenzene in their main chain [320], and Blair and McArdle studied two azobenzene-containing polymers as photochromic materials [521]. Polymeric phthalocyanines belong, in principle, to this class of main-chain polymers. A most significant contribution in this area comes from Wegner *et al.* [330, 522], and we discussed it earlier. We note that the introduction of chromophores and other groups to the polymer backbone has attracted very little attention. It seems, however, that it is an important approach in the engineering of advanced materials for different applications, since these polymers do not contain long alkyl chains, and, therefore, the concentration of the active group in the monolayer is maximized.

A different class of polymers is where a functional moiety (e.g., an NLO chromophore) is attached as a pendent group. In most cases, this requires that an alkyl chain will be a part of the side chain. Vickers *et al.* studied liquid–crystal side-chain polymeric LB films, with major emphasis on waveguide properties (next diagram) [331].

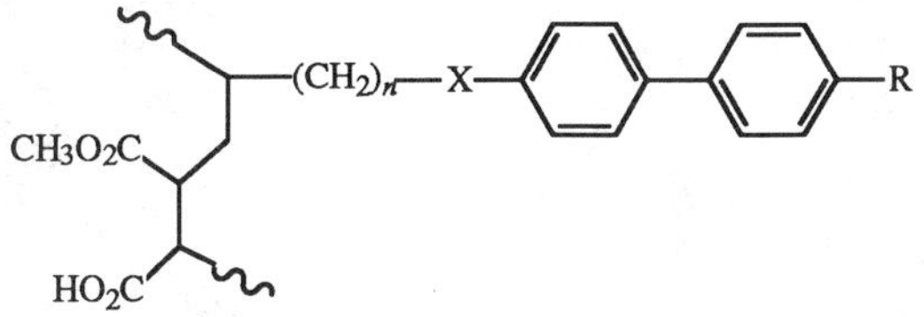

A number of different studies of polymeric LB films for NLO are discussed in Section 5.1.B.1 [335, 523–528]. The formation of LB films of poly(vinyl alcohol) having azobenzene side chains was described by Seki and Ichimura (next diagram) [529]. There are two points of interest in this work. The first is that while monomeric azobenzene amphiphiles in the *cis* form do not form stable monolayers, the polymer backbone makes the formation of stable monolayers possible even when the azobenzene side chains are in the *cis* form. Second, the azobenzene units in the polymeric monolayer did not aggregate, leading to reversible and efficient photoisomerizations of the chromophore groups.

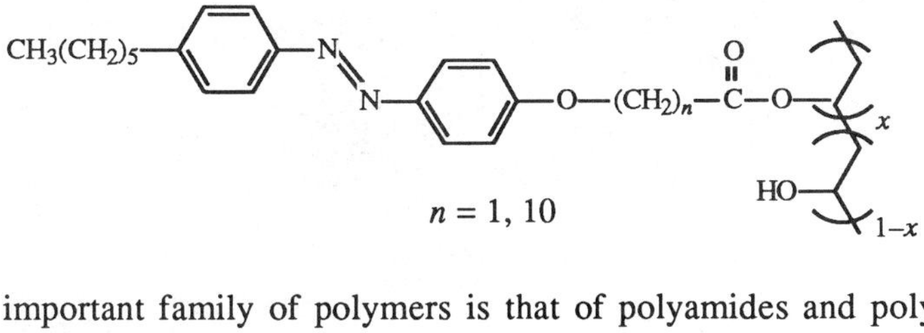

$n = 1, 10$

An important family of polymers is that of polyamides and polyimides. Suzuki *et al.* studied a large series of monolayers of aromatic polyamic acid alkylamine salts, which are precursors of polyimide films (see next diagram) [530–532].

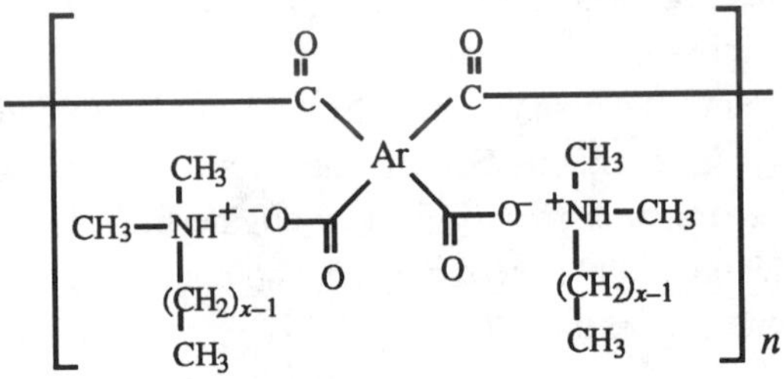

Nishikata *et al.* described the preparation and characterization of polyimide LB films of different kinds [533]. They treated the LB films with acetic anhydride to promote the amine-to-imide reaction. (See next diagram, top).

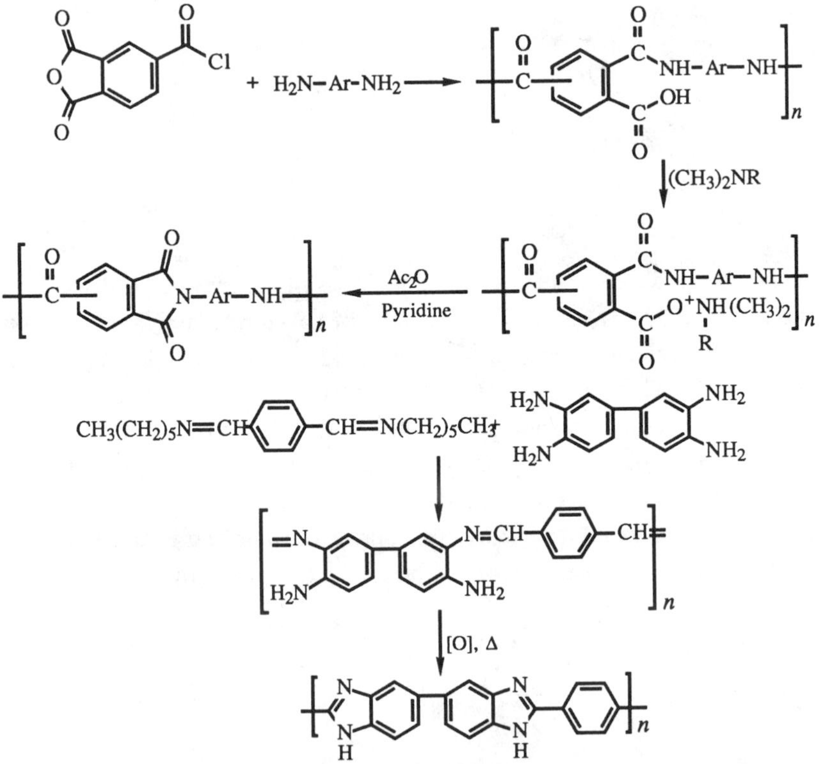

Angel *et al.* reacted dihexylterphthalimide and 3,3'-diaminobenzidine at the water–air interface and formed monolayers of polyanil. (See previous diagram, bottom) [534–536]. The monolayers were transferred onto CaF₂, or activated

202

fused silica substrates and heated in air for 10 min at 250°C, whereby the cyclization reaction was completed. The resulting polyimide films were stable at 420°C in air. These films were used by Flower *et al.* as a thermally stable LB film insulator in an InP metal-insulator-semiconductor (MIS) device [537].

Uekita *et al.* prepared new precursors of polyimide films [538]. (See next diagram.) One important feature in these new materials is that the long-chain C_{18} alkyl group is a part of the leaving group in the cyclization reaction. Thus, while helping in the solubility and LB film formation, it is removed as an octadecanol, leaving only the aromatic parts, with much higher thermal stability (300°–500°C). These heat-stable polymers were found pinhole-free even if the thickness was < 1000 Å. The development of pinhole-free polyimide films, with controlled film thickness and excellent electrical properties, should have an impact not only on the immediate areas where these materials may be useful, but also as an important lesson in materials science.

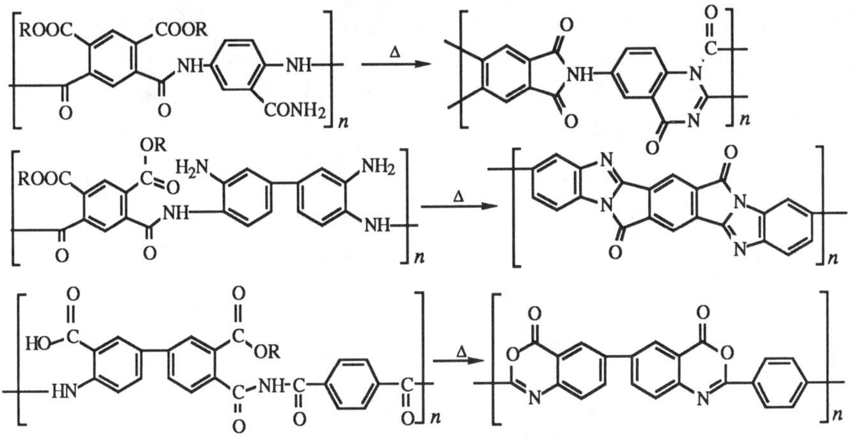

$$R = CH_3(CH_2)_{17} -$$

We already have mentioned the thermal stability of polymeric LB films developed by Ringsdorf *et al.*, and here we have another example where polymers have superior properties. It seems that the future is in polymeric LB films, where principles developed both by the Ringsdorf group and in the Japanese polyimide research are combined. Such advanced organic materials also should have controlled refractive indices; they will be very important in nonlinear optics.

203

2.9. LB Films of Phospholipids

It is truly beyond the scope of this book to cover the richness of the self-organization of lipids, and the formation of liposomes and vesicles, that are issues requiring a separate book by themselves. The reader is referred to the work published by Ringsdorf's group [446, 447]. Here, we discuss LB films of phospholipids, and focus on the chemistry and structural aspects of these films. Monolayers of phospholipids in the water–oil interface have been discussed in Section 2.1.F.

Lipids are a major component (40–50%) of membranes in plant and animal cells, where their bilayers are the matrix of the cell membrane. These membranes are quite flexible and contain ~50–60% proteins, some on the surface (in the hydrophobic part) and some in the interior. The classical structure for the cell membrane has been proposed by Singer and Nicolson, and is called the *fluid-mosaic model* [539]. The picture becomes even more complicated because other cellular functions — such as recognition, fusion, endocytosis, intercellular interactions, transport, and osmosis — all are membrane-mediated processes. The challenge to mimic Nature using synthetic membranes has been a driving force in this area for the last 10 years [446, 447].

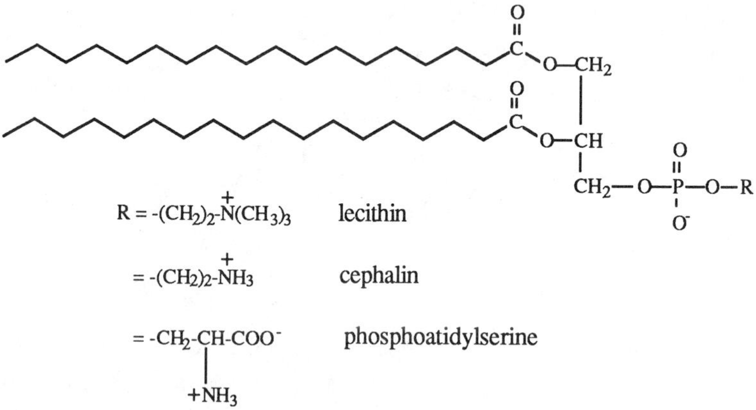

$$R = -(CH_2)_2-\overset{+}{N}(CH_3)_3 \quad \text{lecithin}$$

$$= -(CH_2)_2-\overset{+}{N}H_3 \quad \text{cephalin}$$

$$= -CH_2-CH-COO^- \quad \text{phosphoatidylserine}$$
$$\underset{+NH_3}{|}$$

The most common lipids in the cell membrane are *phospholipids*, where two hydroxy-groups in a glycerine molecule are esterified by long-chain fatty acids, which serve as the hydrophobic part. The third hydroxy-group is connected to a phosphate group, which in turn is connected to an ammonium group, thus forming a zwitterionic hydrophilic part. The most common biomembrane phospholipids are illustrated in the preceding diagram.

Colacicco *et al.* published a series of papers on lipid monolayers [540–544]. The interest in this work is that very early, they suggested the monolayer at the water–air interface is not continuous and consists of lipid clusters. They also proposed that as the pressure increases, these clusters coalesce into large ones, and that this coalescence is mediated by the water structure. Some of their proposals were verified experimentally and theoretically in recent years.

Lösche *et al.* studied the liquid–gel transition in monolayers of DMPA and DLPA, and the shift in this transition caused by the presence of monovalent and divalent ions in the subphase [545].

$$CH_3—(CH_2)_n—\overset{\overset{O}{\|}}{C}—O—CH_2$$
$$CH_3—(CH_2)_n—\overset{\overset{O}{\|}}{C}—O—CH$$
$$CH_2—O—\overset{\overset{O}{\|}}{P}—OH$$
$$O^-$$

$n = 12$ 1,2-dimyristoyl-*sn*-glycero-3-phosphotidic acid — DMPA
$n = 16$ 1,2-dilauroyl-*sn*-glycero-3-phosphotidic acid — DLPA

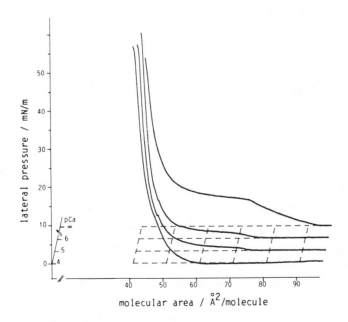

Figure 2.40. π–A isotherms for DMPA monolayers for various concentrations of CaCl$_2$. (From Lösche *et al.* [545], © 1985, Elsevier Science Publishers.)

Figure 2.40 presents π–A isotherms for DMPA monolayers for various concentrations of $CaCl_2$. It is clear that increasing the concentration of Ca^{2+} ions (which appears as pCa in Figure 2.40, with the same being found also for Mg^{2+} ions) had a profound effect on the onset of the condensed phase. Thus, it led to the disappearance of the phase transition. This effect was found to be enhanced further if monovalent ions also were present. Using fluorescence microscopy, they could show the strong effect of the ionic conditions on the geometry of the monolayer, where, in high ionic strength, films could be prepared as regular arrays of single crystalline domains with dimensions up to ~100 μm. A detailed ATR FTIR study of DPPC monolayers (as in next diagram) was published by Okamura *et al.* [546]. It was found that the alkyl chains were oriented vertically to the Ge-surface with all-*trans* configuration, irrespective of the surface pressure on the film transfer. This is in agreement with the existence of surface islands, where the different phases represent different sizes of the individual island and in their packing.

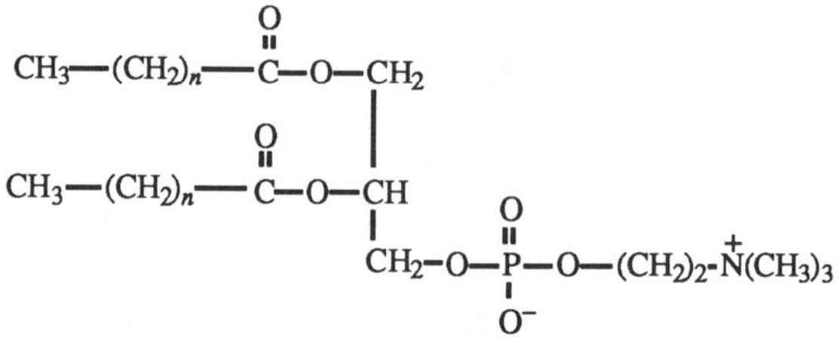

n = 12 1,2-dimyristyl-*sn*-glycero-3-phosphotidylcholin — DMPC
n = 14 1,2-dipalmiloyl-*sn*-glycero-3-phosphotidylcholin — DPPC
n = 16 1,2-distearoyl-*sn*-glycero-3-phosphotidylcholin —DSPC

Early reports discuss the deposition of Y-type multilayers of phospholipids [145, 153, 155, 158]. However, Lotta *et al.* reported the first observation of genuine Z-type multilayers of phospholipids (DPPC), and used ATR FTIR to study their structure [547]. This observation suggested that three methyl groups on ammonium nitrogen may play an important role in reducing the hydrophilic character of the monolayer surface in the subphase, which results in a better free energy match with the hydrophobic surface, and thus the formation of stable Z-type films. The FTIR results showed a wavenumber shift and a decrease in half bandwidth when the number of deposited layers increased, indicating that some kind of an ordering process took place. Similar behavior was observed by Bonnerot for docosanoic and ω-tricosenoic acids [280], and by Kopp for tripalmitin multilayers, where a spontaneous transition from the liquid–crystalline to microcrystalline state took place [548].

Vogel and Möbius use surface potential measurements of lipid monolayers to calculate the effective local dipole moments of polar groups at the water–lipid interface [549]. They showed that the effective dipole moment of polar or charged groups in aqueous solutions becomes negligibly small if the dielectric constant of water is introduced.

Dluhy et al. measured in situ FTIR spectra of DMPC and DSPC monolayers at the water–air interface [550]. Using water as the reflecting substrate, they collected external reflection spectra of both the liquid-expanded and liquid-condensed phases. Their results show that while the chains in the liquid-condensed phase are rigid and mostly in the all-trans configuration, those in the liquid-expanded phase are highly fluid.

The most extensive studies of the microstructure of lipid monolayers were carried out in Möhwald's groups [262, 551–556]. In these studies, they used a combination fluorescence microscopy, synchrotron x-ray diffraction, surface potential and electron diffraction measurements in connection with thermodynamics, to investigate the different ordering processes. Surface plasmon studies of lipid monolayers were carried out by Knoll's group [557].

To start this discussion, we need to consider a schematic phase diagram for lipids having a long hydrocarbon chain (Figure 2.41) [14]. We already have discussed a similar picture in connection with our discussion on the π–A isotherm. Here, the picture is similar, but we discuss the nature of phase transitions from the gas to the liquid phase, and from the liquid to the solid phase. These transitions may be of first or second order, and may involve intermediate fluid phases. If these transitions are first order, then there should be phases at intermediate pressures or temperatures.

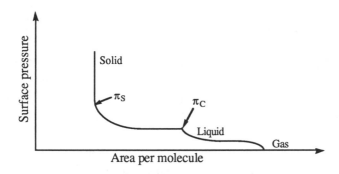

Figure 2.41. A schematic π-A isotherm for a phospholipid of fatty acid.

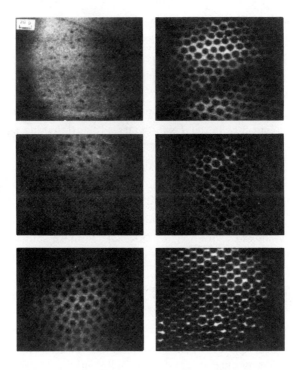

Figure 2.42. Fluorescence micrographs of a DMPA monolayer at pH 11 containing 1 mol% of the dye dipalmitoyl nitrobenzoxadiazol phosphotidylethanolamine (DPNBDPE). The surface pressure π ($\pi_C < \pi < \pi_S$) was increased from the upper left to lower right. (From Möhwald [552], © 1988, Elsevier Science Publishers.)

The visualization of a gas–fluid coexistence phase, using fluorescence spectroscopy, has been demonstrated in different cases [558–560]. In the cases of phospholipids [558] and fatty acids [561], the domains are nearly circular in shape. Once the film is compressed, the coexistence disappears and the monolayer is a homogeneous fluid phase. In most cases, there is a first order transition to a more ordered phase, and it can be visualized because they contain less dye than the fluid phase. The onset of the transition appears as a break in the isotherm and is indicated on the isotherm above (π_C). Figure 2.42 shows, from top left to bottom right, fluorescence micrographs obtained when increasing the surface pressure in a DMPA monolayer. One can see that at initial stages of compression, it is the size of the domains and not their number that increased. The more ordered phase has been called liquid-condensed [562], crystalline-tilted [563], or gel.

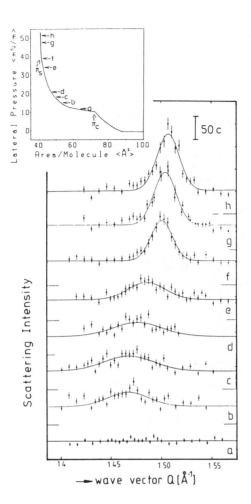

Figure 2.43. (a) Normalized x-ray intensity as a function of the wavevector transfer Q (bottom) for increasing surface pressures (curves a–h) corresponding to the arrows in the π–A isotherm (top), for a DMPA monolayer, pH 5.5; 10^{-2} M NaCl; 5×10^{-5} M EDTA; T = 19 ± 2°C. (From Kjaer *et al.* [262], © 1988, Elsevier Science Publishers.)

The shapes and arrangement of phospholipid islands at the water–air interface are determined in part by electrical forces (dipole–dipole repulsion) [247, 564–567], and in the case of optically active phospholipids, by the driving force to reduce the line tension between the solid and fluid phases [566]. For further discussions on the shape and growth of domains, the reader is referred to References 552, 553, and 555.

209

The first reports on *in situ* synchrotron x-ray studies of lipid monolayers at the water–air interface were published simultaneously by Kjaer *et al.* [261], and Dutta *et al.* [268], but almost at the same time, Richardson and Roser [568] published a similar report. The scattered x-ray intensity as a function of in-plane diffraction angle for various surface pressures is presented in Figure 2.43.

The abrupt change in intensity and line width on exceeding the pressure above π_S indicates the phase transition from a less ordered to a more ordered phase. For DMPA, and also in other phospholipids, the positional coherence lengths below π_S are ~5–10 lattice spacings, whereas for surface pressures above π_S, they vary between 20 and 100 lattice spacings, indicating the large difference in order between the two phases.

2.10. Energy Transfer in LB Films

The idea that the physical properties of organic materials can be engineered at the molecular level was introduced by Kuhn in the 1960s. Although they used the phrase *synthetic molecular organizate* [404, 569, 570], Kuhn and co-workers provided the foundation for today's area of molecular electronics. In the early experiments of light harvesting, a monolayer of a host dye (D) was doped by traces of a guest dye (A). (See next diagram.) Since the cross-section area of the dye was much larger than that of the two C_{18} chains, an equivalent amount of octadecane ($C_{18}H_{38}$) was added. As a result, the octadecane chains filled the free volume and a compact arrangement both in the chromophore and the hydrocarbon parts was formed (Figure 2.44) [15, 571].

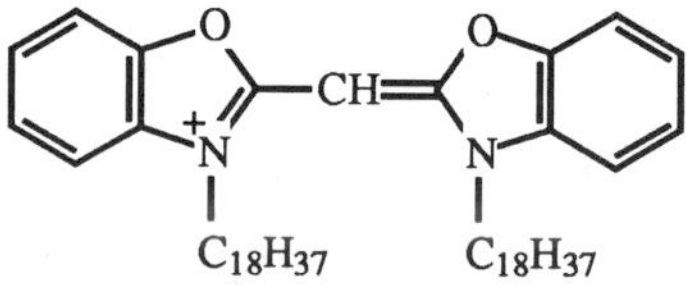

Energy donor (D) 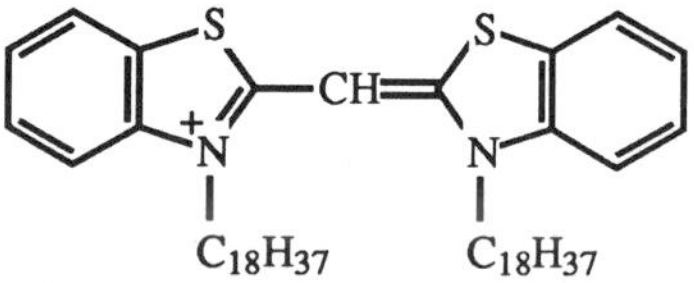   Energy acceptor (A)

The host dye absorbs the light, and the guest, which absorbs at a longer wavelength, is trapping the energy. The energy transfer within the layer depends on the electronic interaction of the molecules. It was calculated that the exciton hopping time in such a compact assembly is $\Delta t = 10^{-13}$ s, which in terms of the uncertainty principle corresponds to a bandwidth of 5 nm [570]. Indeed, it was found that the bandwidth of the fluorescence of D in the monolayer is 8–9 nm. Since a single D molecule has a fluorescence time that is 10^4 longer than Δt, the exciton will have a long diffusion path, and therefore, the energy absorbed by the host can be trapped even if the guest concentration is one molecule in 10^4 host molecules. This predicted effect was demonstrated by Möbius in a classical experiment illustrated in Figure 2.44 (top) [324].

A considerable quenching of the fluorescence of D and a sensitized fluorescence of A was observed even if only one among 50,000 molecules of D was exchanged with a molecule of A. The difference between the two cases can be rationalized on the basis of the extended dipole moment model [572]. Thus, if I and I_0 are the intensity of the fluorescence of D in the presence and absence of the guest, respectively, then

$$I = I_0 \frac{k_d}{k_d + k_{en}} , \qquad 2.1$$

where k_d is the rate of deactivation by fluorescence and thermal collisions and k_{en} is the rate of energy transfer. It is clear that k_{en} is proportional to the concentration of A molecules (which is inversely proportional to N, where N is the number of D molecules per molecule of A); therefore,

$$\frac{I}{I_0} = \frac{1}{1 + \text{const}/N} . \qquad 2.2$$

The experimental data could be fitted to this equation using the values const = 10^4, and const = 300 for cases (a) and (b), respectively. The full and broken lines in Figure 2.45 represent this fit. A quantitative treatment of the resonance energy transfer efficiency between donor and acceptor molecules in parallel layers was derived by Kuhn [573].

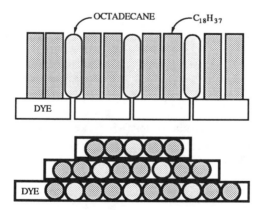

Figure 2.44. A side view (top), and a top view (bottom) of a monolayer of dye D and octadecane, molar mixing ratio 1 : 1.

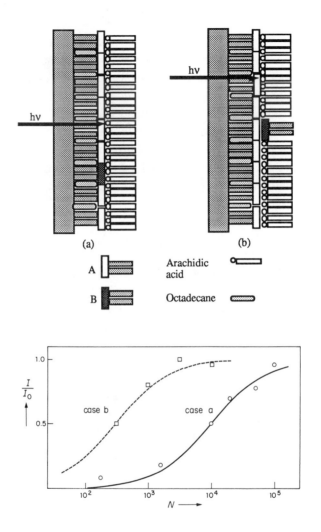

Figure 2.45. Harvesting of light energy by D and trapped by A. The trap is 30 times more efficient in case (a) than in case (b). The graph (bottom) gives a plot of I/I_0 (relative fluorescence intensity of D) against N (number of A molecules per molecule of A) in cases (a) (circles) and (b) (squares). (From Kuhn [570], © 1979, Elsevier Science Publishers.)

When the donor excitation is localized (i.e., there is no aggregation phenomena), the efficiency of energy transfer has been predicted to decrease as the fourth power of the interlayer distance,

$$1 - \phi_{et} = \frac{\phi_D}{\phi_D^0} = \frac{1}{1 + (d_0/d)^4} \, , \qquad 2.3$$

212

where ϕ_{et} is the efficiency of electron transfer, ϕ_D is the donor fluorescence quantum yield with the acceptor layer present, ϕ_D^0 is the donor fluorescence quantum yield with the acceptor layer absent, d is the separation distance between the donor and acceptor layers, and d_0 is the critical transfer distance, which is the separation distance for which the rate of energy transfer is equal to the rate of excited donor deactivation through other paths (e.g., a thermal deactivation). Since d is known from molecular dimensions and tilt angles, a measurement of the donor fluorescence quantum yields (ϕ_D and ϕ_D^0) allows d_0 to be determined.

It also was shown by Kuhn and coworkers that the Förster theory is adequate to obtain the critical distance d_0 from spectroscopic data without directly measuring energy transfer [404, 569, 574].

$$d_0 = \alpha \, \frac{\lambda_D}{n} (q_D q_{AD})^{1/4} , \qquad 2.4$$

where q_D is the probability of fluorescence emission by the donor in the emitting state in the absence of the acceptor; $q_D \geq \phi_D^0$, the fluorescence quantum yield of donor in the absence of acceptor; λ_D, the maximum donor fluorescence wavelength, $n = 1.56$, the index of refraction; A_{AD}, the overlap of donor fluorescence and acceptor absorption defined by the integral below; and α an orientation factor. For monolayers with statistical distribution of the transition moments in the plane of the layers, $\alpha = 0.098$.

The overlap of donor fluorescence and acceptor absorption is calculated by

$$A_{AD} = \int_0^\infty f_D(v) a_A(v)(v_D/v)^4 dv , \qquad 2.5$$

where $a_A(v)$ is the acceptor absorption spectrum, v is the frequency of light, v_D is the frequency of maximum donor fluorescence, and $f_D(v)$ is the instrument corrected donor fluorescence spectrum normalized to unity

$$\int_0^\infty f_D(v) dv = 1 . \qquad 2.6$$

A comparison of the values of d_0 obtained from energy transfer experiments to those obtained from Förster theory shows, in general, quite consistent agreement. However, as pointed out by Drexhage, the experimental values of d_0 usually are higher than the calculated ones, and the dimensionality tends to be slightly less than four [574]. This may be an indication that the donor and acceptor are closer than assumed because of imperfections in the monolayers. We note that Kuhn pointed out another possible reason that the power in Equation 2.3 may be less than 4 [570], namely, if there is more than one excited state population of donor molecules with different lifetimes.

It was predicted by Kuhn that if during the excited state lifetime the excitation in the donor layer is delocalized over large distances relative to the interlayer spacing, the fall-off of energy transfer with distance would be quadratic [573],

$$1 - \phi_{et} = \frac{1}{1 + (d/d_0)^2} \cdot \qquad 2.7$$

The theoretical value of d_0 now will be given by:

$$d_0 = \frac{\beta}{l} (\frac{\lambda_D}{n})^2 (q_n A_{AD})^{1/2} , \qquad 2.8$$

where β is an orientation factor ($\beta \neq \alpha$), which for the present case is equal to 0.016, and l is the distance of excitation delocalization in the donor. Since there is no information on l, the experimental value of d_0 can be used to calculate this delocalization distance (Equation 2.7). One example for a quadratic behavior was shown by Kuhn [8], and the other by Nakahara *et al.* [575].

Fromherz and Reinbold used monolayers of the same cyanine dyes (D and A, described earlier), separated by a single and a double-bilayer of the cadmium salt of C_{14}–C_{24} fatty acids, to study the quenching of the donor fluorescence [576]. They reported that the dependence of energy transfer on the distance of the dye layers differed significantly from the Förster theory, and provided no explanation for these results.

Yonezawa *et al.* used the same two dyes and studied energy transfer as a function of temperature [577]. Lowering the temperature from 293 K to 77 K increased the intensity ratio of the sensitized fluorescence to the donor fluorescence. This result indicates that the rate of thermal deactivation of the excited donor molecule is proportional to temperature, in spite of a temperature-insensitive energy transfer rate. The suggested mechanism for the energy transfer is a direct dipole–dipole interaction.

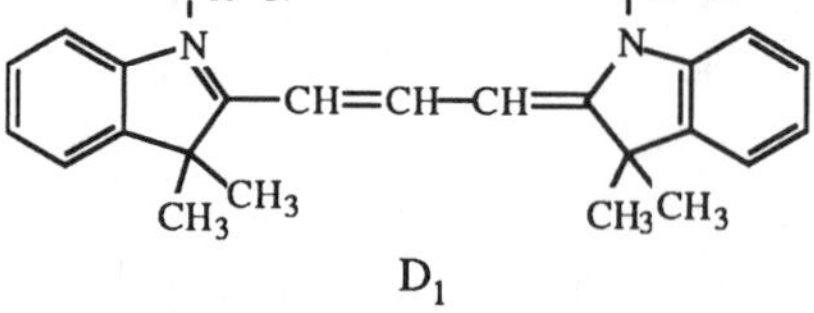

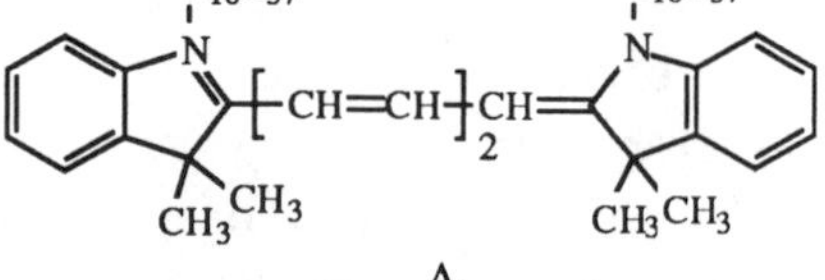

Ruaudel-Teixier and Vandevyver studied a very similar arrangement of other dyes (D_1 and A_1, see diagram), and could calculate that the radius of the excitonic cross section of the acceptor for energy transfer in the plane of the monolayer was

~60 Å [578]. Leitner *et al.* used mixed monolayers of D_1 and A_1 in arachidic acid and reported that fluorescence quenching took place by two-dimensional radiationless energy transfer from donor monomers to donor dimers and then to acceptors [579]. The nonexponential fluorescence decays were in agreement with the Förster theory, while the decay curves in the presence of acceptor layers did not fit the theory.

The question of why some results do not agree with the Förster theory has not been resolved yet. It seems that assemblies of alkyl chains cannot be treated simply as inert spacers between energy donors and acceptors, but more systematic studies are called for before further discussion can be initiated.

Dye aggregation is characterized by close packing of the chromophores in the two-dimensional assembly, which results in a strong coupling of the π-systems, distinct spectral changes, and photophysical behavior [580–582]. Two typical examples for such aggregation are the *J*- and *H*-aggregates of cyanine dyes [583]. In the *J*-aggregates [584, 585], a relatively large (*ca.* 2000 cm^{-1}) bathochromic shift in the lowest absorption transition relative to the non aggregated dye is observed. This shift also is manifested by a narrowing of the band and increase in extinction, but with conservation of oscillator strength [586].

Clearly, LB monolayers are especially suited for studies of dye aggregates, since they provide a highly ordered environment for the molecules, similar to that of a solid matrix, where aggregation of aromatic molecules is observed frequently [587]. In such molecular assemblies, aggregation is enhanced, and can be avoided only by dilution with inert molecules that serve to separate the chromophores. Kuhn and co-workers studied dye aggregation in monolayers [15, 588, 589], and proposed a coherent exciton model for the excited state [590]; however, a delocalized excited state in *J*-aggregates was proposed by Scheibe *et al.* [591]. Although the details at the molecular level of the transitions involved in the excitonic couplings are not fully understood [592, 593], the excitonic nature of the excited state is recognized [594, 595].

Nakahara and Möbius showed that the orientation and aggregation of long-chain merocyanine dyes in mixed monolayers transferred to glass plates depended strongly upon the matrix used [596]. Thus, mixed films with arachidic acid, methyl arachidate, and *n*-hexadecane showed the typical absorption and fluorescence spectra of the *J*-aggregates.

Penner measured energy transfer for the first time between two LB films under conditions where both layers are highly concentrated and *J*-aggregated [597]. The dyes used are shown in the diagram following Figures 2.46–2.48.

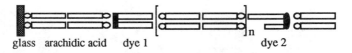

glass   arachidic acid   dye 1   dye 2

Figure 2.46. A schematic representation of the two-dye combination used to study the distance dependence of energy transfer. (From Penner [597], © 1988, Elsevier Science Publishers.)

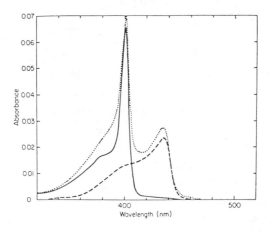

Figure 2.47. Absorption spectra of *J*-aggregated dyes individually and in combination in LB films. ———, dye 1; ---, dye 2; ······, combination. (From Penner [597], © 1988, Elsevier Science Publishers.)

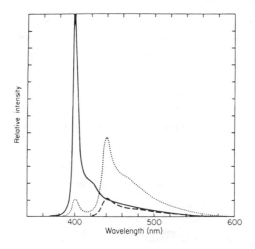

Figure 2.48. Fluorescence spectra of LB films, excited at 350 nm. ———, dye 1; ----, dye 2; ·····, combination. (From Penner [597], © 1988, Elsevier Science Publishers.)

216

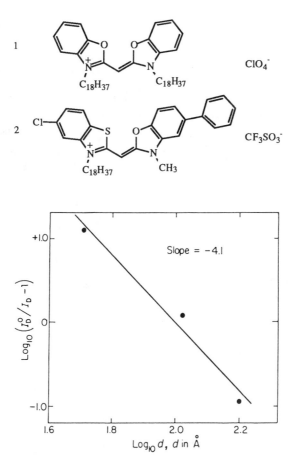

Figure 2.49. Dependence of fluorescence intensity of donor dye 1 on separation from layer of acceptor 2: ——— , least-square fit of Eq. 6. (From Penner [597], © 1988, Elsevier Science Publishers.)

Figure 2.46 presents a schematic representation of the orientation of the amphiphilic molecules, where monolayers of dyes 1 and 2 were prepared with m = 0, 1, and 2 (m being the number of cadmium arachidate bilayers). The separation with $m = 0$ is 51 Å, and each cadmium arachidate bilayer adds 54 Å to the distance. Figure 2.47 shows the absorbance spectra for dyes 1 and 2 individually deposited in the same configuration used in the combinations, and the spectrum of a combination monolayer, in which $m = 0$. Figure 2.48 shows the fluorescence spectra excited at 350 nm. Both absorption and emission spectra did not change by interdeposition of bilayer spacers. Equations 2.4 or 2.7 can be

217

rearranged into a logarithmic form with the power of d (2 or 4) replaced by p so that,

$$\log(\phi_D^0/\phi_D - 1) = -p \log d + p \log d_0 . \qquad 2.9$$

This is a more convenient form, since by plotting $\log(\phi_D^0/\phi_D - 1)$ vs. $\log d$ the power value can be obtained from the slope of the linear fit. Figure 2.48 presents the dependence of fluorescence intensity of donor 1 on separation from a layer of acceptor 2. The slope was found to be -4.1; thus, using the intercept in Equation 2.9, the value of $d_0 = 100 \pm 9$ Å was calculated. From the experimental data, Penner concluded that equation 2.4 is valid and calculated the value of $l = 147$ Å for dye 1, which is in agreement with the assumption that $l >> d$, but rather small, and suggests, therefore, a low degree of exciton delocalization. (Note that for J-aggregated merocyanine dyes, Kuhn suggested for l the range of values 120 Å $\leq l \leq$ 500 Å [573].) In this work, as well as in a more recent paper [598], Penner described experiments in LB films containing three layers of J-aggregated dyes. The reader is referred to References 597, and 598 for further discussion.

Recently, Penner concluded that with $l \sim 150$ Å, the condition $l >> d$ is *not* met, and that is why the 4th power law holds [597a]. As to why this is so, his view is not that l is smaller in this case than in Kuhn's [573]. Thus, Kuhn gave the range of 120 Å $\leq l \leq$ 500 Å, but calculated it assuming that $1 \geq \phi_D \geq \phi_D^0$. By assuming that $\phi_D = 1$, one can get the upper limit (500 Å), while if $\phi_D = \phi_D^0$, the lower limit (~120 Å) is calculated. Penner assumes in his analysis that $\phi_D = \phi_D^0$, since the emitting state for cyanine dyes is almost surely the lowest singlet, which is also the energy transfer state. Therefore, in fact, the excitonic delocalization in Penner's experiment is very similar to that of Kuhn (and of Nakahara [575]). Thus, the failure of $l >> d$ is not due to differences in l (which is supported by the similarity of the donor J-aggregate bandwidth), but, in fact, is due to d_0 being much larger in this experiment. The reason d_0 is larger in this case is becasue the acceptor concentration is high (it is also aggregated); i.e., although the donor excitation is delocalized, a large resonance coupling exists between dyes 1 and 2.

Yamazaki *et al.* also published on a sequential energy transfer in stacking monolayers [599]. Nakahara *et al.* studied J-aggregates of unsymmetrical chromophores in mixed monolayers [575]. In this case, they found $d_0 = 115$ Å, and calculated a value of 51 Å $\leq l \leq$ 130 Å, which is not in agreement with the assumption in the derivation of Equation 2.4, i.e., that $l >> d$, however, the experimental results were in agreement with a quadratic dependence. Yonezawa *et al.* studied luminescence and energy transfer using J-aggregated dye A

218

(structure on p. 210), and suggested that molecular excitation can travel over 3000 molecules within its lifetime [600].

Other papers that are related to the area of energy transfer are those dealing with fluorescence decay in LB films of pyrene–fatty-acid-containing phospholipids [601–604] and pyrene-containing fatty acid [603, 605, 606]. Tamai *et al.* studied energy transfer between carbazole and anthracene in LB films [607]. Biesmans *et al.* studied the fluorescence of anthracene in LB films [608]. In all these cases the photoactive system is in the hydrophobic part of the monolayer. Finally, Flörsheimer and Möhwald studied energy transfer and aggregation in monolayers containing porphyrins and phthalocyanines [609].

References

1. Tabor, D. *J. Colloid Interface Sci.* **1980**, *75*, 240.
2. Aristotle *Problemata Book 23, No 38*; Hett, W. S. translator; Heineman: New York, 1937.
3. Franklin, B. *Phil. Trans. R. Soc.* **1774**, *64*, 445.
4. Pockels, A. *Nature* **1891**, *43*, 437.
5. Pockels, A. *Nature* **1892**, *46*, 418.
6. Pockels, A. *Nature* **1893**, *48*, 152.
7. Pockels, A. *Nature* **1894**, *50*, 223.
8. Rayleigh, L. *Phil. Mag.* **1899**, *48*, 321.
9. Devaux, H. *Smithsonian Institute Ann. Rep.* **1913**, 261.
10. Hardy, W. B. *Proc. R. Soc. A.* **1912**, *86*, 610.
11. Langmuir, I. *J. Am. Chem. Soc.* **1917**, *39*, 1848.
12. Blodgett, K. A. *J. Am. Chem. Soc.* **1935**, *57*, 1007.
13. Blodgett, K. A. *Phys. Rev.* **1937**, *51*, 964.
14. Gaines, G. L., *Insoluble Monolayers Liquid –Gas Interfaces*; Interscience: New York, 1966.
15. Kuhn, H.; Möbius, D.; Bucher, H. *Techniques of Chemistry*, Weissberger, A.; Rositer, B. W., Eds.; Wiley: New York, 1973; Part 3B, Vol. 1.
16. Vincett, P. S.; Roberts, G. G. *Thin Solid Films* **1980**, *68*, 135.
17. Breton, M. *Macromol. Sci.-Rev. Macromol. Chem.* **1981**, *C21*, 61.
18. Roberts, G. G.; Vincett, P. S.; Barlow, W. A. *Ohys. Technol.* **1981**, *12*, 69.
19. Peterson, I. R.; Girling, I. R. *Sci. Prog. Oxford* **1985**, *69*, 533.
20. Sugi, M. *J. Mol. Electron.* **1985**, *1*, 3.
21. Roberts, G. G. *Adv. Phys.* **1985**, *34*, 475.
22. L'vov, Yu. M.; Feigin, L. A. *Sov. Phys. Crystallogr.* **1987**, *32*, 473.
23. Tredgold, R. H. *Rep. Prog. Phys.* **1987**, *50*, 1609.
24. Smith, T. *J. Colloid Interface Sci.* **1968**, *26*, 509.
25. Albrecht, O. *Thin Solid Films* **1983**, *99*, 227.
26. Fox, H. W.; Zisman, W. A. *J. Phys. Chem.* **1955**, *59*, 1233.
27. Wilhelmy, J. *Ann. Physik.* **1863**, *119*, 177.
28. Allan, A. J. G. *J. Colloid. Sci.* **1958**, *13*, 273.
29. Jordan, D. O.; Lane, J. E. *Australian J. Chem.* **1964**, *17*, 7.
30. Bull, H. B. in *Advances in Protein Chemistry*, Anson, M. L. and Edsall, J. T., Eds.; Academic Press: New York, 1947; Part III, p 95.

31. Vonnegut, B. *Rev. Sci. Instr.* **1942**, *13*, 82.
32. Dervichian, D. G. *J. Phys. Rodium [7]* **1935**, *6*, 221.
33. Harkins, W. D.; Anderson, T. F. *J. Am. Chem. Soc.* **1937**, *59*, 2189.
34. Mauer, F. A. *Rev. Sci. Instr.* **1954**, *25*, 598.
35. Rabinovich, W.; Robertson, R. F.; Mason, S. G. *Can. J. Chem.* **1960**, *38*, 1881.
36. Stewart, F. H. C. *Australean J. Chem.* **1961**, *14*, 159.
37. Trapeznikov, A. A. *Doklady Akas. Nauk. USSR* **1941**, *30*, 321.
38. Matuura, R. *Mem. Fac. Sci. Kyushu Univ. Ser. C* **1949**, *1*, 47.
39. Gaines, G. L. *J. Colloid. Sci.* **1960**, *15*, 321.
40. Abribat, M.; Dognon, A. *Compt. Rend.* **1939**, *208*, 1881.
41. LaMer, V. K.; Robbins, M. L. *J. Phys. Chem.* **1958**, *62*, 1291.
42. Ruyssen, R. *Rec. Trav. Chim.* **1946**, *65*, 580.
43. Davis, J. T.; Rideal, E. K. *Interfacial Phenomena*, Academic Press: New York, 1961.
44. Adamson, A. W.; Zebib, A. *J. Phys. Chem.* **1980**, *84*, 2619.
45. Neumann, A. W.; Tanner, W. *Tenside* **1967**, *4*, 220.
46. Princen, H. M. *J. Colloid. Sci.* **1963**, *18*, 178.
47. Gaines, G. L. Jr. *Surface Chemistry and Colloids (MTP International Review of Science)*, Kerker, M., Ed., University Park Press: Baltimore, 1972, Vol 7.
48. Gaines, G. L. Jr. *J. Colloid. Sci.* **1977**, *62*, 191.
49. Askew, F. A.; Danielli, J. F. *Trans. Faraday Soc.* **1940**, *36*, 785.
50. Ellison, A. H.; Zisman, W. A. *J. Phys. Chem.* **1956**, *60*, 416.
51. Cheesman, D. J. *Arkiv Kemi, Mineral. Geol.* **1946**, *B22*, 8.
52. Brooks, J. H.; Pethica, B. A. *Trans. Faraday Soc.* **1964**, *60*, 208.
53. Adam, N. K.; Jessop, G. *Proc. Royal Soc. (London)* **1926**, *A110*, 423.
54. Guastalla, J. *Compt. Rend.* **1929**, *189*, 241.
55. Marcelin, A. *Ann. Physique [10]* **1925**, *4*, 459.
56. Puddington, I. E. *J. Colloid. Sci.* **1946**, *1*, 505.
57. Malcolm, B. R.; Davies, S. R. *J. Sci. Instr.* **1965**, *42*, 359.
58. Guastalla, J. *Compt. Rend.* **1939**, *208*, 973.
59. Moss, S. A.; Rideal, E. K. *J. Chem. Soc.* **1933**, 1525.
60. Guastalla, J. *Compt. Rend.* **1938**, *206*, 993; *Cahiers Phys.* **1942**, *10*, 30.
61. Kalousek, M. *J. Chem. Soc.* **1949**, 894.
62. Tvaroha, B. *Chem. Listy.* **1954**, *48*, 183; *Chem. Abstr*. **1954**, 6494h.
63. Langmuir, I.; Schaefer, V. J. *J. Am. Chem. Soc.* **1937**, *59*, 2400.
64. Adam, N. K. *Proc. Roy. Soc. (London)* **1937**, *B122*, 134.
65. Norris, A.; Taylor, T. W. J. *J. Chem. Soc.* **1938**, 1719.
66. Pak, C. Y. C.; Arnold, J. D. *J. Colloid. Sci.* **1961**, *16*, 513.
67. Sher, I. H.; Chanley, J. D. *Rev. Sci. Instr.* **1955**, *26*, 266.
68. Adamson, A. W. *Physical Chemistry of Surfaces;* Wiley: New York, 1982.
69. den Engelsen, D.; Hengst, J. H. Th.; Honig, E. P. *Philips Tech. Rev.* **1976**, *36*, 44.
70. Peterson, I. R. *Thin Solid Films* **1985**, *134*, 135.
71. Dutta, P.; Halperin, K.; Ketterson, J. B.; Peng, J. B.; Schaps, G.; Baker, J. P. *Thin Solid Films* **1985**, *134*, 5.
72. Bikerman, J. J. *Trans. Faraday Soc.* **1938**, *34*, 800.
73. Barraud, A.; Leloup, J.; Gouzerh, A.; Palacin, S. *Thin Solid Films* **1985**, *133*, 117; *Fr. Patent 8,303,578*, 1983.
74. Girling, I. R.; Milverton, D. R. J. *Thin Solid Films* **1984**, *115*, 85.

75. Holcroft, B.; Petty, M. C.; Roberts, G. G.; Russell, G. J. *Thin Solid Films* **1985**, *134*, 83.
76. Daniel, M. F.; Dolphin, J. C.; Grant, A. J.; Kerr, K. E. N.; Smith, G. W. *Thin Solid Films* **1985**, *133*, 235.
77. Lin, B.; Peng, J. B.; Dutta, P.; Ketterson, J. B.; Wong, G. *Rev. Sci. Instr.* **1988**, *59*, 2623.
78. Miyano, K.; Maeda, T. *Rev. Sci. Instr.* **1987**, *58*, 428.
79. Kumehara, H.; Tasaka, S.; Miyata, S. *Nippon Kagaku Kaishi* **1987**, 2330.
80. Yun, W. B.; Bloch, J. N. *Rev. Sci. Instr.* **1989**, *60*, 214.
81. Lösche, R.; Möhwald, H. *Rev. Sci. Instr.* **1984**, *55*, 1968.
82. Seul, M.; McConnell, H. M. *J. Phys. E: Sci. Instr.* **1985**, *18*, 193.
83. Miyano, K.; Mori, A. *Jpn. J. Appl. Phys.* **1989**, *28*, 252.
84. Mingins, J.; Owens, N. F. *Thin Solid Films* **1987**, *152*, 9.
85. Peterson, I. R. *J. de Chimie. Phys.* **1988**, *85*, 997.
86. Richard, J.; Barraud, A.; Vandevyver, M.; Ruaudel-Teixier, A. *Thin Solid Films* **1988**, *159*, 207.
87. Wolstenholme, G. A.; Schulman, J. H. *Trans. Faraday Soc.* **1950**, *46*, 475.
88. Gains, Jr., G. L. *J. Phys. Chem.* **1959**, *63*, 1322.
89. Schenkel, J. H.; Kitchener, J. A. *Nature (London)* **1958**, *182*, 131.
90. Harkins, W. D. *Physical Chemistry of Surface Films;* Reinholds: New York, 1952, p 79.
91. Harkins, W. D.; Alexander, A. E. *Physical Methods of Organic Chemistry* ; Vol. 1, Wiley- Interscience: New York, 3rd Edn., 1959.
92. Pallas, N. R.; Pethica, B. A. *J. Colloids Surf.* **1975**, *71*, 1161.
93. Peterson, I. R.; Veale, G.; Montgomery, C. M. *J. Colloid Interface Sci.* **1986**, *109*, 527.
94. Saint Pierre, M.; Dupeyrat, M. *Thin Solid Films* **1983**, *99*, 205.
95. Roberts, G. G.; McGinnity, M.; Barlow, W. A.; Vincett, P. S. *Solid State Commun.* **1979**, *32*, 683.
96. Tieke, B.; Lieser, G.; Weiss, K. *Thin Solid Films* **1983**, *99*, 95.
97. Sugi, M.; Saito, M.; Fukui. T.; Iizima, S. *Thin Solid Films* **1983**, *99*, 17.
98. Daniel, M. F.; Lettington, O. C.; Small, M. *Thin Solid Films* **1983**, *99*, 61.
99. Schoeler, U.; Tews, K. H.; Kuhn, H. *J. Chem. Phys.* **1974**, *61*, 5009.
100. Tredgold, R. H.; Jones, S. D.; Evans, S. D.; Williams, P. I. *J. Mol. Electr.* **1986**, *2*, 147.
101. Handy, R. M.; Scala, L. C. *J. Electrochem. Soc.* **1966**, *113*, 109.
102. Pintchovski, F.; Pricw, J. B.; Tobin, P. J.; Peavey, J.; Kobold, K. *J. Electrochem. Soc.* **1979**, *126*, 1428.
103. Wasserman, S. R.; Yu-Tai, T.; Whitesides, G. M. *Langmuir* **1990**, *6*, 1074
104. Carim, A. H.; Dovek, M. M.; Quate, C. F.; Sinclair, R.; Vorst, C. *Science* **1987**, *237*, 630.
105. Zhuravlev, L. T. *Langmuir* **1987**, *3*, 316.
106. Madeley, J. M.; Richmond, C. R. *Z. Anorg. Allg. Chem.* **1972**, *389*, 92.
107. West, R. C.; Astle, M. J., Eds.; *Handbook of Chemistry and Physics*, 60th Ed., Chem. Rubber Co.: Cleavland, 1980, p. F-26.
108. Evans, S. D.; Ulman, A., unpublished results.
109. Gains, G. L., Jr. *J. Colloid. Interface Sci.* **1981**, *79*, 295.
110. Schneegans, M.; Menzel, E. *J. Colloid Interface Sci.* **1982**, *88*, 97.
111. Tredgold, R. H.; El-Badawy, Z. I. *J. Phys. D.* **1985**, *18*, 103.
112. von Tscharner, V.; McConnell, H. M. *Biophys. J.* **1981**, *36*, 421.

113. Honig, E. P.; Hengest, J. H. T.; den Engelson, D. *J. Colloid Interface Sci.* **1973**, *45*, 92.
114. Highfield, R. R.; Thomas, R. K.; Cummins, P. G.; Gregory, D. P.; Mingis, J.; Hayter, J. B.; Schärpf, O. *Thin Solid Films* **1983**, *99*, 165.
115. Chen, S. H.; Frank, C. F. *Langmuir* **1989**, *5*, 978.
116. Peng, J. B.; Ketterson, J. B.; Dutta, P. *Langmuir* **1988**, *4*, 1198.
117. Chidsy, C. E. D.; Loiacono, D. N.; Sleator, T.; Nakahara, S. *Surf. Sci.* **1988**, *200*, 45.
118. Adam, N. K. *The Physics and Chemistry of Surfaces;* 3rd ed., Oxford University Press: London, 1941.
119. Blinov, L. M. *Russ. Chem. Rev.* **1983**, *52*, 713.
120. Milner, S. T.; Joanny, J. F.; Pincus, P. *Europhys. Lett.* **1989**, *9*, 495.
121. Puggelli, M.; Gabrielli, G.; Caminati, G. *Colloid Polym. Sci.* **1989**, *267*, 65.
122. Hiemenz, P. C. *Principles of Colloid and Surface Chemistry;* Marcel Dekker: New York, 1986.
123. Riegler, J. E. *Rev. Sci. Instrum.* **1988**, *59*, 2220.
124. Riegler, J. E.; LeGrange, J. D. *Phys. Rev. Lett.* **1988**, *61*, 2492.
125. Blodgett, K. A.; Langmuir, I. *Phys. Rev.* **1937**, *51*, 964.
126. de Gennes, P. G. *Colloid Polym. Sci.* **1986**, *264*, 463.
127. Petrov, J.; Kuhn, H.; Möbius, D. *J. Colloid Interface Sci.* **1980**, *73*, 66.
127a. Grundy, M. J.; Musgrove, R. J.; Richardson, R. M.; Roser, S. J.; Penfold, J. *Langmuir* **1990**, *6*, 519.
128. Buha enko, M. R.; Richardson, R. M. *Thin Solid Films* **1988**, *159*, 231.
129. Blake, T. D.; Haynes, J. M. *J. Colloid Interface Sci.* **1969**, *30*, 421.
130. Petrov, J. G.; Radeov, B. P. *J. Coll. Polym. Sci.* **1981**, *259*, 753.
131. Buhaenko, M. R.; Goodwin, J. W.; Richardson, R. M.; Daniel, M. F. *Thin Solid Films* **1985**, *134*, 217.
132. Peterson, I. R.; Russell. G. J.; Roberts, G. G. *Thin Solid Films* **1983**, *109*, 371.
133. Barton, S. W.; Thomas, B. N.; Flom, E. B.; Rice, S. A.; Lin, B.; Peng, J. B.; Ketterson, J. B.; Dutta, P. *J. Chem. Phys.* **1988**, *89*, 2257.
134. Blodgett, K. A. *J. Phys. Chem.* **1937**, *41*, 975.
135. Kamel, A. M.; Weiner, N. D.; Felmeister, A. *Coll. Inter. Sci.* **1971**, *35*, 163.
136. Biddle, M. B.; Rickert, S. E.; Lando, J. B. *Thin Solid Films* **1985**, *134*, 121.
137. Langmuir, I.; Schaefer, V. J. *J. Am. Chem. Soc.* **1936**, *58*, 284.
138. Bagg, J.; Abramson, M. B.; Fichman, M.; Haber, M. D.; Gregor, H. P.; Langmuir, I.; Schaefer, V. J. *J. Am. Chem. Soc.* **1964**, *86*, 2759.
139. Sasaki, T.; Matuura, R. *Bull. Jpn. Chem. Soc.* **1951**, *24*, 274.
140. Miyano, K.; Abraham, B. M.; Xu, S. Q.; Ketterson, J. B. *J. Chem. Phys.* **1982**, *77*, 2190.
141. Veale, G.; Peterson, I. R. *J. Colloid Interface Sci.* **1985**, *103*, 178.
142. Spink, J. A.; Sanders, J.V. *Trans. Faraday Soc.* **1951**, *51*, 1154.
143. Goddard, E. D.; Kao, O.; Kung, H. C. *J. Colloid Interface Sci.* **1967**, *24*, 297.
144. Goddard, E. D.; Ackilli, J. A. *J. Coll. Sci.* **1963**, *18*, 585.
145. Stephens, J. F. *J. Colloid Interface Sci.* **1972**, *38*, 557.
146. Neuman, R. D. *J. Colloid Interface Sci.* **1975**, *53*, 161.
147. Tanaka, C.; Fukutome, *J. Colloid Interface Sci.* **1978**, *66*, 492.
148. Hasmonay, H.; Vincent, M.; Dupeyrat, M. *Thin Solid Films* **1980**, *68*, 21.
149. Ohshima, H.; Ohki, S. *J. Colloid Interface Sci.* **1984**, *102*, 85.

150. Fukuda, K.; Shiozawa, T. *Thin Solid Films* **1980**, *68*, 55.
151. Popovitz-Biro, R., Hill, K.; Shavit, E.; Hung, D. J.; Lahav, M.; Leiserowitz, L.; Sagiv, J.; Hsiung, H.; Meredith, G. R.; Vanherzeele, H. *J. Am. Chem. Soc.* **1990**, *112*, 2498.
152. Enkelmann, V.; Lando, J. B. *J. Poly. Sci. Polymer Chem. Ed.* **1977**, *15*, 1843.
153. Ehlert, R. C. *J. Colloid Interface Sci.* **1965**, *20*, 387.
154. Holley, C.; Bernstein, S. *Phys. Rev.* **1936**, *49*, 403.
155. Holley, C.; Bernstein, S. *Phys. Rev.* **1937**, *52*, 525.
156. Clark, G. L.; Sterrett, R. R.; Leppla, P. W. *J. Am. Chem. Soc.* **1935**, *57*, 330.
157. Bernstein, S. *J. Am. Chem. Soc.* **1938**, *60*, 1511.
158. Langmuir, I. *Science* **1938**, *87*, 493.
159. Charles, M. W. *J. Appl. Phys.* **1971**, *42*, 3329.
160. Stephens, J. F. *J. Colloid Interface Sci.* **1972**, *38*, 557.
161. Honig, E. P. *J. Colloid Interface Sci.* **1973**, *43*, 66.
162. Neumann, A. W. *Adv. Coll. Interf. Sci.* **1978**, *4*, 105.
163. Gaines, G. L. *J. Colloid Interface Sci.* **1977**, *59*, 438.
164. Neuman, R. D. *J. Colloid Interface Sci.* **1974**, *63*, 106.
165. Neuman, R. D.; Swanson, J. W. *J. Colloid Interface Sci.* **1979**, *74*, 244.
166. Clint, J. H.; Walker, T. *J. Colloid Interface Sci.* **1974**, *47*, 172.
167. Neumann, A. W. *Z. Phys. Chem.* **1964**, *41*, 339.
168. Neumann, A. W. *Z. Phys. Chem.* **1964**, *43*, 71.
169. Neumann, A. W.; Good, R. J., in *Surface and Colloid Science*, Good, R. J.; Stromberg, R. R., Eds.; Plenum: New York, 1979, Vol. 11.
170. Peng, J. B.; Abraham, B. M.; Dutta, P.; Ketterson, J. B. *Thin Solid Films* **1985**, *134*, 187.
170a. Robinson, I.; Sambles, J. R.; Cade N. A. *Thin Solid Films* **1989**, *178*, 125.
171. Kobayashi, K.; Takaoka, K. *Bull. Chem. Soc. Jpn.* **1986**, *59*, 933.
172. Peterson, I. R.; Russell, G. J. *Br. Polym. J.* **1985**, *17*, 364.
173. Veale, G.; Girling, I. R.; Peterson, I. R. *Thin Solid Films* **1985**, *127*, 293.
174. Billinger, F. W., Jr. *Textbook of Polymer Science*, Wiley: New York, 1971.
175. Blinov, L. M.; Davydova, N. N.; Lazarev, V. V.; Yudin, S. G. *Sov. Phys. Solid State* **1982**, *24*, 1523.
176. Blinov, L. M.; Duninin, N. V.; Mikhnev, L. V.; Yudin, S. G. *Thin Solid Films* **1984**, *120*, 161.
177. Daniel, M. F.; Smith, G. W. *Mol. Cryst. Liq. Cryst.* **1984**, *102*, 193.
178. Smith G. W.; Daniel, M. F.; Barton, J. W.; Ratcliffe, N. *Thin Solid Films* **1985**, *132*, 125.
179. Christie, P.; Roberts, G. G.; Petty, M. C. *Appl. Phys. Lett.* **1986**, *48*, 1101.
180. Christie, P.; Jones, C. A.; Petty, M. C.; Roberts, G. G. *J. Phys. D: Appl. Phys.* **1986**, *19*, L167.
181. Smith, G. W.; Evans, T. J. *Thin Solid Films* **1987**, *146*, 7.
182. Davies, G. H.; Yarwood, J. *Mikrochim. Acta* **1988**, *11*, 305.
183. Smith, G. W.; Matcliffe, N.; Roser, S. J.; Daniel, M. F. *Thin Solid Films* **1987**, *151*, 9.
184. Jones, C. A.; Petty, M. C.; Roberts, G. G. *Thin Solid Films* **1988**, *160*, 117.
185. Davies, G. H.; Yarwood, J.; Petty, M. C.; Jones, C. A. *Thin Solid Films* **1988**, *159*, 461.
186. Roberts, G. G.; Holcroft, B.; Richardson, T.; Colbrook, R. *J. de Chim. Phys.* **1988**, *85*, 1094.

187. Jones, C. A.; Petty, M. C.; Roberts, G. G. *IEEE Trans. on Ultrasonic Ferroelect. Freq. Control* **1988**, *35*, 736.
188. Langmuir, I.; Schaefer, V. J. *J. Am. Chem. Soc.* **1938**, *57*, 1007.
189. See for example, Terada, T.; Yamamoto, R.; Watanabe, T. *Sci. Pap. Inst. Phys. Chem. Res. Jpn.* **1934**, *23*, 173.
190. Kato, T. *Jpn. J. Appl. Phys. Part 2* **1988**, *27*, L1358.
191. Kato, T.; Arai, M.; Ohshima, K. *Nippon Kagaku Kaishi* **1988**, 1774.
192. Kato, T. *Jpn. J. Appl. Phys. Part 2* **1988**, *27*, L2128.
193 Dean, R. B.; Gatty, O.; Rideal, K. K. *Trans. Faraday Soc.* **1940**, *36*, 166.
194. Lange, E. *Z. Elektrochem.* **1951**, *55*, 76.
195. Pearsons, R., in *Modern Aspects of Electrochemistry* , Bockris, J. O'M., and Conway, B. E., Eds.; London: Butterworths (1954).
196. Davies, J. T. *Trans. Faraday Soc.* **1953**, *49*, 683.
297. Davies, J. T.; Rideal, E. K. *Can. J. Chem.* **1955**, *33*, 947.
198. Davies, J. T.; Rideal, E. K. *Interfacial Phenomena*, London: Academic Press (1963).
199. Dean, R. B. *Trans. Faraday Soc.* **1940**, *36*, 166.
200. Alexander, A. E. *Trans. Faraday Soc.* **1941**, *37*, 120.
201. Alexander, A. E.; Teorell, T. *Trans. Faraday Soc.* **1939**, *35*, 727.
202. Powell, B. D.; Alexander, A. E. *J. Colloid Sci.* **1952**, *7*, 482.
203. Powell, B. D.; Alexander, A. E. *J. Colloid Sci.* **1952**, *7*, 493.
204. Dupéyrat, M. *Mémorial Service Chim. État.* **1955**, *40*, 51.
205. Mingis, J.; Zobel, F. G. R.; Pathica, B. A.; Smart, C. *Proc. Roy. Soc. Lond. A* **1971**, *324*, 99.
206. Pal, R. P.; Chattoraj, D. K. *J. Colloid Interface Sci.* **1975**, *52*, 56.
207. Robb, I. D.; Alexander, A. E. *J. Colloid Interface Sci.* **1968**, *28*, 1.
208. Taylor, J. A. G.; Mingis, J.; Pethica, B. A.; Yan, B. Y. J.; Jackson, C. M. *Biochim. Biophys. Acta* **1973**, *323*, 157.
209. Llerenas, E.; Mingis, J. *Biochim. Biophys. Acta* **1976**, *419*, 381.
210. Bell, G. M.; Mingis, J.; Taylor, J. A. G. *J. Chem. Soc. Faraday II* **1978**, *74*, 223.
211. Ohki, S.; Ohki, C. B. *J. Theor. Biol.* **1976**, *62*, 389.
212. Bell, G. M.; Wilson, G. L. *Physica* **1982**, *115A*, 85.
213. Yue, B. Y.; Jackson, C. M.; Taylor, J. A. G.; Mingis, J.; Pethica, B. A. *J. Chem. Soc. Faraday I* **1976**, *72*, 2685.
214. Taylor, J. A. G.; Mingis, J.; Pethica, B. A. *J. Chem. Soc. Faraday I* **1976**, *72*, 2694.
215. Hayashi, M.; Kobayashi, T.; Seimiya, T.; Muramatsu, T.; Hara, I. *Chem. Phys. Lipids* **1980**, *27*, 1.
216. Hayashi, M.; Kobayashi, T.; Seimiya, T *Chem. Phys. Lipids* **1981**, *29*, 289.
217. Mingis, J.; Taylor, G. J. A.; Pathica, B. A.; Jackson, C. M.; Yue, B. Y. T. *J. Chem. Soc. Faraday I* **1982**, *38*, 232.
218. Desai, N.; Stroeve, P. *Chem. Eng. Commun* **1981**, *11*, 113.
219. Davies, R. J.; Jones, M. N. *Biochim. Biophys. Acta* **1986**, *858*, 135.
220. Menter, J. W.; Tabor, D. *Proc. Roy. Soc. London* **1950**, *A204*, 514.
221. Nasselli, C.; Rabolt, J. F.; Swalen, J. D. *J. Chem. Phys.* **1985**, *82*, 2136.
222. Nasselli, C.; Rabe, J. P.; Rabolt, J. F.; Swalen, J. D. *Thin Solid Films* **1985**, *134*, 173.
223. Cohen, S. R.; Naaman, R.; Sagiv, J. *J. Phys. Chem.* **1986**, *90*, 3045.
224. Barraud, A.; Ruaudel-Teixier, A.; Rosilio, C. *Ann. Chim.* **1975**, *10*, 195.
225. Barbaczy, E.; Dodge, F.; Rabolt, J. F. *Appl. Spect.* **1987**, *41*, 176.

226. Rabe, J. P.; Swalen, J. D.; Rabolt, J. F. *J. Chem. Phys.* **1987**, *86*, 1601.
227. Rabe, J. P.; Novotny, V.; Swalen, J. D.; Rabolt, J. F. *Thin Solid Films* **1988**, *159*, 359.
228. Cho, Y.; Kobayashi, M.; Tadokoro, H. *J. Chem. Phys.* **1986**, *84*, 4636.
229. Cho, Y.; Kobayashi, M.; Tadokoro, H. *J. Chem. Phys.* **1986**, *84*, 4643.
230. Rothberg, L.; Higashi, G. S.; Allara, D. L.; Garoff, S. *Chem. Phys, Lett.* **1987**, *133*, 67.
231. Riegler, J. E. *J. Phys. Chem.* **1989**, *93*, 6475.
232. Richardson, W.; Blasie, J. K. *Phys. Rev. B* **1989**, *39*, 12165.
233. Chidsey, C. E. D.; Liu, G. Y.; Rowntree, P.; Scoles, G. *J. Chem. Phys.* **1989**, *91*, 4421.
234. Nuzzo, R. G.; Korenic, E. M.; Dubois, L. H. *J. Chem. Phys.* **1990**, *93*, 767.
235. Shnidman, Y.; Eilers, J. E.; Ulman, A., in preparation.
236. Hautman, J.; Klein, M. L. *J. Chem. Phys.* **1989**, *91*, 4994.
237. Evans, S. D.; Sharma, R.; Ulman, A. *Langmuir* **1990**, *6*, 0000.
238. Hautman, J.; Bareman, J. P.; Mar, W.; Klein, M. L. *J. Chem. Phys.*, submitted.
239. Tillman, N.; Ulman, A.; Penner, T. L. *Langmuir* **1989**, *5*, 101.
240. Claesson, P. M.; Berg, J. M. *Thin Solid Films* **1987**, *176*, 157.
241. Israelachvili, J.; Adam, G. E. J. *J. Chem. Soc. Faraday Trans. 1* **1978**, *74*, 975.
242. Laxhuber, L. A.; Rothenhausler, B.; Schneider, G.; Möhwald, H. *Thin Solid*
243. Jones, J. P.; Webby, G. M. *J. Phys. D: Appl. Phys.* **1987**, *20*, 226.
244. Laxhuber, L A.; Möhwald, H. *Surf. Sci.* **1987**, *186*, 1.
245. Tippmann-Krayer, P.; Laxhuber, L. A.; Möhwald, H. *Thin Solid Films* **1988**, *159*, 387.
246. Harkins, W. D.; Copeland, L. E. *J. Chem. Phys.* **1942**, *10*, 272.
247. Andelman, D.; Brochard, F.; de Gennes, P. G. *C.R. Acad. Sci.* **1985**, *301*, 675.
248. Andelman, D.; Brochard, F.; Joanny, J. F. *Proc. Natl. Acad. Sci.* **1987**, *84*, 4717.
249. Iwahashi, M.; Machara, N.; Kaneko, Y.; Middleton, S. R.; Pallas, N. R.; Pathica, B. A. *J. Chem. Soc. Faraday Trans. I* **1985**, *81*, 973.
250. Hawkins, G. A.; Benedek, G. B. *Phys. Rev. Lett.* **1974**, *32*, 524.
251. Kim, M. W.; Cannel, D. S. *Phys. Rev. Lett.* **1975**, *35*, 889.
252. Kim, M. W.; Cannel, D. S. *Phys. Rev. A* **1976**, *13*, 411.
253. Middleton, S. R.; Iwahashi, M.; Pallas, N. R.; Pathica, B. A. *Proc. R. Soc. Lond, Ser. A* **1984**, *396*, 143.
254. Pallas, N. R.; Pathica, B. A. *J. Chem. Soc. Faraday Trans. I* **1987**, *83* 585.
255. Pallas, N. R.; Pathica, B. A. *Langmuir* **1985**, *1*, 509.
256. Bell, G. M.; Combs, L. L.; Dunne, L. J. *Chem. Rev.* **1981**, *81*, 15.
257. Legré, J. -P.; Albinet, G.; Firpo, J. L.; Tremblay, A. -M. S. *Phys. Rev. A* **1984**, *30*, 2720.
258. Dutta, P.; Peng, J. B.; Lin, B.; Ketterson, J. B.; Prakash, M.; Georegopoulos, P.; Ehrlich, S. *Phys. Rev. Lett.* **1987**, *58*, 2228.
259. Unger, G. *J. Phys. Chem.* **1983**, *87*, 689.
260. Lin, B.; Shih, M. C.; Bohanon, T. M.; Ice, G. E.; Dutta, P. *Phys. Rev. Lett.* **1990**, *65*, 161.
261. Kjaer, K.; Als-Nielsen, J.; Helm, C. A.; Laxhuber, L. A.; Möhwald, H. *Phys. Rev. Lett.* **1987**, *58*, 2224.
262. Kjaer, K.; Als-Nielsen, J.; Helm, C. A.; Tippmann-Krayer, P.; Möhwald, H. *Thin Solid Films* **1988**, *159*, 17.

263. Kjaer, K.; Als-Nielsen, J.; Helm, C. A.; Tippmann-Krayer, P.; Möhwald, H. *J. Phys. Chem.* **1989**, *93*, 3200.
264. Doucet, J.; Denicolo, I.; Craiveich, A. *J. Chem. Phys.* **1981**, *75*, 1523.
265. Doucet, J.; Denicolo, I.; Craiveich, A. *J. Chem. Phys.* **1981**, *75*, 5125.
266. Doucet, J.; Denicolo, I.; Craiveich, A. *J. Chem. Phys.* **1983**, *78*, 1465.
267. Ulman, A.; Eilers, J. E.; Tillman, N. *Langmuir* **1989**, *5*, 1147.
268. Takenaka, T.; Nogami, K.; Gotoh H.; Gotoh, R. *J. Colloid Interface Sci.* **1971**, *35*, 395.
269. Takenaka, T.; Nogami, K.; Gotoh H.; Gotoh, R. *J. Colloid Interface Sci.* **1971**, *40*, 409.
270. Ohnishi, T.; Ishitani, A.; Ishida, H.; Yamamoto, N.; Tsubomura, H. *J. Phys. Chem.* **1978**, *82*, 1989.
271. Chollet, P. A. *Thin Solid Films* **1978**, *52*, 343.
272. Chollet, P. A.; Messier, J. *Chem. Phys.* **1982**, *73*, 235.
273. Chollet, P. A.; Messier, J. *Thin Solid Films* **1983**, *99*, 197.
274. Allara, D. L.; Swalen, J. D. *J. Phys. Chem.* **1982**, *86*, 2700.
275. Rabolt, J. F.; Burns, F. C.; Schlotter, N. E.; Swalen, J. D. *J. Electron Spectr. Related Phenomena* **1983**, *30*, 29.
276. Rabolt, J. F.; Burns, F. C.; Schlotter, N. E.; Swalen, J. D. *J. Chem. Phys.* **1983**, *78*, 946.
277. Prakash, M.; Dutta, P.; Ketterson, J. B.; Abraham, B. M. *Chem. Phys. Lett.* **1984**, *111*, 395.
278. Prakash, M.; Ketterson, J. B.; Dutta, P. *Thin Solid Films* **1985**, *134*, 1.
279. DeGennes, P. G.; Sarma, G. *Phys. Lett.* **1972**, *A38*, 219.
280. Bonnerot, A.; Chollet, P. A.; Frisby, H.; Hoclet, M. *Chem. Phys.* **1985**, *97*, 365.
281. Vogel, V.; Wöll, C. *J. Chem. Phys.* **1986**, *84*, 5200.
282. Kimura, F.; Umemura, J.; Takenaka, T. *Langmuir 1986*, *2*, 96.
283. Garoff, S.; Deckman, H. W.; Dunsmuir, J. H.; Alvarez, M. S. *J. Physique* **1986**, *47*, 701.
284. Moncton, D. E.; Pindak, R. *Phys. Rev. Lett.* **1979**, *43*, 701.
285. Skita, V.; Richardson, A.; Filipkowski, M.; Garito, A.; Blasie, J. K. *J. Physique* **1986**, *47*, 1849.
286. Outka, D. A.; Stöhr, J.; Rabe, J. P.; Swalen, J. D.; Rotermund, H. H. *Phys. Rev. Lett.* **1987**, *59*, 1321.
287. Outka, D. A.; Stöhr, J.; Rabe, J. P.; Swalen, J. D. *J. Chem. Phys.* **1988**, *88*, 4077.
288. Rabe, J. P.; Swalen, J. D.; Outka, D. A.; Stöhr, J. *Thin Solid Films* **1988**, *159*, 275.
289. Guyot-Sionnest, P.; Hunt, J. H.; Shen, Y. R.; *Phys. Rev. Lett.* **1987**, *59*, 1597.
290. Dote, J. L.; Mowery, R. L. *J. Phys. Chem.* **1988**, *92*, 1571.
291. Dote, J. L.; Mowery, R. L. *Mikrochim. Acta* **1988**, *II*, 69.
292. Mizushima, K.; Nakayama, T.; Azuma, M. *Jpn. J. Appl. Phys.* **1987**, *26*, 772.
293. Mizushima, K.; Egusa, S.; Azuma, M. *Jpn. J. Appl. Phys.* **1988**, *27*, 715.
294. Fischetti, R. F.; Filipkowski, M.; Garito, A. F.; Blasie, J. K. *Phys. Rev. B* **1988**, *37*, 4714.
295. Chapman, J. A.; Tabor, D. *Proc. Roy. Soc. London A* **1957**, *242*, 96.
296. Nakahara, H.; Miyata, S.; Wang, T. T.; Tasaka, S. *Thin Solid Films* **1986**, *141*, 165.

297. Takahara, A.; Morotomi, N.; Hiraoka, S.; Higashi, N.; Kunitake, T.; Kajiyama, T. *Macromol.* **1989**, *22*, 617.
298. Nasselli, C.; Swalen, J. D.; Rabolt, J. F. *J. Chem. Phys.* **1989**, *90*, 3855.
299. Rabolt, J. F.; Fanconi, B. *Polymer* **1977**, *18*, 1258.
300. Case, G. D.; Vanderkooi, J. M.; Scarpa, A. *Arch. Biochem. Biophys.* **1974**, *162*, 174.
301. Kafka, M. S.; Holz, R. W. *Biochim. Biophys. Acta* **1976**, *426*, 31.
302. Hyono, A.; Hendriks, Th.; Daemen, F. J.; Bonting, S. I.; Wulf, J. *Biochim. Biophys. Acta* **1975**, *389*, 34.
303. Wulf, J.; Pohl, W. G. *Biochim. Biophys. Acta* **1977**, *426*, 471.
304. Foreman, J. C.; Mongar, J. L.; Gomperts, B. D. *Nature (London)* **1973**, *245*, 249.
305. Grinstein, S.; Erlij, D. J. *Membrane Biol.* **1976**, *29*, 313.
306. Ahkong, Q. F.; Tampion, W.; Lucy, J. A. *Nature (London)* **1975**, *256*, 208.
307. Luckasen, J. R.; White, J. G.; Kersey, J. H. *Proc. Nat. Acad. Sci. USA* **1974**, *71*, 5088.
308. Caspers, J.; Landuyt-Caufriez, M.; Ferreira, J.; Goodmaghtigh, E.; Ruysschaert, J. M. *J. Colloid Interface Sci.* **1981**, *81*, 410.
309. Ferreira, J.; Caspers, J.; Brasseur, R.; Ruysschaert, J. M. *J. Colloid Interface Sci.* **1981**, *81*, 158.
310. Winter, H.-J.; Manecke, G. *Makromol.Chem.* **1985**, *186*, 1979.
311. Matsunura, H.; Furusawa, K.; Inokuma, S.; Kuwamura, T. *Chem. Lett.* **1986**, 453.
312. Matsunura, H.; Watanabe, T.; Furusawa, K.; Inokuma, S.; Kuwamura, T. *Bull. Chem. Soc. Jpn.* **1987**, *60*, 2747.
313. Gold, J. M.; Teegarden, D. M.; McGrane, K. M.; Luca, D. J.; Falcigno, P. A.; Chen, C. C.; Smith, T. W. *J. Am. Chem. Soc.* **1986**, *108*, 5827.
314. Kelker, H.; Hatz, R. *Handbook of Liquid Crystals*, Verlag Chemie: Weinhein, 1980.
315. Dörfler, H. D.; Kerscher, W.; Sackmann, H. *Z. Phys. Chem.* **1972**, *251*, 314.
316. Diep-Quang, H.; Überreiter, K. *Coll. Polym. Sci.* **1980**, *258*, 1055.
317. Diep-Quang, H.; Überreiter, K. *Polym. J.* **1981**, *13*, 623.
318. Rondelez, F.; Koppel, D. *J. Physique* **1982**, *43*, 1371.
319. Daniel, M. F.; Lettington, O. C.; Small, S. M. *Mol. Cryst. Liq. Cryst.* **1983**, *96*, 373.
320. Suresh, K. A.; Blumstein, A.; Rondelez, F. *J. Physique* **1985**, *46*, 453.
321. Okahata, Y.; Kunitake, T. *Ber. Bunsenges Phys. Chem.* **1980**, *84*, 550.
322. Steven, J. H.; Hann, R. A.; Barlow, W. A.; Laird, T. *Thin Solid Films* **1983**, *99*, 71.
323. Mooney III, W. F.; Brown, P. E.; Russel, J. C.; Costa, S. B.; Pederson, L. G.; Whitten, D. G. *J. Am. Chem. Soc.* **1984**, *106*, 5659.
324. Möbius, D. *Ber Bunsenges Phys. Chem.* **1978**, *82*, 848.
325. Heesemann, J. *J. Am. Chem. Soc.* **1980**, *102*, 2167.
326. Shimomura, M.; Ando, R.; Kunitake, T. *Ber. Bunsenges Phys. Chem.* **1983**, *87*, 1134.
327. Kunitake, T.; Okahata, Y. *J. Am. Chem. Soc.* **1980**, *102*, 549.
328. Sprintschnik, G.; Sprintschnik, H. W.; Kirsch, P. P.; Whitten, D. G. *J. Am. Chem. Soc.* **1977**, *99*, 4947.
329. Gaines, G. L.; Behnken, P. E.; Valenty, S. J. *J. Am. Chem. Soc.* **1978**, *100*, 6549.
330. Orthmann, E.; Wegner, G. *Angew. Chem. Int. Ed. Engl.* **1986**, *25*, 1105.

331. Vickers, A. J.; Tredgold, R. H.; Hodge, P.; Khoshdel, E.; Girling, I. *Thin Solid Films* **1985**, *134*, 43.
332. Albrecht, O.; Cumming, W.; Kreuder, W.; Laschewsky, A.; Ringsdorf, H. *Coll. Polym. Sci.* **1986**, *264*, 659.
333. Laschewsky, A. *Angew. Chem. Int. Ed. Engl. Adv. Mater.* **1989**, *28*, 1574.
334. Richardson, T.; Roberts, G. G.; Polywka, M. E. C.; Davies, S. G. *Thin Solid Films* **1988**, *160*, 231.
335. Carpenter, M. M.; Prasad, P. N.; Griffin, A. C. *Thin Solid Films* **1988**, *161*, 315.
336. Vandevyver, M.; Richard, J.; Barraud, A.; Veber, M.; Jallabert, C.; Strzelecka, H. *J. de Chim. Phys.* **1988**, *85*, 385.
337. Sakuhara, T.; Nakahara, H.; Fukuda, K. *Thin Solid Films* **1988**, *159*, 345.
338. Matsumoto, M.; Sekiguchi, T.; Tanaka, H.; Tanaka, M.;Nakamura, T.; Tachibana, H.; Manda, E.; Kawabata, Y. *J. Phys. Chem.* **1989**, *93*, 5877.
339. Cook, M. J.; Daniel, M. F.; Harrison, K. J.; McKeown, N. B.; Thomson, A. J. *J. Chem. Soc. Chem. Commun.* **1987**, 1148.
340. McKeown, N. B.; Cook, M. J.; Thomson, A. J.; Harrison, K. J.; Daniel, M. F.; Richardson, R. M.; Roser, S. J. *Thin Solid Films* **1988**, *159*, 469.
341. Cook, M. J.; McKeown, N. B.; Thomson, A. J. *Chem. Mat.* **1989**, *1*, 287.
342. Alexander, A. E. *J. Chem. Soc.* **1937**, 1813.
343. Bellamy, W. D.; Gaines, G. L.; Tweet, A. G. *J. Chem. Phys.* **1963**, *39*, 2528.
344. Tweet, A. G.; Bellamy, W. D.; Gaines, G. L. *J. Chem. Phys.* **1964**, *41*, 2068.
345. Hopf, F. R.; Möbius, D.; Whitten, D. G. *J. Am. Chem. Soc.* **1976**, *98*, 1585.
346. Horsey, B. E.; Whitten, D. J. *J. Am. Chem. Soc.* **1978**, *100*, 1293.
347. Mercer-Smith, J. A.; Whitten, D. G. *J. Am. Chem. Soc.* **1979**, *101*, 6620.
348. Hopf, F. R.; Whitten, D. G. *J. Am. Chem. Soc.* **1976**, *98*, 7422.
349. Whitten, D. G.; Eaker, D. W.; Horsey, B. E.; Schmehl, R. H.; Worsham, P. R. *Ber. Bunsenges Phys. Chem.* **1978**, *92*, 858.
350. Schmehl, R. H.; Shaw, G. L.; Whitten, D. G. *Chem. Phys. Lett.* **1978**, *58*, 549.
351. Vandevyver, M.; Barraud, A.; Ruaudel-Teixier, A.; Millard, P.; Gianotti, C. *J. Coll. Interf. Sci.* **1982**, *85*, 571.
352. Jones, R.; Tredgold, R. H.; Hodge, P. *Thin Solid Films* **1983**, *99*, 25.
353. Bergerson, J. A.; Gaines, G. L., Jr.; Bellamy, W. D. *J. Colloid Interface Sci.* **1967**, *25*, 97.
354. Ruaudel-Teixier, A.; Barraud, A.; Belbeoch, B.; Roulliay, M. *Thin Solid Films* **1983**, *99*, 33.
355. Lesieur, P.; Vandevyver, M.; Ruaudel-Teixier, A.; Barraud, A. *Thin Solid Films* **1988**, *159*, 315.
356. Bardwell, J.; Bolton, J. R. *Photochem. Photobiol.* **1984**, *39*, 735.
357. Tredgold, R. H.; Vickers, A. J.; Hoorfar, A.; Hodge, P.; Khoshdel, E. *J. Phys. D: Appl. Phys.* **1985**, *18*, 1139.
358. McArdle, C. B.; Ruaudel-Teixier, A. *Thin Solid Films* **1985**, *133*, 93.
359. Callot, M.; *Bull. Soc. Chim. Fr.* **1974**, 7-8, 1492.
360. Miller, A.; Knoll, W.; Möhwald, H.; Ruaudel-Teixier, A. *Thin Solid Films* **1985**, *133*, 83.
361. Möhwald, H.; Miller, A.; Stich, W.; Knoll, W.; Ruaudel-Teixier, A.; Lehmann, T.; Fuhrhop, J. -H. *Thin Solid Films* **1986**, *141*, 261.

362. Ringuet, M.; Gagnon, J.; Leblanc, R. M. *Langmuir* **1986**, *2*, 70.
363. Tredgold, R. H.; Evans, S. D.; Hodge, P.; Hoorfar, A. *Thin Solid Films* **1988**, *160*, 99.
364. Luk, S. Y.; Williams, J. O. *J. Chem. Soc. Chem. Commun.* **1989**, 158.
365. Luk, S. Y.; Mayers, F. R.; Williams, J. O. *J. Chem. Soc. Chem. Commun.* **1987**, 215.
366. Schick, G. A.; Schreiman, I. C.; Wagner, R. W.; Lindsey, J. S.; Bocian, D. F. *J. Am. Chem. Soc.* **1989**, *111*, 1344.
367. Baker, S.; Petty, M. C.; Roberts, G. G.; Twigg, M. V. *Thin Solid Films* **1983**, *99*, 53.
368. Kovacs, G. J.; Vincett, P. S.; Sharp, J. H. *Can. J. Phys.* **1985**, *63*, 346.
369. Hann, R. A.; Gupta, S. K.; Fryer, J. R.; Eyres, B. L. *Thin Solid Films* **1985**, *134*, 35.
370. Aroca, R.; Jennings, C.; Kovacs, G. J.; Loutfy, R. O.; Vincett, P. S. *J. Phys. Chem.* **1985**, *89*, 4051.
371. Kovacs, G. J.; Loutfy, R. O.; Vincett, P. S.; Jenning, C.; Aroca, R. *Langmuir* **1986**, *2*, 689.
372. Hua, Y. L.; Roberts, G. G.; Ahmad, M. M.; Petty, M. C.; Hanack, M.; Rein, M. *Philos. Mag. B* **1986**, *53*, 105.
373. Fujiki, M.; Tabei, H. *Langmuir* **1988**, *4*, 320.
374. Roberts, G. G.; Petty, M. C.; Baker, S.; Fowler, M. T.; Thomas, N. J. *Thin Solid Films* **1985**, *132*, 113.
375. Hirano, M.; Nemori, R.; Nakao, Y.; Yamada, H. *Mikrochim Acta* **1988**, *11*, 43.
376. Aroca, R.; Kovacs, G. J.; Jenning, C. A.; Loutfy, R. O.; Vincett, P. S. *Langmuir* **1988**, *4*, 518.
377. Cook, M. J.; Dunn, A. J.; Gold, A. A.; Thomson, A. J.; Daniel, M. F. *J. Chem. Soc. Dalton Trans.* **1988**, 1583.
378. Yoneyama, M.; Sugi, M.; Saito, M.; Ikegami, K.; Kuroda, S.; Iizima, S. *Jpn. J. Appl. Phys.* **1988**, *25*, 961.
379. Oyanagi, H.; Yoneyama, M.; Matsushita, T. *Thin Solid Films* **1988**, *159*, 435.
380. Kalina, D. W.; Crane, S. W. *Thin Solid Films* **1985**, *134*, 109.
381. Van der Pol, J. F.; Neeleman, E.; Zwikker, J. W.; Nolte, R. J. M.; Drenth, W.; Aerts, J.; Visser, R.; Picken, S. J. *Liq. Cryst.* **1989**, *6*, 577, and references therein.
382. Cook, M. J.; Dunn, A. J.; Daniel, M. F.; Hart, R. C. O.; Richardson, R. M.; Roser, S. J. *Thin Solid Films* **1988**, *159*, 395.
383. Ogawa, K.; Kinoshita, S.; Yonehara, H.; Nakahara, H., Fukuda, K. *J. Chem. Soc. Chem. Commun.* **1989**, 477.
384. Tabei, H.; Fujiki, M.; Imamura, S. *Jpn. J. Appl. Phys. Part I* **1985**, *24*, L658.
385. Fujiki, M. Tabei, H.; Imamura S. *Jpn. J. Appl. Phys. Part I* **1987**, *26*, 1224.
386. Fujiki, M.; Tabei, H.; Kurihara, T. *J. Phys. Chem.* **1988**, *92*, 1281.
387. Troitskiy, V. I.; Bannikov, V. S.; Berzina, T. S. *J. Mol. Electron.* **1989**, *5*, 147.
388. Snow, A. W.; Jarvis, N. L. *J. Am. Chem. Soc.* **1984**, *106*, 4706.
389. Barger, W. R.; Snow, A. W.; Wohltjen, H.; Jarvis, N. L. *Thin Solid Films* **1985**, *133*, 197.
390. Dilella, D. P.; Barger, W. R.; Snow, A. W.; Smardzewski, R. R. *Thin Solid Films* **1985**, *133*, 207.

391. Snow, A. W.; Barger, W. R.; Klusky, M.; Wohltjen, H.; Jarvis, N. L. *Langmuir* **1986**, *2*, 513.
392. Pace, M. D.; Barger, W. R.; Snow, A. W. *J. Mag. Res.* **1987**, *75*, 73.
393. Barger, W.; Dote, J.; Klusty, M.; Mowery, R.; Price, R.; Snow, A. W. *Thin Solid Films* **1985**, *159*, 369.
394. Palacin, S.; Lesieur, P.; Stefanelli, I.; Barraud, A. *Thin Solid Films* **1988**, *159*, 83.
395. Palacin, S.; Barraud, A. *J. Chem. Soc. Chem. Commun.* **1989**, 45.
396. Shutt, J. D.; Batzel, D. A.; Sudiwala, R. V.; Rickert, S. E.; Kenney, M. E. *Langmuir* **1988**, *4*, 1240.
397. Ko, W. H.; Fu, C. W.; Wang, H. Y.; Batzel, D. A.; Kenney, M. E.; Lando, J. B. *Sensors and Materials* **1990**, *2*, 39.
398. Wang, H. Y.; Ko, W. H.; Batzel, D. A.; Kenney, M. E.; Lando, J. B. *Sensors and Actuators* **1990**, *B1*, 138.
399. Gee, G.; Rideal, E. K. *Proc. Roy. Soc. London, Ser. A* **1935**, *153*, 116.
400. Gee, G. *Proc. Roy. Soc. London, Ser. A* **1935**, *153*, 129.
401. Bresler, S.; Judin, M.; Talmud, D. *Acta Physicocchem. URSS* **1941**, *14*, 71.
402. Scheibe, G.; Schuller, H. *Z. Electrochem.* **1955**, *59*, 861.
403. Friedlander, H. *Z. Chem Abstr.* **1962**, *57*, 14008.
404. Bücher, H.; Drexhage, K. H.; Fleck, M.; Kuhn, H.; Möbius, D.; Schafer, F. P.; Sondermann, J.; Sperling, W.; Tillmann, P.; Wiegand, J. *Mol. Cryst.* **1967**, *2*, 197.
405. Hill, T. L. *J. Polym. Sci.* **1961**, *54*, 58.
406. Blumstein, A.; Ries, H. E., Jr. *Polym. Lett.* **1965**, *3*, 927.
407. Ackermann, R.; Inacker, O.; Ringsdorf, H. *Kolloid-Z. Z. Polym.* **1971**, *249*, 1118.
408. Cemel, A.; Fort, T., Jr.; Lando, J. B. *J. Polym. Sci. A-1* **1972**, *10*, 2061.
409. Puterman, M.; Fort, T., Jr.; Lando, J. B. *J. Colloid Interface Sci.* **1974**, *47*, 705.
410. Ackerman, R.; Naegele, D.; Ringsdorf, H. *Makromol. Chem.* **1974**, *175*, 699.
411. Dubault, A.; Veysiié, M.; Liebert, L.; Strzelecki, L. *Nature* **1973**, *245*, 94.
412. Dubault, A.; Casagrande, C.; Veyssié, M. *J. Phys. Chem.* **1975**, *79*, 2254.
413. Lando, J. B.; Fort, T., Jr., in *Polymerization of Organized Systems* (Midland Macromolecular Monograph, Vol. 3), Elias, H. G., Ed.; Gordon and Breach: New York, 1977.
414. Naegele, D.; Lando, J. B.; Ringsdorf, H. *Macromolecules* **1977**, *10*, 1339.
415. Barraud, A.; Rosilio, C.; Ruaudel-Teixier, A. *Polym. Prepe., Am. Chem. Soc., Div. Polym. Chem.* **1978**, *19*, 176.
416. Banerjie, A.; Lando, J. B. *Thin Solid Films* **1980**, *68*, 67.
417. Fariss, G.; Lando, J. B.; Rickert, S. *Thin Solid Films* **1983**, *99*, 305.
418. Rabe, J. P.; Rabolt, J. F.; Brown, C. A.; Swalen, J. D. *Thin Solid Films* **1985**, *133*, 153.
419. Berkovic, G.; Rasing, Th.; Shen, Y. R. *J. Chem. Phys.* **1986**, *85*, 7374.
420. Miyashita, T.; Yoshida, H.; Murakata, T.; Matsuda, M. *Polymer* **1987**, *28*, 311.
421. Uchida, M.; Tanizaki, T.; Kunitake, T.; Kajiyama, T. *Macromolecules* **1989**, *22*, 2381.
422. Barraud, A.; Rosilio, C.; Ruaudel-Teixier, A. *J. Colloid Interface Sci.* **1977**, *62*, 509.
423. Barraud, A.; Rosilio, A. *Solid State Technol.* **1979**, *22*, 120.
424. Barraud, A. *Thin Solid Films* **1983**, *99*, 317.

425. Barraud, A. *Mol. Cryst. Liq. Cryst.* **1983**, *96*, 353.
426. Ogawa, K.; Tamura, H.; Hatada, M.; Ishihara, T. *Langmuir* **1988**, *4*, 1229.
427. Ogawa, K.; Tamura, H.; Hatada, M.; Ishihara, T. *Coll. Polym. Sci.* **1988**, *266*, 525.
428. Tanaka, Y.; Nakatama, K.; Iijima, S.; Shimizu, T.; Maitani, Y. *Thin Solid Films* **1985**, *133*, 165.
429. Schmidt, G. M. J. *Pure Appl. Chem.* **1971**, *27*, 647.
430. Addadi, L.; Mil, J.; Lahav, M. *J. Am. Chem. Soc.* **1982**, *104*, 3422.
431. Koch, H.; Laschewsky, A.; Ringsdorf, H.; Teng, K. *Makromol. Chem.* **1986**, *187*, 1843.
432. Hirota, S.; Itoh, U.; Sugi, M. *Thin Solid Films* **1985**, *134*, 67.
433. Laschewsy, A.; Ringsdorf, H. *Macromolecules* **1988**, *21*, 1936.
434. Fukuda, K.; Shibasaki, Y.; Nakahara, H. *J. Macromol. Sci. Chem.* **1981**, *A15*, 999.
435. Winful, H. G.; Harburger, J. H.; Garmire, E. *Appl. Phys. Lett.* **1979**, *35*, 379.
436. Seaton, C. T.; Mau, X.; Stegeman, G. I.; Winful, H. G. *Opt. Eng.* **1985**, *24*, 593.
437. Sarid, D. *Opt. Lett.* **1981**, *6*, 552.
438. Lattes, A.; Haus, H. A.; Leonberger, F. J.; Ipen, E. P. *IEEE J. Quant. Elect.* **1983**, *QE-19*, 1718.
439. Seymour, R. J.; Carter, G. M.; Chen, Y. J.; Elman, B. S.; Jagannath, C. J.; Rubner, M. F.; Sandman, D. J.; Thakur, M. K.; Tripathy, S. K. *Proc. SPIE* **1985**, *578*, 137.
440. Rustagi, K. C.; Ducuing, J. *Opt. Commun.* **1974**, *10*, 258.
441. Cojan, C.; Agrawal, G. P.; Flytzanis, C. *Phys. Rev. B* **1977**, *15*, 909.
442. Agrawal, G. P.; Cojan, C.; Flytzanis, C. *Phys. Rev. B* **1978**, *17*, 776.
443. Wegner, G. *Z. Naturfosch., Teil B* **1969**, *24*, 824.
444. Wegner, G. *Pure Appl. Chem.* **1977**, *49*, 443.
445. Baughman, R. H. *J. Polym. Sci. Polym. Phys. Ed.* **1973**, *11*, 603.
446. Bader, H.; Dorn, K.; Hupfer, B.; Ringsdorf, H. *Adv. Polym. Sci.* **1984**, *64*, 1, and references therein.
447. Ringsdorf, H.; Schlarb, B.; Venzmer, J. *Angew. Chem. Int. Ed. Engl.* **1988**, *27*, 113, and references therein.
448. Tieke, B.; Wegner, G.; Naegele, D.; Ringsdorf, H. *Angew. Chem. Int. Ed. Engl.* **1976**, *15*, 764.
449. Tieke, B.; Wegner, G., in *Topics in Surface Chemistry*, Kay, E.; Bagus, P. S., Eds.; Plenum Press: New York, 1978, p 121.
450. Day, D.; Ringsdorf. H. *J. Polym. Sci., Polym. Lett. Ed.* **1978**, *16*, 205.
451. Day, D.; Ringsdorf. H. *Makromol. Chem.* **1979**, *180*, 1059.
452. Day, D.; Lando, J. B.; Ringsdorf, H. *Polym. Prepr., Am. Chem. Soc., Div. Polym. Chem.* **1978**, *19*, 176.
453. Lochner, K.; Bässler, H.; Tieke, B.; Wegner, G. *Phys. Stat. Solidi B* **1978**, *82*, 633.
454. Enkelmenn, V.; Tieke, B.; Kapp, H.; Lieser, G.; Wegner, G. *Ber. Bunsenges. Phys. Chem.* **1978**, *82*, 876.
455. Bloor, D.; Preston, F. H.; Ando, D. J.; Batchelder, D. N., in *Structural Studies of Macromolecules by Spectroscopic Methods*, Ivin, K. J., Ed.; Wiley: New York, 1976, p. 91.
456. Lewis, W. F.; Batchelder, D. N. *Chem Phys. Lett.* **1979**, *60*, 232.
457. Tieke, B.; Bloor, D. *Makromol. Chem.* **1979**, *180*, 2275.
458. Day, D.; Lando, J. B. *Macromolecules* **1980**, *13*, 1478.

459. Lieser, G.; Tieke, B.; Wegner, G. *Thin Solid Films* **1980**, *68*, 77.
460. Tieke, B.; Lieser, G. *J. Colloid Interface Sci.* **1982**, *88*, 471.
461. Bubeck, C.; Weiss, K.; Tieke, B. *Thin Solid Films* **1983**, *99*, 103.
462. Kajzar, F.; Messier, J. *Thin Solid Films* **1983**, *99*, 109.
463. Tieke, B.; Weiss, K. *J. Colloid Interface Sci.* **1984**, *101*, 129.
464. Yoshioka, Y.; Nakahara, H.; Fukuda, K. *Thin Solid Films* **1985**, *133*, 11.
465. Chen, Y. J.; Carter, G. M.; Tripathy, S. K. *Solid State Commun.* **1985**, *54*, 19.
466. Chen, Y. J.; Tripathy, S. K.; Carter, G. M.; Elman, B. S.; Koteles, E. S.; George, J., Jr. *Solid State Commun.* **1986**, *58*, 97.
467. Tieke, B.; Weiss, K. *Coll. Polym. Sci.* **1985**, *263*, 576.
468. Laxhuber, L. A.; Scheunemann, U.; Möhwald, H. *Chem. Phys. Lett.* **1986**, *124*, 561.
469. Burzynski, R.; Prasad, P. N.; Biegajski, J.; Cadenhead, D. A. *Macromolecules* **1986**, *19*, 1059.
470. Shutt, J. D.; Rickert, S. E. *Langmuir* **1987**, *3*, 460.
471. Dipoto, E. P. MS Thesis, Case Western Reserve University, Cleveland, Ohio, 1986.
472. Nakahara, H.; Fukuda, K.; Seki, K.; Asada, S.; Inokuchi, H. *Chem. Phys.* **1987**, *118*, 123.
473. Kaneko, F.; Dresselhaus, M. S.; Rubner, M. F.; Shibata, M.; Kobayashi, S. *Thin Solid Films* **1988**, *160*, 327.
474. Göbel, H. D.; Möhwald, H. *Thin Solid Films* **1988**, *159*, 63.
475. Sotnikov, P. S.; Bannikov, V. S.; Troitskiy, V. I.; Iichenko, A. Ya. *J. Mol. Electron.* **1989**, *5*, 155.
476. Ozaki, M.; Ikeda, Y.; Nagoya, I. *Synth. Met.* **1989**, *28*, C801.
477. Delaney, P. A.; Johnstone, R. A. W.; Eyres, B. L.; Hann, R. A.; McGrath, I.; Ledwith, A. *Thin Solid Films* **1985**, *123*, 353.
478. Boothroyd, B.; Delaney, P. A.; Hann, R. A.; Johnstone, R. A. W.; Ledwith, A. *Br. Polym. J.* **1985**, *17*, 360.
479. Crisp, D. J. in *Surface Phenomena in Chemistry and Biology* , Danielli, J. F.; Pankhurst, K. G. A.; Riddiford, Eds.; Pergamon: New York, 1958, p 23.
480. Crisp, D. J. *J. Colloid Sci.* **1946**, *1*, 161.
481. Goddard, E. D., Ed.;*Monolayers* (Adv. in Chem. Ser. 144), American Chemical Society: Washington, D. C., 1975, Ch. 10.
482. Puggelli, M.; Gabrielli, G. *J. Colloid Interface Sci.* **1977**, *61*, 420.
483. Gabrielli, G.; Guarini, G. G. T. *J. Colloid Interface Sci.* **1978**, *64*, 185.
484. Gabrielli, G.; Maddii, A. *J. Colloid Interface Sci.* **1978**, *64*, 19.
485. Mumby, S. J.; Rabolt, J. F.; Swalen, J. D. *Thin Solid Films* **1985**, *133*, 161.
486. Mumby, S. J.; Rabolt, J. F.; Swalen, J. D. *Macromolecules* **1986**, *19*, 1054.
487. Munn, R. W. *Chem Br.* **1984**, *27*, 518.
488. Nutting G. C.; Harkins, W. D. *J. Am. Chem. Soc.* **1939**, *61*, 1180.
489. Duda, G.; Schouten, J. A.; Arndt, T.; Lieser, G.; Schmid, G. F.; Bubeck, C.; Wegner, G. *Thin Solid Films* **1988**, *159*, 221.
490. Naito, K. *J. Colloid Interface Sci.* **1989**, *131*, 218.
491. Tredgold, R. H.; Winter, C. S. *J. Phys. D: Appl. Phys.* **1982**, *15*, L55.
492. Winter, C. S.; Tredgold, R. H. *IEE Proc.* **1983**, *130*, 256.
493. Winter, C. S.; Tredgold, R. H.; Vickers, A. J.; Khoshdel, E.; Hodge, P. *Thin Solid Films* **1985**, *134*, 49.

494. Hodge, P.; Khoshdel, E.; Tredgold, R. H.; Vickers, A. J.; Winter, C. S. *Br. Polym. J.* **1985**, *17*, 368.
495. Tredgold, R. H. *Thin Solid Films* **1987**, *152*, 223.
496. Tredgold, R. H.; Young, M. C. J.; Hodge, P.; Khoshdel, E. *Thin Solid Films* **1987**, *151*, 441.
497. Jones, R.; Winter, C. S.; Tredgold, R. H.; Hodge, P.; Hoorfar, A. *Polymer* **1987**, *28*, 1619.
498. Tredgold, R. H.; Allen, R. A.; Hodge, P.; Khoshdel, E. *J. Phys. D: Appl. Phys.* **1987**, *20*, 1385.
499. Tredgold, R. H. *J. de Chim. Phys.* **1988**, *85* 1079.
500. Kawaguchi, T.; Nakahara, H.; Fukuda, K. *Thin Solid Films* **1985**, *133*, 29.
501. Lovelock, B. J.; Grieser, F.; Sanders, J. V. *J. Colloid Interface Sci.* **1985**, *108*, 297.
502. Ringsdorf, H.; Schmidt, G.; Schneider, J. *Thin Solid Films* **1987**, *152*, 207.
503. Elbert, R.; Laschewsky, A.; Ringsdorf, H. *J. Am. Chem. Soc.* **1985**, *107*, 4134.
504. Biddle, M. B.; Lando, J. B.; Ringsdorf, H.; Schmidt, G.; Schneider, J.*Colloid Polym. Sci.* **1988**, *266*, 806.
505. Schneider, J.; Ringsdorf, H.; Rabolt, J. F. *Macromolecules* **1989**, *22*, 205.
506. Fox, H. W.; Taylor, P.; Zisman, W. A. *Ind. Eng. Chem.* **1947**, *39*, 1401.
507. Fox, H. W.; Solomon, E. M.; Zisman, W. A. *J. Phys. Colloid Chem.* **1950**, *54*, 723.
508. Newing, M. J. *Trans. Faraday Soc.***1950**, *46*, 755.
509. Noll, W.; Steinbach, H.; Sucker, C. *Ber Bunsenges Phys. Chem.* **1963**, *67*, 407.
510. Noll, W.; Steinbach, H.; Sucker, C. *Kolloid Z. Z. Polym.* **1965**, *204*, 94.
511. Garrett, W. D.; Zisman, W. A. *J. Phys. Chem.* **1970**, *74*, 1796.
512. Jarvis, N. L. *J. Phys. Chem.* **1966**, *70*, 2037.
513. Nakihara, Y.; Himmelblau, D. M.; Schechter, R. S. *J. Colloid Interface Sci.* **1969**, *30*, 200.
514. Noll. W.; Steinbach, H.; Sucker, C. *J. Polym. Sci. C* **1971**, *34*, 123.
515. Willis, R. F. *J. Colloid Interface Sci.* **1971**, *35*, 1.
516. Bornett, M. K.; Zisman, W. A. *Macromolecules* **1971**, *4*, 47.
517. Watanabe, M.; Kosaka, Y.; Sanui, K.; Ogata, N.; Oguchi, K.; Yoden, T. *Macromolecules* **1987**, *20*, 452.
518. Watanabe, M.; Kosaka, Y.; Oguchi, K.; Sanui, K.; Ogata, N. *Macromolecules* **1988**, *21*, 2997.
519. Oguchi, K.; Yoden, T.; Kosaka, Y.; Watanabe, M.; Sanui, K.; Ogata, N. *Thin Solid Films* **1988**, *161*, 305.
520. Oguch, K.; Yoden, T.; Sanui, K.; Ogata, N. *Polym. J.* **1986**, *18*, 887.
521. Blair, H. S.; McArdle, C. B. *Polymer* **1984**, *25*, 1347.
522. Duda, G.; Schouten, A. J.; Arndt, T.; Lieser, G.; Schmidt, G. F.; Bubeck, C.; Wegner, G. *Thin Solid Films* **1988**, *159*, 221.
523. Kowel, S. T.; Mayden, L. M.; Selfridge, R. H. *SPIE* **1986**, *682*, 103.
524. Kowel, S. T.; Ye, L.; Zhang, Y.; Hayden, L. M. *Opt. Eng.* **1987**, *26*, 107.
525. Stroeve, P.; Srinivasan, M. P.; Higgins, B. G.; Kowel, S. T. *Thin Solid Films* **1987**, *146*, 209.
526. Hayden, L. M.; Anderson, B. L.; Lam, J. Y. S.; Higgins, B. G.; Stroeve, P.; Kowel, S. T. *Thin Solid Films* **1988**, *160*, 379.
527. Anderson, B. L.; Hall, R. C.; Higgins, B. G.; Lindsay, G.; Stroeve, P.; Kowel, S. T. *Synth. Met.* **1989**, *28*, D683.

528. Carr, N.; Goodwin, M. J. *Makromol. Chem., Rapid Commun.* **1987**, *8*, 487.
529. Seki, T.; Ichimura, K. *Polym. Commun.* **1989**, *30*, 109.
530. Suzuki, M.; Kakimoto, M.; Konishi, T.; Imai, Y.; Iwamoto, M.; Hino, T. *Chem. Lett.* **1986**, 395.
531. Kakimoto, M.; Suzuki, M.; Imai, Y.; Iwamato, M.; Hino, T. *ACS Symp. Ser.* **1987**, *346*, 484.
532. Kakimoto, M.; Morikawa, A.; Nishikata, Y.; Suzuki, M.; Imai, Y. *J. Coll. Interf. Sci.* **1988**, *121*, 599.
533. Nishikata, Y.; Kakimoto, M.; Morikawa, A.; Imai, Y. *Thin Solid Films* **1988**, *160*, 15.
534. Angel, A. K.; Yoden, T.; Sanui, K.; Ogata, N. *Polym. Mater. Sci. Eng.* **1986**, *54*, 119.
535. Angel, A. K.; Yoden, T.; Sanui, K.; Ogata, N. *J. Am. Chem. Soc.* **1985**, *107*, 8308.
536. Angel, A. K.; Yoden, T.; Sanui, K.; Ogata, N. *Proc. Am. Chem. Soc. Div. Polym. Mater. Sci. Eng.* **1986**, *54*, 119.
537. Flower, M. T.; Suzuki, M.; Angel, A. K.; Asano, K.; Itoh, T. *J. Appl. Phys.* **1987**, *62*, 3427.
538. Uekita, M.; Awaji, H.; Murata, M. *Thin Solid Films* **1988**, *160*, 21.
539. Singer, S. J.; Nicolson, G. L. *Science* **1972**, *175*, 720.
540. Colacicco, G.; Rapport, M. M. *Adv. Chem. Ser.* **1968**, *84*, 157.
541. Colacicco, G. *J. Colloid Interface Sci.* **1969**, *29*, 345.
542. Colacicco, G. *Lipids* **1970**, *5*, 636.
543. Colacicco, G. *Ann NY Acad. Sci.* **1972**, *195*, 224.
544. Colacicco, G.; Buckelew, A. R., Jr.; Scarpello, E. M. *J. Colloid Interface Sci.* **1974**, *46*, 147.
545. Lösche, M.; Helm, C.; Mattes, H. D.; Möhwald, H. *Thin Solid Films* **1985**, *133*, 51.
546. Okamura, E.; Umemura, J.; Takenaka, T. *Biochim. Biophys. Acta* **1985**, *812*, 139.
547. Lotta, T. I.; Laakkonen, L. J.; Virtanen, J. A.; Kinnunen, P. K. J. *Chem. Phys. Lipids* **1988**, *46*, 1.
548. Kopp, F.; Fringeli, U. P.; Mühlethaler, K.; Günthard, Hs. H. *Z. Naturforsch.* **1975**, *30c*, 711.
549. Vogel, V.; Möbius, D. *Thin Solid Films* **1988**, *159*, 73.
550. Dluhy, R. A.; Wright, N. A. ; Griffiths, P. R. *Appl. Spectr.* **1988**, *42*, 138.
551. Heckl, W. M.; Miller, A.; Möhwald, H. *Thin Solid Films* **1988**, *159*, 125.
552. Möhwald, H. *Thin Solid Films* **1988**, *159*, 1.
553. Möhwald, H.; Kristin, S.; Haas, H.; Flörsheimer, M. *J. de Chim. Phys.* **1988**, *85*, 1009.
554. Hörber, J. K. H.; Lang, C. A.; Hänsch, T. W.; Heckl, W. M.; Möhwald, H. *Chem. Phys. Lett.* **1988**, *145*, 151.
555. Möhwald, H. *J. Mol. Electron.* **1988**, *4*, 47.
556. Heckl, W. M.; Thompson, M.; Möhwald, H. *Langmuir* **1989**, *5*, 390.
557. Hickel, W.; Kamp, D.; Knoll, W. *Nature* **1989**, *339*, 186.
558. Lösche, M.; Sackmann, E.; Möhwald, H. *Ber. Bunsenges. Phys. Chem.* **1983**, *87*, 848.
559. Weiss, R. M.; McConnell, H. M. *Nature* **1984**, *310*, 5972.
560. Peters, R.; Beck, K. *Proc. Natl. Acad. Sci. USA* **1983**, *80*, 7183.
561. Moore, B.; Knohler, C. M.; Broseta, D.; Rondelez, F. *J. Chem. Soc., Faraday Trans.* **1986**, *2*, 82.

562. Cadenhead, D. A.; Müller-Landau, F.; Kellner, B. M. J. in *Ordering in Two Dimensions*, Sinha, S. K., Ed., Elsevier: New York, 1980.
563. Albrecht, O.; Gruler, H.; Sackman, E. *J. Phys. (Paris)* **1978**, *39*, 301.
564. Andelman, D.; Brochard, F.; Joanny, J. F. *J. Chem. Phys.* **1987**, *86*, 3673.
565. Keller, D. J.; McConnell, H. M.; Moy, V. T. *J. Phys. Chem.* **1986**, *90*, 2311.
566. McConnell, H. M.; Keller, D. J. *Proc. Natl. Acad. Sci. USA* **1987**, *84*, 4706.
567. Fischer, A.; Lösche, M.; Möhwald, H.; Sackmann, E. *J. Phys. Lett.* **1984**, *45*, L-785.
568. Richardson, R. M.; Roser, S. J. *Liq. Cryst.* **1987**, *2*, 797.
569. Kuhn, H. *Verh. Schweiz Naturforsch. Ges.* **1965**, 245.
570. Kuhn, H. *J. Photochem.* **1979**, *10*, 111.
571. Kuhn, H.; Mobius, D. *Angew. Chem. Int. Ed. Engl.* **1971**, *10*, 620.
572. Czikely, V.; Försterling, H. D.; Kuhn, H. *Chem. Phys. Lett.* **1970**, *6*, 11, 207.
573. Kuhn, H. *J. Chem. Phys.* **1970**, *53*, 101.
574. Drexhage, K. H. *Prog. Opt.* **1974**, *12*, 163.
575. Nakahara, H.; Fukuda, K.; Möbius, D.; Kuhn, H. *J. Phys. Chem.* **1986**, *90*, 6144.
576. Fromherz, P.; Reinbold, G. *Thin Solid Films* **1988**, *160*, 347.
577. Yonezawa, Y.; Möbius, D.; Kuhn, H. *J. Appl. Phys.* **1987**, *62*, 2016.
578. Ruaudel-Teixier, A.; Vandevyver, M. *Thin Solid Films* **1980**, *68*, 129.
579. Leitner, A.; Lippitsch, M. E.; Draxler, S.; Riegler, M.; Aussenegg, F. R. *Thin Solid Films* **1985**, *132*, 55.
580. Kuhn, H., in *Light-Induced Charge Separation in Biology and Chemistry*, Gerischer, H.; Katz, J. J., Eds.; Dahlem Konferenzen: West Berlin, 1979, pp. 151-169.
581. Gilman, P. B. in *Photographic Sensitivity*, Cox R. J., Ed., Academic Press: London, 1973, p 187.
582. Sturmer, D. M., in *Special Topics in Heterocyclic Chemistry*, Weissberger, A.; Taylor, E. C., Eds.; Wiley: New York, 1977, p 540.
583. Herz, A. H. *Photogr. Sci. Eng.* **1974**, *18*, 323.
584. Jelley, E. E. *Nature* **1936**, *138*, 1009.
585. Scheibe, G. *Z. Angew. Chem.* **1936**, *49*, 563.
586. Pandolfe, W. D.; Bird, G. R. *Photogr. Sci. Eng.* **1974**, *18*, 340.
587. *Organic Molecular Aggregates: Electronic Excitation and Interaction Processes*, Reineker, P.; Haken, H.; Wold, H. C., Eds.; Springer: Berlin, 1983.
588. Kuhn, H.; Möbius, D. *Angew. Chem.* **1971**, *83*, 672.
589. Kuhn, H.; Möbius, D. *Angew. Chem. Int. Ed. Engl.* **1972**, *10*, 620.
590. Kuhn, H. *Isr. J. Chem.* **1979**, *18*, 375.
591. Scheibe, G.; Schontag, A.; Katheder, F. *Naturwissenschaften* **1939**, *29*, 499.
592. Tanaka, J.; Tanaka, M.; Hayakawa, M. *Bull. Chem. Soc. Jpn.* **1980**, *53*,3109.
593. Delany, J.; Morrow, M.; Eckhardt, C. J. *Chem. Phys. Lett.* **1985**, *122*, 347.
594. McRae, E. G.; Kasha, M. J. *Chem. Phys.* **1958**, *28*, 721.
595. Scherer, P. O. J.; Fischer, S. F. *Chem. Phys.* **1984**, *86*, 269.
596. Nakahara, H.; Möbius, D. *J. Colloid Interface Sci.* **1986**, *114*, 363.
597. Penner, T. L. *Thin Solid Films* **1988**, *160*, 241.
597a. Penner, T. L., private communication.
598. Penner, T. L. *J. de Chim. Phys.* **1985**, *85*, 1081.
599. Yamazaki, I.; Tamai, N.; Yamazaki, T.; Murakami, A.; Mimuro, M.; Fujita, Y. *J. Phys. Chem.* **1988**, *92*, 5035.

600. Yonezawa, Y.; Möbius, D.; Kuhn, H. *J. Appl. Phys.* **1987**, *62*, 2022.
601. Kinnunen, P. K. J.; Virtanen, J. A.; Tulkki, A. P.; Ahuja, R. C.; Möbius, D. *Thin Solid Films* **1985**, *132*, 193.
602. Kinnunen, P. K. J.; Tulkki, A. P.; Lemmetynen, H.; Paakkola, J.; Virtanen, J. A. *Chem. Phys. Lett.* **1987**, *136*, 539.
603. Bohorquez, M.; Patterson, L. K. *Thin Solid Films* **1988**, *159*, 133.
604. Bohorquez, M.; Patterson, L. K. *J. Chem. Phys.* **1988**, *92*, 1835.
605. Fujuhura, M.; Nishiyama, K.; Hamaguchi, Y. *J. Chem. Soc. Chem. Commun.* **1986**, 823.
606. Yamazaki, I.; Tamai, N.; Yamazaki, T. *J. Phys. Chem.* **1987**, *91*, 3572.
607. Tamai, N.; Yamazaki, T.; Yamazaki, I. *J. Phys. Chem.* **1987**, *91*, 841.
608. Biesmans, G.; Verbeek, G.; Verschuere, vsn der Auweraer, M.; de Schryver, F. C. *Thin Solid Films* **1989**, *169*, 127.
609. Flörsheimer, M.; Möhwald, H. *Thin Solid Films* **1988**, *159*, 115.

PART THREE

SELF–ASSEMBLED MONOLAYERS

Self-assembled (SA) monolayers are molecular assemblies that are formed *spontaneously* by the immersion of an appropriate substrate into a solution of an active surfactant in an organic solvent [1, 2]. There are several types of SA methods that yield organic monolayers. These include organosilicon on hydroxylated surfaces (SiO_2 on Si, Al_2O_3 on Al, glass, etc.) [3–15]; alkanethiols on gold [16–24]; silver [25]; and copper [26–28]; dialkyl sulfides on gold [29]; dialkyl disulfides on gold [30]; alcohols and amines on platinum [29]; and carboxylic acids on aluminum oxide [231–33] and silver [34].

From the energetics point of view, a self-assembling surfactant molecule can be divided into three parts (Figure 3.1). The first part is the head group that provides the most exothermic process, i.e., chemisorption on the substrate surface. The very strong molecular–substrate interactions result in an apparent pinning of the head group to a specific site on the surface through a chemical bond. This can be a covalent Si–O bond in the case of alkyltrichlorosilanes on hydroxylated surfaces; a covalent, but slightly polar, Au-S bond in the case of alkanethiols on gold; or an ionic $-CO_2^-Ag^+$ bond in the case of carboxylic acids on AgO/Ag. The energies associated with the chemisorption are at the order of tens of kcal/mol (e.g., ~40–45 kcal/mol for thiolate on gold [35, 36]). As a result of the exothermic head group–substrate interactions, molecules try to occupy every available binding site on the surface, and in this process they push together molecules that have already adsorbed.

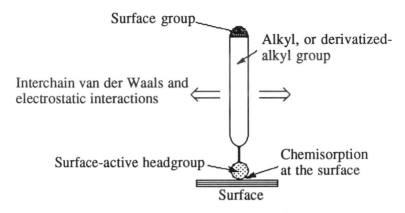

Figure 3.1. A schematic view of the forces in a self-assembled monolayer.

This implies that we assume some surface mobility prior to final pinning, otherwise we cannot explain the formation of crystalline molecular assemblies. The chemisorption exothermicity may be compared, in principle, to the pressure that the barrier in a Langmuir trough applies to the amphiphilic molecules at the water–air interface. It is this spontaneous molecular adsorption that brings molecules close enough together and allows for the short-range, dispersive, London-type, van der Waals forces to become important.

The second molecular part is the alkyl chain, and the energies associated with its interchain van der Waals interactions are at the order of few (<10) kcal/mol (exothermic). It is clear that self-assembly of amphiphilic hydrocarbon molecules cannot be possible given only the interactions among the alkyl chains, and the reader should bear in mind that *the first and most important process is chemisorption*. Only after molecules are put in place on the surface can formation of an ordered and closely packed assembly start. Van der Waals interactions are the main forces in the case of simple alkyl chains (C_nH_{2n+1}). On the other hand, when a polar bulky group is substituted into the alkyl chain (Section 4.2.B), there also are long-range electrostatic interactions that, in some cases, are energetically more important than the van der Waals attractions.

The third molecular part is the terminal functionality, which, in the case of a simple alkyl chain, is a methyl (CH_3) group. These surface groups are thermally disordered at room temperature, as is apparent from helium diffraction [26] and FTIR studies [37] in the case of methyl-terminated monolayers, and from surface reorganization studies in the case of hydroxy-terminated monolayers [38]. The energies associated with this process are at the order of a few kTs, where k is the

Bolznmann constant, and T is the absolute temperature. Usually, a value of ~0.7 kcal/mol is assigned to every *gauche* bond. (*Trans–gauche* isomerization is an endothermic process.)

We note that the terms *order* and *close-packing* used here are relative and depend upon the point of reference. Usually, we use the term *order* to describe molecular assemblies with translational symmetry, i.e., a two-dimensional crystalline-like monolayer. The term *close-packing* usually is associated with the density of crystalline polyethylene. However, a solid-state physicist who compares SA monolayers to inorganic crystals will conclude that these are *highly disordered* systems because of the number of defects. On the other hand, one who compares SA monolayers to polymer glasses or liquid paraffin would suggest that these are *highly ordered* systems. I feel that from the chemistry point of view, and especially when taking into account electron diffraction and FTIR studies, SA monolayers can be considered ordered, closely packed molecular assemblies. This does not mean that SA monolayers are pinhole- and defect-free. We have discussed in Section 2.3.A the structure of LB monolayers at the water–air interface, and the molecular aggregation even at a very low (as low as zero) surface pressure. Similar aggregation of surfactant molecules can occur, in principle, at the solution–substrate interface, prior to or parallel to chemisorption, which would lead to a Swiss-cheese-like structure. We mentioned this possibility in Section 1.1.A in the discussion on ellipsometry. We note, again, that structures containing defects that are larger than very few molecular diameters cannot be stable. Such structures should collapse to either completely disordered, glass-like systems, or to systems where ordered molecular assemblies are embedded in a bulk of a glassy material, depending on chain length. A recent work by Evans *et al.* may provide some support for this suggestion. Their study indicated that OH surfaces (in alkanethiol monolayers on gold) reorganize with kinetics that is strongly dependent on chain length [38]. They suggested that the reorganization process may start from pinholes and grain boundaries, which in my view is where any disorder process in the monolayer starts.

3.1. Monolayers of Fatty Acids

Spontaneous adsorption of long chain *n*-alkanoic acids has been studied in the last few years. Allara and Nuzzo [31, 32] and Ogawa *et al.* [33] studied the adsorption of *n*-alkanoic acids on aluminum oxide, while Schlotter *et al.* studied the spontaneous adsorption of such acids on silver [34]. Huang and Tao studied self-assembled monolayers of long-chain diacetylene amphiphiles [39].

$$CH_3-(CH_2)_n-C{\equiv}C-C{\equiv}C-(CH_2)_m-X$$

I. $n = 13$, $m = 0$, $X = COOH$
II. $n = 13$, $m = 8$, $X = COOH$
III. $n = 8$, $m = 8$, $X = COOH$
IV. $n = 13$, $m = 1$, $X = NH_2$

Note that the rigid, although rod-like, diacetylene π-system breaks the cylindrical symmetry of the all-*trans* alkyl chain in a way similar to that of a phenoxy or a polar aromatic group. Table 3.1 presents the contact angles of hexadecane on various monolayer surfaces. There are two interesting points in these results. The first is that there were hexadecane contact angles of above 46°–47° which have never been reported for a methyl surface, and cannot be correct. (See Part One.) The second, and more important, point is the clear trend of contact angle decrease from I to II to III, which indicates a similarity to what we also found in SA monolayers containing a polar aromatic group [40], i.e., that the position of a π-system in an alkyl chain has an important effect on the overall packing and orientation in the two-dimensional assembly. Note that the highest contact angle was recorded for the shortest molecule in the series, but where $m = 0$, meaning that the diacetylene group was connected to the carboxylic head group. To understand this trend, the reader should consider these diacetylene molecules as composed of three different parts, each contributing to the overall packing and order in the monolayer. (See Part Four.) In solution at room temperature, these alkyl chains are not in an all-*trans* configuration, because of the thermal motion. Therefore, the formation of a solid-like all-*trans* assembly requires rotation motions in the alkyl chains to achieve the *gauche–trans* isomerizations. The alkyl

Table 3.1. Contact angle of hexadecane on various monolayer surfaces. (From Huang and Tao [39], © 1986, Institute of Chemistry, Academia Sinica.)

Compound	Glass		Quartz	Silicon Wafer	
			Solid Substrate		
I	42	(47)	26	31	(45)
II	38	(39)	wet	wet	
III	wet	(22)	wet	wet	
IV	29	(52)	20	27	(50)

Values in parentheses are for surfaces rinsed with dilute NaOH solution (or HCl solution in the case of IV).

chains below the diacetylene are anchored to the surface and to the π-system, and the required rotation motions are thus considerably more restricted.

On the other hand, the alkyl chains facing the solution are anchored only at one side (to the π-system), and can easily squeeze out the *gauche* bonds. As a result, the alkyl chains above the π-systems are more ordered than those below them. The overall packing and order in the assembly are a result of the overall energy, taking into account the van der Waals attractions and the *gauche bonds* (which are assumed to increase energy by ~0.5–0.8 kcal/mol). Now, when there is *no* methylene group below the π-system (I), the total stabilization energy comes from the π–π interaction and the dispersive forces among the C_{13} chains.

The addition of the disordered $(CH_2)_8$ chain below the π-system costs stabilization energy and results in a less solid-like assembly (II). A further reduction of attractive forces by cutting five methylene units from the C_{13} chain results in the most liquid-like assembly (III). This discussion describes important considerations in the engineering of monolayers at the molecular level, and should be studied further in other systems, in a detailed fashion.

The most systematic study on the dynamics of SA adsorption using unpolarized transmission IR or ATR IR, and contact angle measurements, was published recently by Chen and Frank [41].

SA monolayers of fatty acids on glass slides could be prepared on carefully and rigorously cleaned surfaces. (See Section 2.1.B.3.) The authors reported that glass slides used in their studies contains in addition to ~72.6% silica, 14.1% Na_2O, 7.1% CaO, 3.6% MgO, and 1.8% Al_2O_3. These cations promote anchoring of the carboxyl head group, probably via salt formation (as indicated by IR spectra).

Film thicknesses of C_{14}, C_{16}, C_{18}, C_{20}, and C_{22} carboxylic acid monolayers on aluminum oxide substrates showed consistent agreement with the all-*trans* extended molecular length, and are in agreement with those reported by Allara and Nuzzo [31, 32]. Film thickness values for monolayers on glass were not reported. However, wetting data indicates that monolayers on Al_2O_3 are far more stable and ordered than the corresponding samples on glass. Thus, while static water and hexadecane (HD) contact angles for monolayers on Al_2O_3 were 97° (± 2°) and 47° (± 2°), respectively, water contact angles on monolayer samples on glass substrates were 5°–10° smaller, indicating that the examined surfaces are more ordered in the former than in the latter systems. While the qualitative conclusion is correct, the high value of a hexadecane contact angle (47 ± 2°) is suspicious. For this value to be correct, the density of the CH_3 groups in the monolayer surface has to be greater than that in a (111) surface of a $C_{36}H_{74}$ single crystal, which is hard to believe. Another problem with the wetting data is the disagreement between the hydrophilic (H_2O), and hydrophobic (HD) contact angles. Thus, while an HD contact angle of ~47° is a typical value for a highly ordered methyl surface, a water contact angle of 97° is lower than that reported on polyethylene for which the HD contact angle is 0°. (See Section 1.2.A.) We cannot suggest a reasonable explanation for this data.

The difference between the two surfaces is further emphasized by the kinetics results. Figure 3.2 presents the results for stearic acid ($C_{17}H_{35}COOH$) monolayers on both surfaces. There are two important conclusions from the kinetic studies: (*a*) The time required to reach maximum non-wettability (high contact angles) increases with decreasing concentration (similar to adsorption kinetics of alkanethiols on gold; see Figure 3.31); and (*b*) it takes much higher concentration (x 100) to produce good monolayers on glass than on Al_2O_3 (10^{-2} vs. 10^{-4} M). This concentration dependence is in agreement with the expected affinity (i.e., exothermicity of chemisorption) of the COOH groups to the Al_2O_3 and glass surfaces, and to the expected surface concentrations of the salt-forming oxides in the two substrates.

Chen and Frank also developed adsorption kinetics and transient Langmuir kinetic models. They assumed that the plateau adsorption at the highest solution concentration corresponds to full coverage, and calculated fractional coverage from the normalized IR peak intensities. The authors used a configuration in which two substrates with adsorbed films were pressed against the two ATR crystal surfaces, so that the adsorbate was sampled by the internally reflected IR beam. We note that this approach is correct only for atomically smooth surfaces. On the other hand, rough Al_2O_3, glass, and polished Si surfaces are not expected to form good contact on a centimeter-length scale, even at relatively high pressure. Moreover, differences in surface roughness apparently will result in intensity variations.

While neglecting diffusional mass-transfer resistance for adsorption [42], Chen and Frank suggested that adsorption can be thought of as a surface-site filling procedure, with the adsorption and desorption steps counteracting each other. Also, since surface concentration is very small compared to solution concentration, the adsorption step is always rate-controlling [42–44], thus it follows that

$$\frac{d\theta}{dt} = \frac{k_a}{N_o}c\,(1-\theta) - \frac{k_d}{N_o}\theta\,, \qquad\qquad 3.1$$

where θ is the fractional surface coverage, t is the adsorption time, k_a and k_d are the adsorption and desorption rate constants, N_o is the surface adsorbate concentration at full coverage, and c the solution concentration of adsorbate. Integration of Equation 3.1 with the initial condition $\theta = 0$ at $t = 0$ leads to

$$\theta = \frac{k_a c}{k_a c + k_d}\{1 - \exp[-\frac{k_a}{N_0}(c + \frac{k_d}{k_a})t]\} , \qquad 3.2$$

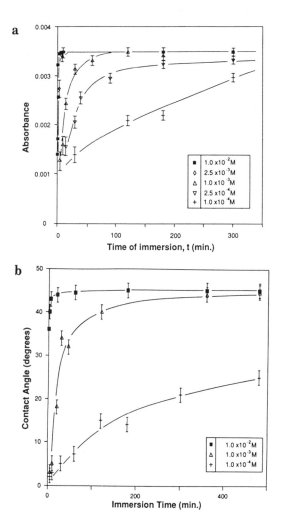

Figure 3.2. Transient adsorption behavior of C_{18} from HD solutions monitored by IR spectroscopy and contact angles: (a) glass slides, 2920 cm^{-1} transmission IR peak intensity; (b) glass slides, HD contact angle; (c) Al$_2$O$_3$ substrates, 2920 cm^{-1} peak intensity; and (d) Al$_2$O$_3$ substrates, H$_2$O and HD contact angles. (From Chen and Frank [41], © 1989, Am. Chem. Soc.)

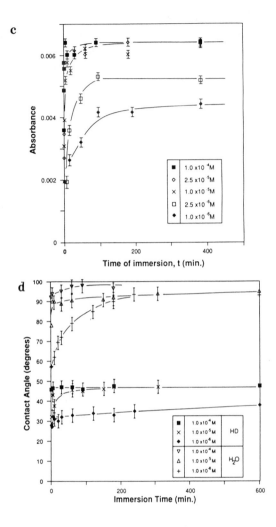

Figure 3.2. *Continued.*

which is reduced to the Langmuir adsorption isotherm at equilibrium (as $t \to \infty$),

$$\theta_{eq} = \frac{k_a c}{k_a c + k_d} = \frac{c}{c + \kappa} , \qquad 3.3$$

where

$$\kappa \equiv k_d/k_a \propto \exp(\Delta G_a^0/RT) , \qquad 3.4$$

244

$$\kappa \equiv k_d/k_a \propto \exp(\Delta G_a^0/RT) \,,$$

and ΔG_a^0 is the free energy of adsorption at infinite dilution. From the kinetics data fitting, the free energy of adsorption $\Delta G_a^0 = -7.3 \pm 0.1$ kcal/mol was calculated for the C_{18}/glass, and $\Delta G_a^0 = -9.2 \pm 0.1$ kcal/mol for the C_{18}/Al$_2$O$_3$ samples. These values are in complete agreement with monolayer order and stability data (as monolayers on glass were lifted off by water), which apparently is the result of differences in the monolayer–substrate chemistry.

One of the first accounts on the structure of SA monolayers of fatty acids is that of Chapman and Tabor [45]. They studied the electron diffraction of a series of fatty acids ($C_nH_{2n+1}COOH$, where $n = 1$–45), octadecanol ($C_{18}H_{37}OH$), octadecylamine ($C_{18}H_{37}NH_2$), and perfluorodecanoic acid ($C_9F_{19}COOH$), deposited by retraction from solution or from the melt onto freshly evaporated metal surfaces. It is quite interesting to read this paper and to realize how much structural information could be extracted from these studies. Since there is no study more current on the structure of corresponding SA monolayers, we provide some of their conclusions here. First, they concluded that in the fatty acid monolayers, the area per molecule (on silver and palladium) is 21.3 Å, quite different from that in the bulk fatty acids and paraffin crystals (18.8 Å) [46, 47], and from the area per molecule in monolayers at the water–air interface (20.4 Å) [48]. Second, they concluded that the alkyl chains are perpendicular to the surface, which, as we already have discussed in Part Two, cannot be correct when such free volume exists. The molecular spacing in monolayers of octadecylamine was 4.4 Å, and in those of octadecanol, 4.3 Å, indicating a much closer packing, probably due to the smaller size of the head group. In the fluorocarbon, the spacing was 5.0 ± 0.1 Å, quite close to the 4.85 Å calculated for a hexagonal close-packing from the known dimensions of fluorocarbon chains [49], indicating a probable vertical geometry of the chains.

3.2. Monolayers of Organosilicon Derivatives

A. Self-Assembled Monolayers and Multilayers: Methodology

The reaction of alkylsilane derivatives — RSiX$_3$, R$_2$SiX$_2$ or R$_3$SiX — where X is chloride or alkoxy, and R is a carbon chain that can bear different functionalities (e.g., amine or pyridyl) with hydroxylated surfaces (e.g., SiO$_2$, SnO$_2$, TiO$_2$), is not new. The electrochemical reduction of methylpyridinum immobilized via a Si–O bond was observed by Moses and Murray in 1976 [50]. Extensive literature also exists on reactions of R$_2$SiX$_2$ and R$_3$SiX organosilanes with hydroxylated surfaces, primarily silica-based, for bonded phases for liquid and gas chromatography, and in enzyme immobilization. In most of these early works, surface polymerization was invoked deliberately by the addition of moisture. This polymerization, mostly in the bulk, probably also was at the film–substrate interface, as later suggested for OTS monolayers [3]. The reader may find accounts of these surface modifications in reviews [51–53], as well as in original papers [54–60]. Weetall *et al.* were the pioneers in enzyme immobilization on controlled pore glass, conducting silanization using $RSiX_3$ reagents in organic solvents [61, 62] and aqueous media [63]. Another important application was surface silanization with $(CH_3)_3SiCl$ or $(CH_3)_2SiCl_2$ for the preparation of hydrophobic substrates for LB films [64–66]. The first SA monolayer of OTS was reported by Sagiv [3].

In a typical procedure, a hydroxylated surface is introduced into a solution of alkyltrichlorosilane in an organic solvent. It usually is preferred to use a *p*-doped, test-grade, polished silicon wafer as the substrate. (See the section on substrates in Part Two for detailed discussion on the cleaning and storage of silicon substrates.)

The substrate is immersed in a ~5 x 10^{-3} M solution of the alkyltrichlorosilane in a mixture of 80/20 Isopar-G/CCl_4 for 2–3 min. After immersion, the substrate is rinsed with ligroin, methanol, distilled water, and finally dried in a steam of nitrogen. One of the most immediate apparent effects of monolayer coating is the drastic change in wettability of the surface. Clean silicon surfaces are completely hydrophilic ($\theta_{water} = 0°$), while the coated surfaces are hydrophobic (not wetted by water, $\theta_{water} > 0°$), and in most cases, oleophobic (not wetted by hydrocarbons such as hexadecane) and autophobic (emerge dry from the amphiphile solution).

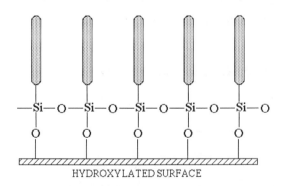

Figure 3.3. An SA monolayer of alkyltrichlorosilane on a hydroxylated surface.

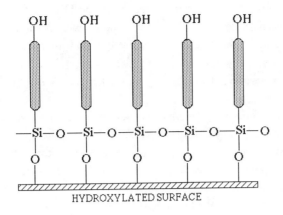

Figure 3.4. A hydroxy-terminated SA monolayer.

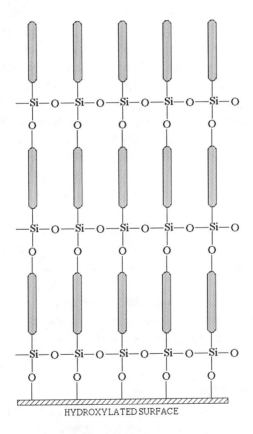

Figure 3.5. SA multilayers.

247

During adsorption, Si–Cl bonds react with the OH groups present on the surface of the substrate [67] — and with trace water, which we presume is adsorbed on the hydrophilic silanol surface [68] — to form a network of Si–O–Si bonds [69]. Recent x-ray photoelectron spectroscopy (XPS) results confirm the complete surface reaction of the SiCl$_3$ groups, since no chlorine could be detected in the monolayers [70].

The result is a monolayer in which the molecules are connected both to each other and to the surface by strong chemical bonds. A schematic view of such a monolayer is suggested in Figure 3.3. (A more detailed discussion on possible siloxane structures can be found later.)

The formation of the SA monolayer essentially is an *in situ* formation of polysiloxane, which is connected to surface silanol groups via Si–O–Si bonds. Thus, the construction of an SA multilayer requires that the monolayer surface can be modified to give a hydroxylated surface, so that another SA monolayer can be adsorbed (Figure 3.4). Such hydroxylated surfaces can be prepared by a chemical reaction and the conversion of a nonpolar terminal group to a hydroxyl group. Examples of such reactions are a reduction of a surface ester group [15], a hydrolysis of a protected surface hydroxy group, and a hydroboration–oxidation of a terminal double bond [10,13]. We note that the use of vinyl-terminated alkyltrichlorosilane (CH$_2$=CH–(CH$_2$)$_{16}$–SiCl$_3$) [71], and a vinyl derivative of silicon phthalocyanine [72] were mentioned in the literature as adhesion layers for polymers (e.g., to glass). Once a subsequent monolayer is adsorbed on the "activated" monolayer, multilayer films may be built by repetition of this process (Figure 3.5).

A close inspection of Figure 3.5 reveals that the SA multilayer structure is similar to a Z-type LB multilayer. (See Section 2.1.E.2.) However, there is a significant difference between the SA system and an LB Z-film. Here, the film is connected to the surface, the molecules in the monolayers are connected to each other, and the monolayers are bonded to each other by strong chemical bonds. In other words, an SA multilayer is a three-dimensional *polymer network*, which explains its excellent material properties. Furthermore, these properties are ideal for NLO and molecular electronics applications, since such a film would be expected to be highly ordered, durable, and to have a density approaching that of organic crystals. In fact, if some of the practical problems are solved (as described in the concluding remarks in this section), NLO may be one of the first technologies based on SA films. Thus, if an NLO chromophore is incorporated into the alkylchain, the resultant film would have polarizable dipoles that are oriented in a noncentrosymmetric arrangement, which is the requirement for a good $\chi^{(2)}$ material (Figure 3.6).

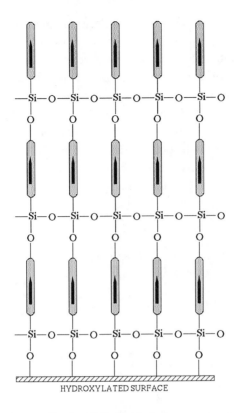

Figure 3.6. An SA multilayer film for NLO; the black arrow represents a molecular dipole.

However, the monolayers in these films would not have to contain dipoles in one particular direction throughout the film. In practice, organic synthesis should allow one to introduce the chromophore into the alkyl chain with either the donor or the acceptor group facing the substrate. This would allow changing the direction of the dipole in the film (in a similar way to the alternating LB films described before).

We have discussed (Section 2.1.E.2) the construction of superlattices from LB monolayers (ABABAB, etc.). However, there are chemical and thermal stability problems in LB multilayers (maybe with the exception of polymeric LB films, discussed in Section 2.8.B). In the case of SA monolayers, on the other hand, the advantage from the material point of view is clear. Thus, the SA technique offers the possibility to design stable organic superlattices — for example, with electron donor and electron acceptor groups, separated by well-

249

defined distances — that can exchange electrons following optical excitation. This may allow the construction, for example, of an electronic shift register memory based on molecular electron transfer reactions, as described by Hopfield *et al.* [73]. Figure 3.7 presents a possible SA shift register system.

Some of the constraints in the design of an electronic shift register memory unit (an $\alpha-\beta-\gamma$ unit, α being an electron donor, and β and γ electron acceptors) include (*a*) low vibronic overlap between anion and cation radicals to decrease the rate of back electron transfer reactions; (*b*) a rigid linking structure so that molecules are not likely to fold back on themselves; and (*c*) good covalent linking pathway for fast electron transfers.

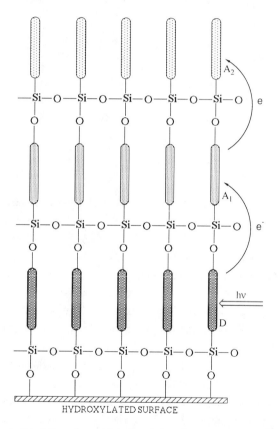

Figure 3.7. A schematic SA shift register unit. D is an electron donor, and A_1 and A_2 electron acceptors. Upon irradiation of D, there is an electron transfer to A_1, and upon irradiation of A_1 an anion radical electron is transferred to A_2.

250

The SA technique seems to address these requirements. A control of the distance between the donor and acceptor, which allows for control of the vibronic coupling, could be achieved, in principle, by introducing the appropriate number of methylene units between the donor and acceptor moieties in two adjacent monolayers. The SA multilayer film also would provide a covalent and rigid linkage to stabilize films, help prevent molecules from folding back, and facilitate fast electron transfer between layers. However, these advantages, although important, have to be discussed within the constraints of the real world, where substrates are not atomically smooth. If we try to imagine a two-dimensional assembly on a rough surface, where the length-scale of roughness is of molecular scale (few tens of ångstroms), it is clear that in some cases an electron donor will be adjacent to an electron acceptor, thus introducing an error to the system. With this in mind, it should be emphasized that atomically smooth mica surfaces are readily available, and that the evaporation of gold on mica can give atomically smooth surfaces with length-scales of thousands of ångstroms [74].

To conclude this discussion, we note that while emphasizing exciting possibilities and pointing out future directions, we should avoid the notion that technologies based on SA monolayers of alkyltrichlorosilanes are just around the corner. There are major material problems in this area, and we shall indicate them as we describe recent results.

B. Kinetics of Formation of Alkyltrichlorosilane Monolayers

The kinetics of the formation of SA monolayers is important for establishing a protocol for the construction of reproducible, closely packed, complete monolayers. This is essential especially where multilayer construction is concerned. One of the first experiments in adsorption kinetics was carried out by Maoz and Sagiv, who studied adsorption isotherms for OTS [4]. They plotted the peak absorbance at ~2918 cm^{-1} [$v_a(CH_2)$] vs. the concentration of the OTS in solution (Figure 3.8). Equilibrium with the solution was assumed to be completed within three min adsorption time, and a plateau was reached in the isotherm, corresponding to complete coverage of the area of the substrate by tightly packed monolayers [3]. This was concluded from a comparison of the absorbance per methylene group in the plateau region of the isotherm with that of a closely packed LB film deposited on the same substrate. The measured contact angles followed the general trend of the isotherm curve, and reached maximal values characteristic of closely packed OTS monolayers exposing methyl groups. However, while the plateau is reached for the isotherm at 1 x 10^{-3} M concentration, it is reached at 2 x 10^{-3} M concentration for the contact angles.

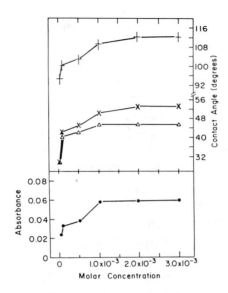

Figure 3.8. Adsorption isotherm and advancing contact angles for OTS/Si, at 20°C, from 80% bicyclohexyl (BCH) + 12% CCl_4 + 8% $CHCl_3$. Lower curve: dependence of the ATR peak absorbance at ~ 2918 cm^{-1} (•) on the molar concentration of OTS. Upper curves: advancing contact angles of H_2O (+), hexadecane (Δ), and BCH (x). (From Maoz and Sagiv [4], © 1984, Academic Press.)

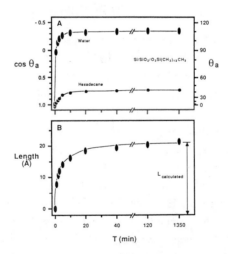

Figure 3.9. Kinetics of formation of TTS monolayer on silicon at 20°C and 30% relative humidity. A. Advancing contact angles of water (●) and hexadecane (•). B. Thickness estimated by ellipsometry. $L_{calculated}$ = 21.1 Å, based on all-*trans* configuration, and chain axis perpendicular to the substrate surface. (From Wasserman *et al.* [70], © 1989, Am. Chem. Soc.)

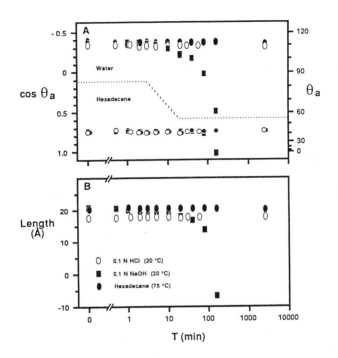

Figure 3.10. Stability of TTS/Si monolayers (●) and exposed to 0.1 N HCl at 20°C (o) 0.1 N NaOH at 20°C (■) or hexadecane at 75°C (●). (A) Water (above) and hexadecane (below) contact angles. (B) Thickness estimated by ellipsometry (with the negative values for 0.1 NaOH are probably due to substrate etching). (From Wasserman *et al.* [70], © 1989, Am. Chem. Soc.)

Thus, it seems that contact angles are more sensitive than FTIR ATR spectroscopy for detection in variations of molecular density and orientation in OTS monolayers.

The kinetics of the formation of long-chain alkyl trichlorosilane monolayers was studied by Wasserman *et al.* (Figure 3.8) [70]. They used 0.01 M solution of tetradecyltrichlorosilane ($CH_3-(CH_2)_{13}-SiCl_3$, TTS) in hexadecane at 20°C and 30% relative humidity. Ellipsometry was used to estimate thickness, and contact angle measurements to establish surface structure. It was found that after ~10 minutes, a plateau was reached in the contact angle measurements, while it took ~40 minutes for a plateau to be reached in the monolayer thickness.

We conclude, therefore, that contact angle measurements cannot serve as the sole analytical tool for the determination of a complete monolayer formation, and that thickness always should be estimated using ellipsometry.

C. Stability of Self-Assembled Monolayers

1. Chemical Stability

As was discussed earlier, SA monolayers of alkyltrichlorosilane essentially are a polysiloxane backbone cross-linked to the surface. Therefore, it can be expected that the stability of these monolayers will be much higher than that of any LB monolayer. Indeed, we found that OTS/Si monolayers can be washed with 1% detergent solution and hot tap water and with organic solvents without any apparent detectable damage to the system [14, 15].

Wasserman *et al.* looked in more detail into the chemical stability of SA monolayers [70]. They used TTS monolayers and immersed them in 0.1 N HCl and 0.1 N NaOH at room temperature and in hexadecane at 75°C. Their contact angle and ellipsometric results are presented in Figure 3.10.

The stability of these monolayers in organic solvents and in acidic media is remarkable. There was no apparent degradation in the monolayers after 44 h. On the other hand, there was a considerable deterioration of the monolayer in basic solution. Immediately after immersion (five min) they noticed loss of material, and after 80 minutes in solution, ~50% of the film was removed. Long contact times have shown etching on the silicon substrate, presumably due to attacking the native SiO_2. These results are not surprising since a Si–O bond is known to undergo hydrolysis in basic media [75].

In conclusion, alkyltrichlorosilane monolayers provide structures that are stable to chemical conditions that most LB films could not stand. However, we should qualify this statement, and add that photopolymerized LB films seem to have considerable stability in organic solvents.

Table 3.2. Advancing contact angles (deg)^a measured before, and after heating and cooling to ambient temperature. (From Cohen *et al.* [76], © 1986, Am. Chem. Soc.)

	Before			After		
Film Type	H_2O	HD	BCH	H_2O	HD	BCH
CA/CA/OTS/Al	115 ± 2	47 ± 1	53 ± 1	109 ± 2	42 ± 1	48 ± 1
CA (LB)	108 ± 2	43 ± 1	49 ± 1	100 ± 2	32 ± 2	41 ± 2
AA (SA)	105 ± 2	45 ± 1	51 ± 1	97 ± 2	30 ± 2	39 ± 2
OTS (SA)	113 ± 2	46 ± 1	51 ± 1	113 ± 2	46 ± 1	51 ± 1
OTS (SA, 60%)	102 ± 2	42 ± 1	42 ± 1	102 ± 2	42 ± 1	42 ± 1

^{a}HD - hexadecane, BCH - bicyclohexyl.

2. Thermal Stability

The thermal stability of OTS monolayers on aluminum was studied by Cohen *et al.* [76]. They used both contact angles and FTIR to examine the monolayers prior to and after a heating process. They heated the monolayer to ~125°C and then cooled it back to room temperature. A partial (60%) SA monolayer of OTS on aluminum, an SA monolayer of arachidic acid (AA, $C_{20}H_{41}COOH$) on aluminum, an LB monolayer of cadmium arachidate (CA), and a trilayer made as an LB bilayer of cadmium arachidate on OTS/Al, were used for comparison. The contact angle results are presented in Table 3.2. Apparently, there is a substantial difference between the OTS/Al monolayer and the other systems.

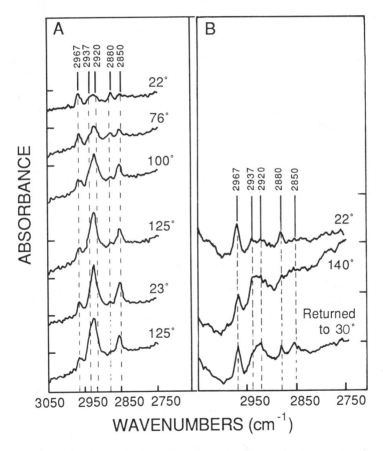

Figure 3.11. Spectral changes in the C–H stretch region of complete OTS/Al monolayers (A), and in cadmium (II) arachidate monolayer on Al (B), upon heating and recooling. (Adapted from Cohen *et al.* [76], © 1986, Am. Chem. Soc.)

255

In all systems except for the monolayer OTS/Al, there was a reduction in both the hydrophilic (H_2O) and hydrophobic (HD and BCH) contact angles. This is a strong indication for a randomization of the surface, i.e., from a close-packed methyl surface to one where there is a mixture of methyl and methylene groups. A process of this nature should be accompanied by changes in the molecular orientation from the all-*trans* extended chain structure, which is associated with a solid monolayer, to a liquid-like system with considerable contribution of *gauche* conformations. FTIR data provided additional information on the thermal-induced disorder in these monolayers. Figure 3.11 present the grazing-angle FTIR spectra of the complete OTS/Al monolayer (A), and of a CA bilayer on OTS (B), before heating, at 120°C, and after cooling to ambient temperature. Again, there is a striking difference between the two systems. In the CA bilayer where the monolayer–substrate interaction is the weakest (van der Waals), Cohen *et al.* observed moderate changes up to ~100°C, but apparent randomization between 100° and 130°C, in both the C–H and the C=O stretching regions. In the SA monolayer of AA and LB monolayers of CA, where the monolayer substrate interaction also has ionic character [2], they observed changes only in the C–H, but not in the C=O, stretching region. The OTS/Al systems showed the highest stability, and they could detect only slight changes in the IR spectra. It can be concluded, therefore, that even the ionic substrate–monolayer interactions in the SA AA/Al monolayer are not enough to provide orientational stability, and only where the head groups are immobilized due to the polysiloxane backbone is a true thermal stability observed. For information on thermal stability of LB films, the reader is referred to Section 2.2.

D. Structural Issues of Self-Assembled Monolayers

The structure of SA monolayers of alkyltrichlorosilane has implications beyond a scientific understanding of these systems. This is because a closely packed and ordered assembly of molecules usually exhibits a closely packed and ordered surface, which, in turn, allows the deposition of a second closely packed and ordered monolayer. (See later in this part.) The issue of monolayer order is connected with the order at the monolayer–substrate interface, which at this time is still an open question.

The small molecular tilt in SA monolayers of OTS ($\leq 15°$), as well as their stability, suggest that polysiloxane structures in some form (linear and/or cyclic) should be the bonding arrangement at the surface. Moreover, it can be expected that oligomeric units formed in solution will adsorb on the surface faster than monomeric trichlorosilane (or partially hydrolyzed monomers). Therefore, the issues of one-dimensional vs. two-dimensional surface polymerization, and of the cross-linking to the surface and its effect on polymer conformation, are important. Figure 3.12 presents a possible cyclic trimer (in a chair conformation) where the alkyl chains are connected at the axial positions (left), and the alkyl chains are connected at the equatorial positions (right).

There are two major differences between the two isomers. The first is the C···C distance, i.e., the distance between the methylene groups connected to the silicons (marked by dark circles in Figure 3.12), which defines the free volume between the alkyl chains. This distance is 4.25–4.35 Å for the axial isomer (in agreement with the tilt angle of $\leq 15°$), and 4.90–5.00 Å for the equatorial isomer (which would require a tilt angle of $\geq 30°$) [78].

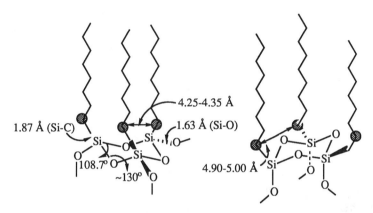

Figure 3.12. A siloxane trimer: left, alkyl chains at axial positions, and right, alkyl chains at equatorial positions.

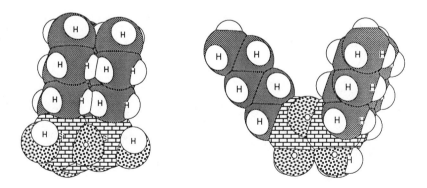

Figure 3.13. Space-filling models of siloxane trimers: left, alkyls at axial positions, and right, alkyls at equatorial positions.

257

The second difference is the possibility of exocyclic Si–O bonds to cross-link with other polysiloxane units on the surface, and to connect the polymer to the surface. In the axial isomer, the Si–O bonds are equatorial, and therefore connection to other polymers, as well as to surface Si–OH groups, is easy and does not require major distortion of the groups around the silicon atom. In the equatorial isomer, on the other hand, the Si–O bonds are axial, and both cross-linking and chemisorption to the surface should be accompanied by a severe distortion. Of course, a silicon atom that is connected to another chain cannot be connected to the surface at the same time. The difference between the two isomers is even more striking in space-filling models (Figure 3.13).

In reality, there probably is a mixture of the two possible isomers in solution. However, the contact angles for water, diiodomethane, and hexadecane for a good OTS monolayer on SiO_2 are 112° (advancing), 73°, and 45° (both sessile drops), respectively. Even more important is the low surface tension of 20 dyn/cm [14], which is associated with a highly ordered CH_3 surface. This data, together with the experimental tilt angle of the alkyl chain ($\leq 15°$), [14, 79] suggests that most (if not all) of the alkyl chains should be connected to axial positions.

It can be assumed further that the two isomers should have very different adsorption kinetics, with a strong preference for the axial isomer, since the alkyl chains in the equatorial isomer mask the hydrophilic siloxane bonds from the hydrophilic SiO_2 surface. Also, it can be expected that cross-linked oligomers in solution — i.e., those species where one alkylsilane molecule is connected to three other molecules — may adsorb to form an ordered assembly, but only if alkyl chains are at axial positions.

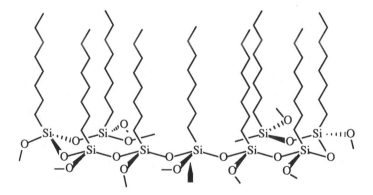

Figure 3.14. A schematic description of a polysiloxane at the monolayer-substrate surface. The arrow points to an equatorial Si–O bond that can be connected either to another polysiloxane chain or to the surface.

258

If some of these trimers are fused together (as in the case of a steroid-like structure), and inner bonds are cut out, a polysiloxane structure is formed. A proposed structure of such polysiloxane, where the alkyl chains are at axial positions, is suggested in Figure 3.14. This general picture of *in situ* polysiloxane formation at the substrate–solution interface — where monomers, dimers, and possibly oligomers adsorb on the silicon surface — suggests that monolayers of alkylsilanes may be *inherently more disordered* than those of alkanethiols, where molecules have more freedom to establish a long-range order. (See later in this part.) Ellipsometric measurements of OTS monolayers on silicon surfaces suggest that these are well formed, relatively dense assemblies, with mean thickness equal to the calculated theoretical length of fully extended alkane chains that are tilted from the surface normal by $\leq 15°$ [3, 4, 14, 80]. However, a comparison of grazing-angle FTIR of OTS monolayers on aluminized surfaces [14], and on HUT/Au [81] (see Section 3.4), reveals that the alkyl chains in the latter are less tilted, and therefore, the free volume in the former should be larger. It is not clear at this point what the effect of surface Si–OH group concentration is on the packing and order of alkyltrichlorosilane monolayers.

Wasserman *et al.* estimated from x-ray reflection data an area per RSi— group of 21 ± 3 Å in the monolayers on SiO_2 [80]. The thickness of an OTS monolayer on silicon wafer is 25 ± 2 Å, which is in close agreement with the predicted value of 26.2 Å. This value was calculated assuming an all-*trans* chain conformation perpendicular to the surface. In this molecular orientation, every methylene unit (CH_2) contributes 1.26 Å to the thickness (for a C–C bond length of 1.50 Å and a tetrahedral angle of 109.5°), the Si–C bond contributes 1.52 Å, the Si–O bond 1.33 Å, and the terminal methyl group 1.50 Å [82]. Tillman *et al.* estimated previously that the alkyl chains in the OTS monolayer tilt ~14° from the normal [14]. This tilt would cause a reduction of ~0.8 Å in film thickness estimated for the perpendicular arrangement.

The structure of partially formed (incomplete) monolayers also is of importance, since it should help us understand the mechanism of monolayer formation. Cohen *et al.* suggested that partially formed monolayer films consist of islands of close-packed, straight, fully extended molecules that are oriented normal to the surface [76]. Such an assembly should contain clean surface areas to which more alkyltrichlorosilane molecules can adsorb, without considerable interference from the already adsorbed ones. The islands structure was disputed recently by Wasserman *et al.* [80], who used x-ray reflectivity and concluded that the structure of incomplete monolayers is described best by a uniform model, in which disordered molecules are distributed uniformly over the substrate, yielding a liquid-like structure with film thickness less than that found in complete monolayer samples (Figure 3.15). Further and indirect information on the packing in SA monolayers of alkyltrichlorosilane can be obtained from a comparison with SA monolayers of alkanethiols. For example, it was found that hydroxy-terminated monolayers on silicon (see Section 3.2.F) reorganize when exposed to a low dielectric constant liquid such as chloroform, which results in an increase of the advancing water contact angle (θ_a) from ~31° to ~60° [11]. However, the hydroxylated surface of 11-hydroxyundecane-1-thiol on silver surfaces (HUT/Ag) can be exposed to boiling chloroform without any detectable change in its advancing water contact angle ($\theta_a = 20$–$21°$) [78]. This difference suggests that the trichlorosilane monolayers are packed less closely (i.e., contain more free volume) than those of alkanethiol on silver. These results bring up the issues of surface mobility during monolayer formation, and of the kinetics of monolayer-to-surface Si–O–Si bond formation. Thus, if alkyltrichlorosilane molecules and siloxane oligomers first physisorb onto the surface, and are free to move prior to being anchored (chemisorbed), then further adsorption can be accompanied by increasing packing and order in the two-dimensional assembly. However, if the monolayer-to-surface Si–O–Si bond formation is fast, and its kinetics competes with that of the adsorption, then there will be many defects in the monolayer. Of course, if the chemical reaction with the surface is reversible, then such defects can be repaired, depending upon the kinetics of the back-reaction. It can be suggested, however, that any repair mechanism, if possible, should be very slow because of the number of bonds associated with each molecule in the assembly.

The quality and structure of OTS monolayers on silicon surfaces suggest that there should be enough molecular mobility prior to any surface pinning. In fact, both ATR and grazing-angle FTIR spectra of OTS monolayers confirm this assumption [3, 5–7, 14, 76, 83]. These spectra, especially the progression of weak twist and wag bands between 1150 and 1250 cm^{-1}, are characteristic of well-oriented and densely packed all-*trans* hydrocarbon chains.

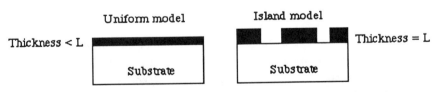

Figure 3.15. Uniform and island models for incomplete monolayers. (Adapted from Wasserman *et al.* [80], © 1989, Am. Chem. Soc.)

E. Self-Assembled Monolayers that Contain Aromatic Groups

We turn now to a discussion on SA monolayers that contain aromatic groups. The ultimate goal of using SA films is the construction — of multilayer films that contain useful physical properties — in a layer-by-layer fashion. Useful films may contain electron donor or electron acceptor groups, NLO chromophores, moieties with unpaired spins, etc. Therefore, there is a need to construct and study monolayer films with progressively more complicated polar and interesting planar bulky π-systems. Such detailed experimental works, when complemented by modeling efforts (Part Four), will provide the systematic understanding of organization in two dimensions, which is the foundation for any SA-based material science. In this section, we discuss experimental work, and in Part Four, we mention preliminary results of modeling efforts.

1. Self-Assembled Monolayers that Contain Phenoxy Groups

Tillman *et al.* prepared SA monolayers from trichlorosilanes 1–4, each containing a benzene ring as a model aromatic functional group [14]. The phenoxy group was the preferred choice mainly because of synthetic convenience.

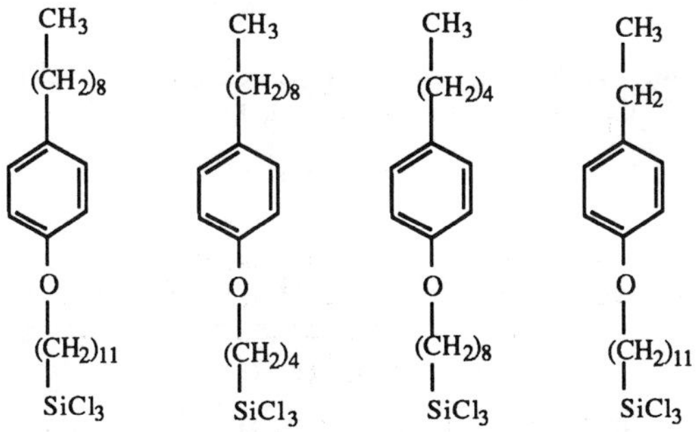

CH_3	CH_3	CH_3	CH_3
$(CH_2)_8$	$(CH_2)_8$	$(CH_2)_4$	CH_2
(benzene ring)	(benzene ring)	(benzene ring)	(benzene ring)
O	O	O	O
$(CH_2)_{11}$	$(CH_2)_4$	$(CH_2)_8$	$(CH_2)_{11}$
$SiCl_3$	$SiCl_3$	$SiCl_3$	$SiCl_3$
1	**2**	**3**	**4**

It was found that monolayer films of each of these materials could be prepared. However, while compound 1 formed monolayers rapidly (in ≤ 5 min), compounds 2–4 required multiple immersions to increase coverage, as judged by ellipsometry and contact angle measurements. Consequently, substrates were treated with the trichlorosilane several times, until maximum contact angles and film thicknesses were observed (Figure 3.16). Table 3.3 presents contact-angle data for four test liquids and critical surface tensions (γ_c) for the monolayers. (See Section 1.2.A for discussion on wetting [85–87].)

261

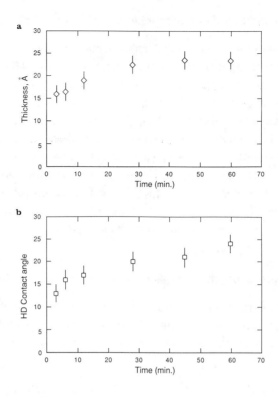

Figure 3.16. (a) Film thickness and (b) HD contact angle vs. total immersion time for monolayer of 3 on a silicon wafer. (From Tillman *et al.* [14], © 1988, Am. Chem. Soc.)

Table 3.3. Advancing contact angles of various test liquids and critical surface tensions for monolayers on silicon wafers. (From Tillman *et al.* [14], © 1988, Am. Chem. Soc.)

monolayer[a]	θ, deg ($\pm$ 2)				γ_c, dyn/cm ($\pm$ 1)[d]
	H_2O	CH_2I_2	BCH[b]	HD[c]	
OTS	111	73	48	45	20
TTS	110	72	45	45	20
UTS	109	69	41	32	22
1	111	74	48	44	20
2	110	70	42	36	22
3	107	66	36	27	24
4	100	60	30	26	26

[a]OTS = octadecyltrichlorosilane (C_{18}), TTS = tridecyltrichlorosilane (C_{13}), UTS = undecyltrichlorosilane (C_{11}). For structures 1-4 see text. [b]BCH = bicyclohexyl. [c]HD = *n*-hexadecane. [d]Critical surface tension measured using a series of homologous *n*-alkanes.

Note that if monolayers of compounds 1–4 are formed with an arrangement of the closely packed alkyl chains, with their axes either parallel to or slightly tilted from the surface normal, then each of these monolayers would be formed with a surface consisting of closely packed methyl groups. Tillman *et al.* observed for compound 1 contact angles of 111° for water, 45° for hexadecane, and $\gamma_c = 20$ dyn/cm, values that are typical for very hydrophobic and oleophobic surfaces [84–86]. Furthermore, these values are identical to what was found for OTS monolayers, which do not contain the phenoxy group and have been well characterized [13, 14]. (See also Table 3.3.) Thus, wettability results already point out that the phenoxy group does not prevent good monolayer formation. However, a close inspection of Table 3.3 reveals that the reduced total number of carbons lowers the level of ordering in the monolayer, which is reflected in the surface wettability. Note that cutting a $(CH_2)_7$ unit from the alkyl chain below the phenoxy group (compound 2), hardly affects the water contact angle (110° and 111° for compounds 1 and 2, respectively), while the HD contact angle shows a considerable reduction (44° and 36° for compounds 1 and 2, respectively). This is not surprising, since HD is more sensitive to the order of the monolayer surface. Indeed, while cutting the same number of methylene units from OTS (to yield UTS) shows only a 2° decrease in the water contact angle, the decrease in the HD contact angle is 13° (Table 3.3).

The thickness of a monolayer of compound 1 was 31 ± 3 Å, close to the 32 Å value calculated by assuming perpendicular alkyl chain axes and normal bond lengths and bond angles. However, while this result may appear convincing, it is difficult to understand in one respect. This is because, given the cross-section area of the phenyl ring (Figure 3.18), a configuration consisting of alkyl chains perpendicular to the surface would leave much free volume in the assembly. FTIR studies revealed that the alkyl chains, in fact, are tilted, which suggests that the lower end of the thickness mentioned earlier (~28 Å) may be a more accurate value.

FTIR spectroscopy detected major effects of phenoxy group substitution on monolayer structure. Figure 3.17 presents grazing-angle FTIR spectra in the C–H stretching region for monolayers of compound 1 and OTS adsorbed on aluminum films. In the bulk spectrum (Figure 3.17a) of these long-chain compounds, the methylene stretching vibrations ($v_a(CH_2)$ and $v_s(CH_2)$ at ~2920 cm^{-1} and ~2855 cm^{-1}, respectively) are extremely intense and bury the methyl peaks ($v_a(CH_3)$ at ~2965 cm^{-1} and $v_s(CH_3)$ at 2880 cm^{-1}). In the grazing-angle spectra, by contrast, the methylene vibrations are diminished greatly in relative intensity.

263

Table 3.4. Molecular orientation in monolayers adsorbed on silicon measured by FTIR ATR linear dichroism. (From Tillman *et al.* [14], © 1988, Am. Chem. Soc.)

	Apparent Orientation Angle, deg [a]	
Monolayer[b]	Aromatic Ring [c]	Alkyl Chain [d]
OTS	--	15
TTS	--	39
1	20	26
2	32	51
3	30	55
4	32	38

[a]The orientation angle is defined here as the measured, average angle between the alkyl chain or aromatic ring axis and the direction normal to the substrate surface. [b]OTS = octadecyltrichlorosilane (C_{18}), TTS = tridecyltrichlorosilane (C_{13}). For structures 1–4 see text. [c]Tilt of the aromatic ring axis lying along the $C_1(O)$–$C_4(C)$ direction. [d]Tilt of the alkyl chain axis. For a fully extended chain, the alkyl chain axis bisects the C–C bonds.

We have discussed in Section 1.1.B that, for perpendicular alkyl chains, both the $v_a(CH_2)$ and the $v_s(CH_2)$ will be oriented parallel to the substrate surface; consequently, their interaction with the electric field vector should approach zero [32, 87–89]. Tillman *et al.* estimated the orientation of the chains in monolayers of compound 1 and OTS by consideration of the measured $v_s(CH_2)/v_s(CH_3)$ intensity ratio, suggesting that the transition dipole moments of both the methyl and methylene vibrations are equal on a per-hydrogen basis. This suggestion should be taken with caution, since the dielectric constant in the bulk of the film (i.e., around the CH_2 groups) is different from that at the interface (i.e., around the CH_3 groups). However, this assumption can be used as a first approximation. The results suggest average alkyl chain tilts of ~15° for OTS and ~26° for compound 1. This tilt is the angle between the chain axis, which bisects the C–C bonds, and the surface normal.

The IR results for the monolayers of compounf 1 are consistent and suggest that the alkyl chains are more tilted than observed for OTS. This likely is due to the steric bulk of the rigid phenoxy π-system, which increases the chain–chain spacing. Consequently, the chains tilt to maintain van der Waals contact. The observation of a ~20° tilt for the aromatic ring is interesting, and can be justified by the following argument.

Let us assume a model of a molecule perpendicular to the surface (Figure 3.18). Here, the alkyl chains are normal to the surface and the phenyl ring is tilted 35° from that normal.

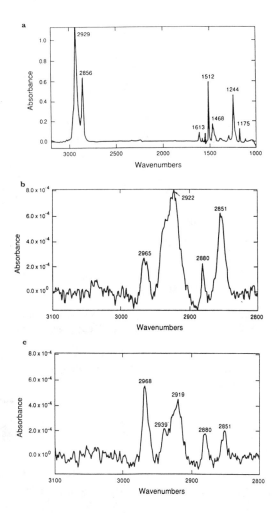

Figure 3.17. (a) FTIR spectrum of compound 1 in CCl_4, (b) C–H stretching region of the grazing-angle external specular reflection IR spectrum of a monolayer of compound 1 on an aluminized silicon wafer, and (c), as in (b), for a monolayer of OTS on an aluminized silicon wafer. (From Tillman *et al.* [14], © 1988, Am. Chem. Soc.)

Thus, while the cross-section area of an all-*trans* alkyl chain is ~20 $Å^2$ [90], simple calculations show that the cross-section area of a phenyl ring is from 21.8 to 25.2 $Å^2$ (up to 25% larger, depending on the phenyl orientation in the monolayer). At ~20° tilt, the cross-section area of the alkyl chain is larger than at 0°, and that of the phenyl ring is smaller than at 35°.

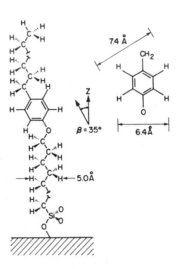

Figure 3.18. (a) Representation of compound 1 in a monolayer with fully extended alkyl chains that are perpendicular to the surface. The length of the alkyl chains has been shortened for clarity. The C_1–C_4 axis of the phenyl ring has a 35° tilt (β) from the substrate surface normal (Z-axis). (b) Dimensions of the phenyl ring, calculated from the usual bond lengths and van der Waals radii [82]. Not shown is the thickness of the π-cloud, 3.4 Å. (From Tillman *et al.* [14], © 1988, Am. Chem. Soc.)

Therefore, to achieve more complete close-packing, the molecules tilt and the benzene rings are squeezed into a more perpendicular orientation. This is an example of how different parts of the molecule affect the overall tilt, and the reader is referred to the discussion at the beginning of Part Four.

2. Self-Assembled Monolayers Containing a Sulfone Group

The introduction of a sulfone group as an electron acceptor for nonlinear optical materials was pioneered at Eastman Kodak Company [91]. There, it also was introduced into an SA monolayer for the first time [92]. The importance of this bifunctional group is apparent, since it allows the introduction of a π-system into the alkyl chain, while controlling the number of methylene units on both its sides.

Tillman *et al.* introduced polar aromatic chromophores into SA monolayers and studied their effects on wettability, packing, and order [92]. They prepared the sulfone-containing compound 6, and as a comparison studied the ester-containing compound 5. However, nonyl (11-trichlorosilylundecyloxy)benzoate (compound 5) yielded low-quality monolayers, with a very low hexadecane contact angle of 22° (≥ 20° lower than expected from comparison with OTS/Si), and high critical surface tension (γ_c = 23 dyn/cm), indicative of a fairly disordered surface. Moreover, the 5/Si monolayer reacted instantly with $LiAlH_4$ in THF, although the carbonyl group is buried under a ~12 Å-long alkyl chain. On the other hand, the preparation of dodecyl 4-(11-trichlorosilylundecyl–oxy)phenyl sulfone monolayers on silicon (6/Si), was straightforward.

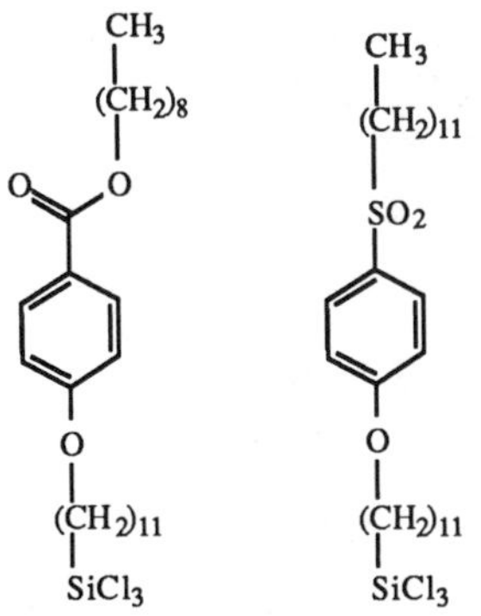

CH3
(CH2)8
O
O
O
(CH2)11
SiCl3
CH3
(CH2)11
SO2
O
(CH2)11
SiCl3

5 **6**

The hexadecane contact angle was 42° (20° higher than observed for 5/Si monolayer), and the surface tension (γ_c = 20 dyn/cm) was identical to the value measured for OTS/Si and 1/Si monolayers. Note that the advancing water and hexadecane contact angles (104° and 42°, respectively) are reminiscent of values obtained by Maoz and Sagiv for partial OTS monolayers at 65% surface coverage [4]. Indeed, measurements of CH_2 stretching intensities in 6/Si monolayer indicated ~70% molar surface coverage relative to OTS. This is in agreement with the alkyl chain having a cross-section area of ~80% that of the aromatic moiety. Thus, a 100% coverage based on the aromatic part essentially is ~80% coverage based on the alkyl chains. Further studies of 6/Si monolayers by ATR FTIR linear dichroism suggested an average tilt angle of the alkyl chains to be ~31°, or ≥ 15° more than observed for OTS/Si. Figure 3.19 presents XPS results for 6/Si monolayer, and Figure 3.20 shows the variable angle XPS results for the same sample [93]. The electron takeoff angle was defined in the variable angle experiments as the angle between the emitted electron and the surface, and it is expected that elements that are present at higher concentrations at the surface should show increasing atomic percent with decreasing takeoff angle, while the reverse should be the case for elements that are concentrated more deeply (~15 Å) in the monolayer/substrate [93–95]. The sulfur and silicon (as SiO_x) each showed an initial increasing intensity with increasing takeoff angle, followed by a drop in intensity, with the maximum for SiO_x occurring at a higher takeoff angle than that observed for sulfur. as is expected for a monolayer where the molecules are tilted slightly from the surface normal. Thus, it is evident that the sulfone group is sandwiched between two alkyl chains.

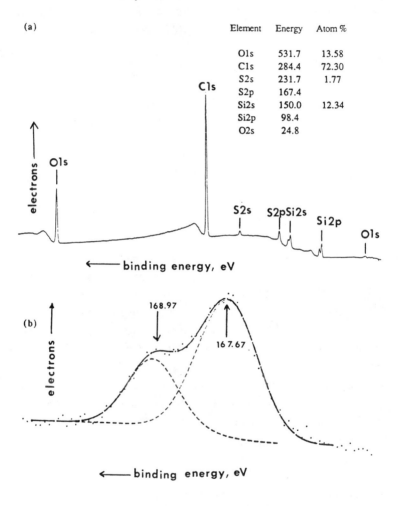

Figure 3.19. XPS results for 6/Si monolayers. Presented is (a) a survey scan, showing the following peaks: O1s (531.7 eV, 13.58 atom %); C1s (284.4 eV, 72.30 atom %); S2s (231.7 eV, 1.77 atom %); S2p (167.4 eV); Si2s (150.0 eV, 12.34 atom %); Si2p (98.4 eV); O2s (24.8 eV); and (b) a high resolution scan of the S2p peak, with the $S2p_{3/2}$ peaks and $S2p_{1/2}$ at 167.67 and 168.97 eV, respectively. (From Tillman *et al.* [92], © 1988, Am. Chem. Soc.)

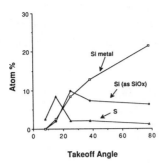

Figure 3.20. Variable angle XPS results for a 6/Si monolayer: silicon metal, silicon (as the oxide), and sulfur percentage, respectively. (From Tillman *et al.* [92], © 1988, Am. Chem. Soc.)

In summarizing the discussion on SA monolayers that contain aromatic groups, I would like to reiterate that the importance of these works is in the structural information they provide. This information may help in the design of future amphiphiles for other self-assembly processes.

F. Formation of Multilayers by Self-Assembly

The first published reports on self-assembled multilayers suggested that the quality of monolayers formed by self-assembly of trichlorosilane derivatives rapidly degrades as the thickness of the films increases [11–13]. Pomerantz *et al.* published results for an x-ray diffraction study of methyl 23-trichlorosilyl-tricosanoate ($H_3CO_2C–(CH_2)_{22}–SiCl_3$, MTST, 7) multilayers. It was found that a three-layer sample could be fit to the data best by assuming as a model a 55% two-layer and 45% three-layer sample [79]. Furthermore, hexadecane contact angles for the unreduced ester surface rapidly deteriorated, from 28° for the initial monolayer on the silicon surface to 12° for the third layer.

Tillman *et al.* reported that the formation of single monolayers of MTST on silicon substrates ($MeO_2CC_{22}Si/Si$) was straightforward [15]. The average thickness for 14 samples of $MeO_2CC_{22}Si/Si$ monolayers was 34 ± 2 Å, using an approximate value for the film refractive index, $n_f = 1.50$. In fact, they independently measured n_f with the ellipsometer on relatively thick (650–700 Å) multilayer films formed from MTST, and established values of $n_f = 1.50 \pm 0.01$. This, however, contradicts the results presented by Wasserman *et al.* [80], where the density of the monolayer was measured, and n_f was assigned a value of

1.45. We can suggest two possible factors contributing to this discrepancy: (*a*) that the multilayers are inherently more dense due to their three-dimensional polymer network, and (*b*) that the polar Si–O bond increases the refractive index.

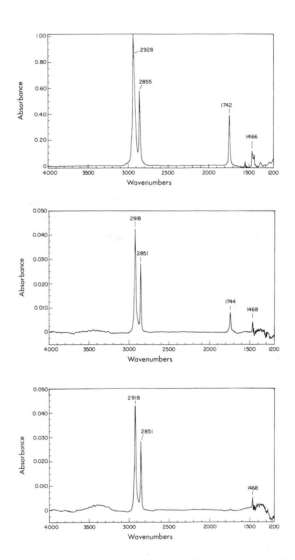

Figure 3.21. FTIR spectrum of MTST in CCl_4 solution (top), an ATR spectrum of a $MeO_2CC_{22}Si/Si$ monolayer (middle), and an ATR FTIR spectrum of the $MeO_2CC_{22}Si/Si$ monolayer reduced 2 x 2 min with $LiAlH_4$ to form a $HOC_{23}Si/Si$ monolayer (bottom). ATR spectra presented are *s*-polarized. (From Tillman *et al.* [15], © 1988, Am. Chem. Soc.)

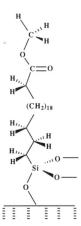

Figure 3.22. Representation of the structure of $MeO_2CC_{22}Si/Si$ monolayers, assuming all-*trans* alkyl chains with a chain axis perpendicular to the substrate surface, showing the silane head group connected to the substrate and to adjacent silicon atoms via Si–O bonds. (From Tillman *et al.* [15], © 1988, Am. Chem. Soc.)

The measured film thickness of $MeO_2CC_{22}Si/Si$ is quite consistent with a close-packed monolayer structure of slightly tilted, fully extended alkyl chains. In fact, film thickness in this study (~34 Å) is greater by 6 Å than the value of 28 Å found using x-ray diffraction techniques [79]. This may indicate that in this case, the monolayers were packed more densely. Contact angle measurements (69 ± 2° for water, 52 ± 2° for diiodomethane, and 25 ± 3° for *n*-hexadecane) are in essential agreement with published values [79]. When complete monolayers of MTST were immersed in a solution of $LiAlH_4$ in THF, and then rinsed with 20% HCl (aq) followed by large volumes of Millipore water, a reduction in the advancing water contact angle to 30 ± 3° and a decrease in the film thickness to 32 ± 2 Å were observed. Receding water contact angles varied from 18° to 27°.

IR data indicated $\geq$ 90% disappearance of the ester carbonyl band at 1743 cm^{-1} (Figure 3.21). Further immersion in the reducing solution did not result in a significant additional change in the water contact angle. This result can be attributed to the reduction of the ester group to a hydroxyl group, forming a new, alcohol-surface monolayer, $HOC_{23}Si/Si$.

In light of the report of a 50° advancing water contact angle for the $HOC_{23}Si/Si$ monolayer [79], the observation of a 30° contact angle was quite puzzling at first. Apparently, this was due to differences in sample handling in the two reports. Pomerantz *et al.* routinely cleaned the monolayer surfaces after

271

reduction by Soxhlet extraction with chloroform [79]. Tillman *et al.* reported that when a sample monolayer initially showing an advancing contact angle of 31° and a thickness of 34 Å was immersed for 10 min in boiling chloroform, the contact angle rose to 60°, while the film thickness increased by 1 Å [15]. This was a marked effect that never was observed with other more hydrophobic surfaces such as OTS/Si. In their paper, Tillman *et al.* provide experimental evidence for a mechanism of surface reorganization. Thus, upon exposure to chloroform, $HOC_{23}Si/Si$ monolayers restructured and minimized the excess interfacial free energy of the system by burying the hydroxyl groups as much as possible, thus exposing more CH_2 groups to the surface. Such "surface reconstruction" has been postulated to occur for the surface of oxidized polyethylene, which consists largely of exposed carboxylic acid groups, with smaller amounts of ketone and possibly aldehyde groups [96]. Recently, Evans *et al.* confirmed this mechanism in a detailed study on surface reorganization in hydroxy-terminated alkanethiol monolayers on gold [38]. The packing (density of surface coverage), of monolayers of $MeO_2CC_{22}Si/Si$ reported, was calculated from ATR FTIR spectra to be *ca.* 100% of the packing in the reference OTS monolayer. Alkyl chain axis tilts were calculated from ATR spectra to be ≤15° (equal to measured values for OTS control samples), and the $v(C=O)$ stretch transition dipole moment to tilt 70°–85° from the substrate normal. Taken together, these data are supportive of an ordered monolayer film with slightly tilted alkyl chains, and surface-exposed methyl ester groups with C=O bonds parallel to the substrate surface, as diagrammed in Figure 3.22.

The formation of multilayer films from MTST could be achieved by a continued application of the procedures outlined previously for the chemisorption of monolayers of MTST on silicon (Figure 3.23).

Contrary to previous reports [79], Tillman *et al.* reported that multilayer samples were fabricated easily on silicon wafers in which a linear relationship between the film thickness and the layer number is observed (Figure 3.24). The linear regression line through the data points showed a slope of 35 Å/layer. There was a tendency toward decreasing precision of thickness measurements with increasing layer numbers: The standard deviation of film thickness at various positions across the face of the sample increased from ~±1 Å for only a few monolayers to ~±8 Å for very thick films of *ca.* 20 monolayers.

ATR FTIR spectra for a multilayer sample are presented in Figure 3.25. Note that the seven-layer sample (Figure 3.25 top) showed approximately 7 x the intensity of the initial monolayer sample (Figure 3.21, bottom). There were small impurity peaks at 1741 and 1721 cm^{-1} in the seven-layer spectrum. These probably correspond to residual ester carbonyl bands resulting from incomplete

272

reduction of the MeOC(O)— group, especially the peak at 1741 cm^{-1}; perhaps the 1721 cm^{-1} band also is an ester carbonyl band, which is localized in a different chemical environment. These impurity bands are too intense to be solely the result of incomplete reduction of the seventh layer, and must be present throughout the entire bulk of the sample. When this seven-layer sample was treated with MTST in the usual fashion, an eight-layer sample was obtained, yielding a thickness of 39 ± 6 Å for the eighth layer, and advancing contact angles of 69° for water and 28° for HD (vs. 69° and 24° for the initial monolayer).

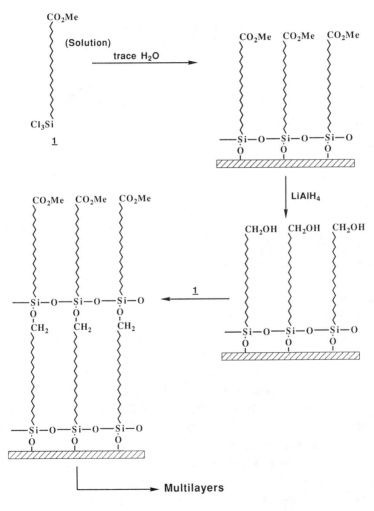

Figure 3.23. Buildup of multilayers from MTST by self-assembly.

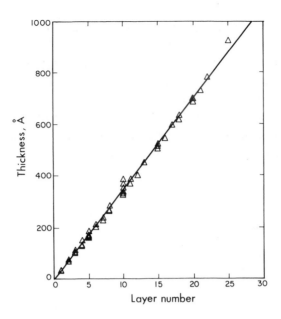

Figure 3.24. Film thickness vs. layer number. (From Tillman *et al.* [15], © 1988, Am. Chem. Soc.).

Reduction of the surface ester group resulted in a monolayer with almost exactly 100% of the expected intensity for the methylene stretching vibrations (based on the unreduced eighth layer). The ester carbonyl band in the IR is not reduced as completely as was observed for the initial monolayer; the residual peak at 1741 cm^{-1} is almost twice that found for the initial $HOC_{23}Si/Si$ monolayer. The resulting water contact angle was 44°, compared with 35° for the $HOC_{23}Si/Si$ monolayer. A chain axis tilt of ~15°–22° was measured from the ATR spectra of both the unreduced and reduced eighth layer samples.

The ellipsometry data, absorbance intensities, and dichroic ratios for the multilayer samples all suggested that the multilayer samples clearly were composed of distinct monolayers, and not layers of bulk film. The IR data indicated that there may be more tilting or disordering of the alkyl chains in the seven-layer sample than for the monolayer samples. These data did not answer unambiguously the important question of whether disorder tends to increase with increasing film thickness, which is a critical issue. In trying to answer this question, the contact angles for both the ester and the reduced alcohol surfaces were monitored during the process of multilayer fabrication. No apparent deterioration of either the advancing hexadecane or water contact angles on the

274

ester surface with increasing layers was found (Figure 3.26). In fact, the tendency was for contact angles to increase after the initial monolayer, and after three or four monolayers the advancing contact angles appeared to stabilize at about $69 \pm 2°$ for water and $30 \pm 2°$ for hexadecane, as compared with $68 \pm 2°$ and $27 \pm 2°$ for the initial monolayers employed in the multilayer work.

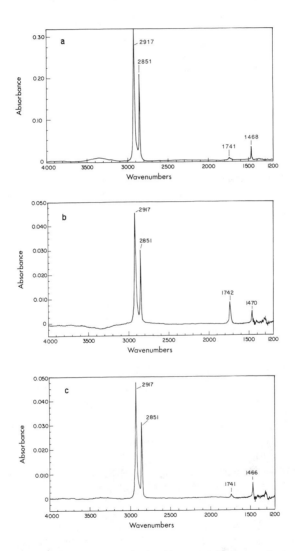

Figure 3.25. ATR FTIR s-polarized spectra for (a) a seven-layer sample, terminated by an $HOC_{23}Si$ monolayer (i.e., reduced); (b) the $MeO_2CC_{22}Si$ eighth layer adsorbed on the sample in (a) ; (c) the $HOC_{23}Si$ eighth layer after treatment of the $MeO_2CC_{22}Si$ eighth layer sample in with $LiAlH_4$. (From Tillman *et al.* [15], © 1988, Am. Chem. Soc.)

275

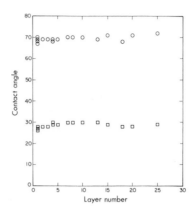

Figure 3.26. Water (circles) and *n*-hexadecane (squares) advancing contact angles on MeO$_2$CC$_{22}$Si/Si monolayers vs. layer numbers. (From Tillman *et al.* [15], © 1988, Am. Chem. Soc.)

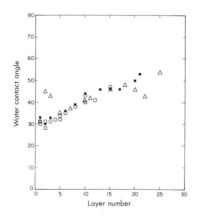

Figure 3.27. Advancing water contact angles on HOC$_{23}$Si/Si monolayers *vs.* layer number. (From Tillman *et al.* [15], © 1988, Am. Chem. Soc.)

However, the same cannot be said for the water contact angles on the reduced, hydrophilic, alcohol surfaces (Figure 3.27). There was a significant general increase in the advancing water contact angle, so that by about 20 monolayers adsorbed on silicon wafers, the contact angle for water on the

276

reduced 20th layer has increased to ~49° from the initial value of ~30°. This result may reflect a general tendency towards increasing disorder in the monolayers with increasing layer numbers. If this is indeed the case, the water contact angle appears to be the most sensitive and obvious indicator of disorder in the monolayer structure, although it is still unknown what the nature and origin of this disordering is.

To investigate further the question of developing disorder, it was necessary to be able to introduce disorder into the monolayers in some kind of controlled fashion, and to measure the resulting contact angles. This was achieved by the formation of partial monolayers (measured by ellipsometry) on silicon wafers. Although partial monolayer formation represents only one of several kinds of disorder that might be expected to be possible in monolayer assemblies, it probably is representative. The alkyl chains would be expected to bend and intertwine to fill the free volume. It also is quantifiable, since partial coverage should lead to reduced film thicknesses, which can be detected with a reasonable level of confidence by the ellipsometer. This model of partial monolayers is in essential agreement with that proposed by Wasserman *et al.* [80]. (See Section 3.1.D.)

The results, plotted as contact angle vs. film thickness, are presented in Figures 3.28 and 3.29. From these data, it appears that the less sensitive indicators of disorder were the water and hexadecane contact angles on the ester surface, while the most sensitive contact angle data, again, were the water contact angles on the reduced alcohol surfaces. Above about 25 Å thickness, there was little change in the contact angles on the ester surface; for the alcohol surface, on the other hand, the water contact angle changed from ~40° at 25 Å thickness to ~30° at the close-packed thickness of 34–35 Å.

This represents a clear trend and justifies the assertion made earlier that there is a general tendency towards increasing disorder in the multilayer samples with increasing number of layers. The idea that increasing disorder would increase the water contact angle is intuitively plausible, since disorder effectively would expose methylene units to the surface, as discussed in Section 1.2.A.

Yet, despite this generally increasing level of monolayer disorder, Tillman *et al.* demonstrated the preparation of a multilayer film with thickness approaching 0.1 μm. Moreover, despite the presence of 5–10% unreduced carbonyl groups, successful adsorptions of 25 layers were accomplished. Thus, it is likely that the *in situ* formation of a polysiloxane backbone at the substrate–solution interface allows the monolayer to "bridge over" defects, such as pinholes and unreduced carbonyl groups. Proof of such a repair mechanism would be very significant, since it is difficult to imagine the construction of very thick films (1–2 mm, 250–500 layers) by self-assembly if defects inevitably propagate and grow.

277

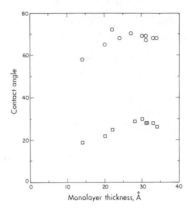

Figure 3.28. Advancing water (circles) and *n*-hexadecane (squares) contact angles for $MeO_2CC_{22}Si/Si$ monolayers and partial monolayers vs. monolayer thickness. (From Tillman *et al.* [15], © 1988, Am. Chem. Soc.)

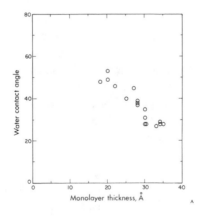

Figure 3.29. Advancing water contact angles for $HOCC_{22}Si/Si$ monolayers and partial monolayers vs. monolayer thickness. (From Tillman *et al* [15], © 1988, Am. Chem. Soc.)

An examination of the surfaces by optical microscopy at 500–1000X revealed that particulate matter of 1–10 μm size was present on two multilayer samples, of five- and 10-monolayer thicknesses. From microscopy, ATR FTIR spectra, and ellipsometry results, it was suggested that adsorbed impurities may be present in the surfaces as both a thin film, and as gel slugs, dust, or other particulate matter. Other gross features, such as cracks in the film, could not be detected under the

278

microscope. As the films grew thicker (> 20 layers), they acquired a blue color due to interference effects, and the surface could be evaluated visually. Except for small regions, usually at the wafer edges, where the process of multilayer formation evidently had been disrupted, the wafers appeared to be coated evenly.

The conclusion of the discussion on SA monolayers of alkyltrichlorosilane derivatives is that, so far, the results are encouraging for the eventual fabrication of relatively thick films by self-assembly from *simple alkyltrichlorosilanes*. However, much remains to be learned regarding the effects of molecular structure on the packing and order, and on the extent and nature of defects, in the two-dimensional assembly.

The conclusion of the discussion on SA monolayers of alkyltrichlorosilane derivatives is that, so far, the results are encouraging for the eventual fabrication of relatively thick films by self-assembly from *simple alkyltrichlorosilanes*. However, much remains to be learned regarding the effects of molecular structure on the packing and order, and on the extent and nature of defects, in the two-dimensional assembly. Thus, issues such as the chromophore size (one, two, or more aromatic rings), its shape (e.g., bipenyl, naphthyl, or stilbene), its position along the alkyl chain, and its dipole moment are still awaiting systematic studies. Also remaining to be addressed is the problem of impurity adsorption. Ellipsometry and IR showed that there was a tendency for excess material to adhere to the monolayer surface. Both also suggested that the nature of this impurity in the multilayer process is a combination of solvent and ester-containing material, presumably polysiloxanes derived from MTST adsorbed during the silanation step. It is possible that in the latter case, the water-saturated hexadecane used as a solvent in the silanation solutions apparently caused severe hydrolysis and polymerization in solution. On the other hand, Wasserman *et al.* reported that highly dried silanation solutions gave very slow and impractical kinetics (~5 h/monolayer) [70]. It is essential, therefore, that adsorption kinetics is tuned so that relatively thick films can be constructed in reasonable time periods, while at the same time, good solvents are used to avoid precipitation. Other improvements in experimental procedures may be to use a clean room for the construction of multilayer films, thus eliminating dust contamination.

3.3. Monolayers of Alkanethiols on Gold

In 1983, Nuzzo and Allara published the first paper in this area, showing that dilkyldisulfides (RS-SR) form oriented monolayers on gold surfaces [97]. Later, it was found that sulfur compounds coordinate very strongly to gold [16–24, 29,

30], silver [25], copper [26–28], and platinum surfaces [29]. However, most of the work to date has been made on gold surfaces, mainly because of the fact that gold does not have a stable oxide [98], and thus it can be handled in ambient conditions.

A fresh, clean, hydrophilic gold substrate (a general discussion on substrates appears in Section 2.1.B.3) is usually immersed into a dilute solution (10^{-3} M) of the organosulfur compound in an organic solvent. Immersion times vary from several minutes to several hours for alkanethiols, while for sulfides and disulfides, immersion times of several days are needed. The result is a close-packed, oriented monolayer (Figure 3.30).

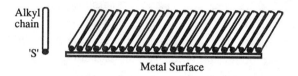

Figure 3.30. A self-assembled monolayer of alkanethiols on a gold surface.

Table 3.5. Adsorption of terminally functionalized alkyl chains from ethanol onto gold. (From Bain *et al.* [99], © 1989, Am. Chem. Soc.)

	$\theta_a(H_2O)^a$	$\theta_a(HD)^b$	Thickness (Å) Obsd^c	Calcd^d
$CH_3(CH_2)_{17}NH_2$	90	12	6	22–24
$CH_3(CH_2)_{16}OH$	95	33	9	21–23
$CH_3(CH_2)_{16}CO_2H$	92	38	7	22–24
$CH_3(CH_2)_{16}CONH_2$	74	18	7	22–24
$CH_3(CH_2)_{16}CN$	69	0	3	22–24
$CH_3(CH_2)_{21}Br$	84	31	4	28–31
$CH_3(CH_2)_{14}CO_2Et$	82	28	6	h
$[CH_3(CH_2)_9C\equiv C]_2Hg$	70	0	4	17–19
$[CH_3(CH_2)_{15}]_3P^e$	111	44	21	21–23
$CH_3(CH_2)_{22}NC$	102	28	30	29–33
$CH_3(CH_2)_{15}SH^f$	112	47	20	22–24
$[CH_3(CH_2)_{15}S]_2$	110	44	23	22–24
$[CH_3(CH_2)_{15}]_2S^g$	112	45	20	22–24
$CH_3(CH_2)_{15}OCS_2Na$	108	45	21	24–26

^{a}Advancing contact angle of water. ^{b}Advancing contact angle of hexadecane. ^{c}Computed from ellipsometric data using $n = 1.45$. ^{d}Assumed that the chains are close-packed, *trans*-extended and tilted between 30° and 0° from the normal to the surface. ^{e}Adsorbed from acetonitrile. ^{f}Reference 15. ^{g}Reference 12. ^{h}An ester group is too large to form a close-packed monolayer.

The effect of the head group on monolayer formation on gold was studied by Bain et al. [99] They examined a variety of head groups and also conducted competition experiments in dilute solutions containing the components, $A(CH_2)_nX$ and $B(CH_2)_mY$, in a range of molar fractions but with the same total concentration. Their criteria for a formation of a good monolayer were the advancing contact angles of water and hexadecane, and the thickness estimated by ellipsometry. They considered a well-packed monolayer for long-chain, methyl-terminated molecules when $\theta_a(H_2O) > 100°$, and $\theta_a(HD) > 40°$. (Note that while the suggested value for HD is acceptable, the one for water is ~14° lower than that suggested for a close-packed methyl surface. (See Section 1.II.1.) The results are summarized in Table 3.5. Clearly, it is only the sulfur and phosphorus, strong interaction with the gold substrate that promoted the formation of a close-packed, ordered monolayer. The isonitrile, although known to coordinate to gold [29], formed poorly packed monolayers compared to those formed by the thiols or phosphines. In a competition experiment, Bain et al. immersed a gold slide in an acetonitrile solution containing a 3:1 mixture of triocty phosphine, $[CH_3(CH_2)_7]_3P$ and 11-hydroxyundecane-1-thiol, $HO–(CH_2)_{11}–SH$ (HUT), respectively. They measured a 47° water contact angle on the resulting monolayer, which suggests an approximate 1:1 ratio of methyl (or methylene) and hydroxyl groups on the surface. This indicates that there is a preference for the adsorption of the thiol, since the methyl–hydroxyl ratio in solution was 9:1. It thus can be concluded that *the thiol group forms the strongest interaction with the gold surface* over all the head groups studied.

Ulman et al. adsorbed octadecanethiol ($C_{18}H_{37}SH$, ODT) and dodecanethiol ($C_{12}H_{25}SH$, DDT) onto silver surfaces. The reaction of the thiol group with the silver was very fast and the monolayers were stable. They had samples of ODT/Ag that were over one year old and had not tarnished [25]. The film thickness was 26 ± 2Å and 15 ± 2Å for ODT/Ag and DDT/Ag, respectively. Advancing contact angles for water and hexadecane were found, within experimental error, to be similar to those measured for the corresponding monolayers on gold ($112 \pm 1°$ and $44 \pm 1°$, respectively).

A. Kinetics of Formation of Monolayers of Alkanethiols on Gold

The kinetics of the formation of alkanethiol monolayers on gold was studied by Bain et al. [16]. At relatively dilute solutions (10^{-3} M), they could observe two distinct adsorption kinetics: a very fast step, which takes a few minutes, by which the contact angles are close to their limiting values and the thickness about 80–90% of its maximum, and a slow step, which lasts several hours, at the end of which the thickness and contact angles reach their final values (Figure 3.31).

We note that while these results suggest that 1 mM is a convenient concentration for most experimental work, Ulman found that a higher concentration, e.g., 10 mM, can be used for simple alkanethiols. However, functionalized thiols that form surfaces with high excess free energy — for example, HUT — tend to yield monolayers (HUT/Au) with excessive thicknesses, probably due to adsorption of the alcohol on the surface [81].

The effect of chain length on the kinetics of formation of alkanethiol monolayers on gold was studied by Bain *et al.* for C_{10} and C_{18} alkanethiols [1]. It was found that the kinetics is faster for longer alkyl chains. This is due to the fact that van der Waals interactions are a function of the chain length.

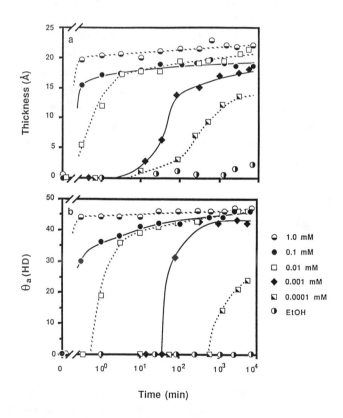

Figure 3.31. Kinetics of adsorption of octadecanethiol from ethanol as a function of concentration: (a) ellipsometric thickness, (b) advancing contact angles. (From Bain *et al.* [16], © 1989, Am. Chem. Soc.)

282

The same kinetics behavior reported in the preceding also is typical for alkanethiols containing more bulky groups, such as phenyl rings. The kinetics in this case is a function of the total number of alkyl carbons in the molecule, but also of the position of the phenyl ring in the chain. Faster kinetics are observed when the phenyl is closer to the thiol and thus, the larger portion of the alkyl chain faces the solution [40].

B. The Effect of Chain Length on Monolayer Formation

Porter *et al.* [17] were the first to study the effect of chain length on the properties on monolayers in a series of *n*-alkanethiols, $CH_3(CH_2)_nSH$, and Bain *et al.* repeated these studies and report water and HD contact angles as a function of the chain length [16]. Figure 3.32 shows film thickness estimated by ellipsometry [17]; Figure 3.33 shows the same measurements, and the advancing water and HD contact angles, all as a function of the chain length [16]. The measured thickness by Porter *et al.* shows two different regions of dependence on *n*: the first between $n = 1$ and $n = 9$, and the second between $n = 9$ and $n = 21$.

While in the first, a linear regression between $n = 1$ and $n = 5$ results in a slope of 0.56 Å/CH$_2$ and intercept of 4.6 Å, in the second, the region between $n = 9$ and $n = 21$ is linear and results in a slope of 1.5 Å /CH$_2$ and an intercept of 3.8 Å. Note that this behavior has not been observed by Bain *et al.* (Figure 3.33) [16], or by Evans and Ulman [100]. Figure 3.33 does not show such pronounced differences between the short ($n < 9$) and long ($n > 9$) alkanethiols, and the data could be fitted to a straight line. The estimated thickness calculated for all-*trans* fully extended *n*-alkanethiols perpendicular to the surface were calculated in both cases using known bond lengths and bond angles [101]. According to these calculations the estimated slope and intercept in Figure 3.32 should be 1.3 Å /CH$_2$ and 5.6 Å, respectively.

Note that the intercept corresponds to the length of the CH_3SH molecule. Of course, if a tilt of 25° is taken into account (and FTIR data suggests an average tilt of 20°–30° from the surface normal; see later in Section 3.3.E), the estimated slope is only 1.1 Å/CH$_2$. Three factors stand out as possible contributors to the considerable discrepancy between the expected calculated thicknesses and that estimated by ellipsometry: the refractive index of the gold substrate, the refractive index of the adsorbed film, and the surface coverage (and therefore, the packing of the monolayer).

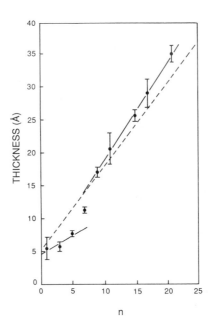

Figure 3.32. Film thickness (•) of n-alkanethiols adsorbed on gold; the ellipsometric-determined thickness (——) and estimated thickness for fully extended chain normal to the surface in Å (----) are plotted against the number of methylene groups in the chain, n. (From Porter *et al.* [17], © 1987, Am. Chem. Soc.)

It seems that one may not be able to obtain a true optical response of a bare, unreacted gold substrate unless ultrahigh vacuum is used. This is because even with the most careful operation, there always will be film of contaminants on the gold surface, i.e., chemisorbed CO, O_2, H_2O, and hydrocarbons [102–104], and physisorbed H_2O, hydrocarbons, and other organic compounds that exist in the laboratory [104–106]. Apparently, this contaminant-film affects only the kinetics of adsorption, and is displaced from the surface by the thiols in the process of monolayer formation [16]. Therefore, when the thickness is estimated by ellipsometry, a monolayer that is adsorbed on a "bare" gold substrate — and for which there are no optical constants — is examined. Furthermore, even if true optical constants for the gold substrate were available, it is not clear if they are not altered by the interaction between the sulfur and the gold. Moreover, even greater discrepancies between calculated and observed thicknesses can be expected for metal substrates such as copper and silver, due to the existence of native oxides on their surfaces.

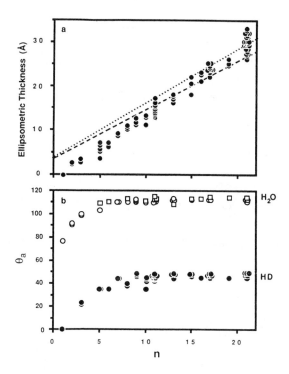

Figure 3.33. Monolayers of n-alkanethiols, $CH_3-(CH_2)_n-SH$, on gold. (a) Ellipsometric thickness. The dotted (dashed) line represents the thickness expected theoretically for a close-packed monolayer oriented normal (tilted 30° from the normal) to the surface. (b) Advancing contact angle (•) hexadecane, (o) water. (From Bain *et al.* [16], © 1989, Am. Chem. Soc.).

Thus, if some of the metal oxide has been reduced by the thiol, the refractive index of the substrate following adsorption may be different. Moreover, differences in substrate refractive index may be related to the kinetics of the oxide reduction, and therefore, may be a function of the chemical properties of metal substrates.

Bain *et al.* used the value of $n_f = 1.45$ (similar to that used by Porter *et al.*), and suggested that film thickness is "only moderately sensitive" to the value of n_f [16]. They calculated a decrease of ~2.5% in the thickness for $n_f = 1.47$, and noted that for loose packing, the uncertainty in their model should be greater. On the other hand, Porter *et al.* found that if the value of $n_f = 1.50$ was used for a monolayer of docosanethiol (C_{21}), a thickness of 31.6 Å was measured; however, when the value $n_f = 1.45$ was used, the thickness estimated was ~33 Å (a difference of ~4.4%). The calculated maximum thickness for an all-*trans* chain [101], with 25° tilt was ~29 Å. Wasserman *et al.* suggested the value of n_f = 1.45 for monolayers of alkyltrichlorosilane on silicon [80], but Tillman *et al.* measured a value of n_f = 1.50 for SA multilayers of MTST [15]. It may be possible that the Au-S dipole sheet increases the value of n_f, since surface potential measurements showed that even a C_{22}-length alkyl chain did not screen the effect of this dipole sheet [100].

The arguments discussed so far do not suggest a clear explanation for the differences among the aforementioned reports [16, 17, 100]. To consider the effect of chain length, one can use FTIR data, where it is suggested that SA monolayers of short alkanethiols ($n \leq 9$) are liquid-like [17]. Such systems are more sensitive to thermally induced disorder, which is a function of chain length [38]. Of course, differences in coverage could explain the differences in thickness measurements. However, in all three experiments mentioned [16,17,100], substrates were left for long periods of time in the thiol solution, and differences in coverage thus should be very small.

One parameter that will have an apparent effect on molecular order in the monolayers is the crystallographic face of gold in contact with the thiol solution. Strong and Whitesides showed that the S···S distance on (111) gold is 4.97 Å, while that distance on (100) gold is only 4.54 Å [21]. Moreover, the symmetry of the two-dimensional assembly was pseudohexagonal in the former and square lattice in the latter case. (See a more detailed discussion in Section 3.3.D.) If the gold substrate is a mixture of crystallographic faces, the assembly structure at the molecular level will become a mixture of the aforementioned structures. The overall monolayer packing in that case will be a function of the ratio of these phases and of their domain size. Porter *et al.* did not discuss the structure of their gold substrates; however, we and others found that evaporation of gold on silicon yields the (111) phase. (See Section 2.1.B.3 for further discussion on gold substrates.)

In conclusion, it seems that none of the factors just discussed can explain, by itself, the results reported by Porter *et al.* for the short alkanethiol monolayers. However, I brought up this detailed discussion here for the reader to appreciate the complexity of a molecular level discussion, where the available analytical tools are far from being sensitive enough to deal with such problems.

C. Solvent Effects on Monolayer Formation

Most of the work reported in the literature suggests the use of ethanol as the preferred solvent. However, other solvents also can be used, and sometimes are preferred. Practically, the solubility properties of the alkanethiol derivative dictate the choice of solvent. Figure 3.34 presents the effect of solvent on the formation of hexadecanethiol monolayers. Apparently, there is no considerable solvent effect, but Bain *et al.* mentioned in their paper that hexadecanethiol monolayers adsorbed from hexadecane had abnormally low contact angles, presumably due to the incorporation of the solvent into the monolayers [16]. Therefore, it is recommended to use a solvent that does not show a tendency to incorporate into the two-dimensional system (ethanol, THF, acetonitrile, etc.).

Ulman *et al.* found that in the preparation of mixed monolayers of different alkanethiols (e.g., HUT and DDT), THF solutions yield very close to a linear relationship between the solution and surface OH concentrations [107]. This probably is because THF is not very polar, but still can form hydrogen bonds with the HUT molecules, thus stabilizing both thiols in a similar way. On the other hand, using alcohol in similar mixtures yielded preferential adsorption of the nonpolar component at low concentrations of the polar one [23]. Thus, it should be possible to control the concentration of different surface functionalities in mixed monolayers by the choice of solvent (or solvent mixture).

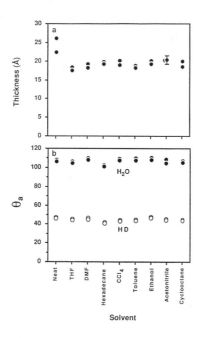

Figure 3.34. The effect of solvent on the formation of hexadecanethiol monolayers: (a) Ellipsometric thickness; (b) advancing contact angles: (●) water, (o) hexadecane. (From Bain *et al.* [16], © 1989, Am. Chem. Soc.)

287

Table 3.6. Atomic composition of a monolayer of $HS-(CH_2)_{10}-CO_2CH_3$ on gold, derived from angle-dependent XPS. (From Bain *et al.* [16], © 1989, Am. Chem. Soc.)

Element (%) or Ratio of Elements	Takeoff Angle		
	90°	35°	15°
Au	44	31	17
C	44	55	66
O	8	12	17
S	4	2	1
C/O	5.8	4.7	4.1
C/S	10	30	60

D. The Structure of Alkanethiol Monolayers on Gold

Chemisorption of alkanethiols or dialkyldisulfines on gold(0) surfaces yields the gold(I) thiolate (RS^-) species [108]. The reaction of dialkyldisulfides with gold(0) is an oxidative addition one,

$$RS\text{-}SR + Au_n^0 \Rightarrow 2RS^- - Au^+ + Au_n^0 \ ,$$

while the mechanism involved in the reaction of alkanethiols with gold(0) is not completely understood. Thus, the formation of $RS^- - Au^+$ requires the loss of the SH hydrogen, but it has not been determined yet whether this proton is lost as H_2, either via the reductive elimination reaction of the gold(II) hydride ($RS^- - Au^{2+} - H^-$, formed by oxidative addition of the alkanethiol) or by another unknown reaction,

$$RS\text{-}H + Au_n^0 \Rightarrow RS^- - Au^+ + \tfrac{1}{2}H_2 + Au_n^0 \ ,$$

or as H_2O, either via the reaction of either the gold(II) hydride or another unknown species with traces of oxidant,

$$RS\text{-}H + Au_n^0 + \text{oxidant} \Rightarrow RS^- - Au^+ + \tfrac{1}{2}H_2O + Au_n^0 \ .$$

It was found that the exchange of a chemisorbed thiolate with alkanethiols of dialkyl disulfides in solution is slow (from hours to days) [16], and that heating these monolayers under vacuum results in desorption of disulfides [109].

Bain *et al.* [16] studied monolayers of alkanethiols on gold using x-ray photoelectron spectroscopy (XPS) [110]. It should be emphasized that XPS is not recommended as an analytical tool for the calculation of atomic composition in monolayers. This is due to the fact that these calculated compositions are sensitive to various factors, such as the energy of the primary x-ray beam,

288

variations in photoionization cross section with chemical structure, the takeoff angle, and the elemental distribution perpendicular to the surface. (For a more detailed discussion on XPS, see Section 1.2.L.) However, elemental compositions obtained by XPS indicate qualitatively the elements present in the monolayer, which is very useful in the study of mixed monolayers. Furthermore, the use of angular dependence allows obtaining information on the depth profile for the surface, which is useful for the study of complex systems in which, for example, the monolayer properties are studied as a function of the position of a specific group in the chain [92]. Table 3.6 presents the results of a study on a monolayer of $HS-(CH_2)_{10}-CO_2CH_3$ on gold.

The takeoff angle is the angle between the surface and the photoelectrons accepted by the analyzer. Therefore, as the takeoff angle decreases, it will sample atoms on the surface, and as the takeoff angle increases, it will sample atoms deeper in the monolayer [110]. The preceding results are in agreement with a monolayer structure in which the ester group is at the surface and the sulfur is on the substrate.

Strong and Whitesides studied monolayers of docosanethiol using electron diffraction [24]. They found that the symmetry of sulfur atoms in a monolayer of docosanethiol on (111) gold is hexagonal with an S···S spacing of 4.97 Å, and calculated area per molecule of 21.4 Å^2. This spacing also is found for a stearic acid monolayer on the (111) face of silver [111], and is close to the spacing between the next nearest gold atoms on the (111) gold surface (4.99 Å). Thus, while the chain spacings are equivalent to the second nearest-neighbor gold spacing, the monolayer lattice vectors lie in the same direction as the principal gold lattice directions {110}, and therefore, they concluded that the monolayer is not epitaxial to the gold lattice. Later, after the helium diffraction studies by Chidsey et al. (detailed in Section 1.2.C) [22], it was suggested that alkanethiols and dialkyl disulfides chemisorb epitaxially on Au(111) [20].

Recently, Sellers et al. performed ab initio geometry optimization of HS— and CH$_3$S— groups on cluster models of Au(111) surface, at the PECR Hartree-Fock + electron correlation (MBPT2) level [112]. The structure of SH in the on-top position on the Au$_{16}$ cluster model (Figure 3.35) was optimized, and then the SH fragment was allowed to move off the on-top site. This process took took the SH group to the hollow site and to an equilibrium geometry having a linear Au(111)–S–H angle. The hollow equilibrium site is 6.03 kcal/mol more stable than the on-top optimized configuration.

The bonding between the Au(111) surface and the sulfur atom in the hollow site has both σ- and π-character. The σ-bonding involves primarily the S p-orbitals and the Au 6s-orbitals with significant contribution from the Au p- and d-orbitals. The Au p-orbitals contribute significantly to the π-bonding orbitals as well. It appears that the π-interaction is responsible for the linear structure in the hollow site. The equilibrium surface–S distance is 1.978 Å which gives 2.586 Å for the Au–S distance. The S–H distance is 1.353 Å. In the CH_3S on Au(111) case, the equilibrium geometry was linear also (Au(111)–S–CH_3 angle = 180°). The Au(111)–S distance is 1.905 Å which gives an Au–S distance of 2.530 Å.

The chemisorption arrangement emerging from these calculations is consistent with the picture suggested by Chidsey and Loiacono [21], which is based on the Strong and Whitesides data [24]. The calculations confirm that the hollow site on the Au(111) surface is the more stable binding site, and that the chemisorption is epitaxial. Figure 3.35 shows the arrangement of metal atoms (the open circles) of the (111) surface. The black (filled) circles denote hollow sites that are arranged in a hexagonal relationship with respect to each other as is indicated by the dark lines connecting six of them. With the gold at their bulk value (2.884 Å) these hollow sites are 4.99 Å apart, which is in an excellent agreement with the electron diffraction [24] and helium diffraction data [22]. Recent STM studies confirmed the hexagonal symmetry [113].

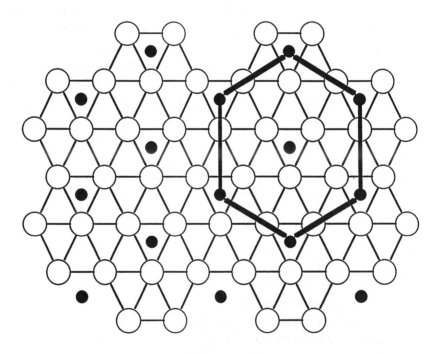

Figure 3.35. Proposed coverage scheme for alkanethiols on Au(111) surface.

The symmetry of sulfur atoms in a monolayer of docosanethiol on (100) gold is a base-centered square with an S···S spacing of 4.54 Å, and calculated area per molecule of 20.6 Å^2 [24]. This S···S spacing is problematic if the chemisorbed sulfur atoms are assumed to occupy the same binding sites, and hence, have the same distance from the Au(100) surface. The reader is referred to Section 4.4.2.A for a detailed discussion, however for the present discussion we ask the reader to turn to Figure 4.10. There, it is clear that assemblies where S···S distances are 4.54 Å are highly unstable. This is because the interlocking of hydrogen atoms in neighboring alkyl chains (Figure 4.9) cannot be achieved. However, if the sulfur atoms are chemisorbed on both on-top and hollow sites (in alternating fashion), neighboring sulfur atoms would have different distances from the solid surface. If the difference between the two Au(100)–S distances is right, the alkyl chains would form a stable assembly with all hydrogen atoms interlocking as in Figure 4.9, but with a tilt angle much smaller than 32°. A possible alkanethiol arrangement on a gold(100) surface is presented in Figure 3.36 [112]. In this figure, the top layer gold atoms are shaded and the small black dots represent sulfur chemisorbed sites. Thus, looking at the Au(100) surface as a chess board, the coverage scheme works as a knight moves in chess.

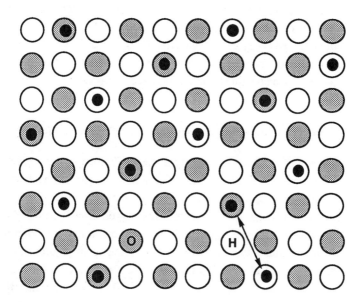

Figure 3.36. Proposed coverage scheme for alkanethiolates on Au(100) surfaces. The on-top and hollow sites are denoted by O and H, respectively. The distance between the on-top and hollow sites (represent by the arrow) is 4.56 Å.

Another interesting observation made by Strong and Whitesides is the translational correlation distance, which is indicative of the crystalline order [24]. In the monolayer of docosanethiol on (111), this translational correlation has the value of 60 Å, while in the monolayer on (100), it is 130–200 Å (in the direction of the nearest-neighbor vectors). These results may suggest that translational correlation in two-dimensional, crystalline monolayers are a function of the spacing between the head groups (S⋯S). This is because the spacing defines the free volume, and hence the molecular tilt angle in the assembly. The tendency of small clusters, with molecular tilt angles in different directions (in the xy plane), to merge into a larger cluster (larger translational correlation) in the adsorption process decreases with the increasing tilt angle, since overlap between such clusters at their boundaries is expected to be very small. As the molecular tilt angle decreases, the difference between tilts in different directions decreases, and the merging of neighboring small clusters becomes possible.

The fact that a (111) gold surface directs an assembly of the alkanethiols in a hexagonal symmetry, while on the (100) gold surface the symmetry of the alkanethiol assembly is base-centered square, is remarkable. We have discussed the two chemisorption schemes but have not addressed an important question, i.e., what is the relationship (if any) between the chemisorption at the gold surface and the interchain interactions? In other words, can interchain interactions change the chemisorption scheme calculated for the CH_3S- group, and, for example, reduce the Au(111)–S–C angle from being linear? Such questions are interesting, especially when the interchain interactions are not purely van der Waals in nature but also contain strong electrostatic components.

E. FTIR Studies of Alkanethiol Monolayers on Gold and Silver Surfaces

We discussed in Part One that grazing-angle FTIR spectroscopy represents a useful method for estimating alkyl chain orientations in monolayers adsorbed on reflecting metallic surfaces due to the strong polarization normal to the metal surface that is established. In an oriented monolayer, transition dipoles show varying tilt angles with respect to this polarization vector, and the orientation of various molecular features can be estimated from the relative intensities of the bands observed. The C–H stretch region is quite characteristic for long-chain alkyl monolayers, with relative methylene and methyl C–H stretch intensities markedly different from those found in isotropic states. Nuzzo et al. [30, 97] and Porter et al. [17] used this technique to estimate the orientation of the chains in long alkyl-chain thiols and disulfides on gold. Finklea and Melendrez also reported the grazing-angle spectra for monolayers of alkanethiols on gold [114]. The tilt angle of the chain axis from the normal in thiol monolayers on gold was estimated to be on the order of 20°–35° [17].

Ulman used grazing-angle FTIR spectroscopy to compare monolayers of alkanethiols on gold and on silver surfaces [25]. Figures 3.37 and 3.38 present the spectra in the C–H stretch region of 2800–3100 cm^{-1} for monolayers ODT/Au, and ODT/Ag. The $v_a(CH_2)$ and $v_s(CH_2)$ peak positions for ODT/Ag agreed well with those of condensed-phase alkanes [115], as well as with those of CH_3–$(CH_2)_{21}$–SH in KBr [17]. (see Part One for further discussion.) These spectra suggest that the alkyl chains in the monolayers on silver are less tilted (~6–7° tilt) than the corresponding monolayers on gold (~30° tilt). Furthermore, the spectrum of the ODT/Ag monolayer resembles, for example, the spectrum of cadmium arachidate on an Si (111) surface, where a tilt angle of $0 \pm 15^\circ$ was proposed [116].

It is apparent that there are differences in the alkyl chain orientation for monolayers prepared on the two metal surfaces. However, since Ag and Au lattices are very similar (only ~0.3% difference) [117], and the organic molecules employed are identical, these results lead to the conclusion that the tilt of the alkyl chains may be controlled by the bonding and/or spacing of the head group sulfur atoms at the metal surface [118]. (See Section 4.2.A for further discussion.)

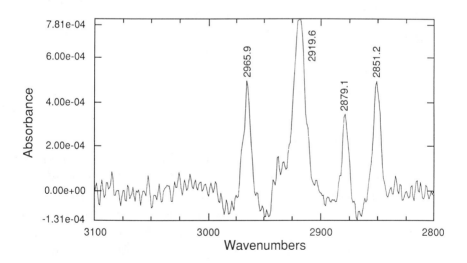

Figure 3.37. A grazing-angle FTIR spectrum of ODT/Au.

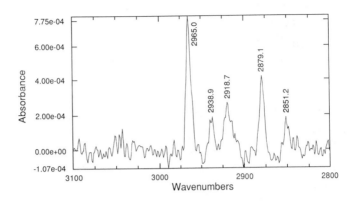

Figure 3.38. A grazing-angle FTIR spectrum of ODT/Ag.

Recently, Harris *et al.* studied the bonding of methanethiolate on Ag(111) using sum-frequency generation spectroscopy [119]. They reported that only the symmetric methyl stretch, $v_s(CH_3)$, could be observed. The absence of any of the two $v_a(CH_3)$ vibrations (Figure 1.6) suggests that the C–S bond in this system is parallel to the surface normal (Figure 3.39); however, no conclusion could be drawn regarding the Ag–S–C angle. If the Ag–S bond also is parallel to the surface normal, it requires the Ag–S–C angle to be ~180°, which was the value calculated for CH_3S– on Au(111) [112]. For the same phenomenon to occur in longer alkyl chains, i.e., that the $v_s(CH_3)$ transition dipole is normal to the surface, the Ag–S–C angle has to be close to the ~110° conventionally used for the Au–S–C angle, provided that the Ag–S bond is parallel to the surface normal (Figure 3.40) [18, 36]. The understanding of the interplay of chemisorptive bonds and alkyl chain packing in monolayers is not completely understood; however, it is clear that both contribute to the final structure of the molecular assembly.

The above structure, however, is true only for the even case (an even number of carbon atoms in $C_nH_{2n+1}SH$). In the odd case, on the other hand, given the same bonding geometry at the surface, the $v_s(CH_3)$ transition dipole is more parallel to that surface (Figure 3.39). The same arguments can be made for the $v_a(CH_3)$ vibrations, where both the in-plane and out-of-plane dipoles are more parallel to the surface in the even case, and more perpendicular to the surface in the odd case. Thus, it is clear that a careful analysis of the methyl vibrations may give valuable information on molecular orientation. Note that in both the odd and even cases, the $v_s(CH_2)$ vibration has a component that is normal to the surface, since the alkyl chains are tilted from the surface normal.

294

Figure 3.39. A methanethiolate ion on Ag(111) surface.

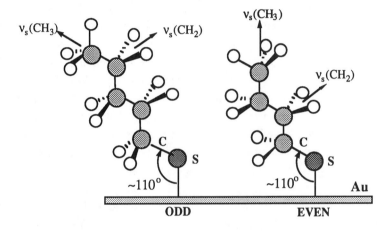

Figure 3.40. A schematic view of odd and even cases for a tilted alkanethiol on gold.

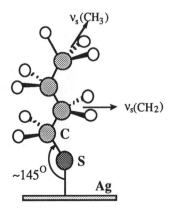

Figure 3.41. A schematic view of a perfectly perpendicular alkanethiol molecule on Ag.

295

This really is the case of alkanethiol monolayers on gold and not on silver. For the $v_s(CH_2)$ transition dipoles to be perfectly parallel to the surface, and hence undetectable by the p-polarized IR light, this angle should be larger ($\sim 145°$, Figure 3.41), assuming again an Ag–S bond normal to the substrate surface. It is more reasonable, however, to assume that the C–S–surface angle is not 145°, and that that alkyl chain is not perfectly perpendicular to the surface.

To summarize, the preceding arguments point to an interesting question regarding the nature of alkanethiol monolayers on gold and silver surfaces, and the mechanism of their organization. We already have discussed the electron diffraction studies (see Section 3.3.A.4), and elaborated on FTIR spectra of alkanethiols on gold and silver surfaces. (See also Section 1.1.B.2.) The picture emerging from these studies is that alkanethiol monolayers are close-packed assemblies with crystalline-like order. However, for a two-dimensional alkanethiol crystal to be formed, molecules should have enough lateral mobility on the surface, either before or after chemisorption. Three scenarios of thiol–metal interactions and molecular assembly can be proposed. In the first and most unlikely one, chemisorbed alkanethiols (i.e., R–S⁻) are organized on a Au⁺/Au surface, and have lateral mobility (i.e., alkanethiolates can "hop" on the gold surface). In the second scenario, which is more reasonable, alkanethiolates are bonded to the surface through a covalent, although polar, Au–S bond [100], and the assembling units are gold mercaptide molecules (R–S–Au) that have enough lateral surface mobility. This may be correct only if the thiolate is chemisorbed at the on-top position (see earlier discussion), or can move back to that position from the hollow site. In the third and most likely mechanism, alkanethiols physisorb on the surface, and are chemisorbed after a brief lateral motion to form gold mercaptide units, as just described. This would explain why as the alkyl chain becomes longer, the assembly becomes more ordered, since organization is driven by van der Waals attraction forces. We note, however, that the second scenario may contribute to ordering, but at a longer time scale. We deduce this assumption from the aging of stearic acid monolayers on Al_2O_3 surfaces observed by Dote and Mowery [120]. In that case, it was found that monolayers became less tilted after one week.

Finally, we refer the reader to Section 1.1.B.2, where the detailed characterization of alkanethiol monolayers on gold by grazing-angle FTIR, made by Porter *et al.* is discussed [3].

3.4. Formation of Double Layers on Gold

We have discussed SA monolayers of alkyltrichlorosilanes on hydroxylated surfaces, and of alkanethiols on gold. In trying to merge these two chemistries, Ulman and Tillman used HUT adsorbed onto gold as an adhesion layer for OTS [81]. Treatment of HUT/Au monolayers (thickness 14 ± 1 Å) with OTS resulted in the formation of ordered, oleophobic monolayers. The resulting bilayer thickness measured $40–41 \pm 2$ Å, thus giving a thickness value of $26–27$ Å for the OTS second layer. The advancing contact angles measured for the second OTS layer were $112°$ for water and $44°$ for HD [4, 14].

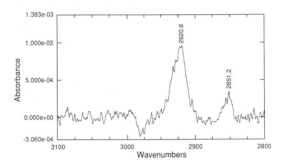

Figure 3.42a. A grazing angle FTIR spectrum of an HUT/Au monolayer. (From Ulman and Tillman [81], © 1989, Am. Chem. Soc.)

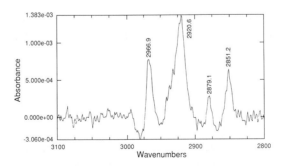

Figure 3.42b. A grazing angle FTIR spectrum of OTS on HUT/Au. (From Ulman and Tillman [81], © 1989, Am. Chem. Soc.)

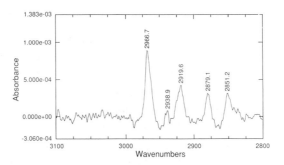

Figure 3.42c. A grazing angle FTIR spectrum of the OTS *second* layer on HUT/Au. (From Ulman and Tillman [81], © 1989, Am. Chem. Soc.)

The grazing-angle FTIR spectrum of HUT/Au is presented in Figure 3.42a. The bilayer spectrum on gold is presented in Figure 3.42b, and the spectrum of the OTS second layer on HUT/Au (obtained by subtraction of the HUT monolayer spectrum from the bilayer spectrum) is presented in Figure 3.42c.

The OTS monolayers had spectral features that are quite similar to those of ODT/Ag (Figure 3.36). From the $v_a(CH_2)/v_a(CH_3)$ intensity ratio, it was estimated that the alkyl chains in the OTS monolayer on HUT/Au were tilted about the same as those in ODT/Ag, and in cadmium arachidate on a Si (111) [121].

Previously published results for OTS monolayers on aluminum [14] and gold surfaces [7], showed more canted alkyl chains, i.e., ~15°–20° tilt (coupled with a 45° rotation). On the other hand, these results are similar to the tilt angle of ≤ 10° proposed by Pomerantz *et al.* on the basis of x-ray diffraction data [79].

We conclude that IR results, together with the ellipsometry and contact angle data, indicate that the OTS second layers are ordered assemblies comparable to OTS monolayers on Si. Thus, the thiol and trichlorosilane techniques in combination yield ordered bilayer structures.

3.5. Orthogonal Self-Assembled Monolayers

One of the questions still open in this area of self-assembly is that of the lateral organization of molecules in two dimensions. It is clear that solving this problem will require the use of molecular recognition in the design of surfactant molecules for SA. However, a very recent report suggested the use of modified substrates to achieve some degree of control on the lateral structure and composition [122].

Taking advantage of the difference in the interaction of carboxylic acids with Al_2O_3, and that of alkanethiols with gold, Liabinis *et al.* prepared substrates by evaporation of ~500 Å of aluminum through a mask onto gold substrates, and exposed them to mixtures of carboxylic acids and thiols. Using series of alkane carboxylic acids [Y–$(CH_2)_n$–COOH] and alkanethiols [X–$(CH_2)_m$–SH], they found that the carboxylic acids adsorbed preferentially on the aluminum oxide, as expected, while the thiols adsorbed preferentially on the gold (Figure 3.43).

To establish further the structure of these assemblies, the authors prepared samples from CF_3–$(CF_2)_8$–COOH and Cl–$(CH_2)_{11}$–SH and studied the composition distributions by introducing them into a scanning Auger microprobe spectrometer. One limitation should be mentioned, however. It was found that $HO(CH_2)_{15}COOH$ adsorbed on aluminum oxide with both polar functionalities to form looped structures. This, of course, prevented the adsorption of a second layer (e.g., OTS) and thus the further demonstration of the utility of this approach for the construction of three-dimensional molecular assemblies. There is no doubt, however, that further developments in the area of orthogonal monolayers may help establish the SA technique as a construction tool for certain molecular devices.

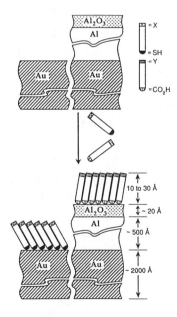

Figure 3.43. Schematic illustration of the formation of orthogonal self-assembled monolayers by adsorption of alkanethiol [X–$(CH_2)_m$–SH] and alkyl carboxylic acid [Y–$(CH_2)_n$–COOH] from a common solution onto a patterned gold-aluminum oxide surface. (From Liabinis *et al.* [122], © 1989, Nature)

299

The demonstration of orthogonal SA monolayers is an interesting advancement in this area. One of the most apparent contributions of such structures may be to the understanding of wetting of molecularly heterogeneous, rough surfaces. Using well-designed surfaces may help in the systematic study of problems such as surface roughness and contact angle hysteresis.

3.6. Concluding Remarks

Self-assembled monolayers of alkyltrichlorosilane may have important applications in the engineering of surface properties or as adhesion layers for catalysts, biomolecules, etc. However, because of the polymeric nature of these assemblies, they are not as ordered as monolayers of alkanethiols on metal surfaces, and, therefore, will not be good as model systems for fundamental studies. We have mentioned that alkyltrichlorosilanes containing polar aromatic groups are practically impossible to purify, and that their oligomers tend to precipitate out of the organic solutions a short time after the solutions are prepared. Thus, although appealing from the material engineering point of view, it is not expected that multilayers of alkyltrichlorosilanes will become a viable technology. This, of course, may change if critical chemical engineering studies are carried out and eliminate the aforementioned problems.

A very recent contribution from Li *et al.* suggests an interesting utilization of alkyltrichlorosilanes in the fabrication of SA multilayers for NLO [123]. There, they used 2-(4-iodophenyl)ethyltrichlorosilane, $I-C_6H_4-CH_2-CH_2-SiCl_3$) as a binding layer in a novel hemicyanine-containing SA film. This material is a liquid and therefore can be distilled. It is also soluble in solvents such as CCl_4 and HD and therefore would not cause severe problems due to precipitation.

Self-assembled, chemisorbed monolayers of alkanethiols on metal substrates open exciting new possibilities of engineering smooth surfaces with their chemical properties fine-tuned at the molecular level. These novel systems exhibit a rich variety of packing and ordering phenomena, as a result of competing steric, dispersive, chemisorption, and intramolecular elastic potentials (Part Four). Hence, these thin films can serve as excellent model systems for evaluating modern theories of wetting, spreading, adhesion, friction, molecular recognition, and related phenomena.

References

1. Bigelow, W.C.; Pickett, D. L.; Zisman, W. A. *J. Colloid Interface Sci.* **1946**, *1*, 513.
2. Zisman, W. A. *Adv. Chem. Ser.* **1964**, *43*, 1.
3. Sagiv, J. *J. Am. Chem. Soc.* **1980**, *102*, 92.
4. Maoz, R.; Sagiv, J. *J. Colloid Interface Sci.* **1984**, *100*, 465.
5. Gun, J.; Iscovici, R.; Sagiv, J. *Ibid.* **1984**, *101*, 201.
6. Gun, J.; Sagiv, J. *Ibid.* **1986**, *112*, 457.
7. Finklea, H. O.; Robinson, L. R.; Blackburn, A.; Richter, B.; Allara, D.; Bright, T. *Langmuir* **1986**, *2*, 239.
8. Cohen, S. R.; Naaman, R.; Sagiv, J. *J. Phys. Chem.* **1986**, *90*, 3054.
9. Maoz, R.; Sagiv. J. *Thin Solid Films* **1985**, *132*, 135.
10. Maoz, R.; Sagiv, J. *Langmuir* **1987**, *3*, 1045.
11. Netzer, L.; Iscovici, R.; Sagiv. J. *Thin Solid Films* **1983**, *99*, 235.
12. Netzer, L.; Iscovici, R.; Sagiv, J. *Ibid.* **1983**, *100*, 67.
13. Netzer, L.; Sagiv, J. *J. Am. Chem. Soc.* **1983**, *105*, 674.
14. Tillman, N.; Ulman, A.; Schildkraut, J. S.; Penner, T. L. *J. Am. Chem. Soc.* **1988**, *111*, 6136.
15. Tillman, N.; Ulman, A.; Penner, T. L. *Langmuir* **1989**, *5*, 101.
16. Bain, C. D.; Troughton, E. B.; Tao, Y. -T.; Evall, J.; Whitesides, G. M.; Nuzzo, R. G. *J. Am. Chem. Soc.* **1989**, *111*, 321.
17. Porter, M. D.; Bright, T. B.; Allara, D. L.; Chidsey, C. E. D. *J. Am. Chem. Soc.* **1987**, *109*, 3559.
18. Nuzzo, R. G.; Dubois, L. H.; Allara, D. L. *J. Am. Chem. Soc.* **1990**, *112*, 558.
19. Rubinstein, I.; Steinberg, S.; Tor, Y.; Shanzer, A.; Sagiv, J. *Nature* **1988**, *332*, 426.
20. Whitesides, G. M.; Laibinis, P. E. *Langmuir* **1990**, *6*, 87.
21. Chidsey, C. E. D.; Loiacono, D. N. *Langmuir* **1990**, *6*, 709.
22. Chidsey, C. E. D.; Liu, G. -Y.; Rowntree, P.; Scoles, G. *J. Chem. Phys.* **1989**, *91*, 4421.
23. Bain, C. D.; Whitesides, G. M. *J. Am. Chem. Soc.* **1989**, *111*, 7164,
24. Strong, L.; Whitesides, G. M. *Langmuir* **1988**, *4*, 546.
25. Ulman, A. *J. Mat. Ed.* **1989**, *11*, 205.
26. Stewart, K. R.; Whitesides, G. M.; Godfried, H. P.; Silvera, I. F. *Surf. Sci.* **1986**, *57*, 1381.
27. Blackman, L. C. F.; Dewar, M. J. S. *J. Chem. Soc.* **1957**, 171.
28. Blackman, L. C. F.; Dewar, M. J. S.; Hampson, H. *J. Appl. Chem.* **1957**, *7*, 160.
29. Troughton, E. B.; Bain, C. D.; Whitesides, G. M.; Nuzzo, R. G.; Allara, D. L.; Porter, M. D. *Langmuir* **1988**, *4*, 365.
30. Nuzzo, R. G.; Fusco, F. A.; Allara, D. L. *J. Am. Chem. Soc.* **1987**, *109*, 2358.
31. Allara, D, L.; Nuzzo, R. G. *Langmuir* **1985**, *1*, 45.
32. Allara, D, L.; Nuzzo, R. G. *Langmuir* **1985**, *1*, 52.
33. Ogawa, H.; Chihera, T.; Taya, K. *J. Am. Chem. Soc.* **1985**, *107*, 1365.
34. Schlotter, N. E.; Porter, M. D.; Bright, T. B.; Allara, D. L. *Chem. Phys. Lett.* **1986**, *132*, 93.
35. Dubois, L. H.; Zegarski, B. R.; Nuzzo, R. G. *Proc. Natl. Acad. Sci. U.S.A.* **1987**, *84*, 4739.

36. Dubois, L. H.; Zegarski, B. R.; Nuzzo, R. G. *J. Am. Chem. Soc.* **1990**, *112*, 570.
37. Nuzzo, R. G.; Korenic, E. M.; Dubois, L. H. *J. Chem. Phys.* **1990**, *93*, 767.
38. Evans, S. D.; Ulman, A.; Sharma, R. *Langmuir* **1991**, *7*, 0000.
39. Huang, D. Y.; Tao, Y. T. *Bull. Inst. Chem., Academia Sinica* **1986**, *33*, 73.
40. Evans, S. D.; Urankar, E.; Ulman, A.; Ferris, N. *J. Am. Chem. Soc.*, submitted.
41. Chen, S. H.; Frank, C. F. *Langmuir* **1989**, *5*, 978.
42. Grow, D. T.; Shaeiwitz, J. A. *J. Colloid Interface Sci.* **1982**, *86*, 239.
43. Trogus, F. L.; Sophany, T.; Wade, W. H.; *Soc. Pet. Eng. J.* **1977**, *17*, 337.
44. Dobiás, B. *Colloid Polym. Sci.* **1978**, *256*, 465.
45. Chapman, J. A.; Tabor, D. *Proc. R. Soc. A* **1957**, *242*, 96.
46. Müller, A. *Proc. R. Soc. A* **1928**, *120*, 437.
47. Verma, A. R. *Proc. R. Soc. A* **1955**, *228*, 34.
48. Adam, N. K. *The Physics and Chemistry of Surfaces*, Oxford University Press: Oxford, 1941.
49. Bunn, C. W.; Howells, E. R. *Nature, London* **1954**, *174*, 549.
50. Moses, P. R.; Murray, R. W. *J. Am. Chem. Soc.* **1976**, *98*, 7435.
51. Walton, *Anal. Chem.* **1976**, *48*, 52.
52. Grushka, E., Ed. *Bonded Stationary Phases in Chromatography*, Ann Arbor Science Publication: Ann Arbor, 1974.
53. Locke, D. C. *J. Chromatogr.* **1973**, *11*, 120.
54. Aue, W. A.; Hastings, C. R. *J. Chromatogr.* **1969**, *42*, 319.
55. Hastings, C. R.; Aue, W. A.; Augl, J. M. *J. Chromatogr.* **1970**, *53*, 487.
56. Hastings, C. R.; Aue, W. A.; Larsen, F. E. *J. Chromatogr.* **1971**, *60*, 329.
57. Kirkland, J. J.; DeStafano, J. J. *J. Chromatogr. Sci.* **1970**, *8*, 309.
58. Kirkland, J. J. *J. Chromatogr. Sci.* **1971**, *9*, 206.
59. DeStafano, J. J.; Kirkland, J. J. *J. Chromatogr. Sci.* **1974**, *12*, 337.
60. Majors, R. E.; Hopper, M. J. *J. Chromatogr. Sci.* **1974**, *12*, 767.
61. Weetall, H. H. *Science* **1969**, *160*, 615.
62. Weetall, H. H.; Hersh, L. S. *Biochim. Biophys. Acta* **1970**, *206*, 44.
63. Weetall, H. H.; Havewala, N. B., in *Enzyme Engineering*, Wingard, L. B., Ed., Wiley: New York, 1972, Vol. 3, p. 241.
64. von Tscharner, V.; McConnell, H. M. *Biophys. J.* **1981**, *36*, 421.
65. Honig, E. P.; Hengest, J. H. T.; den Engelson, D. *J. Colloid Interface Sci.* **1973**, *45*, 92.
66. Highfield, R. R.; Thomas, R. K.; Cummins, P. G.; Gregory, D. P.; Mingis, J.; Hayter, J. B.; Schärpf, O. *Thin Solid Films* **1983**, *99*, 165.
67. Abel, E. W.; Pollard, F. H.; Uden, P. C.; Nickless, G. *J. Chromatogr.*, **1966**, *22*, 23.
68. van Roosmalen, A. J.; Mol, J. C. *J. Phys. Chem.* **1979**, *83*, 2485.
69. Kallury, K. M. R.; Krull, U. J.; Thompson, M. *Anal. Chem.* **1988**, *60*, 169.
70. Wasserman, S. R.; Tao, Y-T; Whitesides, G. M. *Langmuir* **1989**, *5*, 1074.
71. Lando, J. B.; Rogers, C. E.; O'Brien, K.; Rahnenfuehere, E., AFOSR-78-3692, November 30, 1980.
72. Ishida, H.; Koenig, J. L.; Asumoto, B.; Kenney, M. E. *Polym. Comp.* **1981**, *2*, 75.
73. Hopfield, J. J.; Onuchic, J. N.; Beratan, D. N. *Science* **1988**, *241*, 817.
74. Chidsey, C. E. D.; Loiacono, D. N.; Sleator, T.; Nakahara, S. *Surf. Sci.* **1988**, *200*, 45.

75. Cotton, F. A.; Wilkinson, G. *Advanced Inorganic Chemistry*, 4th ed.; John wiley and Sons: New York, 1980; p. 374.
76. Cohen, S. R.; Naaman, R.; Sagiv, J. *J. Phys. Chem.* **1986**, *90*, 3054.
77. Barraud, A.; Ruaudel-Teixier, A.; Rossilo, C. *Ann. Chim.* **1975**, *10*, 195.
78. Ulman, A., unpublished results.
79. Pomerantz, M.; Segmuller, A.; Netzer, L.; Sagiv, J. *Thin Solid Films* **1985**, *132*, 153.
80. Wasserman, S. R.; Whitesides, G. M.; Tidswell, I. M.; Ocko, B. M.; Pershan, P. S.; Axe, J. D. *J. Am. Chem. Soc.* **1989**, *111*, 5852.
81. Ulman, A.; Tillman, N. *Langmuir* **1989**, *5*, 1418.
82. *Handbook of Chemistry and Physics;* Weast, R. C.; Ed.; CRC: Boca Raton, FL, 1984.
83. Sabatani, E.; Rubinstein, I., Maoz, R.; Sagiv, J. *J. Electroanal. Chem.* **1987**, *219*, 365.
84. Zisman, W. A. In *Adhesion and Cohesion;* Weiss, P. Ed.; Elsevier: New York, 1962.
85. Adam, N. K. *Adv. Chem. Ser.* **1964**, *43*, 52.
86. Blake, T. D. In *Surfactants*; Tados, Th. F. Ed.; Academic Press: New York, 1984.
87. Greenler, R. G. *J. Chem. Phys.* **1966**, *44*, 310.
88. Rabolt, J. F.; Jurich, M.; Swalen. J. D *J. Appl. Spectrosc.* **1985**, *39*, 269.,
89. Rabolt, J. F.; Burns, F. C.; Schlotter, N. E.; Swalen, J. D. *J. Chem. Phys.* **1983**, *78*, 946.
90. Kuhn, H.; Möbius, D. In *Physical Methods in Chemistry* , A. Weissberger, B. Rossiter, Eds.; Wiley: New York, 1972; Part 3B, Vol. 1, p. 577.
91. Ulman,A.; Williams, D. J.; Penner, T. L.; Robello, D. R.; Schildkraut, J. S.; Scozzafava, M.; Willand, C. S.; US Patent 4,792,208, Dec. 20, 1988.
92. Tillman, N.; Ulman, A.; Elman, J. *Langmuir* **1990**, *6*, 1512.
93. *Handbook of X-Ray Photoelectron Spectroscopy*; Muilenberg, G. E.; Ed; Perkin-Elmer Corporation, Physical Electronics Division: Eden Prairie, MN, 1979; p. 56.
94. Andrade, J. D. in *Surface and Interfacial Aspects of Biomedical Polymers*; Andrade, J. D. Ed; Plenum: New York, 1987; Vol. 1, p. 175.
95. Paynter; R. W.; Ratner, B. D. *Ibid.*, Vol. 2, pp. 198-9.
96. Holmes-Farley, S. R.; Whitesides, G. M. *Langmuir* **1987**, *3*, 62.
97. Nuzzo, R. G.; Allara, D. L. *J. Am. Chem. Soc.* **1983**, *105*, 4481.
98. Somorjai, G. A. *Chemistry in Two Dimensions* – *Surfaces;* Cornell University Press: Ithaca, New York, 1982.
99. Bain, C. D.; Evall J.; Whitesides, G. M. *J. Am. Chem. Soc.* **1989**, *111*, 7155.
100. Evans, S. D.; Ulman, A. *Chem. Phys. Lett.* **1990**, *170*, 462.
101. Bond length and angles used were: C–C = 1.545 Å, C–C–C = 110.5 °, C–S = 1.81 Å, C–H = 1.1 Å, contribution from S $^-$ estimated at 1.5 Å [16].
102. Canning, N. D.; Outka, D.; Madix, R. J. *Surf. Sci.* **1984**, *141*, 240.
103. Chester, M. A.; Somotjai, G. A. *Surf. Sci.* **1975**, *52*, 21.
104. Trapnell, B. M. W. *Proc. R. Soc. London, A* **1953**, *218*, 566.
105. Richter, J.; Stolberg, L.; Lipkowski, J. *Langmuir* **1986**, *2*, 630.
106. Krim, J. *Thin Solid Films* **1986**, *137*, 297.
107. Ulman, A.; Evans, S. D.; Sharma, R.; Shnidman, Y.; Eilers, J. E. *J. Am. Chem. Soc.* **1991**, *113*, 0000.
108. Widring, C. A.; Chung, C.; Porter, M. D. *J. Electroanal. Chem.*, in press.

109. Nuzzo, R. G.; Zegarski, B. R.; Dubois, L. H. *J. Am. Chem. Soc.* **1987**, *109*, 733.
110. Bussing, T. D.; Holloway, P. H. *J. Vac. Sci. Tech., A* **1985**, *3*, 1973.
111. Mathieson, R. T. *Nature*, **1960**, *186*, 301.
112. Sellers, H., Ulman, A.; Shnidman, Y.; Eilers, J. E., in preparation.
113. Widrig, C. A.; Alves, C. A.; Porter, M. D., preprint.
114. Finklea, H. H.; Melendrez, J. A. *Spectroscopy* **1986**, *1*, 47.
115. Snyder. R. G.; Strauss, H. L.; Ellinger, C. A. *J. Phys. Chem.* **1982**, *86*, 5145.
116. Rabe, J. P.; Swalen, J. D.; Outka, D. A.; Stöhr, J. *Thin Solid Films*, 1988, **159**, 275.
117. Kittel, C. *Introduction to Solid State Physics*, 5th ed., John Wiley: New York, 1976.
118. Ulman, A.; Eilers, J. E., Tillman, N. *Langmuir* **1989**, *5*, 1147.
119. Harris, A. L.; Tothberg, L.; Duboid, L. H.; Levinos, N. J.; Dhar, L. *Phys. Rev. Lett.* **1990**, *64*, 2086.
120. Dote, J. L.; Mowery, R. L. *J. Chem. Phys.* **1988**, *92*, 1571.
121. Outka, D. A.; Stöhr, J.; Rabe, J. P.; Swalen, J. D.; Rotermund, H. H. *Phys. Rev. Lett.* **1987**, *59*, 1321.
122. Laibinis, P. E.; Hickman, J. J.; Wrighton, M. S.; Whitesides *Science* **1989**, *245*, 845.
123. Li, D.; Ratner, M. A.; Marks, T. J.; Zhang, C.; Yang, J.; Wong, G. K. *J. Am. Chem. Soc.* **1990**, *112*, 7389.

PART FOUR

MODELING OF MONOLAYERS

4.1. General Considerations

The understanding of the interrelationships between the molecular structure of amphiphiles and their organization on different surfaces is a fundamental problem. The packing and orientation of such molecules affect the surface chemistry of the monolayer, and play an important role in the phenomena of boundary lubrication, corrosion inhibition, adhesion, and catalysis [1–3]. The two-dimensional ordering results from intermolecular interactions such as van der Waals and electrostatic interactions [4–7]. Their strength is a function of the spacing between the molecular head groups. The reader may find specific mechanisms of ordering in References 8 and 9. Such molecular level understanding is essential for any successful molecular engineering of amphiphiles with useful properties. For example, we already have noted that an organic film for nonlinear optics should have a noncentrosymmetric arrangement of molecular dipoles, preferably perpendicular to the monolayer surface. Hence, for these purposes, it is desirable to maximize the average orientational order parameter of the molecular assembly. (See also Part Five.) Packing and ordering considerations also are important in applications such as conductivity (and photoconductivity), electron transfer, ferromagnetic materials, etc. In other words, our capability to produce useful thin organic films is the result of our ability to engineer two-dimensional molecular assemblies. This requires

305

understanding packing and ordering in these systems on the most detailed, molecular level. Computer modeling starting from individual molecules, and going all the way to assemblies and materials, is an indispensable tool for the design of advanced materials at the molecular level. I will get back to this point in my final note at the end of this book.

To illustrate the kind of problems one may encounter, let us consider here the case of a bulky group (in most cases a rigid, planar π-system) that is incorporated into a long alkyl chain (Figure 4.1). This group disrupts the cylindrical symmetry and forms a kink at some position of the alkyl group. This brings up the following structural questions:

(*a*) What is the optimal position of the bulky chromophore in an alkyl chain? In other words, does a chromophore closer to the substrate or closer to the surface give the best two-dimensional packing?

(*b*) What are the effects of the total molecular length on the packing and orientation in the two-dimensional assembly?

(*c*) What are the effects of the size (in Å) and magnitude (in D) of a dipole moment on the two-dimensional packing and orientation of these molecules?

(*d*) Is there a relationship between the order in one part of the molecule and that in another? Thus, is a disorder in the alkyl chain below the π-system will affect the order in the alkyl chain above it?

The answers to these questions provide the understanding of monolayer–bulk structure relationships, which is the foundation for any molecular design program we discussed earlier.

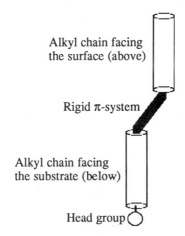

Figure 4.1. A schematic description of an amphiphile containing a π-system.

Molecular modeling is aimed at developing a better understanding of the origin of the macroscopic behavior of materials based on the information about the microscopic (molecular) level. Theoretical methods for modeling monolayers, bilayers, and related systems (such as paraffin crystals) include analytical approaches, such as single-chain mean-field theories in the rotational isomeric states (RIS) approximation [10], various lattice models [11–29], and the Landau–Ginzburg phenomenological approach [30-34]. However, the information provided by this kind of modeling is qualitative in nature, and though helpful in elucidating some general features of these systems, it is not capable of quantitative predictions needed for material engineering through molecular design. To achieve this capability, one has to resort to computer modeling methods that incorporate the chemical information on the atomic level. In this part, we discuss the most common of these techniques.

A. Energy Minimizations

This method provides solutions for local minima of the potential energy function of the molecular assembly, with or without periodic boundary conditions, using standard numerical methods, such as the conjugate–gradients, or the Fletcher–Powell method. For the description and explanation of these and other relevant numerical algorithms, see References 35 and 36. A global minimum corresponds to the expected structure at zero temperature (0 K), but these methods typically find only local minima, and the problem of sorting out the global minimum among them is rather complicated. For an example of a systematic search method for global minima of packing energies for two-dimensional assemblies of rigid organic molecules, see Reference 37. The rigidity constraint then can be relaxed using local minimization methods described earlier. Finding a local minimum usually is needed also as an input for more sophisticated simulations at finite temperatures, such as molecular dynamics (MD) simulations.

In many cases, the problem of getting stuck in an unrepresentative local minimum can be avoided by combining energy minimization using simulated annealing methods [38], in which either Monte Carlo (MC) or molecular dynamics (MD) steps (described in text sections) are employed to get the system out of a local minimum.

B. Simulations at Finite Temperatures

1. Monte Carlo Simulations

Monte Carlo (MC) simulations start from a given system configuration and generate new ones, at discrete time steps, using specific transition probabilities W_{ij}, from configuration i to configuration j. The time-dependent probability P_i, to find the system in configuration i, is described by the so-called master equation,

$$\frac{dP_i}{dt} = \sum_{j\,(\neq i\,)} W_{ji}\, P_j - \sum_{i\,(\neq j\,)} W_{ij}\, P_i \ . \qquad 4.1$$

If the transition probabilities W_{ij} satisfy the detailed balance condition,

$$W_{ij}\, e^{-E_i/kT} = W_{ji}\, e^{-E_j/kT} \ , \qquad 4.2$$

where E_i is the energy of configuration i, the system will equilibrate after a sufficient number of time steps, as follows from the theory of stochastic processes (this being an example of Markov chain dynamics). After reaching equilibrium, MC steps are used to generate representative statistical ensemble, where configurations appear with proportional Boltzmann probabilities [39, 40]. In principle, static equilibrium properties can be obtained by averages over the ensemble. However, this method yields only very limited information about the nonequilibrium kinetics of the system, since the Newtonian dynamics has been replaced by the physically ambiguous Markov chain dynamics. The MC method is too time-consuming for closely packed assemblies of organic chain-like molecules because of high-energy barriers that prevent an efficient sampling of the configuration space (sometimes referred to in the literature as "the problem of attrition"). Therefore, MC is not recommended for systems with crystalline-like packing. However, it has been applied successfully to more loosely packed phases, such as expanded phases of monolayers at the water–air interface, lipid bilayers, and micelles [41–44].

2. Molecular Dynamics Simulations

Molecular dynamics (MD) simulations produce a trajectory of the system in phase space by efficient numerical solution of the Newton equations of motion for all the particles in the system. A long enough trajectory contains a representative ensemble of equilibrium configurations. Different versions of MD exist that correspond to different ensembles of statistical mechanics, such as microcanonical, canonical, constant pressure, etc. In addition, the trajectory also contains dynamic information, e.g., the velocity–velocity time autocorrelation function that yields information about the power spectrum of the system. This method has been applied lately with increasing success to systems such as amphiphilic monolayers [45–49], lipid bilayers [50–54], paraffin crystals [55, 56], and self-assembled monolayers of simple alkanethiols on gold [57], and seems to be a potentially powerful tool to address the four questions mentioned at the beginning of Part Four.

C. The Choice of a Force Field and Calculations of Its Parameters

Both energy minimization and MD methods need a reliable model for intra- and intermolecular interactions as a starting point. For condensed organic molecular assemblies, computational time constraints typically limit our choice to classical atomic force-field models. Examples of popular force fields frequently used in the literature are MM2 [58, 59], MM3 [60, 61], GROMOS [62], OPLS [63], AMBER [64, 65], CHARMM [66], etc. These models represent the total potential energy of the system as a sum of various bond and nonbond interaction terms between point-like atoms or pseudo-atoms as follows (Figure 4.2):

TOTAL POTENTIAL ENERGY = BOND STRETCHES + ANGLE BENDS + TORSIONS
+ VAN DER WAALS + ELECTROSTATIC.

Typically, the bonded terms include bond stretches and bends, which are represented by harmonic potentials (with possible higher-order corrections), and torsions, which are represented by a truncated Fourier expansion in dihedral angles. The nonbonded terms include either a 12–6 or an exp-6 representation of van der Waals pair-interactions, and the electrostatic term is expressed as interactions between partial atomic charges at a given dielectric constant, ε. A simplified collection of such terms for the total potential energy surface, V_{total} is shown (although most force fields include higher-order corrections which are not shown here for the sake of clarity),

$$V_{total} = \sum_{bonds} K_r (r - r_{eq})^2 + \sum_{angles} K_\theta (\theta - \theta_{eq})^2 \qquad 4.3$$

$$+ \sum_{dihedrals} V_n [1 + \cos(n\phi + \gamma)] + \sum_{i < j} \{\varepsilon_{ij} [(\frac{r_{ij}^*}{r_{ij}})^{12} - (\frac{r_{ij}^*}{r_{ij}})^6] + \frac{q_i \, q_j}{\varepsilon r_{ij}}\} ,$$

where r, θ, ϕ, denote intramolecular bond length, bond angle, and dihedral angles. γ stands for a relative phase, by convention taken to be $\gamma = 0$ for $n = 1$ and 3, and $\gamma = \pi$ for $n = 2$, respectively. K_r and K_θ are harmonic constants for bond stretch and angle-bend potentials, respectively, and V_n are torsional barriers of period n. r_{ij} are intermolecular nonbond atomic distances, ε_{ij} are Lennard–Jones potentials, and $r_{ij}*$ the equilibrium intermolecular nonbond atomic distances. q_i are partial atomic charges, and ε is the appropriate dielectric constant.

Different classical force-field models may differ widely in the number of independent degrees of freedom chosen. The choice of the model appropriate for a given system is dictated by the need to balance competing requirements. On the one hand, one would like to have as detailed a model as possible, to faithfully represent the real system. However, limitations of computation time and complexity dictate coarse-graining of the degrees of freedom representing the physical system, and constrain the configurational space available to them. Thus, to speed up the computation, most MC and MD simulations coarse-grain the atomic representation to united atoms. This United Atoms Approach (UAA) represents $-CH_2-$, $-CH_3$, etc. groups as single effective atoms, interacting with effective potentials. The advantages of the UAA are (a) the gain of factor $\approx$ $(N/n)^2$ in pair interactions computation time (where N is the total number of atoms in the real system, and n is the average number of real atoms in a united atom), and (b) the absence of hydrogen vibrations allowing a significant increase in the MD time step. However, the disadvantages of the UAA are (a) the loss of molecular details that may be important for packing and ordering, and (b) since C–H vibrations are missing, IR spectra potentially cannot be calculated, if the force field is good enough, and compared to experimental data. Sometimes, even more coarse-grained representations of the molecular structure are used, such as spheres of influence, etc.

Another way to decrease computation time is the SHAKE method that constrains vibrations of united atoms along the chain backbone for MD [67, 68]. Still another one, useful for MC simulations, limits the configurational space to the so-called Rotational Isomeric States (RIS) that allow only discrete dihedral torsions with rigid bonds and bond angles along the chain [69].

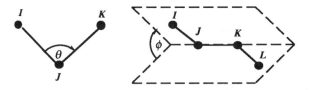

Figure 4.2. Bond angle bending (left, between bonds $I–J$ and $J–K$), and a definition of torsion angle (right, the angle between planes JKL and IJK).

For a tightly packed system such as LB and SA monolayers, it is important to note that the use of united atoms representation for the alkyl CH_2 and CH_3 groups does not allow a full expression of molecular stereochemistry, since these united atoms are spherical and the real CH_2 groups are not. Thus, we believe that an in-depth understanding of packing and molecular orientation in molecular assemblies can be obtained only by an explicit representation of all atoms and charges, including hydrogens in the atomic force-field model.

The various parameters are typically determined by a combination of *ab-initio* and semi-empirical calculations on small molecules, and by fitting to a large database of crystallographic structures. However, most force fields do not contain all the parameters one may need. For example, in our studies of alkanethiol monolayers containing a polar aromatic group, parameters for torsions around the SO_2 (α) and ether (β) bonds (Figure 4.3) were of great importance.

Shnidman *et al.* [70] have derived these parameters, therefore, from a semi-empirical quantum mechanical calculation using the PM3 Hamiltonian, as implemented in the MOPAC program [71], on the parent molecule shown in Figure 4.3. The torsional parameters for the bonds indicated were determined by calculating the variation of the heat of formation of the molecule in various torsionally rotated configurations. Figure 4.4 presents the results of these calculations. Subsequently, the variation of the MM2 internal energy, as a function of the same torsions as in Figure 4.5 (calculated with the torsion for this bond set to zero), was subtracted from the first curve. The resulting energies represent the net torsional contribution around the α-bond to the internal energy of the molecule. These were fitted to the MM2 form of the torsional energy term E_{tor}, as shown (with ϕ a dihedral angle, from Figure 4.5),

$$E_{tor}(\phi) = [V_1(1+\cos(\phi)) + V_2(1-\cos(2\phi)) + V_3(1+\cos(3\phi))]/2 . \qquad 4.4$$

D. Determination of Partial Atomic Charges

In monolayers of polar molecules, the electrostatic energy constitutes a significant portion of the total nonbond energy, and in principle, may play an important role in the packing and order of the molecular assembly. This requires the correct representation of the electrostatics in the system. In the context of classical force-field models, this is done typically by assignment of partial atomic charges to the atomic centers.

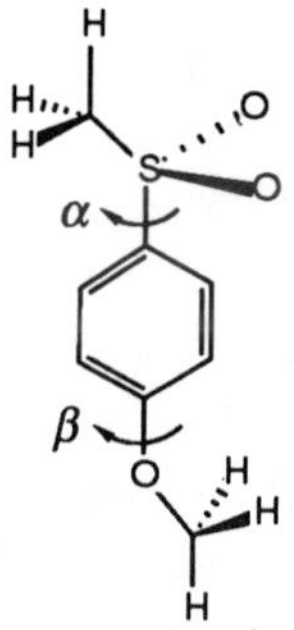

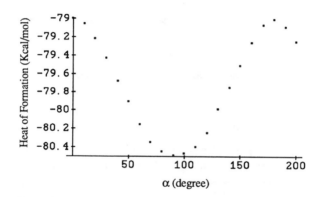

Figure 4.3. 4-Methoxyphenyl methyl sulfone used for calculation of α and β.

Figure 4.4. MOPAC PM3 heat of formation as a function of torsion around the α-bond indicated in Figure 4.3.

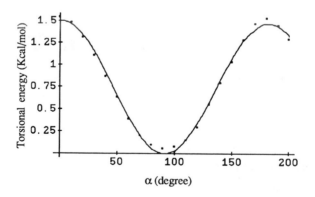

Figure 4.5. Variation of torsional parameters for the α-bond indicated in Figure 4.3.

312

A widely used method for obtaining partial atomic charges is Mulliken's population analysis of quantum orbitals [72, 73], which is based on assignment of a partial charge to a particular atom based on the orbitals centered on that atom. However, this method is problematic; for example, see the discussion in Reference 74. Modifications of this method have been suggested based on the preservation of the molecular dipole and higher moments [75, 76]. Another method to determine partial atomic charges is by fitting charges to the expectation values of the molecular electrostatic potential [77–79]. Common to all the methods just described is that their results are not unique, and are very sensitive to the choice of the basis set, and to the method of the quantum mechanical calculations (i.e., semi-empirical or *ab-initio*). This kind of calculation requires long computation times, and despite significant efforts by many groups in recent years [80–85], a reliable simple method for calculation of partial atomic charges is still not at hand.

4.2. Molecular Mechanics

A. Simple Alkanethiol Monolayers

We start here with a simple example that will illustrate the usefulness of the application of the energy minimization method of molecular mechanics to deduce structural information for the ground-state two-dimensional assembly [86]. The starting point in this study was an electron diffraction study of monolayers of docosanethiol on gold single-crystal foils with an exposed (111) surface [87]. There, it was shown that the alkanethiols have hexagonal packing with S···S spacing of 4.97 Å (Part Three), a distance that would leave substantial free volume between molecules if they "stood up" normal to the plane of the lattice. Since FTIR spectroscopy and electron diffraction experiments suggest that alkanethiol monolayers on gold surfaces have crystal-like periodicity in two dimensions (for alkyl chain length $\geq C_{10}$) [9], all the molecules in the system were treated as having identical conformations and identical orientations relative to a hexagonal close-packed grid of the S atoms (Figure 4.6). A rigid molecule of dodecanethiol ($C_{12}H_{25}SH$) in the all-*trans* conformation, with its geometry being optimized using the MM2 force field [58, 88] in the MACROMODEL program [89], was placed perpendicular to the plane of the lattice, with its S atom at the origin and oriented so that the yz plane bisects the methylene H–C–H bond angles (Figure 4.6). Exact duplicates were placed on all other lattice points under consideration. The first degree of freedom was a twist (α) around the long axis of the molecule (z), the second was an in-plane tilt of the molecule in the yz

plane or, equivalently, a rotation of every molecule around a local x axis by θ, and the third was an out-of-plane tilt or, equivalently, a rotation of each molecule around a local y axis by θ'.

The minimization was performed on a 3 x 3 patch over these three degrees of freedom only, with free boundary conditions, and using standard molecular mechanics methods. The energy of interaction of a molecule with its neighbors was examined as a function of these three angles. The energetics considered only intermolecular two-body terms (van der Waals and electrostatics). All energies in the figures refer to the energy of interaction of the central molecule in the cluster with its near neighbors. If the reader wants to consider the stabilization energy of an infinite sheet, he should divide these energies by two to avoid double counting. The simple model described earlier does not permit the molecules to adjust their geometries to accommodate the presence of their near neighbors. Nevertheless, it afforded considerable insight to the packing and orientation mechanism of alkyl chains in self-assembled monolayers.

The first step in the calculations was to examine closely one row of the assembly. Perhaps the most striking observation at this stage was the correlation between compactness of structure (smaller d) and favorable energetics (more negative). The most energetically favored twist for one row of alkanethiols is $\alpha \approx 0°$. There are minima at $\alpha \approx 70°$ and $\alpha \approx 110°$, which are slightly higher in energy that that at $0°$ (Figure 4.7). A perfect hexagonal symmetry should have yielded identical minima at $0°$, $60°$, and $120°$.

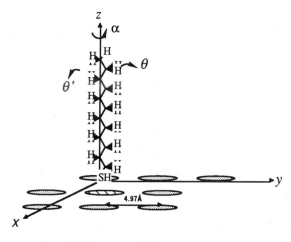

Figure 4.6. Hexagonal arrangement of alkanethiol molecules ($C_{12}H_{25}SH$) on (111) gold surface. (From Ulman [87], © 1989, Am. Chem. Soc.)

314

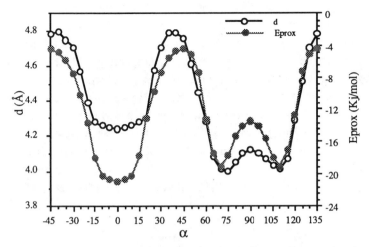

Figure 4.7. Optimal interaction energy and the corresponding spacing as a function of the twist angle (α) for a single row of alkanethiol molecules of (111) gold surface. d is the S$\cdots$S spacing, and E_{prox} is the interaction energy. (From Ulman [87], © 1989, Am. Chem. Soc.)

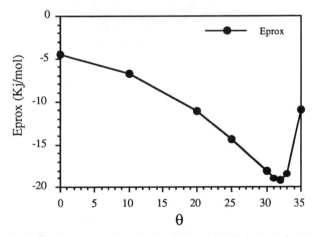

Figure 4.8. The interaction energy in a single row of alkanethiol molecules as a function of the in-plane tilt (θ) for 4.97 Å S$\cdots$S spacing. (From Ulman [87], © 1989, Am. Chem. Soc.)

These minima are crucial for understanding hexagonal close packing, since if the molecules have a twist angle of α relative to one triangular lattice axis, they must have twist angles of $\alpha + 60$ and $\alpha + 120$ relative to the other row directions. On the other hand, optimal close-packing of alkanethiols in a row results in a perpendicular chain orientation, with spacing of 4.24 Å. This optimal packing will only be achieved if the attachment to the underlying substrate is compatible with such a lattice. If the chemisorption to the underlying substrate were to enforce a lattice spacing significantly larger than 4.24 Å, as appears to be the case on gold, there would be no close intermolecular contacts at all for molecules oriented normal to the surface. Hence, the molecules would be expected to "tip" away from the normal in such a way as to establish again optimum van der Waals contacts. Figure 4.8 shows the energy as a function of θ, for $\alpha = 0°$, where the repulsive wall for tilts >35° is quite steep, and there are no barriers to the in-plane tilting. Figure 4.9 shows the interlocking of hydrogen atoms at neighboring alkyl chains (the so-called "racheting effect").

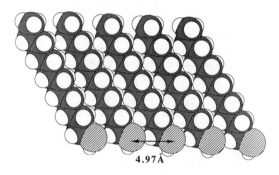

Figure 4.9. The interlocking of hydrogen atoms in neighboring alkyl chains tilted 32°.

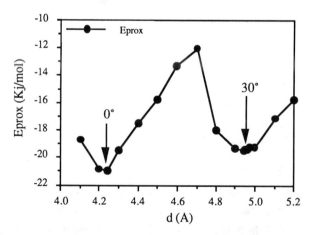

Figure 4.10. The best interaction energy (for optimal in-plane tilt) as a function of the spacing (d). (From Ulman [87], © 1989, Am. Chem. Soc.)

It already is clear that at lattice spacings less than about 4.8 Å, it is not possible for the racheting effect to establish this special interlocking. Figure 4.10 presents the calculated energy in different S⋯S distances, when in each lattice spacing the molecules were allowed to tilt to optimum van der Waals contact. The minimum near 0° requires a nearly vertical orientation, and that near 4.95 Å requires tilts near 30°. Thus, it seems that spacings other than 4.1 to 4.3 Å, and 4.8 to 5.05 Å, and tilt angles other than ~0° or ~30°, are unlikely for rows of alkyl chains. The twist angle being near 0° and the in-plane tilt being about 30° apparently are the two most important factors in establishing good van der Waals contacts. However, if the reader returns to Figure 4.7, it should become clear that a monolayer with $\{\alpha = 0°, \theta = 30°, \text{ and } \theta' = 0°\}$ would have close-packing within parallel planes perpendicular to the lattice, but still would have appreciable free volume between the adjacent planes of packed molecules. A simultaneous tilt (in the x direction) of these tilted rows closes the gap between these planes and provides the final structure, which is tightly packed in all directions. Calculations on a 3 x 3 cluster showed that the minimum energy is for an out-of-plane tilt of $\theta' = 22°$, and that this tilt provides approximately 25% of the total stabilization energy. These results are equivalent to a total chain-axis tilt of 38° from the normal in a plane containing the z axis and rotated 59° from the xz, followed by a rotation about the alkyl chain axis so that the angle between this plane and the plane bisecting the methylene units becomes ~46°.

The results of this study, however simple, relate to numerous studies, either already mentioned (Parts One and Two), or to be discussed later [46, 90–95]. The reader is referred to the original paper [1] for these discussions.

B. Alkanethiol Monolayers Containing a Polar Aromatic Group

For a close-packed assembly of molecules containing rigid, bulky groups embedded in semi-flexible alkyl chains (Figure 4.1), the assumption of rigidity of the individual molecules cannot be justified. Close-packing of the bulky group results in a large free volume for the semi-flexible chains. The competition between the elastic energy associated with the internal degrees of freedom and the van der Waals interactions responsible for the close-packing may lead to a minimized close-packed structure with distorted molecular conformations. To investigate these questions, the minimization should include the intramolecular, as well as the intermolecular degrees of freedom.

Shnidman et al. started with a single, isolated molecule similar to that presented in Figure 4.3, with different alkyl chains, and with the alkyl group below the π-system terminated by a thiol group [96]. Explicit hydrogen atoms and partial atomic charges were used, and the energy of the isolated molecule was minimized using the conjugate–gradients method, at constant stress, with dielectric constant $\varepsilon = 1$. This results in a straight, all-*trans* molecular conformation. Four such molecules then were placed in an orthogonal simulation cell, with dimensions significantly larger than the contact dimensions, and periodic boundary conditions. The vertical separation between layers was taken large enough (beyond the chosen cutoff of 8 Å for nonbond interactions) to ensure a two-dimensional assembly.

Constant stress minimization that allows self-adjustment of both the *shape* and the *volume* of the unit cell (Figure 4.11), as implemented in the POLYGRAF software package from Biodesign Inc., was used with a constrained fixed vertical dimension of the simulation cell. They found that once minimization was started from this initial configuration with the electrostatics turned on, the system gets trapped in a nonphysical local minimum that stabilizes a loosely packed assembly. However, if the minimization was initiated with the electrostatics turned off, they found two close local minima, corresponding to very different closely packed structures. A subsequent minimization with the electrostatics turned on (with $\varepsilon = 1$) led to large energy separation between these two structures, with a minimum significantly lower than the one corresponding to the aforementioned loosely packed system. Thus, while the van der Waals interactions are the major contribution to the close-packing of the assembly, in a case where two close local minima exist, introduction of electrostatics was found to pick only one of them. Such behavior was predicted by Kitaigorodtskii [97, 98]. The molecules in the minimized close-packed assembly exhibit chain conformations significantly distorted from the conformation of a molecule minimized in isolation.

4.3. Dynamics Simulations of Monolayers

A. Molecular Dynamics

1. Methodology

Classical molecular dynamics (MD) simulations consist of an efficient numerical integration of Newton's equations of motion, with classical intra- and intermolecular pairwise interactions between the atoms. This can be done with or without periodic boundary conditions, depending on the context and computation time limitations. The interactions are described by the classical force-fields as explained earlier, with classical electrostatics between partial atomic charges. For any atom i, Newton's equation of motion is,

318

$$m_i \frac{d^2\mathbf{r}_i}{dt^2} = \mathbf{F}_i \, , \qquad\qquad 4.5$$

where m_i is the atomic mass, $\mathbf{r}_i$ is its position, $\mathbf{F}_i$ is the force on atom i due to the interactions with all the other atoms in the system, and t is the time. Time evolution of the system is found by integration of those equations for any atom i. Since analytical integration of such equations is an impossible task, one has to resort to various algorithms for efficient numerical integration methods. One of the simplest of those is the Standard Verlet Algorithm [99], which approximates the preceding Newton equations,

$$\mathbf{r}_i\,(t+\Delta t) = 2\mathbf{r}_i\,(t) - \mathbf{r}_i\,(t-\Delta t) + (\Delta t)^2 \mathbf{F}_i\,(t) \, , \qquad\qquad 4.6$$

where Δt is the discrete time step, and $\mathbf{r}_i\,(t-\Delta t)$ relates to the velocity $\mathbf{v}_i(t)$ as follows:

$$\mathbf{v}_i\,(t) = [\mathbf{r}_i\,(t+\Delta t) - \mathbf{r}_i\,(t-\Delta t)]/2\Delta t \, . \qquad\qquad 4.7$$

The two preceding equations are called the two-point Verlet algorithm, and are used at all time-steps other than the initial one. To initiate the integration, the so-called one-point Verlet algorithm is used,

$$\mathbf{r}_i\,(\Delta t) = \mathbf{r}_i\,(0) + \Delta t\, \mathbf{v}_i\,(0) + (\Delta t)^2 \mathbf{F}_i\,(0)/2 \, . \qquad\qquad 4.8$$

The initial velocity, $\mathbf{v}_i(0)$, is taken either as the final velocity of a previous trajectory, or from a Maxwell–Boltzmann distribution corresponding to the given temperature.

The inherent inaccuracy in any numerical integration method imposes a limit on the maximum time step possible. A greater time step leads to instabilities in the integration of the equations of motion. Typically, the criterion for choosing a maximum time step for a given system is dictated by the frequency of the fastest motion in the system. The fastest motion in an organic chain molecule is the vibration of the C–H bond ($\sim$3000 cm^{-1}), which corresponds roughly to a period of 0.01 ps. Thus, when hydrogens are represented explicitly in the force field, the maximum time step should be 0.0005–0.001 ps. On the other hand, when united atoms are used, the maximum time step can be increased to $\sim$0.002 ps. Algorithms constraining vibrations along the C–C backbone, such as SHAKE [67, 68], allow a further increase in the time step used.

A long enough trajectory contains a representative ensemble of equilibrium configurations. A trajectory in the $6N$-dimensional phase-space $(\mathbf{r}^N, \mathbf{p}^N)$ will correspond to sampling of the microcanonical ensemble that represents the statistics of the motion of the atoms within an isolated system, i.e., without thermal contact with the surroundings. Here, we use the notation as follows:

$$\mathbf{r}_N = (\mathbf{r}_1, \mathbf{r}_2, ..., \mathbf{r}_N) \, , \qquad\qquad 4.9$$

$$\mathbf{p}^N = (\mathbf{p}_1, \mathbf{p}_2, ..., \mathbf{p}_N) \, , \qquad\qquad 4.10$$

where $\mathbf{p}_i = m_i \mathbf{v}_i$ is the momentum of atom i. In the real world, experiments are performed in contact with a heat reservoir, at a constant temperature. This corresponds to the canonical ensemble of statistical mechanics. Different methods have been suggested to emulate the thermal contact with a heat bath in MD simulations, such as *ad hoc* velocity rescalings [100], Andersen's heat bath method [101], the time relaxation method of coupling to heat bath of Berendsen *et al.* [102], and Nosé's method, in which the velocities' rescaling factor is regarded as an additional degree of freedom in a generalized Lagrangian, coupling the system with a heat bath [103].

As an example, in the velocity rescaling method, a given number of microcanonical dynamics steps are performed, after which a temperature T, defined in terms of the average kinetic energy,

$$3N \, kT \, / 2 = \left\langle \sum_i m_i \, \mathbf{v}_i^2 \, / 2 \right\rangle \, , \qquad\qquad 4.11$$

is calculated, where N is the number of degrees of freedom, k is the Boltzmann constant, and < > denotes an average over the number of microcanonical dynamics steps. The calculated temperature T is compared then to the bath temperature T_0, and if it is outside a prescribed range, velocities are rescaled ,

$$\frac{v_{\text{new}}}{v_{\text{old}}} = \sqrt{\frac{T_0}{T}} \, . \qquad\qquad 4.12$$

In many situations of interest in the subject area of this book, the experimental setups correspond to still other statistical mechanical ensembles. For example, when surface pressure is applied in the Langmuir trough to stabilize a monolayer at the water–air interface, the system is in contact with a pressure or stress reservoir (the moveable barrier). This is in addition to the system being in thermal contact with the ambient. Versions of MD simulations have been developed, producing an MD trajectory that samples statistical ensembles of this kind [101, 102, 104–106]. Such methods also are particularly suited to study closely packed self-assembled monolayers. This is because the simulation-cell dimensions and angles are adjusted continuously throughout the numerical integration by incorporating these extra degrees of freedom within the Lagrangian formulation of the Newtonian mechanics. We describe here one example of a constant stress MD algorithm [104–106].

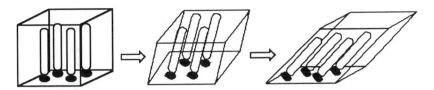

Figure 4.11. Changing box shape and volume. Note that while the assembly is in square lattice symmetry in the beginning, it is in pseudohexagonal arrangement at the end of the simulations.

In this algorithm one introduces scaled coordinates through the equation,

$$\mathbf{r} = \mathbf{Bs}, \qquad\qquad 4.13$$

where $\mathbf{B} = (\mathbf{b}_1, \mathbf{b}_2, \mathbf{b}_3)$ is a transformation matrix whose columns are the three vectors $\mathbf{b}_\alpha$, $\alpha = 1, 2, 3$, representing the sides of the box. V, the volume of the box, is given by

$$V = |\mathbf{B}| = \mathbf{b}_1 \cdot \mathbf{b}_2 \times \mathbf{b}_3. \qquad\qquad 4.14$$

The changing box shape and volume are represented in Figure 4.11. One associates an extra "potential energy" with the box, given by,

$$\mathcal{V}_B = PV, \qquad\qquad 4.15$$

and an extra "kinetic energy" term of the form,

$$\mathcal{T}_B = [Q \sum_\alpha \sum_\beta \dot{\mathbf{B}}_{\alpha\beta}^2]/2 , \qquad\qquad 4.16$$

where $\mathcal{T}_B$ is the "kinetic" energy associated with the box degrees of freedom, and Q is the box "mass." The modified Newton equations of motion are obtained in the usual way from the Lagrangian,

$$\mathcal{L} = \mathcal{T} + \mathcal{T}_B - \mathcal{V} - \mathcal{V}_B, \qquad\qquad 4.17$$

where $\mathcal{T}$, $\mathcal{T}_B$ are the kinetic energies associated with the molecular and box degrees of freedom, and $\mathcal{V}$, $\mathcal{V}_B$ are the potential energies, respectively. The resulting equations of motion are

$$m \frac{d^2 s}{dt^2} \mathbf{B}^{-1} f - m \, \mathbf{G}^{-1} \dot{\mathbf{G}} \frac{ds}{dt} , \qquad\qquad 4.18$$

$$Q \ddot{\mathbf{B}} = (\mathbf{P} - l \, \mathbf{P})V \, (\mathbf{B}^{-1})^{\mathrm{T}} , \qquad\qquad 4.19$$

where $\mathbf{G} = \mathbf{B}^{\mathrm{T}}\mathbf{B}$ is a metric tensor, T stands for transpose, and $\mathbf{P}$ is the stress tensor,

321

$$\mathbf{P}_{\alpha\beta} = \frac{1}{v} \{ \sum_i m \, (\mathbf{B} \frac{ds_i}{dt})_\alpha^{\mathrm{T}} \, (\mathbf{B} \frac{ds_i}{dt})_\beta^{\mathrm{T}} + \sum_i \sum_{j>i} (\mathbf{B}s_i \,)_\alpha \, (f_{ij} \,)_\beta \} \, . \qquad 4.20$$

The constant stress MD simulation method has been extended to molecular systems by Nosé and Klein [107].

Simulations of condensed assemblies usually involve the use of periodic boundary conditions. Thus, the atoms close to the boundaries of the simulation cell interact with external images of atoms within the box. To save computation time, the simulations are frequently performed with a finite cutoff for the nonbond interactions. However, because of the long-range electrostatic and van der Waals nonbond interactions, a more accurate method involves performing Ewald lattice sums for those interactions [108–111]. Thus, calculation of Ewald sums requires a considerable increase in the required computation time, and therefore is not used often.

MD and MC simulations yield statistical information about properties that are explicit functions of the phase-space coordinates of the system, such as internal energy, temperature, stresses, and orientational order parameters. However, they do not provide direct information on thermal properties that depend on the total phase-space volume accessible to the system, such as the entropy, various free energies, and chemical potentials. Nevertheless, methods have been developed to calculate entropy and free energy differences that mimic their thermodynamic definitions. Thus, a reversible path is found first that links the state point of interest A to a state B of known free energy and entropy (e.g., an ideal gas state or an Einstein solid state, the latter being an idealized solid with all atoms vibrating at an identical frequency). One method evaluates the free energy difference, ΔG_{BA}, by the so-called thermodynamic integration method [112],

$$\Delta G_{BA} = \int_{\lambda_A}^{\lambda_B} \langle \frac{\partial H \, (\mathbf{r}^N, \mathbf{p}^N, \lambda)}{\partial \lambda} \rangle_\lambda d\lambda \, , \qquad 4.21$$

while another uses the so-called perturbation technique,

$$\Delta G_{AB} = -kT \, \ln \langle \frac{\exp[-H \, (\mathbf{r}^N, \mathbf{p}^N, \lambda_B \,) - H \, (\mathbf{r}^N, \mathbf{p}^N, \lambda_A \,)]}{kT} \rangle_{\lambda_A} \, . \qquad 4.22$$

In the preceding formulas, $H(\mathbf{r}^N, \mathbf{p}^N, \lambda)$ is the phase space Hamiltonian, where λ denotes its set of coupling parameters. The angular brackets $<...>_\lambda$ denote an ensemble average over the phase space at specified values λ of the coupling parameters. The latter formula (Equation 4.22) is called the perturbation formula because it gives accurate results only when state B is close to state A. For a review discussing these two methods, see Reference 112.

To model properly the interactions at the water–air (oil–water) interface, solvation effects should be included. One possibility is to include solvent molecules explicitly in the simulations. The problem with this approach is that it increases enormously the number of degrees of freedom in the system, and the required computation time grows correspondingly. In most cases, a more practical alternative is to treat the solvent as a statistical continuum. Unless the discreteness of the solvent molecules is essential for the phenomenon under study (e.g., Israelachvili's model of hydration) [113], this approach results in a satisfactory approximation. A recent example of a continuum solvation model is implemented in the MACROMODEL V3.0 software package [89], and will be described briefly.

Still *et al.* considered the solvation free energy (G_{sol}) as consisting of a solvent–solvent cavity term (G_{cav}), a solute–solvent van der Waals term (G_{vdW}), and a solute–solvent electrostatic polarization term (G_{pol}) [114],

$$G_{sol} = G_{cav} + G_{vdW} + G_{pol} \qquad 4.23$$

Based on the observation that G_{sol} for the saturated hydrocarbons in water is related linearly to the solvent accessible surface area (*SA*), it is assumed that

$$G_{cav} + G_{vdW} = \Sigma\, \sigma_k SA_k, \qquad 4.24$$

where SA_k is the total solvent accessible surface area of atoms of type k, and σ_k is an empirical atomic solvation parameter. Still *et al.* used the value $\sigma_k = +7.2$ cal/mol-Å^2 for all atom types to reproduce hydration energies for hydrocarbons. The electrostatics of the atomic system is assumed to be that of spherical particles of charges q_i, radii α_i, and separations r_{ij}, in a medium of dielectric constant ε. The polarization term G_{pol} has been termed the generalized Born equation [114], and is given by,

$$G_{pol} = -166\,(1 - \frac{1}{\varepsilon}) \sum_{i,j=1}^{n} \frac{q_i q_j}{f_{GB}} . \qquad 4.25$$

f_{GB} was chosen in Reference 114 as

$$f_{GB} = \sqrt{r_{ij}^2 + \alpha_{ij}^2\, e^{-D}} . \qquad 4.26$$

Here,
$$\alpha_{ij} = \sqrt{\alpha_i \alpha_j} \qquad 4.27$$

and
$$D = \frac{r_{ij}}{2\alpha_{ij}^2} . \qquad 4.28$$

According to the authors [114], small-molecule hydration energies computed with this method are of comparable accuracy to those obtained from contemporary free energy perturbation methods, but at only a fraction of the computational expense.

2. MD Simulations of LB Films and Lipid Membranes

Van der Ploeg and Berendsen published one of the first papers on molecular dynamics of a bilayer membrane in 1982 [50]. In this study, they used the Verlet algorithm [99], where the bond lengths in the molecules were constrained by the SHAKE method [67, 68]. They built their system from two layers of small number (initially 16, later extended to 64) decane molecules ($C_{10}H_{22}$), each with periodic boundary conditions in two dimensions (x,y). They used a surface area of 25 $Å^2$ per molecule, hence the bilayer thickness was not a parameter. The interaction model for the chain in this study was based on the dihedral potential function, $V(\phi)$, of Ryckaert and Bellemans [115, 116],

$$V(\phi) = \sum_{i=0}^{5} C_i (\cos\phi_i)^i ,$$ 4.29

where ϕ is the dihedral angle, and C_i are constants. They used a *trans-gauche* barrier energy of 12.3428 kJ/mol, and the energy of the *gauche* minimum was 2.9288 kJ/mol. The form of the Lennard–Jones interaction, $V(r_{ij})$, between all intramolecular and intermolecular pairs was

$$V(r_{ij}) = 4\varepsilon [(\frac{\sigma}{r_{ij}})^{12} - (\frac{\sigma}{r_{ij}})^6] ,$$ 4.30

where ε and σ are appropriate coupling constants. Note that this is an alternative but equivalent form of the Lennard–Jones interaction given in Equation 4.3. They calculated ε and σ values for the CH_2, CH_3, and COO^- head group from polarizabilities and van der Waals radii [117, 118], where these groups were treated as pseudo-atoms, i.e., spherical objects with no sense of stereochemistry. The authors used a cutoff radius of 10 Å and the following harmonic function for the interaction, $V(z)$, of the polar group with the water,

$$V(z) = [k_h (z - \overline{z})^2]/2 ,$$ 4.31

where $z - \overline{z}$ denotes the distance from the water–air interface, and k_h is a harmonic constant. Using a time step of 8 x 10^{-3} ps, they reported that it took 10,000 steps to achieve equilibrium. To check their results they carried out runs for 80 ps and calculated the fluctuation in the total energy, which was ~7%. The calculations provided an excellent agreement between the bond orientational order parameter, S^{CD}, determined from deuterium resonance data [119], and that calculated from the molecular dynamics. Another interesting observation was that the rotation around the molecular z axis is anisotropic, which means that there is a difference in the order parameter in the x and in the y directions. (The order parameters are S_{XX}^{CD} and S_{YY}^{CD}, respectively.) They also found a strong correlation between the total molecular tilt of 30° and the tilt of individual methylene units, and concluded that the molecular tilt is a collective property, which fluctuates with time (every 10–20 ps) between a structure with long–range order and one when the directional correlation is low. It may be suggested that this observation is a result of the lack of symmetry in this calculation, i.e., the use of pseudo-atoms, since in such a process there should be an energy barrier for the disordering process.

Cardini *et al.* used molecular dynamics for the characterization of Langmuir–Blodgett monolayers [46]. They used an ensemble of 90 molecules (in a triangular lattice) each with 20 pseudo-atoms, with interchain separation of 4.9 Å and interacting with each other potentials similar to that mentioned earlier [67]. The interaction with the flat surface was via a surface 9-3 potential, $V(z)$,

$$ V(z) = 20\varepsilon \left[(\frac{\sigma}{Z})^9 - (\frac{\sigma}{Z})^3 \right] , \qquad 4.32 $$

where $Z = z - \bar{z}$, the distance from the surface, and ε and σ are coupling constants. Two-dimensional boundary conditions were used and the unit cell was a rectangular box with the dimensions $L_x = 44.1$ Å, and $L_y = 42.4$ Å. The time step was 2.5 fs, and the C–C bonds were kept rigid, with 1.53 Å length [120].

The most important result is the 40° canting angle they observed for their model of arachidate chain, in excellent agreement with the value of $33 \pm 5°$ published by Outka *et al.* for calcium arachidate on Si(111) [121]. This canting angle, of course, is a result of their choice of 4.9 Å for molecular spacing, and is in complete agreement with the molecular mechanics calculations of alkanethiols on gold, described earlier [86]. The authors reported that calculation with even- (20) and odd- (21) carbon atoms gave the same tilt angle, which is contrary to what is known for bulk solid alkanes [122]. At $T = 306$ K, they estimate ~1% of *gauche* bonds, while at $T = 435$ K, the number was ~4%, which is in agreement with what is found for bulk alkanes [123, 124].

Another result that should be mentioned is their calculation of power spectra for intramolecular and external modes of the C_{20} monolayer. The frequencies calculated (0–600 cm^{-1}) are in the low energy IR spectral region. It should be possible, in principle, to calculate frequencies in the region of C–H bond vibrations (2800–3000 cm^{-1}) if the calculations are carried out with explicit hydrogen atoms (rather than with pseudo-atoms), and a high-quality force field is used. Such calculations, however long and tedious, may be compared with actual IR spectra, and therefore, molecular orientation as a function of temperature can be deduced. Force fields of such quality are currently under development (e.g., MM3).

Bareman et al. carried out further molecular dynamics on the C_{20} ensemble [47]. Here, they used the same starting conditions as described above, on a triangular lattice, but assigned different area per molecule values, 21, 26, and 35 $Å^2$, corresponding to what commonly is denoted as solid, liquid, and gas phases, respectively. Figure 4.12 presents the two-dimensional cross sections of the final molecular dynamics configurations. These results indicate that in the case of 26 $Å^2$/chain, there are two coexisting regions (phases), one condensed and one disordered, with domain dimensions of ~20 Å. The results are in agreement with the picture found by Mann et al. in their "melting" experiment, and indicates that in the liquid phase, there may be clusters of molecules with long-range order [125]. We note that the system in simulations of Bareman et al. is relatively small, and thus the observed final configuration may be strongly affected by the boundary conditions.

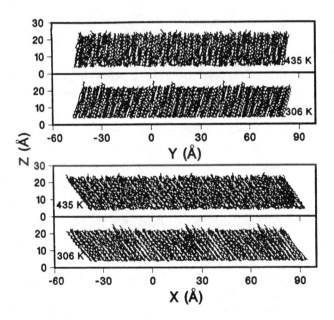

Figure 4.12. Two-dimensional cross sections of final molecular dynamics configurations viewed in the surface plane. $T = 300$ K, (a) 21 $Å^2$/chain, (b) 26 $Å^2$/chain, and (c) 35 $Å^2$/chain. (From Bareman et al. [47], © 1988, Am. Phys. Soc.)

326

In this context, it is interesting to note that nucleation of solid regions in a liquid background was observed first by Lösche *et al.* [126], and then by Weis and McConnell [127]. Riegler and LeGrange were the first to observe this effect in the meniscus region in a monolayer at the water–air interface during deposition (details in Part Two) [128], which also was observed by Möhwald in phospholipids [129, 130]. Here, there is a theoretical indication that an equilibrium between liquid expanded–liquid condensed phases does exist in monolayers where the area per molecule is 26 Å^2, as seen experimentally by Pathica *et al.* [131, 132].

Harris and Rice carried out a molecular dynamics study of the structure of a model Langmuir–Blodgett monolayer of pentadecanoic acid ($C_{14}H_{29}COOH$) on water. (The effect of the water was represented by an effective surface 9-3 potential; no explicit electrostatic and solvation interactions were used [48].) Using 15 pseudo-atoms with parameters similar to those described earlier, they observed one low-density vapor phase and one well-ordered condensed phase in which the molecules were canted at an angle of 30° to the z axis (≤ 23 Å^2/molecule). They could not detect a liquid-expanded phase and a transition from this phase to liquid-condensed and to the crystalline phase.

Although this and other molecular dynamics simulations have contributed to our understanding of the order at the molecular level, the use of united atoms may be the reason for some of their problems. They used spherical united atoms, in a hexagonal arrangement rather then anisotropic bodies in a quasihexagonal (or face-centered orthorhombic) arrangement. Hence, there is no difference between a tilt in a plane that bisects the methylene units, and one that is out of this plane. (See earlier discussion in the section on molecular mechanics.) According to Lin *et al.* [133], a transition from a condensed-liquid to a crystalline phase may be from a free rotor phase (one where the molecules have enough free volume to allow free rotation about their molecular axis) to a phase where there is a strong racheting in both the x and y directions. If this is the case, only those simulations that take the methylene symmetry into account, i.e., molecular dynamics with explicit representation of hydrogen atoms, may be able to predict phase transitions and correlation with FTIR spectra.

3. Modeling of Self-Assembled Monolayers

Hautman and Klein were the first to carry out molecular dynamics simulations of a model monolayer of alkanethiol chains on a gold (111) surface [57]. In their study, the molecules of hexadecanethiol ($CH_3(CH_2)_{15}SH$) consisted of 17

327

spherical united atoms connected by rigid constrains ($d_{cc} = 1.53$ Å, $d_{cs} = 1.82$ Å, Figure 4.13).

Two different approximations were used to model chemisorption. In the first model (I), they left the angle between the C–S bond and the surface normal unrestricted, while in the second (II), the angle was restricted to $\theta_{h'} = 100°$. These simple approximations were used in lieu of the detailed chemisorption potential between the thiols and the gold, which is not currently known. Their molecular dynamics simulations used 90 molecules with periodic boundary conditions, at 300 K, an S···S spacing of 4.97 Å, and an S–Au bond length of 2.4 Å. The most striking difference between the results of model I and II was an extra *gauche* bond near the sulfur in model II (Figure 4.14), and the molecular tilt angle that was 28° for model I and 19.6° for model II.

Let us consider the two scenarios for chemisorption at 0 K, assuming all-*trans* conformations in both cases (Figure 4.15). In case *A*, where the C–S bond is allowed to adopt an angle of 0° with the surface normal, the 15 methylene units contribute optimally to the attractive, intermolecular van der Waals interactions. On the other hand, if this angle is restricted to 100° (*B*), both the first methylene and the methyl groups do not contribute strongly to those interactions. Thus, under these assumptions, model *A* leads to a lower intermolecular energy for the system. This seems to be in agreement with published IR data. (See also the discussion earlier on molecular mechanics.)

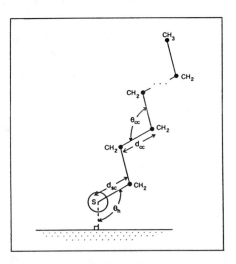

Figure 4.13. Coordinations for hexadecanethiol model molecules on gold. (From Hautman and Klein [57], © 1989, Am. Inst. Phys.)

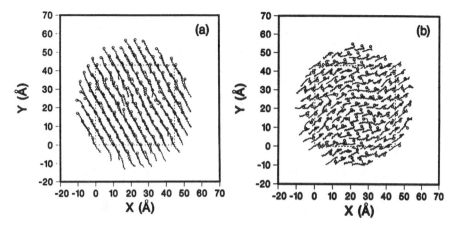

Figure 4.14. Instantaneous configurations at the end of molecular dynamics runs using (a) model I and (b) model II. (From Hautman and Klein [57], © 1989, Am. Inst. Phys.)

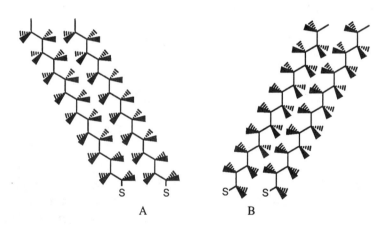

Figure 4.15. Van der Waals interactions in hexadecanethiol molecules with a C–S bond perpendicular, $\theta_h = 0$, to the surface (A), and with the C–S bond at $\theta_h = 100°$ from the surface normal (B).

In model B, at a finite temperature, the molecules can lower their average intermolecular energy at the expense of their intramolecular energy, by rotating the C_{15} fragment by 60° about the C_1–C_2, bond. (C_1 is the carbon atom connected to the SH group.) This operation creates a *gauche* conformation near the sulfur. (A *gauche* conformation in any other position would result in less

329

attractive van der Waals interaction.) However, this operation reduces the free volume, thus allowing for only a 19.6° tilt to occur.

B. Brownian Dynamics

So far, we have discussed molecular dynamics simulations of two-dimensional assemblies *in vacuo*. While this is a reasonable approximation for LB and SA monolayers on solid substrates, an accurate representation of monolayers at the water–air interface should account for the viscous water medium that is in contact with the organic molecules. Such contact leads to nonconservative dynamics due to friction with the viscous medium, which is best modeled by the stochastic Brownian dynamics simulations [134]. The presence of solvation and hydration interactions poses an additional difficulty that should be accounted for in such a medium.

Brownian dynamics is a stochastic version of the molecular dynamics method. Here, instead of solving the deterministic Newton's equations of motion that conserve the total energy of the system, the stochastic Langevin equations of motions are used,

$$m_i \frac{d\,\mathbf{v}_i}{dt} = -\sum_i \xi_{ij} \cdot \mathbf{v}_j + \mathbf{F}_i + \mathbf{F}_{i\,\Sigma} + \delta\,\mathbf{F}_i \ , \qquad 4.33$$

$$\mathbf{v}_i = \frac{d\,\mathbf{r}_i}{dt} \ , \qquad 4.34$$

where m_i in Equation 4.33 is the mass of the ith atom or pseudo-atom, $\mathbf{r}_i$ is the Cartesian coordinate of pseudo-atom i, $\mathbf{v}_i$ is its velocity, t is the time, and ξ_{ij} the friction coefficient tensor, representing velocity damping due to friction with the viscous medium. $\mathbf{F}_i$ is the force that each atom feels as a result of all the intra- and intermolecular interactions with the other pseudo-atoms, and $\mathbf{F}_{i\Sigma}$, if nonzero, represents an additional contribution to the forces, due to effective interactions with the medium (such as solvation and hydration forces). $d\mathbf{F}_i$ is a fluctuating stochastic force, giving rise to the Brownian motion, and representing collisions with the medium molecules (i.e., water in the case of a monolayer at the water–air interface), usually chosen to satisfy

$$\langle \delta\,\mathbf{F}_i(t)\delta\,\mathbf{F}_i(0)\rangle = 2m_i kT\xi\delta\,(t) \ . \qquad 4.35$$

The random force time autocorrelation function on the left side of Equation 4.35, correlating the force at time zero to that at time t, is assumed to be diagonal, and proportional to the temperature and to the mass of the particles. The constant of proportionality ξ is the scalar friction coefficient that is related to the diffusion coefficient D through Einstein's relation,

$$\xi = \frac{kT}{m_i D} .$$
4.36

In addition to the advantages listed previously, the use of Brownian dynamics typically permits a considerable increase in the time step, thus allowing a more efficient exploration of the phase-space. Practically, it means that larger assemblies of molecules can be studied by this technique, as compared to Newtonian MD.

Mann *et al.* used this method in their studies of Langmuir monolayers at the water–air interface [125]. Their model consisted of 64 molecules (an 8 x 8 patch) of hexadecanoic acid ($C_{15}H_{31}COOH$) at the water–air interface. They used pseudo-atoms, or spheres of influence (SI), which, in principle, are similar to the united atoms described earlier (Figure 4.16). The SI represented, in their case, an average over configurations of a C_3 alkyl fragment ($CH_3CH_2CH_2$ or $CH_2CH_2CH_2$), larger than the united atoms representing CH_2 or CH_3 groups, used in the dynamics simulations described earlier [91]. They used a simplified force field for the intra- and intermolecular interactions, where bond lengths were constrained to $2R$, where R is the SI radius. Constant constraints were imposed on bond angles, and torsional interactions were ignored. A hard-core variety of nonbond potentials was used, neglecting electrostatics. This string of beads representation of the amphiphile had been used previously in simulations of polymer motion [135].

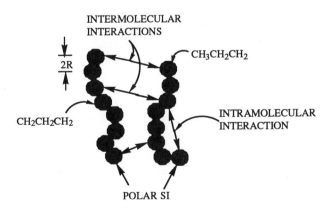

Figure 4.16. The sphere of influence (SI) model for the Brownian motion calculations. (Adapted from Mann *et al.* [125], © 1987, Elsevier Science Publishers.)

In practice, the authors derived the friction coefficient tensor ξ_{ij} by imposing boundary conditions (stick and no stick) giving a value of either $6\pi\eta R$, or $4\pi\eta R$, respectively, where R is the radius of the SI and η is the macroscopic shear viscosity of the solvent medium.

In the context of these calculations, $\mathbf{F}_{i\Sigma}$ represented an effective force exerted on the SI by the medium, due to its density gradient at the water–air interface, and it is maximal for a fragment that is at the interface. In these calculations the authors estimate the density distribution of the amphiphiles, their spreading pressure tensor, and the elastic moduli of the film.

Let us consider now what happens when an amphiphile moves from the vapor phase towards the air-water interface. The polar head group has higher affinity to the water than to the vapor phase, therefore, the potential it responds to is zero (or isotropic) until the polar group is within few molecular dimensions of the interface. At this point the density gradient of the polar group from the liquid phase to the gas phase generates a potential well (~1 kT). It can easily be understood that in the opposite case, where the polar group moves from the liquid phase to the vapor phase, it is repulsed by a barrier. Simulations were carried out where five SIs were designed to represent propyl groups (C_3), and one SI represented a carboxylic group (COOH). Cyclic boundary conditions were used, and since interactions between macroscopic spheres involve a hard core limit at closest approach, a Kihara type potential was used [136].

Figure 4.17 presents the results for surface pressure of 37.5 mN m^{-1}, at 25°C, using time steps of 10 fs. This time step of 10 fs is adequate for the SI model, since the faster C–C and C–H vibrational motions are by definition absent from it. Note the development of tilt correlation as the simulation proceeds. After 50 ps (*a*), there is little correlation between the chain tilts, however, after 200 ps (*b*), one can observe the emergence of domains with different average tilt. After 450 ps (*c*), the tilt becomes more strongly correlated, and further approach towards equilibration (700 ps) involves isomerization of the chains to the all-*trans* configuration, and the emergence of a single tili-domain. Finally at 1500 ps the system is in equilibrium.

In another interesting simulation Mann *et al.* started from a monolayer with 23 Å^2 per molecule and did a "melting" experiment, i.e., reduced surface pressure, and consequently increased area per molecule to 25 Å^2 (liquid phase, see Part Two), and finally to 45 Å^2 (gas phase). Two different expansion processes can occur in principle. In the first, the increased surface area is evenly distributed among the molecules, thus resulting in a homogeneous film with uniform density. On the other hand, in the second, the extra area is unevenly distributed, the area per molecule is smaller then expected, and thus, there is surface area with very few or no surfactant molecules between clusters of molecules that "remember" their packing and orientation in the original solid-like film.

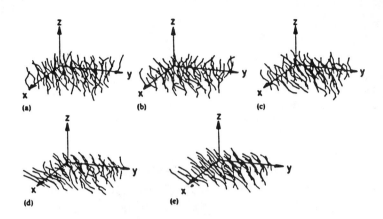

Figure 4.17. Development of equilibrium state for an 8 x 8 patch of hexadecanoic acid molecules: (a) 50 ps; (b) 200 ps; (c) 450 ps; (d) 700 ps; (e) 1500 ps.) (From Mann *et al.* [125], © 1987, Elsevier Science Publishers.)

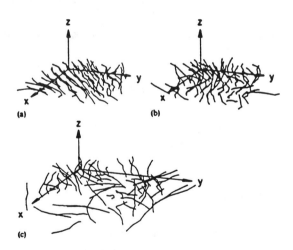

Figure 4.18. Comparison of three equilibrium states measuring (a) 23 Å^2; (b) 25 Å^2; and (c) 45 Å^2 per molecule, and with aging of 720 ps in each case. Note that some molecules are flat on the surface in (c). (From Mann *et al.* [125], © 1987, Elsevier Science Publishers.)

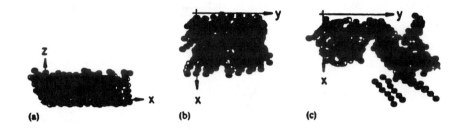

Figure 4.19. The space-filling representatives for two densities and two views. (a) The view down the y axis at $\pi = 37.5$ mN m^{-1}; (b) as in (a), except that the view is down the z axis; (c) 45 Å^2 per molecule viewed down the z axis. Clustering is obvious. (From Mann *et al.* [125], © 1987, Elsevier Science Publishers.)

If the first mechanism underlies the observed phenomenon, then within the limits of this simulation, the formation of a condensed two-dimensional LB film from a molecular film in an ideal-gas phase should be a reversible process. However, if the second mechanism is the one responsible, and clusters thus are formed, the formation of an LB film at the water–air interface should not be reversible, and the density of the organic film should not be uniform. Figure 4.18 presents a comparison of three equilibrium states (23 Å^2, 25 Å^2, and 45 Å^2), with aging of 720 ps in each. Figure 4.19 shows a space-filling model similar to Figure 4.18. From both figures, it is clear that the second mechanism is observed.

These simulations provide information on the development of order, conformational changes and life-time of conformations, *gauche-trans* isomerization, aggregation, desorption of amphiphiles from the water–air interface, pressure-area isotherms, chain tilt, chain order parameter (very important for applications that are sensitive to the order perpendicular to the surface), order parameter of different fragments (from which information on the influence of different groups on the molecular order parameter can be deduced), and two-dimensional elastic moduli.

Brownian dynamics is an interesting approach that, while somewhat limited due to the use of SIs, provides a relatively simple and quick way to investigate new two-dimensional assemblies. In many cases, this method should be considered as a first step before large, time-consuming, and expensive molecular dynamics calculations are carried out, if at all.

334

References

1. Zisman, W. A. In *Friction and Wear*, Davis R., Ed.; Elsevier: New York, 1959; p. 118.
2. Adamson, A. W. *Physical Chemistry of Surfaces* , Wiley: New York, 1976, and references cited therein.
3. Somorjai, G. A. *Chemistry of Two Dimensions: Surfaces* ; Cornell University Press: Ithaca, NY, 1981, and references cited therein.
4. Kitaigorodskii, A. I. *Organic Chemical Crystallography,* Consultants Bureau: New York, pp. 177-217, 1959.
5. Garoff, S. *Proc. Natl. Acad. Sci. USA* **1987**, *84*, 4729, and references cited therein.
6. Langmuir, I *J. Chem. Phys.* **1933**, *1*, 756.
7. Epstein, H. T. *J. Colloi. Chem.* **1950**, *54*, 1053.
8. Safran, S. A.; Robbins, M. O.; Garoff, S. *Phys. Rev. A* **1986**, *33*, 2188.
9. Porter, M. D.; Bright, T. B.; Allara, D. L.; Chidsey, C. F. D. *J. Am. Chem. Soc.* **1987**, *109*, 3559.
10. Ben-Shaul, A.; Gelbart, W. M. *Ann. Rev. Phys. Chem.* **1985**, *36*, 179, and references therein.
11. Nagle, J. F. *J. Chem. Phys.* **1973**, *58*, 252.
12. Nagle, J. F. *J. Chem. Phys.* **1975**, *63*, 1255.
13. Nagle, J. F. *Ann. Rev. Phys. Chem.* **1980**, *31*, 157.
14. Pink, D. A.; Georgallas, A.; Zuckermann, M. J. *Z. Physik. B* **1980**, *40*, 103.
15. Pink, D. A.; Green, T. J.; Chapman, D. *Biochem.* **1980**, *19*, 349.
16. Chaillé, A.; Pink, D.; de Verteuil, F.; Zuckermann, M. J. *Can. J. Phys.* **1980**, *58*, 581.
17. Izuyama, T.; Akutsu, Y. *J. Phys. Soc. Jpn.* **1982**, *51*, 50.
18. Izuyama, T.; Akutsu, Y. *J. Phys. Soc. Jpn.* **1982**, *51*, 730.
19. Pearce, P. A.; Scott, H. L., Jr. *J. Chem. Phys.* **1982**, *77*, 951.
20. Georgallas, A.; Pink, D. A. *J. Colloid Interf. Sci.* **1982**, *89*, 1.
21. Scott, H. L. *J. Chem. Phys.* **1984**, *80*, 2197.
22. Dill, K. A.; Cantor, R. S. *Macromolecules* **1984**, *17*, 380.
23. Cantor, R. S.; Dill, K. A. *Macromolecules* **1984**, *17*, 384.
24. Scott, H. L. *Phys. Rev. A.* **1988**, *37*, 263.
25. Maroncelli, M.; Strauss, H. L.; Snyder, R. G. *J. Chem. Phys.* **1985**, *82*, 2811.
26. Benza, V.; Bassetti, B.; Jona, P. *Phys. Rev. B.* **1987**, *36*, 7100.
27. Bassetti, B.; Benza, V.; Jona, P. *Europhys. Lett.* **1988**, *7*, 61.
28. Mouritsen, O. G.; Ipsen, J. H.; Zuckermann, M. J. *J. Colloid Interf. Sci.* **1989**, *129*, 32.
29. Bassetti, B.; Benza, V.; Jona, P. *J. Phys. France* **1990**, *51*, 259.
30. Doniach, S. *J. Chem. Phys.* **1979**, *70*, 4587.
31. Falkovitz, M. S.; Seul, M.; Frisch, H. L.; McConnell, H. M. *Proc. Natl. Acad. Sci. USA* **1982**, *79*, 3918.
32. Marder, M.; Frisch, H. L.; Langer, J. S.; McConnell, H. M. *Proc. Natl. Acad. Sci. USA* **1984**, *81*, 6559.
33. Andelman, D.; Brochard, F.; Joanny, J. -F. *Proc. Natl. Acad. Sci. USA* **1987**, *84*, 4717.
34. Carlson, J. M.; Sethna, J. P. *Phys. Rev. A.* **1987**, *36*, 3359.

35. Fletcher, R. *Practical Methods of Optimization* , Vol. 1, John Wiley & Sons: New York, 1980.
36. Press, W. H.; Flannery, B. P.; Teukolsky, S. A.; Vetterling, W. T. in *Numerical Recipes, The Art of Scientific Computing* , Cambridge University Press: Cambridge, 1986.
37. Scaringe, R. P. in *Proceedings of the 45th Annual Meeting of the Electron Microscopy Society of America* , Bailey, G. W., Ed.; San Francisco Press: San Francisco, 1987.
38. Kirkpatrick, S.; Gelatt, C. D., Jr.; Vecchi, M. P. *Science* **1983**, *220*, 671.
39. Metropolis, N.; Rosenbluth, A. W.; Rosenbluth, M. N.; Teller, A. H.; Teller, E. *J. Chem. Phys.* **1953**, *21*, 1087.
40. Binder, K. in *Monte Carlo Methods in Statistical Physics* , Binder, K.; Ed., Springer-Verlag: Berlin, 2nd Ed., 1986, Ch. 1.
41. Scott, H. L. *Biochem. Biophys. Acta.* **1977**, *469*, 264.
42. Owenson, B.; Pratt, L. R. *J. Phys. Chem.* **1984**, *88*, 2905.
43. Hann, S. W.; Pratt, L. R. *Chem. Phys. Lett.* **1981**, *79*, 436.
44. Owenson, B.; Pratt, L. R. *J. Phys. Chem.* **1984**, *88*, 6048.
45. Kox, A. J.; Michels, J. P. J.; Wiegel, F. W. *Nature* **1980**, *287*, 317.
46. Cardini, G.; Bareman, J. P.; Klein, M. L. *Chem. Phys. Lett.* **1988**, *145*, 493.
47. Bareman, J. P.; Cardini, G.; Klein, M. L. *Phys. Rev. Lett.* **1988**, *60*, 2152.
48. Harris, J.; Rice, S. A. *J. Chem. Phys.* **1988**, *89*, 5898.
49. Bareman, J. P.; Cardini, G.; Klein, M. L. *Mat. Sci. Soc. Symp. Proc.* **1989**, *141*, 411.
50. van der Ploeg, P.; Berendsen, H. J. C. *J. Chem. Phys.* **1982**, *76*, 3271.
51. Edholm, O.; Berendsen, H. J. C.; van de Ploeg, P. *Mol. Phys.* **1983**, *48*, 379.
52. van der Ploeg, P.; Berendsen, H. J. C. *Mole. Phys.* **1983**, *49*, 233.
53. Pastor, R. W.; Venable, R. M.; Karplus, M. *J. Chem. Phys.* **1988**, *89*, 1112.
54. Pastor, R. W.; Venable, R. M.; Karplus, M.; Szabo, A. *J. Chem. Phys.* **1988**, *89*, 1128.
55. Ryckaert, J. -P.; Klein, M. L. *J. Chem. Phys.* **1986**, *85*, 1613.
56. Ryckaert, J. -P.; Klein, M. L.; McDonald, I. R. *Phys. Rev. Lett.* **1987**, *58*, 698.
57. Hautman, J.; Klein, M. L. *J. Chem. Phys.* **1989**, *91*, 4994.
58. Allinger, N. L. *J. Am. Chem. Soc.* **1977**, *99*, 8127.
59. Burkert, U.; Allinger, N. L. *Molecular Mechanics* ; American Chemical Society: Washington, D.C., 1982, and references cited therein.
60. Allinger, N. L.; Yuh, Y. H.; Lii, J. -H. *J. Am. Chem. Soc.* **989**, *111*, 8551.
61. Lii, J. -H.; Allinger, N. L. *J. Am. Chem. Soc.* **989**, *111*, 8566.
62. van Gunsteren, W. F.; Berendsen, H. J. C. *Groningen Molecular Simulation (GROMOS) Library Manual* , pp. 1–229, Biomos Nijenborgh 16, Groningen, The netherlands.
63. Jorgensen, W. L.; Madura, J. D.; Swenson, C. J. *J. Am. Chem. Soc.* **1984**, *106*, 6638.
64. Weiner, S. J.; Kollman, P. A.; Case, D. A.; Singh, U. C.; Ghio, C.; Alagona, G.; Profeta, S., Jr.; Weiner, P. *J. Am. Chem. Soc.* **1984**, *106*, 765.
65. Weiner, S. J.; Kollman, P. A.; Nguyen, D. T. *J. Comp. Chem.* **1986**, *7*, 230.
66. Brooks, B. R.; Bruccoleri. R. E.; Olafson, B. D.; States, D. J.; Swaminathan, S.; Karplus, M. *J. Comp. Chem.* **1983**, *4*, 187.
67. Ryckaert, J. P.; Ciccotti, G.; Berendsen, H. J. C. *J. Comp. Sci.* **1977**, *23*, 327.
68. van Gunsteren, W. F.; Berendsen, H. J. C. *Mol. Phys.* **1977**, *34*, 1311.

69. Flory, P. *Statistical Mechanics of Chain Molecules* , Academic Press, Wiley-Interscience: New York, 1969.
70. Shnidman, Y.; Eilers, J. E.; Ulman, A.; Evans, S. D., in preparation.
71. Stewart, J. J. P. Quantum Chemistry Program Exchange (QCPE) # 455.
72. Mulliken, R. S. *J. Chem. Phys.* **1955**, *23*, 1833, 1841.
73. Hayes, D. M.; Kollman, P. A. *J. Am. Chem. Soc.* **1976**, *98*, 3335.
74. Thole, B. T.; van Duijnen, P. Th. *Theoret. Chim. Acta* **1983**, *63*, 209.
75. Löwdin, P. O. *Adv. Phys.* **1956**, *5*, 2.
76. Jug, K. *Theort. Chim. Acta* **1956**, *39*, 301.
77. Smit, P. H.; Derissen, J. L.; van Duijneveldt, F. B. *Mol. Phys.* **1979**, *37*, 521.
78. Momany, J. *J. Phys. Chem.* **1978**, *82*, 529.
79. Cox, S. R.; Williams, D. A. *J. Comp. Chem.* **1981**, *2*, 304.
80. Gasteiger, J.; Marsili, M. *Tetrahedron* **1980**, *36*, 3219.
81. Del Re, G. *J. Chem. Phys.* **1958**, *29*, 4031.
82. Del Re, G.; Pullman, B.; Yonezawa, T. *Biochim. Biophys, Acta* **1963**, *75*, 153.
83. Mortier, W. J.; van Genechten, K.; Gasteiger, J. *J. Am. Chem. Soc.* **1985**, *107*, 829.
84. Montier, W. J.; Ghosh, S. K.; Shankar, S. *J. Am. Chem. Soc.* **1986**, *108*, 4315.
85. Rappé, A. K.; Goddard, W. A., III *J. Phys. Chem.*, in press.
86. Ulman, A.; Tillman, N.; Eilers J. *Langmuir* **1989**, *5*, 1147.
87. Strong, L.; Whitesides, G. M. *Langmuir* **1988**, *4*, 546.
88. Allinger, N. L. *J. Org. Chem.* **1987**, *52*, 5162.
89. Still, C., Department of Chemistry, Columbia University, New York, N.Y.
90. Guyot-Sionnest, P.; Hunt, J. H.; Shen, Y. R. *Phys. Rev. Lett.* **1987**, *59*, 1597.
91. van der Ploeg, P.; Berendsen, H. J. C. *J. Chem. Phys.* **1982**, *76*, 2358.
92. Rabe, J. P.; Swalen, J. D.; Outka, D. A.; Stöhr, J. *Thin Solid Films* **1988**, *159*, 275.
93. Dote, J. L.; Mowery, R. L. *J. Phys. Chem.* **1988**, *92*, 1571.
94. Barton, S. W.; Thomas, B. N.; Flom, E. B.; Rice, S. A.; Lin, B.; Peng, J. B.; Ketterson, J. B.; Dutta, P. *J. Phys. Chem.* **1988**, *89*, 2257.
95. Heard, D.; Roberts, G. G.; Holcroft, B.; Goring, M. J. *Thin Solid Films* **1988**, *160*, 491.
96. Shnidman Y.; Eilers, J. E.; Ulman, A.; Evans, S. D., in preparation.
97. Kitaigorodtskii, A. I. *Organic Chemical Crystallography* , Consultants Bureau: New York, 1961.
98. Kitaigorodtskii, A. I. *Molecular Crystals and Molecules* , Academic Press: New York, 1973.
99. Verlet, L. *Phys. Rev.* **1967**, *159*, 98.
100. Andrea, T. A.; Swope, W. C.; Andersen, H. C. *J. Chem. Phys.* **1983**, *79*, 4576.
101. Andersen, H. C. *J. Chem. Phys.* **1980**, *72*, 2384.
102. Berendsen, H. J. C, Postma, J. P. M.; van Gunsteren, W. F.; DiNola, A.; Haak, J. R. *J. Chem. Phys.* **1984**, *81*, 3684.
103. Nosé, S. *Mol. Phys.* **1984**, *52*, 255.
104. Parrinello, M.; Rahman, A. *Phys. Rev. Lett.* **1980**, *45*, 1196.
105. Parrinello, M.; Rahman, A. *J. Appl. Phys.* **1981**, *52*, 7182.
106. Parrinello, M.; Rahman, A. *J. Chem. Phys.* **1982**, *76*, 2662.

107. Nosé, S.; Klein, M. L. *Mol. Phys.* **1983**, *50*, 1055.
108. Ewald, P. *Ann. Phys.* **1921**, *64*, 253.
109. Madelung, E. *Phys. Z.* **1918**, *19*, 524.
110. De Leeuw, S. W.; Perram, J. W.; Smith, E. R. *Proc. R. Soc. London A* **1980**, *373*, 27.
111. Heyes, D. M. *J. Chem. Phys.* **1981**, *74*, 1924.
112. van Gunsteren, W. F. *Protein Eng.* **1988**, *2*, 5, and references cited therein.
113. Israelachvili, J. N. *Acc. Chem. Res.* **1987**, *20*, 415.
114. Still, W. C.; Tempczyk, A.; Hawley, R. C.; Hendrickson, T. *J. Am. Chem. Soc.* **1990**, *112*, 6127.
115. Ryckaert, J. P.; Bellemans, A. *Faraday Discuss. Chem. Soc.* **1978**, *64*, 95.
116. Ryckaert, J. P.; Bellemans, A. *Chem. Phys. Lett.* **1975**, *30*, 123.
117. Gibson, K. D.; Scheraga, H. A. *Proc. Natl. Acad. Sci. U.S.A.* **1967**, *58*, 420.
118. Warms, P. K.; Scheraga, H. A. *J. Comput. Phys.* **1973**, *12*, 49.
119. Saupe, Z. *Naturforrach. Teil A* **1964**, *19*, 161.
120. Ciccotti, G.; Ryckaert, J. -P. *Computer Phys. Rept.* **1986**, *4*, 347.
121. Outka, D. A.; Stöhr, J.; Rabe, J. P.; Swalen, J. D.; Rotermund, H. H. *Phys. Rev. Lett.* **1987**, *59*, 1321.
122. Ungar, G. *J. Phys. Chem.* **1983**, *87*, 689.
123. Maroncelli, M.; Qi, S. P.; Strauss, H. L.; Snyder, R. G. *J. Am. Chem. Soc.* **1982**, *104*, 6237.
124. Maroncelli, M.; Strauss, H. L.; Snyder, R. G. *J. Chem. Phys.* **1985**, *82*, 2811.
125. Mann, J. A.; Tjatjopoulos, G. J.; Azzam, M. -O. J.; Boggs, K. E.; Robinson, K. M.; Sanders, J. N. *Thin Solid Films* **1978**, *152*, 29.
126. Lösche, M.; Rabe, J.; Fischer, A.; Rucha, B.; Knoll, W.; Möhwald, H. *Thin Solid Films* **1984**, *117*, 284.
127. Weis, R. M.; McConnell, H. M. *Nature* **1984**, *310*, 47.
128. Riegler, J. F.; LeGrange, J. D. *Phys. Rev. Lett.* **1988**, *61*, 2492.
129. Lösche, M.; Sackmann, E.; Möhwald, H. *Ber. Bunsenges Phys. Chem.* **1983**, *87*, 848.
130. Miller, A.; Knoll, W.; Möhwald, H. *Phys, Rev, Lett.* **1986**, *56*, 2633.
131. Middleston, S. R.; Iwahashi, M.; Pallas, N. R.; Pathica, B. A. *Proc. R. Soc. Lond. Ser. A* **1984**, *396*, 143.
132. Pallas, N. R.; Pathica, B. A. *Langmuir* **1985**, *1*, 509.
133. Lin, B.; Peng, J. B.; Ketterson, J. B.; Dutta, P.; Thomas, B. N.; Buontempo, J.; Rice, S. A. *J. Chem. Phys.* **1989**, *90*, 2393.
134. Binder, K., Ed. *Monte Carlo Methods in Statistical Physics, Top. Curr. Physics;* Springer-Verlag: Berlin, 1979, Vol 7.
135. Baumgartner, A. *Annu. Rev. Phys. Chem.* **1984**, *35*, 419.
136. Feke, D.; Prabhu, N. D.; Mann, J. A. Jr.; Mann, J. A. III *J.Phys, Chem.* **1984**, *88*, 5735.

PART FIVE

APPLICATIONS OF LB AND SA FILMS

The integrated circuit industry, both in electronic and electro-optic device applications, is in a continuous quest for faster speeds and larger memories. This has led to an impressive refinement of microlithography, and experiments are carried out in x-ray lithography where ultimately the resolution will be of the order of hundreds of Ångstroms. If this is the case, then what are the advantages of organics, especially those of monolayers and thin organic films? In this part, we describe different applications of LB and SA films, and try to address this question.

5.1. Nonlinear Optics

A. Background

Nonlinear optics (NLO) comprises the interaction of light with matter to produce a new light field that is different in wavelength or phase. The reader may find detailed explanations on the origin of the nonlinear effect, as well as on the applications of $\chi^{(2)}$ and $\chi^{(3)}$ materials, in a recent book by P. N. Prasad and D. J. Williams [1] , and in other books and reviews [2–4].

In this part, we discuss equations necessary for the understanding of thin organic films with NLO properties. Let us begin with the interaction of light with molecules. When an electromagnetic wave interacts with a molecule, a

339

polarization, **p**, is induced in the molecule by the electric field component, **E**, of the wave. This polarization can be approximated as a power series,

$$\mathbf{p} = \alpha \cdot \mathbf{E} + \beta \cdots \mathbf{EE} + \gamma \cdots \mathbf{EEE} + \dots \ . \tag{5.1}$$

This equation describes the polarization induced in a molecule as a function of the electric field. Both **p** and **E** are vector quantities, α, β, and γ are tensors. To understand this point, one should remember that if the electric field vector has three components (i.e., $\mathbf{E}_x$, $\mathbf{E}_y$, and $\mathbf{E}_z$), each of these components can contribute to polarization in each of the three directions of the molecular system. Therefore, if we consider the first term $\alpha \cdot \mathbf{E}$, α is a second-rank polarizability tensor with nine elements (α_{xx}, α_{xy},..., α_{zz}). Using the same argument, it is clear that, since in the next term the polarization is proportional to $\mathbf{E}^2$, β is a third-rank polarizability tensor with 27 components (β_{xxx}, β_{xxy},..., β_{zzz}), and γ is a fourth-rank tensor with 81 elements (γ_{xxxx}, γ_{xxxy},..., γ_{zzzz}). In most cases, molecular symmetry reduces the number of significant tensor elements simply because the molecular dipole moment, induced by the intramolecular charge transfer, usually is in a direction that is parallel to the long axis of the molecule (which we call z), and therefore, tensor elements that do not have z-components (e.g., β_{xxx}, γ_{xxxx}) are negligible.

We already have discussed (in Part Two) that noncentrosymmetry of a film is a requirement for second-order hyperpolarizability; however, the same also is true for the individual molecule. In other words, unless the molecular coordinate system lacks an inversion center, the odd-rank tensors such as β will be zero. On the other hand, molecules that are symmetrical but highly conjugated (e.g., polyacetylene, polypyrrole, etc.) can have large α and γ parameters. The 4-nitroaniline, a conjugated donor–acceptor system that does not have a center of symmetry, exhibits a second-order hyperpolarizability coefficient (β),

Turning from the microscopic to the macroscopic domain, a power series relating the macroscopic effective polarizability **P**, to the electric field **E**, can be written similarly,

$$\mathbf{P} = \chi^{(1)} \cdot \mathbf{E} + \chi^{(2)} \cdots \mathbf{EE} + \chi^{(3)} \cdots \mathbf{EEE} + \dots \ , \tag{5.2}$$

where $\chi^{(1)}$, $\chi^{(2)}$ and $\chi^{(3)}$, are tensors with similar meaning to α, β, and γ, but here they describe the polarization induced in the ensemble. It should be emphasized that the nonlinear effect is a ground-state phenomena. Hence, there is no excitation of the molecules, although there are cases where absorption can be used for resonance enhancement. However, the excited states of a molecule also are considered in the calculation of molecular hyperpolarizabilities, since their interaction with the ground state produces these effects. To better understand this last statement, we have to go back to 4-nitroaniline. The ground state of 4-nitroaniline is aromatic, and the charge-separated form, which is a result of an electron transfer from the amine donor to the nitro acceptor group, is quinonic (simply because the aromaticity of the benzene ring is disrupted in this electron transfer process). Spectroscopic studies (e.g., solvatochromism) suggest that the excited state of such molecules is highly polar and should have a contribution of the quinonic, charge-separated form. However, it appears that the ground-state dipole moment also is quite large, which usually is explained by suggesting that a low-lying excited charge-transfer state mixes into the ground state via the charge-transfer resonance.

One of the manifestations of the nonlinear interaction of light with matter is *second-harmonic generation* (SHG). The simplest way to explain SHG is to view it as a combination of two photons at frequency ω to give one proton at a frequency 2ω. Figure 5.1 illustrates some of the effects occurring through $\chi^{(2)}$. We can explain further that the SHG process occurs because the polarization that is created in the material by the interaction with the electric field of the light has a Fourier component at 2ω that acts as a source of an electromagnetic radiation at 2ω. Another important parameter in $\chi^{(2)}$ materials is called *phase matching*. This merely says that if the fundamental wave at ω and the second-harmonic wave at 2ω travel in phase through the material (i.e., ω and 2ω travel at the same speed) the harmonic field can build up considerable amplitude [4].

When instead of using only one frequency (ω), we use two frequencies (ω_a and ω_b) we get $\omega_a \pm \omega_b$. (This process also occurs with ω, but one of the resulting frequencies here is zero) [5]. The case of $\omega_a - \omega_b$ is both interesting and important. Let us consider a case where one of the waves (ω_a) is very intense and the other (ω_b) is weak. Here, the nonlinear interaction leads to the formation of $\omega_a - \omega_b$, but at the same time to the amplification of ω_b [6]. It seems as if the energy of ω_a was divided into two parts, where one part was used to amplify ω_b (Figure 5.1). Therefore, this phenomenon is called *parametric amplification,* and is extremely useful in spectroscopy for the detection of very weak signals, and in optical communication for repeaters [7].

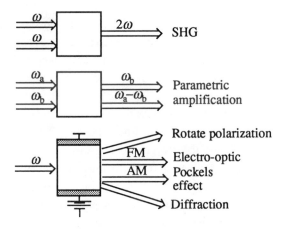

Figure 5.1. Effects occurring through $\chi^{(2)}$.

The last effect that occurs through $\chi^{(2)}$ is the *electro-optic* or *Pockels effect* [8–10]. This is an important effect, since it is used in integrated electro-optics. In such devices the optical signal (usually confined in a waveguide) is modified by the application of an external electric field. This can be used in switching of optical signals in telecommunications and signal processing [11]. An essential material requirement for these devices is a fast change (<1 nsec) of the refractive index by the electric field, which is based on the Pockels effect, where the change in the dielectric constant, $\Delta\varepsilon$, is a linear function of the applied field,

$$\Delta\varepsilon_{ij} = \sum_k 4\pi\chi_{ijk}^{(2)}\mathbf{E}_k(\omega_0) \qquad \text{(esu units)}. \qquad 5.3$$

Another way to write this equation is

$$\Delta(1/\varepsilon)_{ij} = \sum_k r_{ijk}\,\mathbf{E}_k(\omega_0)\,, \qquad 5.4$$

where $\chi^{(2)}$ is the second-order susceptibility, $\mathbf{E}(\omega_0)$ is the applied field at the modulation frequency at ω_0, and r is the electro-optic coefficient. The reader may find that some papers use r in picometers/volt, and others use $\chi^{(2)}$ in cm^5/esu. It is good to remember that the conversion between the two is given by

$$r_{ijk} = -(\chi_{ijk}^{(2)}/\varepsilon_{ii}\,\varepsilon_{jj})\times 4.19\times 10^8 \text{ pm/V}\,. \qquad 5.5$$

In all the devices that are based on the electro-optic effect [12], there is one common requirement, i.e., an electrically induced phase shift of the order of π in the optical fields ($\Delta\phi$) as they pass through the electro-optic guiding region. The required r or $\chi^{(2)}$ may be estimated from

$$\Delta\phi = 2\pi l \, \Delta n \, /\lambda \equiv \pi \,,$$ 5.6

where l is the path length through the medium, Δn is the electrically induced index change (as defined by Equation 5.3, and $\varepsilon = n^2$), and λ is the free-space wavelength.

Let's turn our attention to the $\chi^{(3)}$ effects (Figure 5.2). The first effect that we mention is *third-harmonic generation* (THG). In this case, the input frequency is ω and the output frequency is 3ω. This frequency tripling is useful in the study of new $\chi^{(3)}$ materials and may be useful for deep uv conversion.

The second effect is *optical bistability*. This is similar to the Pockels effect, but here the refractive index is changed as a result of the light intensity in the resonant cavity [13]. The change in the refractive index affects the propagation of the beam in the material. When the light intensity increases, its effect on the refractive index increases, too, and as a result, the material becomes transmitting. The increase of the output light intensity (I_{out}) with the intensity of the input light (I_{in}) continues until a saturation occurs. If we decrease I_{in} we see that there is a hysteresis and bistability in the I_{out} vs. I_{in} characteristics. One of the potential applications of this effect may be the optical analogue of a transistor in which a weak beam is used to control the transmission characteristics of an intense beam [14].

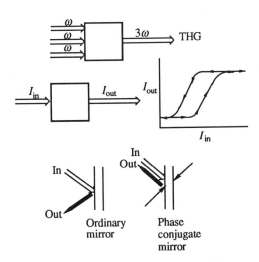

Figure 5.2. Effects occurring through $\chi^{(3)}$.

The third interesting effect is that of *optical phase conjugation*. Without getting into many details, this is basically a four-wave mixing process (the interaction of three electromagnetic waves to produce a fourth one) [9]. To make it more clear, let us start with two beams that intersect to create a phase grating in the material (Figure 5.2). Because of this grating, the complex conjugate of the phase front of the third beam is created as an outgoing beam (fourth beam). As a result, the third and fourth beams are related, where the fourth beam propagates as if it were time-reversed relative to the third beam. This is very significant since any loss of information in the third beam during propagation is recovered in the fourth beam, as it propagated back *through the same path*. Moreover, in the NLO material, the fourth, phase-conjugate, beam can be amplified in the wave mixing process. In summary, phase-conjugate mirrors open interesting possibilities in lensless imaging and image processing.

Before turning to the advantages of thin organic films in NLO applications, let us consider the following equations that may clarify the relationship between intramolecular charge transfer and molecular second-order hyperpolarizability coefficient,

$$\beta_{vec} = \frac{3e^2\hbar^2}{2m} F(\omega) f \Delta\mu_{e,g} , \qquad\qquad 5.7$$

$$F(\omega) = \frac{W}{[W - (2\hbar\omega)^2][W^2 - (\hbar\omega)^2]} , \qquad\qquad 5.8$$

where

$$\beta_{vec} = \beta_{zzz} + (\beta_{zzx} + \beta_{zzy} + 2\beta_{xxz} + 2\beta_{yyz})/3 . \qquad\qquad 5.9$$

β_{vec} is the measured second-order hyperpolarizability coefficient, $\Delta\mu_{e,g}$ is the difference between the ground and excited state dipole moments ($\Delta\mu_{e,g} = \mu_e - \mu_g$), W is the energy of the optical transition from the ground to that excited state for which $\Delta\mu_{e,g}$ is defined, and f is the oscillator strength of that optical transition. It is $\Delta\mu_{e,g}$ and f that we try to maximize by molecular engineering. We usually do so by increasing both the electron donating ability of the donor and electronegativity of the acceptor substituents, which enhances the intramolecular charge transfer. As a result, both $\Delta\mu_{e,g}$ and f are increased, and therefore, β also is enhanced.

Let us turn now to macroscopic nonlinearity. Here, we find that for isotropic cases, $\chi^{(1)}$ and $\chi^{(3)}$ are related to the molecular parameters α and γ by,

$$\chi^{(1)} = N\alpha F^2(\omega) , \qquad\qquad 5.10$$

$$\chi^{(3)} = N\gamma F(\omega_1)F(\omega_2)F(\omega_3)F(\omega_4) , \qquad\qquad 5.11$$

344

where N is the number density of molecules and $F(\omega)$ is a local field factor that provides the correction of the incident field. This local field factor simply is describing the interaction of a molecule with the vibrating dipoles of its neighbors. For simplicity, we assume that α and γ are orientational averages of the tensor quantities in Equation 5.1 so that expressions 5.10 and 5.11 apply only to isotropic media. The local field factors are estimated from the Onsager expression,

$$F(\omega_i) = \frac{(\varepsilon_\infty^i + 2)\varepsilon_\omega}{\varepsilon_\infty^i + 2\varepsilon_\omega} , \qquad 5.12$$

where ε_ω and ε_∞ are the dielectric constants at frequency ω and at the optical frequency limit [15]. For $\chi^{(2)}$ materials, however, since $\chi^{(2)}$ is zero in centrosymmetric media, we have to add an orientational factor $O(\Omega)$ to the relationship between $\chi^{(2)}$ and β,

$$\chi^{(2)} = N\gamma F(\omega_1)F(\omega_2)F(\omega_3)O(\Omega) , \qquad 5.13$$

When we analyze LB and SA monolayers and films, where the chromophore dipoles are oriented parallel to the long axis of the amphiphile, we assume $C_{\infty v}$ symmetry. We can write the following equation for near-perfect orientation,

$$\chi_{zzz}^{(2)} \sim N\beta_z f_z(2\omega)^2 f_z(\omega) , \qquad 5.14$$

and assume any other component of $\chi^{(2)}$ is practically zero. However, as the reader already knows, perfect perpendicular alignment is not common, and the molecules are tilted in the film. As a result, tensor elements other than $\chi_{zzz}^{(2)}$ become nonzero ($\chi_{xzz}^{(2)}$, $\chi_{yzz}^{(2)}$, etc.) and, in some cases, may even have significant values.

Before we discuss some specific LB systems, it is important to discuss the issue of SHG in waveguides, which will emphasize even more the promise in organic thin films for NLO. Here, we start with a crystalline material, where Zyss has shown that the optimum harmonic efficiency (the conversion of ω to 2ω) is given by [16],

$$\eta_{\text{crystal}} = \frac{\omega^2}{n^3}(\frac{\mu_0}{\varepsilon_0})^{3/2}(\chi^{(2)})^2 \lambda^{-1} L P^{(\omega)} F(\frac{\Delta k L}{2}) , \qquad 5.15$$

where λ, ω, η are the wavelength, frequency, and refractive index of the fundamental beam, respectively, $P^{(\omega)}$ is the power, Δk the momentum difference at the fundamental and harmonic frequencies, L is the interaction length, and $F(\Delta k \cdot L/2)$ is the phase mismatch factor (the difference between the propagation of ω and 2ω in the material) defined as $F(x) = \sin^2 x/x^2$ with x

$= \Delta kL$. The quantity $\Delta k \cdot L$ is a phase angle $\Delta \phi$; therefore, $F(x)$ is zero at $\Delta k = 0$, and a maximum at a phase angle $\pi/2$. It is clear that for a phase-matched harmonic generation, the conversion of the fundamental to the harmonic is determined by the interaction length. Ledoux *et al.* showed that in a waveguide geometry, the conversion efficiency is given by [17]

$$\eta_{\mathrm{wg}} = \frac{\omega^2}{n^3}(\frac{\mu_0}{\varepsilon_0})^{3/2}(\chi^{(2)})^2 a^{-1} I_R F(\frac{\Delta kL}{2}) ,$$ 5.16

where a is the wavelength thickness, I_R is the overlap integral and defined as

$$I_R = \int P_{\mathrm{NL}} E_{\mathrm{m}}^{2\omega} d\sigma ,$$ 5.17

where P_{NL} is the nonlinear polarization, $E_{\mathrm{m}}^{2\omega}$ is the harmonic field distribution, and σ is the cross-section coordinate. It can be calculated that I_R is optimal at l/a [17]. Now if we divide Equation 5.16 by 5.15, we can get an estimate on the advantage of the waveguide relative to the crystal,

$$\eta_{\mathrm{wg}} / \eta_{\mathrm{crystal}} \sim L/\lambda .$$ 5.18

In real-life situations, the advantage of a waveguide will be determined most of all by the coherence length; that is, by the degree of mismatch between ω and 2ω within the waveguide. Since the effective refractive index for the guided mode depends both on the thickness and on its uniformity over the length of the waveguide, the phase mismatch, which determines the conversion efficiency, strongly depends on these parameters.

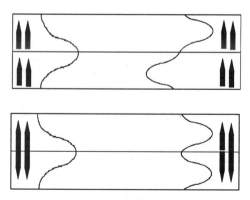

Figure 5.3. $TM_0 \rightarrow TM_1$ conversion: the problem (top) and the solution (bottom). The arrows denote molecular dipole direction in the multilayer film.

Here, the ability to control film thickness and uniformity clearly demonstrates the advantages of the LB and SA techniques in waveguide technology. Furthermore, according to Equations 5.15 and 5.16, the efficiency is a function of $\chi^{(2)}/n^3$, where n is the refractive index. While a typical value for the refractive indices of organic materials is ~1.5, the most commonly used inorganic material for waveguides, LiNbO$_3$, has a refractive index of 2.24. Therefore, for a given value of $\chi^{(2)}$, the ratio between these two values can be calculated as

$$n^3_{\text{LiNbO}_3}/n^3_{\text{organic}} \approx 3.4 \, , \qquad\qquad 5.19$$

which is another important advantage of organic waveguides.

At this point, the reader has become familiar with the preparation and properties of LB and SA monolayers, as well as with some of the theory of nonlinear optics. We have discussed the issues of order parameter in monolayers, and the advantages of molecular and material design. The conclusion that emerges from all these discussions is that, conceptually, the LB and SA techniques may provide preferred design qualities, and, therefore, ideal thin film materials for nonlinear optics. Here, molecularly engineered chromophores can be functionalized into molecularly engineered amphiphiles, which can be incorporated in high concentration into highly ordered films with thicknesses defined by molecular resolution.

The last point that demonstrates the advantage of the layer-by-layer construction of a waveguide is the possibility of a $TM_0 \rightarrow TM_1$ conversion. Figure 5.3 presents both the problem and its solution. The spacial distribution of electric fields for the TM_0 and TM_1 indicates that the overlap integral between them should be zero, because TM_0 is all positive while TM_1 has a node; i.e., TM_1 changes sign. This is true when the dipole moments are oriented in the same direction throughout the film. However, if the molecular dipole (and $\chi^{(2)}$) changes sign at the TM_1 node, the negative part of TM_1 will become positive, and a $TM_0 \rightarrow TM_1$ conversion will become possible and efficient.

We have seen before (in Part Two) how molecular design can provide amphiphiles for noncentrosymmetric Y-type ABABAB films. The LB technique offers a unique solution to the $TM_0 \rightarrow TM_1$ conversion problem, since by simply repeating a deposition of a B molecule at the appropriate thickness, (ABAB...AB|BA...BABA), the direction of the dipole at the node is changed and TM_1 will be all positive, allowing for an efficient $TM_0 \rightarrow TM_1$ conversion.

B. Monolayers and Films with Nonlinear Optical Properties

1. Second–Order Effects

In this section, we use Roman numbers for the different materials under study, since we need to come back to different structures for comparison.

Blinov *et al.* were the first to report on spontaneous polarization of LB films [18, 19]. They used a series of azobenzenes to form X- and Y-type monolayers and studied their piezoelectric, pyroelectric, and electrooptic characteristics. We shall discuss their work in Section 5.2.

I

Aktsipetrov *et al.* in 1983 published the first paper on SHG from LB multilayers [20], using 4-octadecylamino-4'-nitroazobenzene (I) as the amphiphile. The interesting feature in their experiment is that they measured SHG in reflection from the surface, thus learning about molecular orientation in the two dimensions. To better understand this issue, we have to remember that p-polarized light has its electric field perpendicular to the monolayer surface, and hence parallel to the amphiphile dipole, while s-polarized light has an electric field parallel to the surface and thus perpendicular to the molecular dipole. Therefore, for a perfect perpendicular orientation with $C_{\infty v}$ symmetry, i.e., with the polar axis perpendicular to the surface but isotropic in the plane of the film, one would expect large $I^{2\omega}(p \rightarrow p)$, a somewhat smaller $I^{2\omega}(p \rightarrow s)$, and a practically zero $I^{2\omega}(s \rightarrow s)$. Here, we denote $I^{2\omega}$ $(i \rightarrow j)$ the intensity of the reflected harmonic, where i is the polarization of the incident beam, and j the polarization of the reflected beam. It is easy to understand that in a case of a tilted amphiphile, the ratio $I^{2\omega}(p \rightarrow p)/I^{2\omega}(p \rightarrow s)$ is proportional to the tilt angle. (As $I^{2\omega}(p \rightarrow p)/I^{2\omega}(p \rightarrow s)$ decreases, the tilt angle increases.) However, in this case, $I^{2\omega}(s \rightarrow s)$ was found to be significant, and therefore, it was not possible to calculate a unique tilt angle. The authors reported a value of 2.8 x 10^{-28} cm^5/esu for $\chi^{(2)}$ in this monolayer.

In the next paper from this group [21], they reported on the preparation of Y- and Z-type LB films from this molecule (I). The centrosymmetric Y film was prepared by starting with a clean SiO$_2$, hydrophilic surface. For this film, they measured $I^{2\omega}$ (p-polarized incident light, reflection mode), which was an order of magnitude lower than that of the monolayer. However, when they prepared a

Z film by starting with a hydrophobic surface, they found that $I^{2\omega}$ increased with the number of layers (n). Figure 5.4 presents their results.

It should be pointed out that they did not plot $[\chi^{(2)}]^{1/2}$ (which is the usual practice), but rather $I^{2\omega}/(I^{\omega})^2$ as a function of n, and therefore, the result is a parabola [22]. One last note concerning this study is that the authors report on a five-layer film. There are some who argue that it would be impossible to prepare a Z-type, 100-layer film with a strong dipole because of the large charge separation that will be built in the film. There is no direct evidence to support this argument.

Ledoux *et al.* [23] and Loulergue *et al.* [24] improved the amphiphilic properties of I by adding a carboxylate group (which also is an electron acceptor group). Another very important change was to shorten the alkyl chain from 18 to 12 carbons, thus increasing the surface energy of the monolayer. (The polar chromophore is closer to the surface.) As a result the energy difference between the head-group and tail-group surfaces decreased and the stability of a Z film increased. Indeed, they could get Z films with and without Cd^{2+} in the subphase.

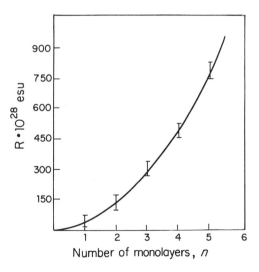

II

Figure 5.4. Conversion coefficient $R = I_{2\omega}/I^2_{\omega}$ vs. the number (n) of monolayers. (From Aktsipetrov *et al.* [4], © 1985, Am. Inst. Phys.)

It was found that Cd^{2+} stabilizes the film (at 24 mN m^{-1}), while without Cd^{2+} in the subphase, the film collapsed at $\approx$18 mN m^{-1}, and was deposited, therefore, at 15 mN m^{-1}. Furthermore, the interference fringe of the second harmonic (SH) was dramatically different between these two films, when for the Cd^{2+}-stabilized film there was well-contrasted fringe pattern with a quasi-vanishing value at a normal incidence. The SHG measurements at an incident angle of 45° yielded a large signal (532 nm), where the intensity was 10 times higher for p- than for s-polarized excitation. The dependence of $I_{2\omega}$ on layer number n was sublinear for the film without Cd^{2+} and subquadratic for the stabilized film. The authors report a value of $\chi^{(2)} = 6.7 \times 10^{-6}$ cm^5/esu for their film, the largest reported so far in the literature. However, they attribute this value to some resonance enhancement at $2\omega = 532$ ($\lambda_{max} = 504$ nm without Cd^{2+} and 464 nm with Cd^{2+}). It seems that the need for an NLO chromophore that is fully transparent to ω and to 2ω, for the conversion of 840 to 520 nm, is becoming clear, and we shall emphasize this issue as we go along. The merocyanine (III) has a very high β value due to the gain of aromatization energy in the intramolecular electron transfer ($\approx 10^{-27}$ cm^5/esu [25]).

Gaines was the first to prepare a monolayer of the pyridine-substituted alkyl merocyanine [26], but Daniel and Smith were the first to report on a noncentrosymmetric multilayer structure of merocyanine amphiphiles [27]. They used the same derivatives reported by Gaines (C_{16} and C_{22}), but decided to introduce long-chain amines ($C_{18}H_{37}NH_2$) [28, 29] as the counter layer in their ABABAB system (due to the protonation problem mentioned earlier).

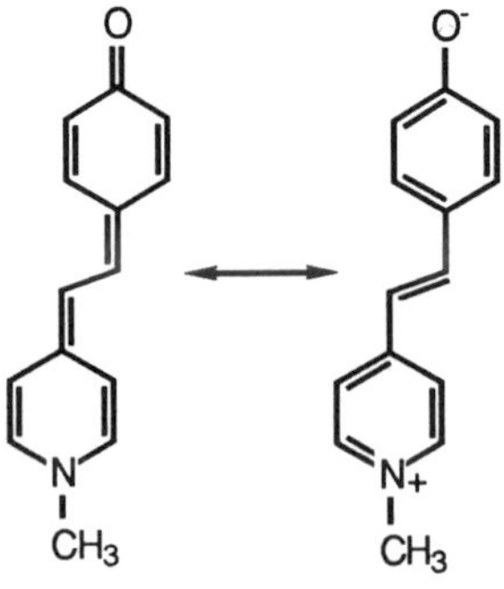

III

Even with these amines, they found that the merocyanine in their monolayers was protonated, and they had to expose the monolayers to ammonia or methylamine vapor or to NaOH solution to deprotonate partially the dyes. A detailed spectral study of the protonation process is described. It seems that a most important statement in this paper is that when monolayers were transferred at a pressure of 28 mN m^{-1}, a film of 55 monolayer did not appear to scatter light.

Girling *et al.* were the first to measure SHG from a merocyanine LB film [30]. They used the C_{22} derivative with ω-tricosenoic acid as an intermediate layer. After modeling the film as uniform axially isotropic dielectric, he could report that β_z for the dye was 2.42 x 10^{-27} cm^5/esu, which is a very high number and may be resonance-enhanced at 2ω (533 nm).

When alternating layers of the dye and arachidic acid were prepared (up to three pairs), the ratio of the harmonic intensity was 1:3.2:6.2, rather than the expected ratio of 1:4:9. This subquadratic ratio can be attributed to layer imperfections, but also to incomplete deprotonation of the dye. It has been shown by IR spectroscopy that when ammonia is used, the ammonium ion is a part of the film. [26] It is not clear how much reorganization is needed to accommodate these ions in the multilayer structure, and if it is at all possible to achieve a complete deprotonation in the film.

One of the most commonly used chromophores is the hemicyanine (IV). Meredith *et al.* did some studies on hemicyanine salts and could show that its β coefficient should be comparable to that of the merocyanine chromophore [31]. Here, as in the case of the merocyanine dye, the acceptor is the alkyl pyridinium ring, but in this case, the donor is an amine, which makes the dye bifunctional. This bifunctionality is essential for any modification, since substitution on both ends makes it possible to incorporate the dye into amphiphiles and polymers, and thus to prepare ABAB multilayer systems. However, while both the phenolate oxygen and dialkylamino nitrogen are good electron donors, a dialkylamino *para* to an alkylpyridinium group is more difficult to protonate.

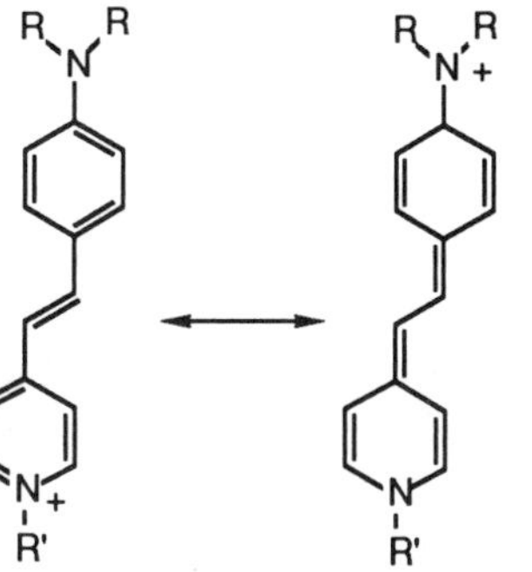

IV

Indeed, Marowsky *et al.* investigated the optical absorption spectra of hemicyanine monolayers at different pH conditions and found that even in substantial concentrations of H_2SO_4 (0.02%, $\approx 2 \times 10^{-3}$ M) in 10^{-3} M dye in propanol, the dye is less than 40% protonated [32]. This finding supports the idea that hemicyanine is more suitable than the merocyanine dye for LB multilayer fabrication.

Girling *et al.* were the pioneers in the studies of LB films from hemicyanine (V) [33]. They prepared both monolayers and ABAB multilayers with ω-tricosenoic acid as the intermediate layer.

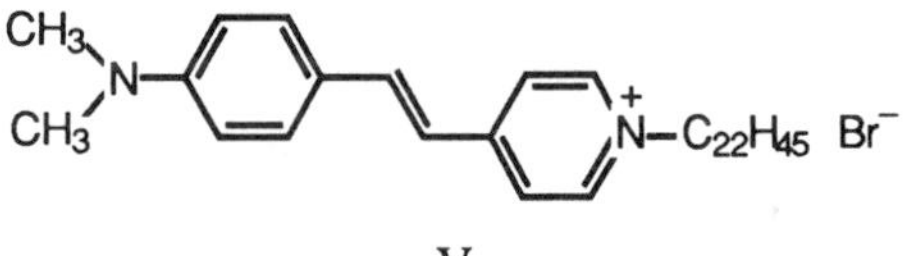

V

There are some very important results in their paper. First, they could estimate that the tilt angle of the chromophore in the monolayer (away from the polar axis) is 37°, that $\beta_z = 2.29 \times 10^{-28}$ cm^5/esu, and that the dependence of the second harmonic on the number of layers is 1.2:2.9:5.9, i.e., subquadratic.

Second, and most important, they noted that $I^{2\omega}(p \rightarrow s)$, which was considerable in the first layer, decreased in the subsequent layer, indicating that the dye becomes more perpendicular with increasing layer number. In a subsequent paper, Neal *et al.* added a new dimension to the preparation of ABAB films, when they used as the intermediate layer between hemicyanine (V) layers a chromophore with an amide group (VI), thus introducing intralayer hydrogen bonding as a stabilization mechanism [34]. Note that while in V the alkyl chain is on the acceptor side, it is substituted in VI on the donor side.

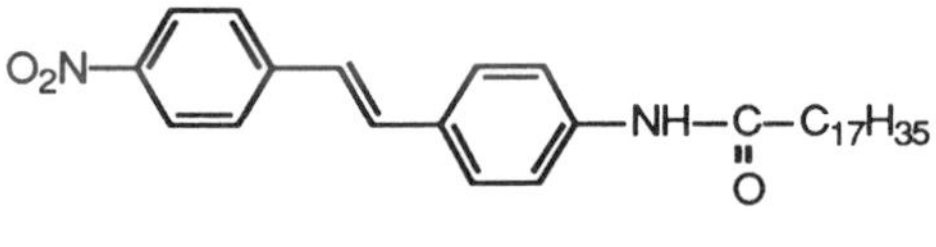

VI

Hence, within the stable Y-type film, the dyes are in a noncentrosymmetric arrangement. They report that monolayers of V gave a β value of 2.8×10^{-28} cm^5/esu, and a tilt angle of 24°, while those of VI gave a β value of 5.2×10^{-29} cm^5/esu and a tilt angle of 30°. On the other hand, the mixed layers gave a β value of 7.64×10^{-28} cm^5/esu and a tilt angle of 23°, a major improvement even over the best V monolayer. This result strongly suggests that there may be an enhancement mechanism in the bilayer, where the neighboring dimethylamino and nitro groups stabilize each other. If this is correct, it should have a major impact on the engineering of NLO materials; however, at this moment, there is no additional unambiguous evidence for such an enhancement mechanism.

At about the same time, Earls *et al.* studied the electron diffraction (reflection and transmission) from LB films of V [35]. They concluded that the monolayer is quasi-crystalline molecular packing, that the chromophore and alkyl chain moieties are parallel to each other, and that the chromophore is tilted 40° away from the surface normal. If the chromophores are arranged indeed like a deck of cards, the question is what kind of interaction (if any) is there in the monolayer, and if an interaction exists, what is its effect on the NLO properties of the two-dimensional assembly. There are two issues that have to be considered; the first is the aggregation, and the second the local field effect. The aggregation is a well-known phenomenon, and dyes can interact to form different types of aggregates depending on the shift in their optical spectra (e.g., *H*-aggregates are blue-shifted, while *J*-aggregates are red-shifted). The local field effect basically is a result of the interaction of a vibration molecular dipole with the vibrating dipoles of neighboring chromophores. (For further reading, see References 36–41.) In a simple schematic way, it says that the microscopic polarizability of a molecule is affected by the electron density distribution in its neighbors. The Onsager model in principle is the same, when taking the dipoles of solvent molecules into account. However, since in an LB film we consider a system with inherent order and with correlation among the molecules in the two dimensions, the neighboring dye molecules are not distributed randomly around the dye but rather in a well-defined symmetry, forming a local field. Therefore, if the dipoles were perpendicular to the surface (and thus parallel to each other), the repulsion between them would cause a decrease in the charge separation, and hence in β_z. Hayden tried to calculate these local field effects for LB films of mixtures of V and behenic acid ($C_{20}H_{41}COOH$), and concluded that the range of dipole–dipole interaction in the mixed monolayers is ~ 10Å [42]. However, this study did not account for aggregation of the hemicyanine dye (discussed later) and therefore may not be correct.

Girling *et al.* studied the SHG ($I^{2\omega}(p{\rightarrow}p)$) on mixed monolayers of V with arachidic acid ($C_{19}H_{39}COOH$) [43]. They showed that the SHG intensity of a pure LB monolayer of V was $\approx$50% of that measured for a monolayer of 1:1 mixture. While this is an interesting experiment, again, it does not distinguish between the aggregation and local field mechanisms.

Schildkraut *et al.* investigated the optical absorption of mixed monolayers of V with arachidic acid. They found that the optical absorption is concentration dependent [44]. Pure monolayers exhibited absorption at 350 nm with a very small absorption at 460 nm, where the dilute dye adsorbed. They studied the SHG of these mixed monolayers and concluded an H-aggregate is formed in the pure dye monolayer, and that the nonresonant value of β is three times larger in the dye than in its aggregate. This observation was confirmed later by Marowsky and Steinhoff, who studied the SHG in these mixed films and found that when they used incident light at 820 nm ($2\omega = 410$ nm), the 2ω signal was much more enhanced than for an incident light of 1060 nm ($2\omega = 530$ nm) [45]. They concluded, therefore, that the aggregate is more NLO-active than the monomeric dye. However, 410 nm is within the absorption band of the hemicyanine chromophore; hence, their result may be due to resonance enhancement in 2ω.

The question of aggregation and its relationship to the structure of the amphiphile is even more interesting in the work of Cross et al. [46]. In this study, they investigated the LB monolayers of V and compared the results to that from an LB monolayer of VII. They report that $I^{2\omega}(p\rightarrow p)$ for V is about an order of magnitude larger than that measured for VII. It thus seems that the aggregation is very sensitive to the position of the alkyl chain and that there is more aggregation in VII than in V. This difference may be a result of the different areas per molecule in the two cases. Thus, while in V there is a dimethylamino group, in VII there is a methylpryridinium group that occupies less space. As a result, there is more free volume in V than in VII, and more aggregation in the latter. Indeed the authors report on a slight difference in the tilt angle ($22°$ for V and $23.7°$ for VII), which is in agreement with the excess free volume in VII.

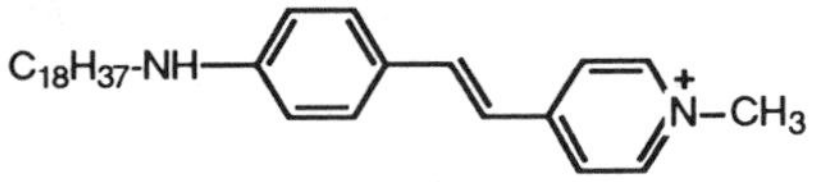

VII

Further support for the suggested relationship between the area per molecule and the degree of aggregation (and hence $\chi^{(2)}$) comes from the work of Marowsky et al. [32], as well as from recent results from our laboratory [47].

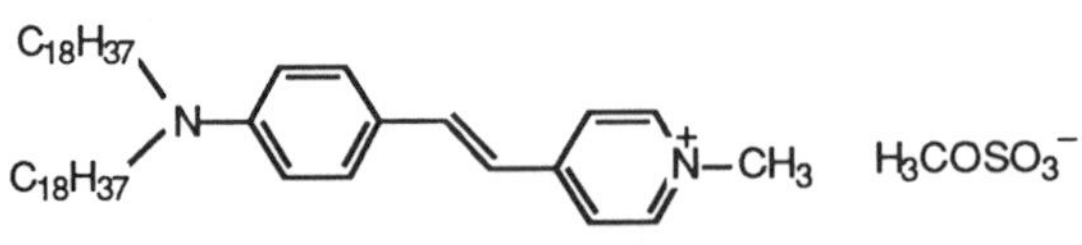

VIII

They studied the optical spectra and SHG of LB monolayers of V and VIII, and report that the extinction coefficient of VIII at 450 nm is much more enhanced compared to that of V. The same trend was found in the SHG, and the authors rightly attribute it to the difference in ε (oscillator strength).

However, the authors suggest a difference in inductive effect rather than in aggregation as the reason for their observations. We suggest that the more reasonable explanation is that the area per molecule in VIII is about twice that in V, and therefore, the distance between the dyes in the monolayer is large and no effective aggregation is detected.

Steinhoff *et al.* studied in a recent work the effect of protonation and aggregation on SHG in monolayers of IX [48]. They found that protonation on pyridine nitrogen leads to a considerable increase in its hyperpolarizability. This is expected, since a strong acceptor is being formed in the molecule.

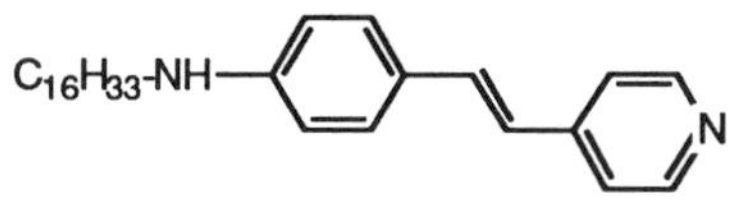

$$C_{16}H_{33}\text{-NH}-\text{(benzene ring)}-\text{CH=CH}-\text{(pyridine ring)}N$$

IX

The maximum susceptibility they measured was 1.6×10^{-14} cm^5/esu, and a linear dependence of $(\chi^{(2)})^{1/2}$ on protonation induced absorption at 435 nm (resonance enhanced). However, further addition of H_2SO_4 caused a dramatic two orders of magnitude decrease in the susceptibility to the value of 2.3×10^{-16} cm^5/esu. The authors suggest that inversion-symmetric *H*-aggregates are formed and that despite the observed decrease in susceptibility, the hyperpolarizability per molecule in the aggregates exceeds the β value of the monomer by at least a factor of six.

The effect of aggregation on NLO activity of organized assemblies is a very important question and is still not well understood. Much more systematic work is needed where series of compounds are prepared and studied with absolutely no resonance enhancement (incident light at 1907 nm) before clear conclusions can be made.

One of the issues in the engineering of systems ordered at the molecular level is to achieve a match between the cross-section area of the chromophore and that of the alkyl chains. The reader may remember that we discussed this issue in the context of SA monolayers of alkyltrichlorosilane containing a phenoxy group (Section 3.2.E). Indeed, it can be expected that since the cross-section area of a hemicyanine dye is larger than that of one alkyl chain, a two-chain system (VIII) improves the order in the two-dimensional assembly. However, it is not clear if two C_{18} chains are really needed, or that shorter ones (e.g., C_{10}) would be sufficient.

This question of chain length in double-chain hemicyanine dyes was studied by Cross *et al.*, who prepared the zwitterionic molecule X with different alkyl chains on the amino group (C_6, C_{10}, C_{14}, and C_{18}) [49]. Although this is a molecule that is much more water-soluble than a simple stilbene, azobenzene, or hemicyanine, it is interesting to note that the most stable monolayer, measured by the rate of decrease in the surface area (ΔA, %/h) is the C_{18} derivative ($\Delta A = 6$, compared to, for example, 32 for the C_{14} analogue).

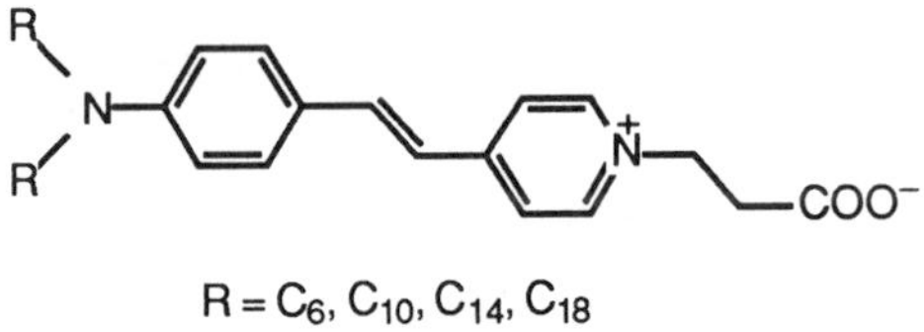

$$R = C_6, C_{10}, C_{14}, C_{18}$$

X

The question of how the different water solubilities of these systems affect the stability of the corresponding monolayers is not completely clear from their study, although their stability results correlate nicely with the order of water solubility of these derivatives. The connection between the head group and molecular orientation was studied by Marowsky *et al.* [50]. They prepared a series of compounds containing *N*-methylpyridinium and compared it to materials containing the nitrophenyl moiety. Their conclusion was that the *N*-methylpyridinium is a much better head group, strongly bonded to the hydrophilic surface, resulting in a very good molecular orientation in the film. This result is not surprising and emphasizes again that the balance between the hydrophobic and hydrophilic parts of the amphiphile is very important.

Lupo *et al.* investigated some hydrazones and compared them to the hemicyanine system [51]. Their results confirm the already known fact that two alkyl groups on the nitrogen donor provide a monolayer with high $\chi^{(2)}$.

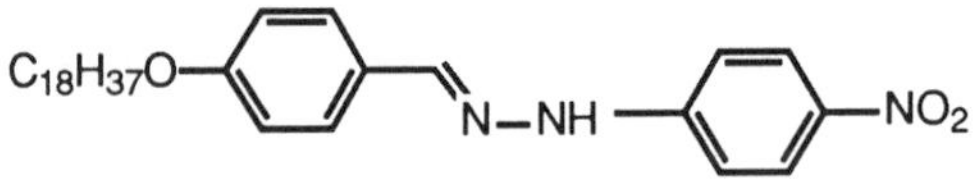

XI

For example, they reported $\beta = 4 \times 10^{-28}$ cm^5/esu for XI, while from their SHG experiments with the C_{16} analogue of VIII, they calculate $\beta = 2 \times 10^{-27}$ cm^5/esu. The β value reported for X is rather surprising, considering the fact that the donor is an alkoxy group, and may indicate that the hydrazone nitrogen also is a donor, contributing to the overall hyperpolarizability.

Roberts *et al.* were the first to report on an organometallic system (XII) in an LB film [52]. The authors noted that the $\chi^{(2)}$ value for XII is well in excess of lithium niobate. Using an alkoxy group as the donor and a ruthenium complex of a cyano group as the acceptor, they could get good Z-type deposition with a good quadratic dependence of the second harmonic on the number of layers.

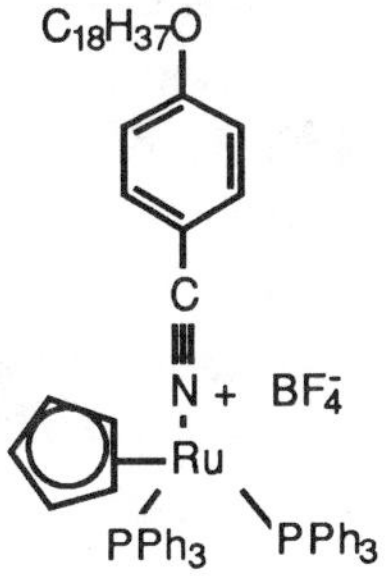

$$C_{18}H_{37}O$$

$$
\begin{array}{c}
C \\
\parallel \\
N \, + \quad BF_4^- \\
\mid \\
Ru \\
\diagup \quad \diagdown \\
PPh_3 \quad PPh_3
\end{array}
$$

XII

Apparently, the "strange" shape of the molecule with a huge ruthenium complex and one alkyl group did not prevent a formation of stable LB films, which is an unexpected result. Thus, the perception that a good amphiphile is a long one with a substantial difference between the hydrophilic and hydrophobic parts may not always be correct. It is quite clear from this study that the area of NLO awaits some innovative syntheses of new organometallic amphiphiles, where the unique electronic and magnetic properties of transition metal ions can be fully utilized.

One of the most innovative experiments is described by Popovitz-Biro *et al.*, where they prepared an amphiphile with amide groups in the alkyl chain, thus introducing hydrogen bonding in addition to the van der Waals interaction (e.g., XIII) [53]. Furthermore, they used a *p*-nitroaniline chromophore at the end of their alkyl group, again allowing for hydrogen bonding to stabilize a Z-type multilayer.

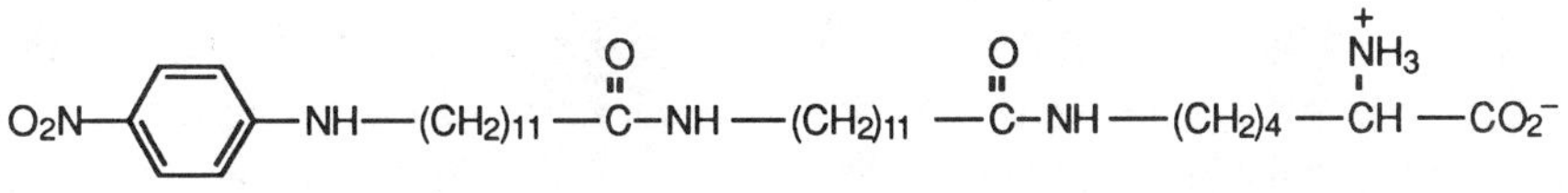

$$O_2N-\!\!\!\bigcirc\!\!\!-NH-(CH_2)_{11}-\overset{O}{\overset{\parallel}{C}}-NH-(CH_2)_{11}-\overset{O}{\overset{\parallel}{C}}-NH-(CH_2)_4-\overset{\overset{+}{NH_3}}{\underset{\mid}{CH}}-CO_2^-$$

XIII

Indeed, they report an excellent noncentrosymmetric, Z-type deposition with an impressive quadratic dependence of the SHG signal with layer number (up to 10 layers). An interesting result of this work is their statement on the requirements for Z-type deposition, i.e., an advancing contact angle of near 90°, and a receding contact angle close to $0°$. This is clearly understood in terms of surface energy, and the reader is referred to our discussion in Section 2.1.E.1 on the three types of LB deposition. The conclusion from the Popovitz-Biro experiments is that a Z-type, stable LB multilayer film can be constructed provided that the amphiphile is designed with careful engineering of surface energies in mind.

So far, we have discussed the formation of LB films from monomeric amphiphiles (where there is no chemical bond between the units in the monolayer). However, the use of polymeric amphiphiles either in the Y- or the Z-type deposition is becoming more attractive because of the thermal stability and, possibly, amorphous nature of the films.

Carr and Goodwin pioneered this area when they prepared the polysiloxane polymer XIV [54]. The interesting features in this system are the lack of a long alkyl chain and the fact that the hydrophilic head group (-OH) is on the pending chromophore, thus allowing for the polymer to serve as a "hydrophobic blanket" in the monolayer. This structure is interesting because a large volume in the film, otherwise occupied by the alkyl chains, is saved. Therefore, more active NLO chromophores can be introduced into a given volume yielding a better performing NLO film. The average area per molecule in this case is 38 $Å^2$ at $\pi = 30$ mN m^{-1}, and when the monolayers were maintained at this surface pressure, no significant change in the surface area was detected over a 64 h period. It thus can be concluded that the polysiloxane system is an excellent monolayer-forming polymer.

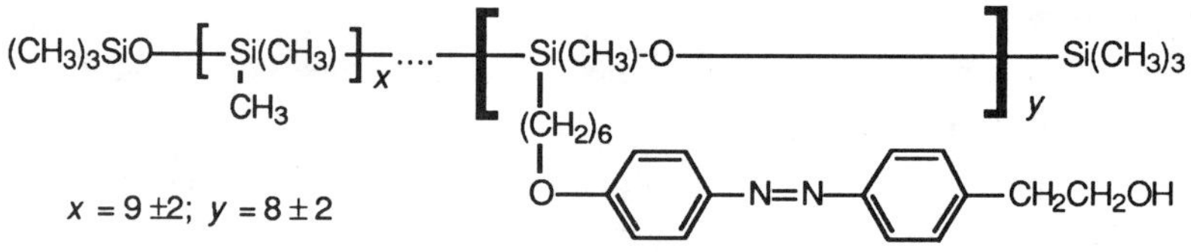

$x = 9 \pm 2; y = 8 \pm 2$

XIV

Another important point in this study is that their SHG signal was predominately p-polarized from both p- and s-polarized incident light ($I^{2\omega}(p{\to}p)$ and $I^{2\omega}(s{\to}p)$, respectively), indicating that the chromophores are perpendicular to the surface. This means that the order parameter in this polymeric monolayer is larger than in any of the LB systems discussed so far (where $I^{2\omega}(p{\to}s)$ was sizeable), which should have an important effect on the performance of any NLO device. The authors report a value of 8.4 x 10^{-29} cm^5/esu for the repeat unit.

Anderson *et al.* prepared the two hydrophilic polyether polymers with hemicyanine as pending groups extending toward the hydrophobic region (XV) [55]. Notice that in polymer A, the alkyl chain is connected to the pyridine acceptor, while in polymer B, it is connected to the oxygen donor. This work is the first demonstration of a true noncentrosymmetric, stable Y-type multilayer of a polymeric amphiphile, where the chromophores are with their dipole pointing in the same direction; that is, they have an additive effect. Both polymers (A and B) are ~50% substituted; i.e., every second Cl in the CH_2Cl groups is substituted with the chromophore.

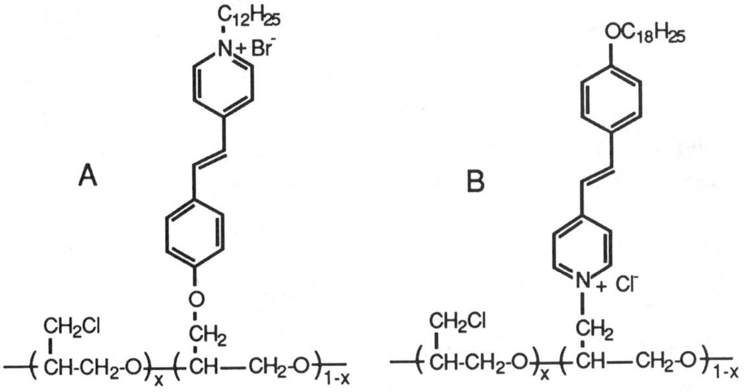

A ... B

$C_{12}H_{25}$ $N^+ Br^-$ $OC_{18}H_{25}$ $N^+ Cl^-$

CH_2Cl CH_2 $-(CH-CH_2-O)_x-(CH-CH_2-O)_{1-x}-$

CH_2Cl CH_2 $-(CH-CH_2-O)_x-(CH-CH_2-O)_{1-x}-$

XV

The area per molecule was 27 $Å^2$ for A and 40 $Å^2$ for B, quite reasonable since A has 43 repeat units with 47% chromophore load, and B has 11 repeat units with 33% load. The authors report an excellent quadratic behavior for up to eight monolayers (four AB bilayers). These results are very encouraging, since it becomes clear that with the right molecular and polymer engineering, it should be possible to construct thick films (≈ 1 µm) for waveguide and other applications.

In concluding this discussion on second-order NLO applications, we note that this is an area where LB films may be a viable technology. From the thermal stability point of view, however, polymeric amphiphiles may be the materials of choice. In general, the use of organic materials has many advantages, but thermal and photostability of the NLO-active moiety should be a major concern when choosing a chromophore. It is important here to emphasize what we already have noted in previous parts of this book, that polymeric materials have higher surface viscosity than monomeric amphiphiles, and therefore can not be transferred at high speeds. With this in mind, using horizontal lifting to transfer monolayers of polymeric amphiphiles with stable chromophores may become the technology that will allow mass production of high-quality NLO devices.

The preparation of SA multilayers of alkyltrichlorosilanes is associated with serious practical problems, and therefore will not become a viable route for the production of NLO devices in the near future. This by no means suggests that self-assembly should not be considered as an option. On the contrary, we feel that self-assembly will be the preferred technology, but that different chemistries should be developed for the construction of ordered thin organic films. We shall elaborate more on this question in our final note at the end of this book.

Thus, while NLO is a most promising area for the use of LB and SA monolayers, much work should be done in the understanding of issues such as the relationship between the structure of an amphiphile and the order of its monolayer in two dimensions. Other problems related to the use of monolayers in NLO applications are the relation of aggregation phenomena to NLO, local field effects in two dimensions, the thermal and photochemical stability of the chromophore as a function of material embodiment, and the relation between the polymer structure and the surface viscosity of its monolayer.

2. Third–Order Effects

Kajzar and Messier studied third-order nonlinear properties of polymerized diacetylenes (deposited on CaF_2) [56]. They reported the value $\chi^{(3)}_{xxxx} = (0.70 \pm 0.12) \times 10^{-14}$ esu for a polymerized monolayer of the ammonium salt of the diacetylene $CH_3-(CH_2)_{17}-C\equiv C-C\equiv C-COO^-NH^{4+}$, and a value of $\chi^{(3)}_{xxxx} = (0.69 \pm 0.20) \times 10^{-12}$ esu for the polymer film with the diacetylene group in the middle of the alkyl chain, $[CH_3-(CH_2)_{15}-C\equiv C-C\equiv C-(CH_2)_8-COO^-]_2Cd^{2+}$. The polymer was in the blue form, and was stable at room temperature. Estimating ~50% polymerization, the estimated value of $\chi^{(3)}$ is $(1.34 \pm 0.4) \times 10^{-12}$ esu.

Using the same cadmium salt, Kajzar et al. built up 90 multilayer films (2700 Å), and reported the value $\chi^{(3)} = (1.33 \pm 0.68) \times 10^{-12}$ esu [57]. In another paper, Kajzar et al. reported that for similar polymerized films of 2700 Å, the square root of the third-harmonic intensity, $I_{3\omega}$, shows a linear dependence on the polymer thickness l in Å [58]. There are two points of interest here; first, the value of $\chi^{(3)}$ is somewhat resonance-enhanced, since 3ω for the YAG Q-switched laser ($\omega = 1064$ nm) is at ~355 nm, in which the optical density in the spectra of both the blue and red forms of the polymer is not zero. Second, the film thickness is well below the coherence length, and as a result, only quadratic dependence (rather than cubic dependence) between the third-harmonic intensity and film thickness is observed, and does not depend on the refractive index dispersion.

360

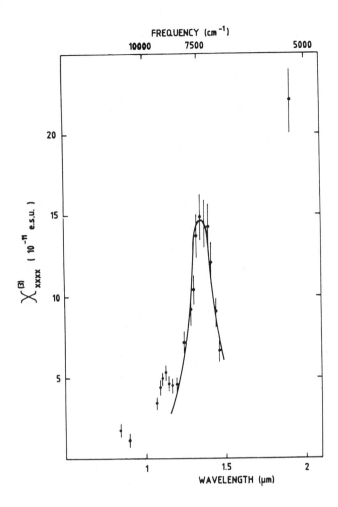

Figure 5.5. Cubic susceptibility $\chi^{(3)}_{xxxx}$ $(-3\omega,\omega,\omega,\omega)$, as a function of the laser fundamental wavelength. The continuous line is a least-square fit. (From Kajzar *et al.* [59], © 1983, Institute of Electrical Engineers.)

Kajzar *et al.* measured the third-harmonic generation from LB multilayers of the C_{22} derivative of III [59]. They used a pH 7.3 (Na_2HPO_4) subphase (dye completely protonated) deposited on silica plates, with the first monolayer deposited by withdrawal. The area per molecule was 26.06 Å^2, corresponding to 3.83×10^{13} molecules/m^2. Samples were prepared of 2 x 23, 2 x 49, and 2 x 79 monolayers, each layer being 38 Å in thickness. The THG was carried out at 1.064 and 1.907 μm, with the first yielding 3ω at ~354 nm (resonance-

361

enhanced), and the second at ~636 nm (off resonance). Based on this experiment, they report $\chi^{(3)}(-3\omega, \omega, \omega, \omega)$ at 1064 μm to be 20, while at 1907 μm to be 10 times that of silica, and their values for $<|\chi^{(3)}(-3\omega, \omega, \omega, \omega)|>$ are $(5.1 \pm 5) \times 10^{-13}$ esu at 1.064 μm, and $(3.1 \pm 3) \times 10^{-13}$ esu at 1907 μm.

The first study on resonance enhancement in third-harmonic generation of LB films of polydiacetylenes was reported by Kajzar and Messier [60]. They measured the wave-dispersed third-harmonic generation of polymerized LB films of the diacetylene $[CH_3-(CH_2)_{15}-C\equiv C-C\equiv C-(CH_2)_8-COO^-]_2Cd^{2+}$, and reported the existence of two resonances in $\chi^{(3)}_{xxxx}(-3\omega, \omega, \omega, \omega)$. The first was at 1.35 mμ, a two-photon resonance, and the second at 1.907 mμ, a three-photon resonance. The reported values for $\chi^{(3)}_{xxxx}(-3\omega, \omega, \omega, \omega)$ were $(1.5 \pm 0.2) \times 10^{-10}$ esu, and $(2.2 \pm 0.2) \times 10^{-10}$ esu. Figure 5.5 presents the cubic susceptibility as a function of the laser fundamental wavelength.

Prasad et al. studied a group of soluble polydiacetylenes having the general formula $R-C\equiv C-C\equiv C-R$, where $R \equiv (CH_2)_n OOCNHCH_2COO(CH_2)_3CH_3$ [61]. The π–A isotherms for both poly 4-BCMU ($n = 4$) and poly 3-BCMU ($n = 3$) are very unusual, and a monolayer-to-bilayer phase transition, similar to that observed for polypeptides, occurs. Using a degenerate four-wave-mixing technique, the subpicosecond response of the electronic $\chi^{(3)}$ was established [62]. The calculated value of susceptibility for the red form of the polymer was $\chi^{(3)} = 2 \times 10^{-10}$ esu (probably resonance enhanced) [63]. Berkovic et al. studied monolayers of 4-BCMU at the water–air interface [64]. A two-dimensional phase transition from the polymer yellow form to the red form was clearly observed when the area of the monolayer on water fell below 100 Å^2/repeating unit. A value of $\chi^{(3)}_{xxxx} \approx 3 \times 10^{-11}$ esu was reported for the red form at 1.06 mμ.

Electrochemically prepared   Chemically prepared

I

Logsdon et al. studied films of poly(3-dodecylthiophene, I) [65]. They prepared the polymer either electrochemically (ECP-PDDT) from the 3-dodecylthiophene) or chemically (CP-PDDT, from the 2,5-diodo-3-dodecyl-thiophene). The area per molecule for ECP-PDDT was 11.2 Å^2, usually low, and not in agreement with the reported value of 26 Å^2 [66]. The only possible explanation for this observation is that, in average, *ca.* one out of two thiophene moieties is in contact with the water subphase. This immediately suggests that no information on the packing, order, and molecular orientation in the monolayer can be obtained and related to the physical properties. Degenerate four-wave-mixing experiment gave a resonance-enhanced (602 nm) $\chi^{(3)}$ value of ~1 x 10^{-9} esu for a sample of 20 monolayers.

Kajzar *et al.* studied third-order hyperpolarizability of centrosymmetric multilayers of hemicyanine [67]. They reported the value of $\chi^{(3)} = (1.4 \pm 0.2)$ x 10^{-12} esu, and (0.90 ± 0.16) x 10^{-12} esu, for 1.064 and 1.907 mμ, respectively.

Ogawa *et al.* studied the THG spectra of LB films of low and high density x-ray-polymerized $CH_3-(CH_2)_{11}-C\equiv C-C\equiv C-(CH_2)_8-COOH$ (10,12-pentacosa-diynoic acid) [68]. While no values were reported for $\chi^{(3)}$, their results indicate that the THG intensity of the high-density polydiacetylene was ~2 times larger than that of the low-density sample, although the difference in density was only ~20%.

5.2. LB Films in Piezoelectric Devices

The *piezoelectric device* is a gravimeter, i.e., a device that can measure small changes in mass. In its simplest form, the piezoelectric device is a capacitor with a dielectric material that has a preferred axis normal to the capacitor plates. In such devices, the resonance frequency depends on the total mass of the gravimeter, where any small change in the mass will shift the resonance frequency,

$$\frac{\delta f_R}{f_R} = -\frac{\delta M}{2M}. \qquad 5.20$$

Let us consider a device as shown in Figure 5.6. We can write the total mass of the gravimeter as

$$M = M_{PG} + m + \delta m, \qquad 5.21$$

where M_{PG} is the mass of the piezoelectric gravimeter, m the mass of the chemically active layer, and δm the mass due to the products of reaction with the layer. The shift in the resonance frequency will be

$$\frac{\Delta f_R}{f_R} = -\frac{\delta m}{2M} = -\frac{\delta m}{2M_{PG}}(1 - \frac{m + \delta m}{M_{PG}}). \qquad 5.22$$

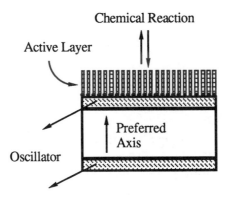

Chemical Reaction

Active Layer

Preferred
Axis

Oscillator

Figure 5.6. Setup for the piezoelectric gravimeter. The chemically active layer is presented as a monolayer.

It is clear that for such devices, $m + \delta m / M_{PG} \ll 1$, and therefore, any change in the mass of the chemically active layer will be observed as a change in Δf_R, and we can write

$$\frac{\Delta f_R}{f_R} = \frac{\delta m}{2M_{PG}} . \qquad 5.23$$

If we assume that the minimum measurable frequency shift is of the order of 1 Hz, then δm can be calculated from equation 5.20, with the assumption that M_{PG} is ~10 mg, to be

$$\delta m \approx 2 \times 10^{-9} g = N M_{mol} , \qquad 5.24$$

where N is the number of molecules, and M_{mol} is the mass of an individual molecule. In other words, for small molecules with a molecular weight of 50 g per mol (or 10^{-22} g per molecule), δm corresponds to the surface molecular density N of 2×10^{-14} cm^{-2}, which is at the order of magnitude of 0.1 monolayer.

Blinov *et al.* published the first paper on spontaneous polarization of LB films [18, 19]. They used alkoxyazobenzene derivative to prepare X- and Z-type multilayer films, and measured spontaneous polarization of the order of 0.6–1.3 x 10^{-6} C·cm^{-2}. Novak and Myagkov studied the piezoelectric effect in noncentrosymmetric multilayers of 4-nitro-4'-octadecylazobenxene [69], Snow *et al.* used the piezoelectric effect to study simultaneous electrical conductivity and mass measurements on iodine-doped phthalocyanine LB films [70], and Okahata and Ariga used piezoelectric effect for *in-situ* weighing of water-deposited LB films [71]. However, the most comprehensive review on opportunities of LB films in piezoelectric and pyroelectric devices is that of Biddle and Rickert [72].

It seems that one of the most promising applications of the piezoelectric device is in biological sensors. Here, due to the high molecular weight of biomolecules (e.g., proteins), a very small number of such molecules will cause a large change in frequency, and hence can be detected. Of course, detection of more complicated systems such as antibodies and even living cells may be accomplished by the modification of the organic film, for example, by receptors or antigens.

5.3. LB Films in Pyroelectric Devices

Pyroelectric devices have considerable importance for the detection of infrared radiation. The advantage of such materials over inorganic crystals (e.g., cadmium mercury telluride, CMT) is their broader spectral sensitivity and efficient operation at room temperature. For a material to be pyroelectric, it should have a noncentrosymmetric structure; i.e., it should have a unique polar axis [73]. As discussed before for NLO materials, LB and SA offer an elegant method for producing noncentrosymmetric structures in ultrathin films, needed as a basis for the pyroelectric detector.

Christie *et al.* proposed a theory that enables the pyroelectric coefficient of an LB film to be calculated from the measured dynamic voltage [74]. In this model, they assume that the film is sufficiently thin so that its contribution to the thermal mass of the device is negligible, and therefore the thermal properties are dominated by the substrate and the blacking layer. According to this theory, the pyroelectric coefficient p is given by

$$p = \frac{2\sqrt{2}v_m \varepsilon_0 \varepsilon_r (\omega_0 K \rho c)^{1/2}}{W \eta N d} , \qquad 5.25$$

where v_m is the measured pyroelectric voltage, ε_r is the relative permittivity of the LB film, ω_0 is the modulation frequency, K is the thermal conductivity, ρ is the density, c is the specific heat capacity of the substrate, W is the incident intensity, η is the emissivity of the blacking layer, N is the total number of monolayers, and d is the average monolayer thickness.

The pyroelectric coefficient p also can be determined by the *static technique*. Here, the device is mounted on a black material (e.g., copper black), and heated or cooled at a constant rate by a thermoelectric element. The pyroelectric current i is measured, and p is calculated from where dT/dt is the rate of change of temperature, and A the device area, so

$$i = pA \frac{dT}{dt} .$$ 5.26

Christie et al. describe the preparation of LB films of ionically terminated polybutadiene [75]. They established that the X-type films possessed a large polarization using surface potential measurements; however, no pyroelectric signal was detected. Blinov et al. reported pyroelectric activity in monolayer and multilayer LB films of a series of azoxy compounds (discussed earlier). Their best pyroelectric coefficient is 0.3 nC cm^{-2} K^{-1} [18, 19, 76, 77].

The use of alternate-layer LB films (Y-type, ABAB) for pyroelectric applications was pioneered in England. The general structure of the ABAB layers is described in Figure 5.7.

Smith et al. used stearylamine (C$_{18}$H$_{37}$NH$_2$) and a series of straight-chain fatty acids, and achieved in a thick film (several hundreds of monolayers) a pyroelectric coefficient of ~0.05 nC cm^{-2} K^{-1} [78].

Christie et al. reported a coefficient of 0.3 nC cm^{-2} K^{-1} for an 11-monolayer sample of 22-tricosenoic acid (CH$_2$=CH–(CH$_2$)$_{20}$–COOH) and docosylamine (C$_{22}$H$_{45}$NH$_2$) [79]. Further studies of 99-layer samples gave a pyroelectric coefficient of 1 nC cm^{-2} K^{-1}. Jones et al. repeated this experiment and reported a static value of 0.2 nC cm^{-2} K^{-1} (static value) [80]. The pyroelectric coefficient was independent of the number of layers, but did depend on the thermal expansion coefficient of the substrate, thus indicating that there was a piezoelectrically induced secondary effect contributing to overall activity.

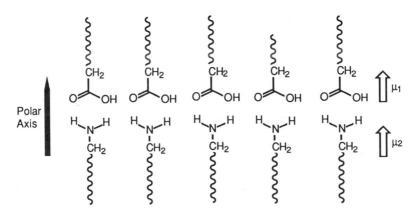

Figure 5.7. Interaction between fatty acids and amines to produce an ABAB film with a polar axis.

366

5.4 Electrical Properties of Monolayers

A. Dielectric Properties and Breakdown

The use of very thin dielectrics in electronic devices has important advantages. In capacitors, for example, while solving the mechanical problem of maintaining two metal surfaces at an extremely small separation but without actual contact, an organic film usually has dielectric breakdown much smaller than that of air. Furthermore, the capacitance of a capacitor of given dimension is larger when there is dielectric material between the plates than when the plates are separated only by air or vacuum. The LB and SA techniques give ultrathin dielectric films with uniform, controllable, and known thicknesses that are of the magnitude of a molecular length, and may have interesting applications. Devices in which such films can be utilized are metal–insulator–semiconductor (MIS), metal–insulator–metal (MIM), electroluminescence cells, metal–insulator–n$^-$ semiconductor–p$^+$ semiconductor that have bistable switching characteristics, and superconductor–insulator–superconductor (SIS or Josephson junction). Note that for these films to become technologically important, they need to be pinhole-free, which is a very difficult specification to meet. The reader may find accounts on early work in this area in a review by Agarwal [81]. Theories explaining breakdown conduction in thin insulating films were proposed by Forlani and Minnaja [82, 83] and by Klein [84].

Agarwal and Srivastava studied the temperature dependence of the dielectric breakdown field in LB films of barium salts of fatty acids [85]. Kapur and Srivastava studied thickness dependence of the static dielectric constant of multilayer films of barium palmitate, stearate, and behenate, and reported that the breakdown field depends weakly on temperature [86]. Honig reported that the current–voltage characteristics of insulation LB multilayer films, 25–400 Å thick, are linear at <105 V/cm, but exponential at high fields [87]. Gupta *et al.* reported on an odd–even effect in the internal voltage in Al–Ba stearate films on Ag substrates [88]. Srivastava studied thickness and temperature dependence of the breakdown field in LB films [89]. The results established the dominance of a Shottky mechanism of electric breakdown. Taylor and Mahboubian-Jones studied the electrical properties of synthetic phospholipid LB films [90]. The thickness-dependent internal voltage appeared to be due to the presence of water in these films. Iwamoto and Kang studied electrical properties of monolayers both at the water–air interface and sandwiched between aluminum electrodes [91].

B. Conduction through LB Films

Long-chain amphiphiles such as fatty acids have very high dielectric constants, and therefore have been considered as very thin insulators for different applications. These systems, because of their high two-dimensional order, also are ideal for studying electron tunneling, provided there are no pinholes and defects in the organic film, which is very difficult to achieve. The issue of electron tunneling is of importance these days with the still existing controversy around imaging of organic monolayers using STM. Figure 5.8 shows a schematic model of a capacitor, where the insulator thickness can be controlled by either changing the chain length or the number of monolayers in the film, or both. In such experiments, an LB film is deposited on a metal electrode, and another electrode then is evaporated on top. If there are defects, atoms of the second electrode will penetrate through the film and some of the observed film may result from shorts and not from tunneling. The question of monolayer aging also is of interest, since it can either increase or decrease conductivity, and we will discuss this issue.

Some of the early works in tunneling were reviewed by Handy and Scala, who also published some of their own results, demonstrating some of the problems mentioned earlier [92]. They reported the best results when a lower Sn/SnO_2 electrode was used; however, although their monolayers were quite resistive, their results were not reproducible. Miles and McMahon used electrodes made of tin–lead (Sn–Pb), and tin–indium (Sn–In) alloys, but could not get very reproducible results [93]. Mann and Kuhn used Al/Al_2O_3 as lower electrodes and Al as the top ones, prepared monolayer capacitors, and demonstrated that the measured current decreased exponentially with d [94].

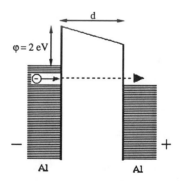

Figure 5.8. A schematic model of a capacitor, where the fatty acid LB film is the tunneling barrier between aluminum electrodes. d is the film thickness, and φ is the metal work function. (From Kuhn [102], © 1979, Elsevier Science Publishers.)

Further investigations were carried out by Polymeropoulos [95], and Polymeropoulos and Sagiv [96], who studied C_{14} to C_{23} fatty acid monolayers. Figure 5.9 shows their results. It is clear that resistance does not increase as expected for Ohmic resistance, since an increase in thickness of ~30% results in an increase by a factor of 10^6 in resistance.

It was suggested, therefore, that the residual current is due to quantum or resonance tunneling. In other words, because of its wave nature, the electron in the metal has a certain probability to penetrate the barrier to the other side. There were other experiments where it could be concluded that tunneling sites must be available that match energetically [97–101]. Other experiments with multilayer capacitors yield similar results [97], which brought the suggestion that electrons may jump from one interlayer to another [102].

In simple chemistry, the tunneling mechanism means that electrons tunnel from one electrode to the other through some states in the organic assembly. One possible way to explain this phenomenon is that weak intermolecular interactions in the two-dimensional organic assembly result in the formation of filled and empty "assembly orbitals" or virtual states, similar to bands in semiconductors. This band structure can be formed by the overlap of filled bonding σ-orbitals, and of either empty d-orbitals or σ^*-orbitals, or both.

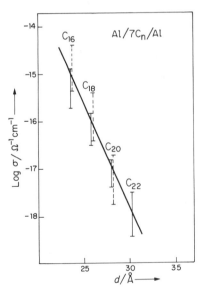

Figure 5.9 Log σ ($\Omega^{-1}cm^{-1}$) against d: broken bard according to Sugi [97], full bars according to Polymeropoulos [95]. (From Kuhn [102], © 1979, Elsevier Science Publishers.)

Table 5.1. Tunneling conductivity potential barrier height and conduction level of fatty acid salt monolayers and carotenoic acid array in the mixed monolayer. (From Yamamoto *et al.* [103], © 1976, Chem. Soc. Jpn.)

Material	σ_t (Ω^{-1}cm^{-1})	ϕ (eV)	V_0 (eV)	d (Å)
Barium stearate	2.70×10^{-16}	2.53	1.00	25.8
Barium arachidate	8.70×10^{-17}	2.26	1.31	28.0
Barium behenate	1.96×10^{-17}	2.05	1.45	30.2
Carotenoic acid	1.95×10^{-14}	1.70	1.90	28.0

Of course, the organic assembly is an insulator, simply because the energy gap between these states is very large. An electron injection from a filled assembly band to the metallic electrode results in hole migration, and the injection of an electron from the metallic electrode to an empty assembly band results in electron migration. This is a similar picture to the valence band, and conduction band conductivity in organic conductors. We note that although not impossible theoretically, one has to exclude carefully any other possibility before resorting to resonance tunneling.

Honig found that conductance of LB films was strongly dependent on the electrode material; e.g., there was a difference of a factor of 10^7 between Al and SnO$_2$ electrodes [87]. Yamamoto *et al.* studied the electrical properties of LB films of fatty acid salts and cartenoic acid [103].

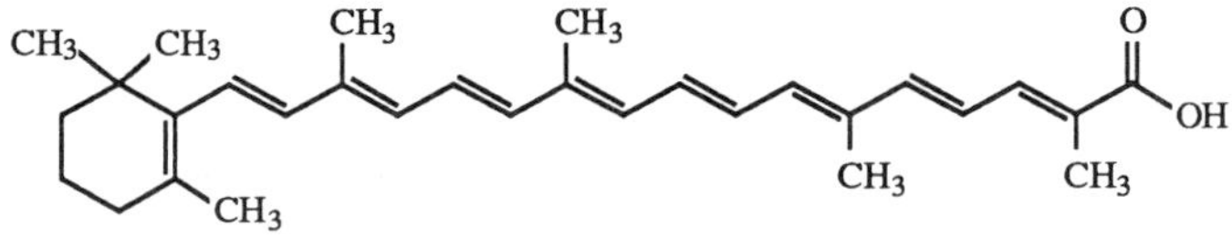

Table 5.1 summarizes their results, where σ_t (Ω^{-1}cm^{-1}) is the tunneling conductivity; ϕ (eV) is the potential barrier height, which is the difference between the work function of the metal electrode and the electron conduction level of the monolayer, V_0 (eV); and d is the monolayer thickness in Ångstroms. The trend is clear and suggests a large contribution to conductivity from the π-system. This is rather expected since the cartenoic acid is a truly conjugated system, where the highest occupied π-molecular orbital (HOMO) is much higher in energy than any of the σ-orbitals in the saturated fatty acids, and the lowest unoccupied π^*-molecular orbital (LUMO) is lower than any of the σ^*-orbitals in those acids. Therefore, even when diluted in a saturated acid, the conjugated system has a profound effect on the electron conductivity.

Peterson studied structural defects in LB films in great detail, especially in ω-tricosenoic acid monolayers [104–110]. This work is important for understanding the alternative mechanism of conductivity through organic assemblies, since it demonstrates the microstructure of the film, with all its complexity. In the studies of conductivity through monolayers of eicosanoic acid (arachidic acid, C_{20}) deposited on nickel, Peterson suggested that the results could be well understood in terms of conduction through defects [104]. We note that the conducting defects in LB films are very much like the defects in semiconductors (introduced by doping), i.e., sites where electrons can reside. Included in this category, besides pinholes, are sites where the density of the two-dimensional assembly is different (e.g., more liquid-like), and grain boundaries. Peterson suggested that the localized defects in LB films are analogous to the disclinations observed in liquid crystals [108–110].

Tredgold and Winter deposited stearic acid monolayers on Al_2O_3/Al and evaporated gold electrode on them [111]. They then stored the samples in dry helium four times up to 40 days and studied their conductivity. They found that over this period, the conductivity decreased by two orders of magnitude. Conductivity measurements in temperatures from 4.2 K to 280 K revealed that the *I-V* curves had the same shape, indicating that the number of conducting defects decreased, but that their characteristics remained the same. This can be understood if larger crystallites grow at the expense of the smaller ones, while the grain boundaries retain their structure. Later, Tredgold *et al.* studied multilayers of cadmium stearate and interpreted the conductivity results as associated mainly with film defects [112]. Hao *et al.* concluded that the defect mechanism always dominates over the tunneling one, even at 4.2 K, based on studies of conductivity in monolayers of oxirane carboxylic acid sandwiched between electrodes of lead–indium alloy [113]. Couch *et al.* studied metallic conduction through LB films in Ag/ω-tricosenoic acid/Ag and Au-Pd/cadmium stearate/Au systems [114]. They concluded that metallic pathways with an estimated average diameter of 6 x 10^{-3} mμ^2 and a number density of 1–10 per mm^2 are responsible for the observed metallic conductivity in samples with LB films as thick as 40 monolayers. On the other hand, in much thicker films (~100 layers), they suggested that the conductivity is mainly through the organic assembly. In another paper, Couch *et al.* studied the electromigration failure through LB films [115]. Gemma *et al.* reported the existence of short circuits in LB films of stearic acid on various kinds of substrates [116]. They concluded that this may be considered as a general phenomenon in LB films. Iwamoto *et al.* reported the first study of the conductivity in an LB film sandwiched between Pb-Bi superconductors [117]. Recently, Matsuda *et al.* investigated conducting defects in LB films [118].

C. Photoinduced Electron Transfer and Photoconductivity

1. Photoinduced Electron Transfer

The common photoinduced electron transfer reaction occurs when an electron donor is excited and, as a result, one electron is promoted from its π to its π^* orbital [119–121]. The energy of this orbital (π^*_D) is higher than the lowest unoccupied molecular orbital of the acceptor (π^*_A), and as a result one electron transfers from π^*_D to π^*_A,

$$D^* + A \rightarrow D^{\cdot+} + A^{\cdot-}. \qquad 5.27$$

The radical cation of the acceptor is produced when the fluorescence of D* is quenched. This was demonstrated in the case where the acceptor was a viologen, and the viologen radical was identified by its absorption spectrum.

$$C_{18}H_{37}-N^+ \underset{}{\underset{}{\bigcirc}} \underset{}{\underset{}{\bigcirc}} {}^+N-C_{18}H_{37}$$

The dark recombination reaction involves an electron transfer from the acceptor anion radical to the donor cation radical,

$$D^{\cdot+} + A^{\cdot-} \rightarrow D + A . \qquad 5.28$$

This is illustrated in Figure 5.10.

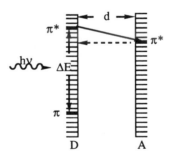

Figure 5.10. Photoinduced electron transfer from an excited donor to an acceptor. The solid arrow represents the forward reaction, and the dashed arrow the recombination reaction. π is an occupied molecular orbital, π^* is an empty molecular orbital, and d is the distance between the donor and acceptor layers. (From Kuhn [102], © 1979, Elsevier Science Publishers.)

372

There are cases where electron transfer reaction is not efficient, and enhancement is needed. This is generally done by physically separating the products of the photoreaction [122–124], neutralizing or removing one of them by means of reactions with added substances [125–127], or a combination of both of them [128, 129]. If a good reducing agent (an electron donor) is added to the reaction, it reacts with $D^{\cdot+}$,

$$R + D^{\cdot+} \rightarrow R^{\cdot+} + D , \qquad\qquad 5.29$$

and the overall reaction is:

$$R + D^* + A \rightarrow R^{\cdot+} + D + A^{\cdot-} . \qquad\qquad 5.30$$

Such a reaction is known as *supersensitization*, and has been discussed in the literature [130–132].

One of the interesting experiments is to compare fluorescence intensity I of a sample with a fatty acid spacer layer of a different length. The ratio $(I_0 - I)/I$, where I_0 is the fluorescence intensity without the acceptor, measures the rate of electron transfer. If the electron transfer is purely tunneling, its rate should be practically temperature-independent, and decrease exponentially with increasing d of spacer layer (the number of CH_2 units in the fatty acid). Thus, the logarithm of $(I_0 - I)/I$ plotted against d should form a straight line. Figure 5.11 shows that this is the case for spacer layers of C_{14}–C_{22} [102].

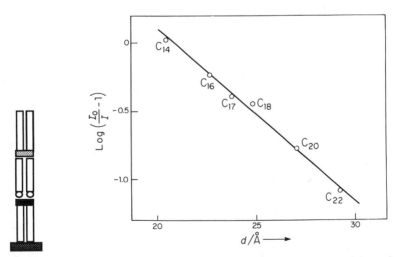

Figure 5.11. Quenching of fluorescence of dye D (see structure earlier) by viologen in the indicated arrangement; $\log[(I_0 - I)/I]$ is plotted against the spacer layer thickness d of fatty acids with 14–22 C atoms. (From Kuhn [102], © 1979, Elsevier Science Publishers.)

We note that the quality of LB films in these cases is of great importance. Residual currents in LB films have been interpreted in some cases as being due to quantum tunneling [94–96, 133, 134]; however, it seems more likely that in some cases, these currents should be associated with structural defects, although tunneling may still be involved [135].

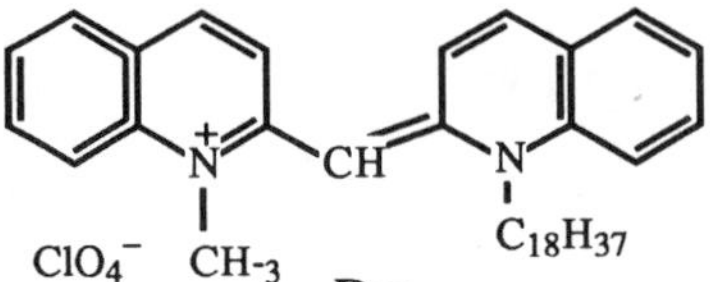

ClO$_4^-$ CH$_3$
Dye

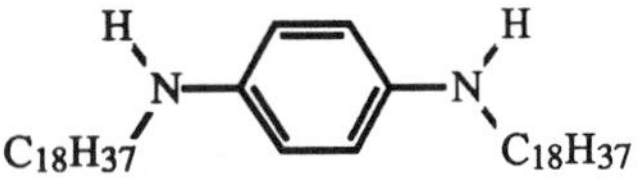

Donor

Penner and Möbius studied a case of supersensitization in three-component LB films containing a cyanine dye, viologen, and an electron donor (structures shown earlier) [136]. They could show that supersensitization could occur only if the dye was organized in *J*-aggregates.

Fugihira *et al.* studied photoinduced electron transfer from sensitized pyrene and Ru(bpy)$_3^{2+}$ derivatives [137, 138].

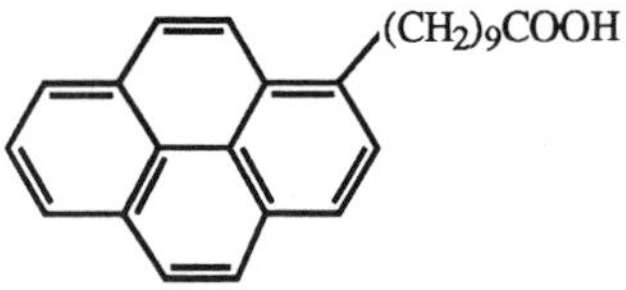

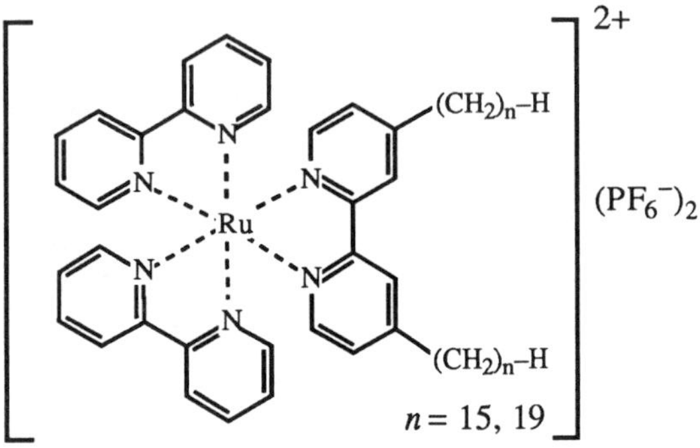

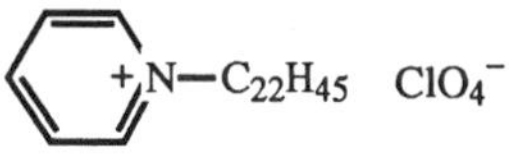

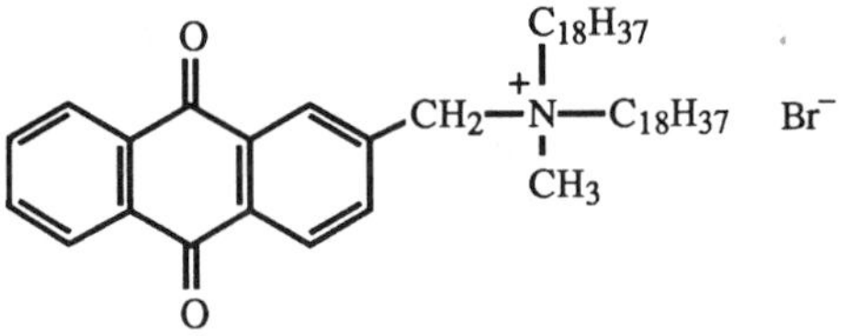

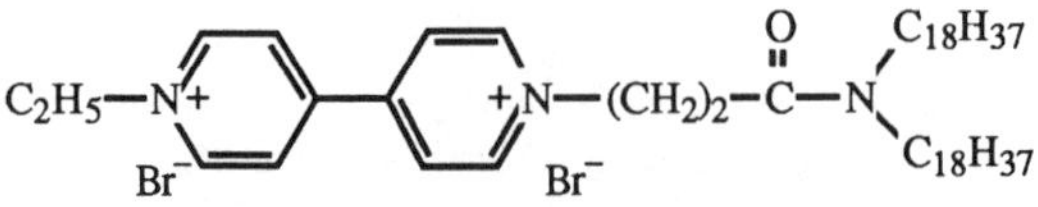

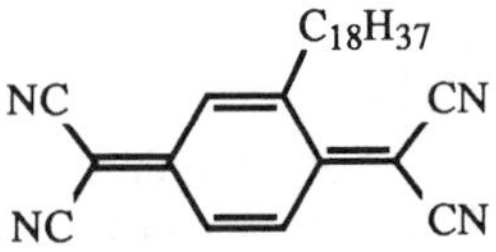

$n = 15, 19$

They prepared mixed monolayers of the sensitizers with arachidic acid and studied the electron transfer to a variety of electron acceptors (structures in previous page). The logarithms of the quenching rate constant corresponding to the different acceptors were plotted as a function of ΔG^0 (calculated from redox potentials). Based on these plots, the authors suggested the possibility of a Marcus-type inverted region at high negative ΔG^0 [139–142].

Murakata *et al.* studied photoinduced electron transfer from LB films of polymers containing carbazole and naphthalene units, with viologen as the electron acceptor [143]. Although there is no unique feature in this study, we note that the use of polymeric amphiphiles in this area has been very limited so far.

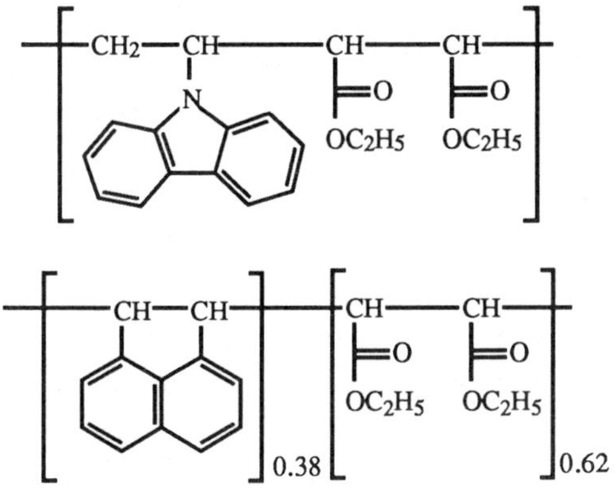

$$\left[\text{CH}_2-\text{CH}\cdots\cdots\text{CH}\cdots\cdots\text{CH}\right]$$

The advantages of polymeric LB films in this case are two: (*a*) thermal stability and (*b*) while the spreading of mixed amphiphiles does not give a uniform distribution at the molecular level, a polymeric amphiphile with different combinations of pendant photoactive groups, in designed free energies (ΔG^0) and distances, usually can be prepared easily.

2. Photovoltaic Effects

The sensitization of semiconductors and the separation of electron-hole pairs in a downhill potential gradient is the basic feature of solar-energy conversion to electrical power [144]. This is useful especially when the semiconductor has a band gap that corresponds to the energy of photons in the uv (e.g., TiO_2, SnO_2, $SrTiO_3$, etc.). In this case, sensitization would make a device more suitable for use with solar energy [145–149]. Figure 5.12 presents a schematic description of semiconductor sensitization. This figure is complementary, in a sense, to our earlier discussion on supersensitization.

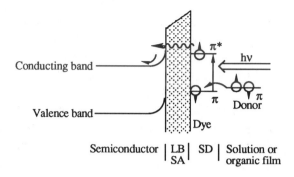

Semiconductor | LB | SD | Solution or
              | SA | | organic film

Figure 5.12. A schematic presentation of supersensitization at semiconductor interface under anodic bias. In this general picture, the sensitizing dye (SD) is adsorbed at the surface, and the donor in solution, or in the organic film, is the supersensitizer. In this picture, the dye either can be adsorbed directly on the semiconductor or on an LB or SA spacer film.

The reader should bear in mind that the supersensitizing donor either can be in solution, as in solar cells, or in a monolayer. Furthermore, this architecture is not limited to one supersensitizing dye. In principle, one could have a series of dyes and construct an electron-hole separation, as long as the electron moves on a downhill energy surface. This can be achieved if a series of electron donors is arranged in a decreasing oxidation potential order. With this stepwise drop in redox potential, one could, in principle, design a photoelectrode with a reduction (oxidation) potential carefully designed to carry out a specific reaction in solution.

Roberts *et al.* demonstrated that MIS solar cells can be produced by depositing fatty-acid-based LB films onto *n*-type CdTe [150–152]. Yoneyama *et al.* studied photovoltaic properties of copper phthalocyanine LB films in the MIM device Al/CuPc/Ag [153]. Willig *et al.* used the surface layer of an anthracene crystal as a third layer in a molecular triad [154], and studied the influence of different parameters on hole injection from LB films into the anthracene crystal [155, 156]. Yoneyama *et al.* used surface-active rhodamine and merocyanine dye to form an organic *p–n* heterojunction by surface pressure control [157]. Sasabe *et al.* studied the photovoltaic properties of purple membrane LB films [158, 159].

The electronic and optical properties of chlorophyll films (next structure shown) have been studied for a long time because of their bioactivity in photosynthesis [160–165]. It was first used in LB films by Gaines [166], and then by others [167–169], for the study of absorption spectra, fluorescence, and energy transfer. Jones *et al.* studied the photovoltaic effect in a chlorophyll *a* film adsorbed on Al_2O_3/Al with a gold electrode [170]. They suggested that the photovoltage was generated at the Al_2O_3–chlorophyll interface, similar to the effect observed in an SnO_2–chlorophyll interface. [147, 148, 171, 172]

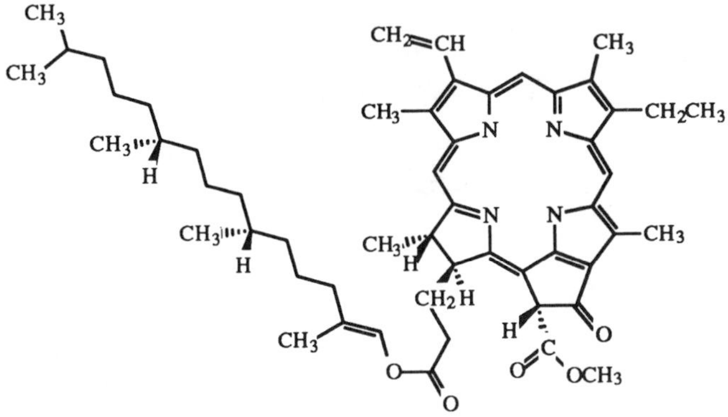

Chlorophyll *a*

Janzen and Bolton studied the photovoltaic effect in a system containing chlorophyll LB films deposited on a transparent Al_2O_3/Al substrate, with mercury as the top electrode [173]. Lawrence *et al.* deposited chlorophyll and chlorophyll–quinone multilayers on Al_2O_3/Al with silver as the top electrode [174]. The photovoltaic effect in LB films of azobenzene deposited on SnO_2 was studied by Blinov *et al.* [175]. Holoyda *et al.* studied photoeffects in chromophore–phospholipid LB films [176]. Saito *et al.* reported the fabrication of Schottky and *p–n* junction diodes from LB films based on merocyanine and triphenylmethane dye derivatives [177]. Iriyama *et al.* reported that LB films of *J*-aggregated merocyanine dye on a transparent SnO_2 electrode tend to act as photocathodes [178]. Sugi *et al.* studied the photoelectric effect in heterojunction LB film diodes and reported on a rectification effect with a voltage-dependent capacitance [179]. Saito *et al.* studied dye-sensitized *p–n* junctions of LB films and reported that two types of photocarriers could be recognized in the system, electron and holes [180].

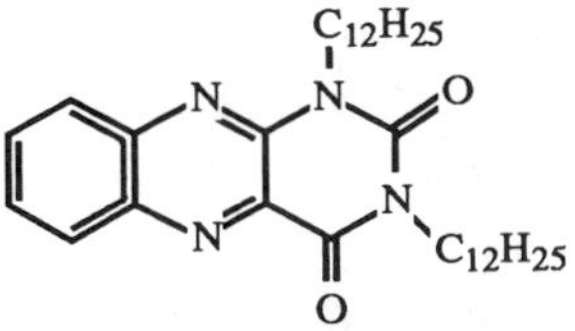

Nagamura *et al.* studied the photoelectric properties of Alloxazine monolayer assemblies (preceding structure) [181]. They reported that the dye aggregated and the in-plane photoconductivity was almost isotropic, and was greater than that of the out-of-plane one by about five orders of magnitude.

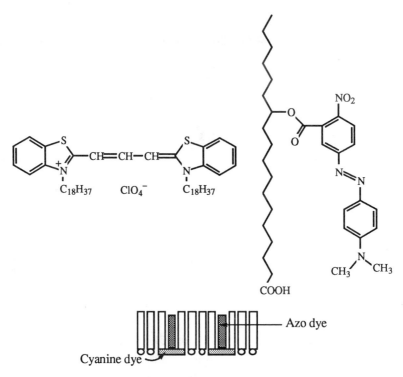

Figure 5.13. Possible coupling of a cyanine dye with an azo dye (structures shown) in a mixed monolayer. (Adapted from Polymeropoulos *et al.* [182], © 1978, Am. Inst. Phys.)

3. Photocurrent through LB Films

We have considered the vectorial separation of charges in monolayer assemblies. However, for a photocurrent to be of any significance, charge separation has to be maintained, and the assembly should be able to support electron migration. Polymeropulos *et al.* used a combination of cyanine and azo dyes to study photoconduction in LB films (Figure 5.13) [182]. The cyanine dye was excited by light, and the azo dye served as a conducting element. The photocurrent was an order of magnitude smaller if the azobenzene π-system was absent, and the conducting mechanism was found to be temperature dependent. In a later paper, Polymeropoulos *et al.* added an acceptor and found that it had a profound effect on the photocurrent [183].

Sugi *et al.* studied the effect of chromophore aggregation of the photocurrent in LB films, and reported that the lateral photoconductivity component $\Delta\sigma_\parallel$ was 10^3–10^4 times larger than the photoconductivity $\Delta\sigma_\perp$ in the normal direction

[184–186]. Moreover, both $\Delta\sigma_\parallel$ and the ratio between the lateral and normal photoconductivities, $\Delta\sigma_\parallel/\Delta\sigma_\perp$, were more than an order of magnitude larger for the aggregated systems.

Yamamoto et al. studied the photocurrents in LB films of cadmium arachidate mixed with cyanine dyes, and reported that their efficiency was dependent on the applied voltage, the temperature and the distance from the electrode [187]. Nagamura et al. studied LB films of amphiphilic porphyrin deposited on cadmium arachidate layers [188]. They reported the first observation of transient photocurrents due to excitation of the porphyrin. This current showed a fast rise and a slow decay, with a characteristic decay length of 180–208 Å, suggesting that the photocarriers decay in the arachidate layers. Steven et al. calculated 50 ns and less than 5 ns for the electron hop time in two differently prepared protoporphyrin IX dimethyl ester films [189]. Tran-Thi et al. used time-resolved absorption spectroscopy for the direct determination of transient charge carriers which are the primary species responsible for photoconduction processes in these films [190]. The transient charge carriers were reported to have a half-life time from 700 to 300 ns with increasing laser energies.

4. Conduction in the Plane of an LB Film

The conduction in a direction normal to the film plane is rather small, and thus may be influenced by defects. On the other hand, conduction in the plane of the film is orders of magnitude larger, and therefore, the results are much less confusing. The first studies of in-plane conductivity in LB films were done using a lightly substituted anthracene derivative.

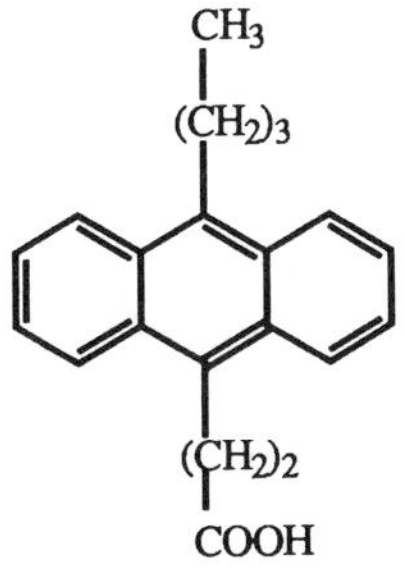

$$
\begin{array}{c}
CH_3 \\
| \\
(CH_2)_3 \\
\text{[anthracene structure]} \\
(CH_2)_2 \\
| \\
COOH
\end{array}
$$

Barlow et al. [191], Vincett et al. [192], Roberts et al. [193, 194], and Vincett and Barlow [195] found an anisotropy in conductivity $(\Delta\sigma_\parallel/\Delta\sigma_\perp)$ of about 10^8. Day and Lando reported that polydiacetylene bilayers are very poor conductors [196].

379

Much work on conductivity of porphyrin LB films has been carried out by Tredgold *et al.* Jones *et al.* reported anisotropy in conductivity of up to 10^5 for LB films of mesoporphyrin derivatives [197–199]. (See Section 2.7.A for a discussion on LB films of porphyrins for structures.) Tredgold *et al.* exposed LB films of copper mesoporphyrin to NO_2 gas (an oxidant) and studied the conductivity as a function of the gas pressure [200]. The system was sensitive to this gas and therefore, can be used as a sensor (Section 5.8). Tredgold *et al.* studied conductivity in alternating layers of porphyrins and fatty acids [201]. McArdle and, Ruaudel-Teixier studied conductivity in LB films of copper cyanoporphyrins [202]. Under iodine pressure, the conductivity of a four-chain dicyanoporphyrin LB film was $10\ \Omega^{-1}\ cm^{-1}$.

Baker *et al.* studied LB films of the free-base phthalocyanine and reported very low conductivity [203]. Hann *et al.* studied LB films of copper and zinc derivatives of tetra-*tert*-butylphthalocyanine and reported that in-plane conduction was ohmic and approximately $10^{-2}\ \Omega^{-1}\ cm^{-1}$ [204]. Hua *et al.* reported that the conductivity in LB films of tetra-*tert*-butylphthalocyaninesilicon dichloride was only $2 \times 10^{-6}\ \Omega^{-1}\ cm^{-1}$ [205]. Kutzler *et al.* reported that exposure of LB films of nickel and copper phthalocyanines to NH_3 gas enhanced their conductivity by two orders of magnitude [206]. Fujiki *et al.* exposed LB films of substituted nickel phthalocyanine to iodine and reported a two orders of magnitude increase in their dark conductivity, with a final value of $10^{-6}\ \Omega^{-1}\ cm^{-1}$ [207]. They suggested that the extremely low conductivity was a result of the presence of insulating alkane chains. Fujiki and Tabei doped LB films of lightly substituted metallophthalo-cyanines and reported values as high as $3.4 \times 10^{-4}\ \Omega^{-1}\ cm^{-1}$ for the dark conductivity [208].

Nakamura *et al.* prepared LB films from a mix-valence complex of $[(C_nH_{2n+1})_2N(CH_3)_2][Ni(dmit)_2]_2$, where H_2dimt = 4,5-dimercapto-1,3-dithiol-2-thione, and n = 10, 12, 14, 16, 18, and 22 [209, 210].

The dark conductivities of the films were from 10^{-3} to $10^{-1}\ \Omega^{-1}\ cm^{-1}$, where the highest conductivity was recorded for the C_{10} chain derivative. This is a promising system, since for complexes where the charge of the metal-$(dmit)_2$ unit was -0.5 (rather than -0.25), the bulk conductivity was higher, i.e., $20-100\ \Omega^{-1}$ cm^{-1} and $60\ \Omega^{-1}\ cm^{-1}$ for $(C_2H_5)_2N(CH_3)[Ni(dmit)_2]_2$ and $(CH_3)_4N[Ni-(dmit)_2]_2$, respectively. Thus, further work is needed to establish experimental conditions where such high conductivities are attained in LB films.

Tetracyanoquinodimethane (TCNQ) and tetrathiafulvalene (TTF) are the electron acceptor and electron donor that made the first organic metal. Later, tetraselenafulvalene (TSeF) and tetratellurafulvalene (TTeF) were made and showed very interesting properties. Some charge-transfer TSeF derivatives even exhibit superconductivity. Thus, it is not surprising to find a large number of reports on conducting LB films using these molecules. We first discuss LB films of either TCNQ or TTF derivatives, and then LB films of the charge transfer complex between them.

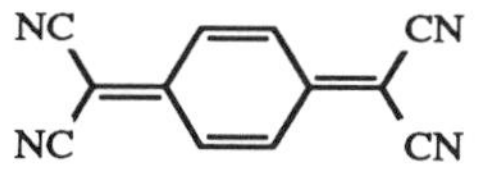

Tetracyanoquinodimethane (TCNQ) 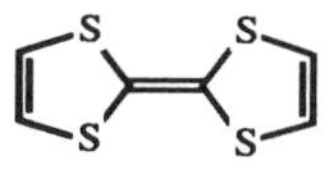   Tetrathiafulvalene (TTF)

Barraud *et al.* [211] and Ruaudel-Teixier *et al.* [212] used an alkylpyridinium as counter ion for TCNQ anion radical and studied its alternating LB films. The original film was an insulator (10^{-5}–10^{-7} Ω^{-1} cm^{-1}), but after doping using iodine vapor, they obtained very high in-plane conductivity (0.1 Ω^{-1} cm^{-1}). For more details on the structure and properties of these films, see References 213–218.

Roberts *et al.* studied the properties of these films using acoustoelectric devices, and reported that the films are sensitive to NO_2 [219]. Nakamura *et al.* found that at a low temperature (4°C), the structure of the monolayer of this material is expanded, and the conductivity is low, while at 13°–17°C, the film was more closely packed and the lateral conductivity, even without iodine doping, was higher [220]. Nakamura *et al.* [221, 222] and Matsumoto *et al.* [223] reported that an LB film of 1:2 pyridinium charge-transfer TCNQ salt, N-docosylpyridinium-(TCNQ)$_2$, exhibited lateral conductivity of 0.1 Ω^{-1} cm^{-1}. This is comparable to the value of iodine-doped 1:1 complex (discussed earlier), and probably results from the fact that TCNQ in the 1:2 complex is present in a non-integral oxidation state. Barraud described the preparation and properties of LB films from a phosphonium TCNQ salt, $C_{18}H_{37}P^+Me_3$-$TCNQ^-$ [224]. The conductivity was not measured directly, but was estimated from the maximum of the charge transfer absorption band to be about 20–50 Ω^{-1} cm^{-1}. Another salt prepared by Barraud *et al.* was the dimethyl octadecylsulfomium TCNQ, $C_{18}H_{37}S^+Me_2$–

TCNQ$^-$ [225]. In this case, however, preliminary dc measurements give conductivity values that are at least an order of magnitude lower than the value estimated based on optical spectra, which may indicate that the same also may be correct for the phosphonium analogue.

Saito wrote an overview of a series of tetrachalcogenafulvalene derivatives, which is a good starting point for a reader seeking synthetic directions [226]. Dhindsa *et al.* reported values of 0.01 Ω^{-1} cm^{-1} for the conductivity of 30 monolayer samples of hexadecanoyltetrathiafulvalene, after exposure to iodine for a few hours [227, 228].

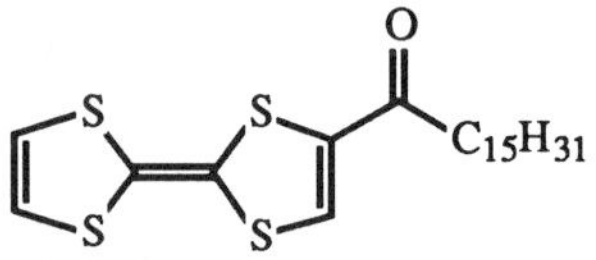

In contrast to this, films doped with bromine remained insulating, probably due to complete oxidation. Richard *et al.* prepared a new TTF derivative (ethylene-dithiodioctadecylthio TTF), and studied its mixed LB films with ω-tricosenoic acid [229]. Samples of up to 100 monolayers were exposed to iodine vapor and exhibited in-plane conductivity of about 0.2 Ω^{-1} cm^{-1}.

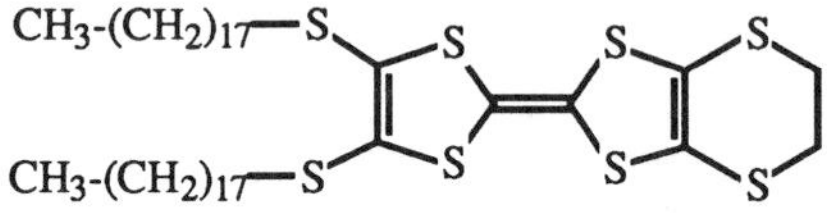

Another interesting TTF derivative was reported by Bertho *et al.* However, no values were given for the in-plane conductivity; the authors mentioned, however, that after bromine doping, all samples showed a 10^5 increase in conductivity [230].

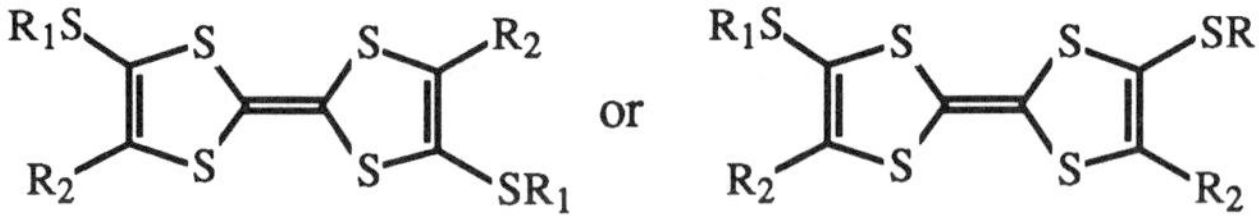

Nakamura *et al.* [231, 232], Ikegami *et al.* [233, 234], Kawabata *et al.* [235], and Matsumoto *et al.* [236] studied LB films of amphiphilic charge-transfer complexes of TTF and TCNQ. In the case of the TTF salt (with C$_{18}$TCNQ, structures on next page), the conductivity was only 10^{-3} Ω^{-1} cm^{-1}; however, by replacing the TTF with TMTTF, the conductivity increased to up to 1 Ω^{-1} cm^{-1}. An interesting observation was made in mixed monolayers of the TMTTF–C$_{18}$TCNQ salt with fatty acids, where the percolation threshold of the conducting monolayer was *ca.* 40% [232]. Multilayer films of TMTTF-C_{18}TCNQ were stable for months.

Fujiki and Tabei used the horizontal lifting method to transfer different TTF_xTCNQ_{1-x} alloys (x = molar composition) onto hydrophobic substrates [237]. Note that both donor and acceptor are not substituted. The conductivity reported was 5.5 and 1.5 Ω^{-1} cm^{-1} at $x = 0.6$ and 0.4, respectively.

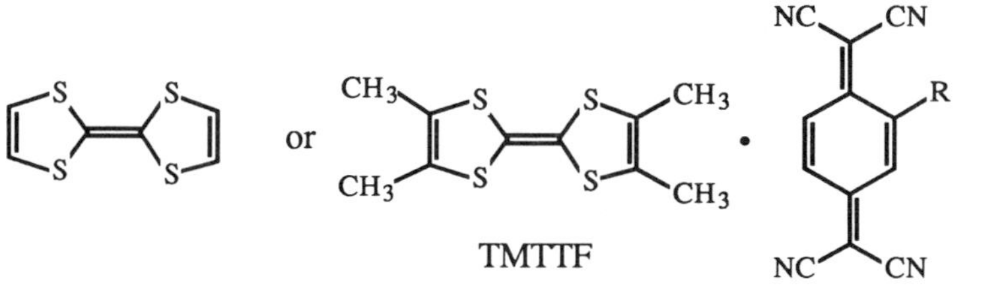

TMTTF

$R = C_{14}, C_{18}, C_{22}$

Richard *et al.* [238] and Morand *et al.* [239] described the preparation of conducting LB films based on ethylenedithiodioctadecylthio TTF and tetrafluoro-tetracyanoquinodimethane.

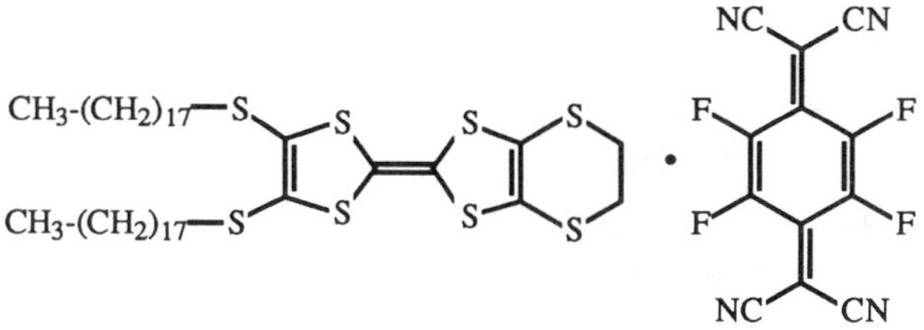

The original films were insulating and were doped by iodine to give conductivity of 0.05 Ω^{-1} cm^{-1}. However, the authors reported that the effect of iodine vanished slowly in ambient (over a few days), and more quickly *in vacuo,* yielding the original insulating film.

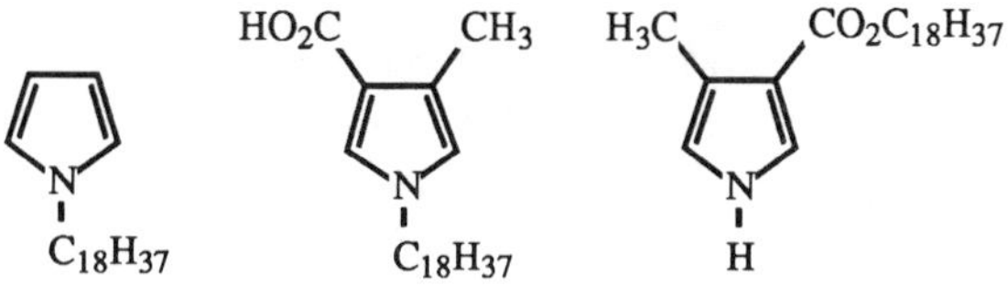

Iyoda *et al.* [240, 241] and Shimidzu *et al.* [242] prepared conducting LB films of polypyrrole by electrochemical polymerization of amphiphilic pyrrole derivatives. (See previous structures.) The electrochemistry of an LB film of 2

with octadecane (in 2:1 ratio) was carried out on a Y-type 120-layer film transferred onto a silane-treated ITO electrode. The highly anisotropic D.C. conductivity of the polymerized film, $\sigma_\parallel = 10^{-1}\ \Omega^{-1}\ cm^{-1}$, and $\sigma_\perp = 10^{-11}\ \Omega^{-1}\ cm^{-1}$ is a strong indication of a layer-structure.

Hong and Rubner used $FeCl_3$ in the subphase to prepare conductive polypyrrole at the water–air interface from a mixture of 3-octadecylpyrrole and pyrrole [243]. No conductivity measurements were presented.

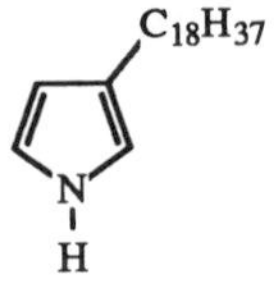

$C_{18}H_{37}$

Tasaka et al. prepared LB films of α-quinquethiophene (QT)/stearic acid, doped it with iodine, and measured conductivity of $0.2\ \Omega^{-1}\ cm^{-1}$ for samples of ≥ 10 layers, containing ≥ 35 mol% QT [66].

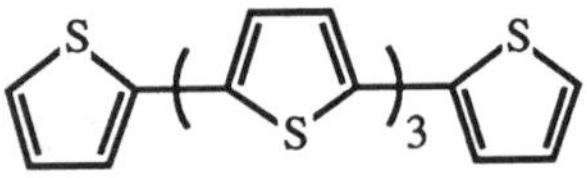

Nakahara et al. prepared LB films from a series of oligo- and polythiophenes, and studied their optical and electrical properties [244]. They reported that LB films of tetrathiophene ester derivatives showed a large dielectric constant and a high conductivity that were enhanced by iodine doping, but did not report conductivity values.

Logsdon et al. polymerized 3-dodecylthiophene electrochemically, and studied conductivity in undoped and iodine-doped LB films of the polythiophene [65]. The conductivity of the undoped film was $4.3 \times 10^{-10}\ \Omega^{-1}\ cm^{-1}$, but increased slowly upon exposure to iodine vapor, and reached a maximum value of $0.51\ \Omega^{-1}\ cm^{-1}$.

Watanabe et al. prepared mixed LB films containing stearic acid and poly(3-octadecylthiophene) [245, 246]. A value of about $5\ \Omega^{-1}\ cm^{-1}$ was reported for the conductivity of an LB film doped with nitrosyl hexafluorophosphate ($NO^+PF_6^-$).

Nishikata et al. studied LB films of poly(p-phenylene vinylene), and reported that the in-plane conductivity of a 300-layer film doped with SO_3 was $0.5\ \Omega^{-1}\ cm^{-1}$ [247].

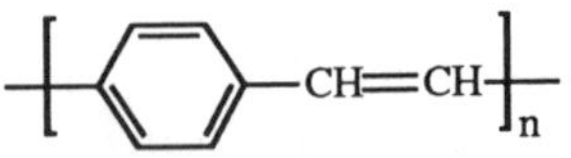

5.5. Chromic Effects In LB Films

A. Electrochromic LB Films

The *electrochromic effect* is the reversible change of an absorption band of a material induced by an external electric field. In this respect, it is similar to the *solvatochromic effect,* where the absorption is shifted due to the dielectric constant of the solvent. The shift of an absorption is determined by the electric field strength at the locus of the absorbing group, and, therefore, depends on the nature of the medium surrounding the molecule (local field effect). Usually, the local field is smaller than the applied one, $F = V/d$, when the ratio between the two is a constant (d). However, the σ-electrons of a chromophore, and of the surrounding hydrocarbon moieties, can be treated as a continuum of dielectric constant $\varepsilon = 2$ [248], and as a result, the π-electrons behave as if they were embedded in a uniform medium and the local field strength is identical with the applied field strength F. Now, the difference in the excitation energy with and without the field is $\Delta E - \Delta E_0$ and is given by

$$\Delta E - \Delta E_0 = h\Delta v_0 = \Delta\mu_z F - \Delta\alpha_{zz}F^2/2 , \qquad 5.31$$

where $\Delta\mu_z = \mu_z{}^* - \mu_z$, and $\Delta\alpha_{zz} = \alpha_{zz}{}^* - \alpha_{zz}$. μ_z is the component of the ground-state dipole moment, and α_{zz} is the polarization in the same direction, in a medium dielectric constant ε. We assume that the direction of the electric field is z. Note that if the main molecular axis is also in the z direction, and the dipole moment of the molecule is parallel to it, the effect is maximized. (See the earlier discussion on NLO.) Accordingly, $\mu_z{}^*$ and $\alpha_{zz}{}^*$ refer to the dipole moment and polarization in the excited state, respectively. All this is correct only as a first approximation. A more detailed treatment should take into account the dipole–dipole interactions in the material, their effect on the absorption spectra, and hence on the effective field that can be calculated from the spectral shift. Bücher and Kuhn measured $\Delta\mu_z$ and $\Delta\alpha_{zz}$ from electrochromism of Scheibe-dye aggregates in monolayers, and compared them with the theoretical calculated values [249]. Their measured values of $\Delta\mu_z = 0.07$ D, and $\Delta\alpha_{zz} = 1.5$ Å^3, are smaller than the calculated values of 0.2 D, and 2.0 Å^3, respectively.

Yamamoto *et al.* studied electrochromism of metal-free phthalocyanine LB films [250]. However, they found electrochromism only in fresh films, while no color change could be observed in the annealed films. Reich and Scheerer studied electrochromism of oriented chlorophyll *b* in lipid layers [251].

B. Thermochromic LB Films

The *thermochromic effect* is the reversible change of an absorption band of a material induced by a temperature change. This color change may be the result of either induced isomerization, i.e., *cis-trans* isomerization of an X = Y double bond (X, Y = CH, N), or of a chemical reaction, e.g., a ring opening. Fukui *et al.* studied the thermochromic behavior of merocyanine LB films [252].

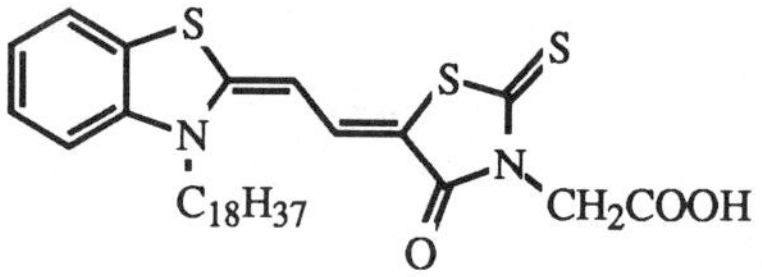

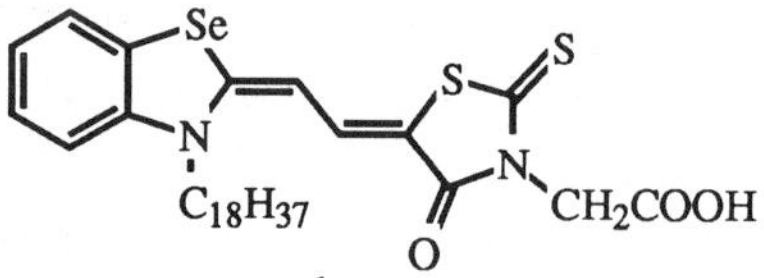

They reported a reversible behavior only in the case of dye c, where the red-shifted band was found to be restored upon cooling.

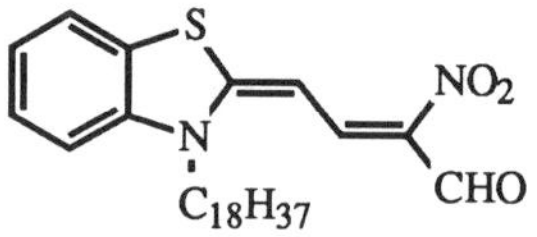

Decher and Tieke studied thermochromic LB films of an amphiphilic azobenzene dye [253]. The molecules formed good monolayers at the water–air interface with an area per molecule of 47 Å^2 (20°C), and x-ray diffraction of deposited multilayers (Y-type) showed a spacing of 52.2 ± 0.3 Å. Figure 5.14 shows the absorption spectra as a function of temperature for a 60-monolayer film.

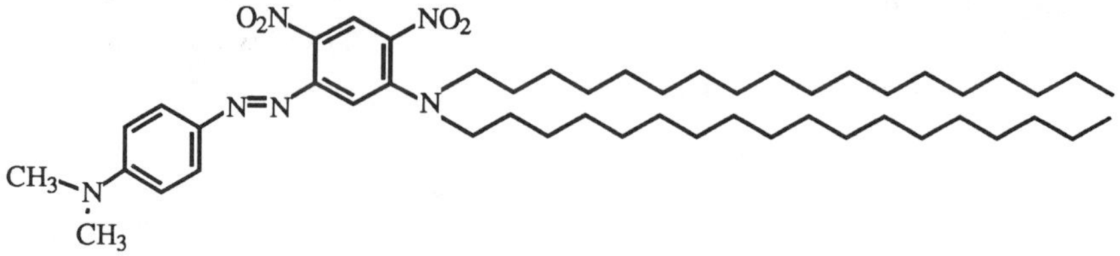

It is clear that this material is very temperature-sensitive, probably due to the large intramolecular charge transfer that weakens the N=N bond, and, therefore, reduces the activation energy for the *cis-trans* isomerization.

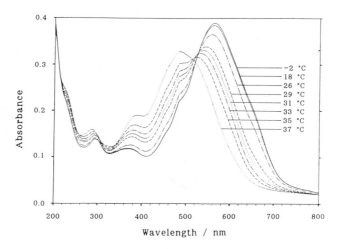

Figure 5.14. Absorption spectra of freshly-prepared 60 layers film. (From Decher and Tieke [253], © 1988, Elsevier Science Publishers.)

C. Photochromic LB Films

The *photochromic effect* is the reversible change of a molecule between two states having different absorption spectra, induced, at least in one direction, by an irradiation. As in previous chromic effects, this is also a result of either isomerization or ring-opening reactions. This effect is interesting because of a number of possible applications, such as recording media.

Gruda and Leblanc prepared a series of spiropyran-indoline molecules containing an *n*-hexadecyl chain ($C_{16}H_{33}$) at the R_2 or R_3, and measured the equilibrium constant between the colorless form (spiro) and the colored one (opened) [254].

It was found that the compounds form stable monolayers at the water–air interface that can be transferred onto quartz [255]. Irradiation at 253.7 nm showed the appearance of a new absorption band at 570 nm, which is the charge transfer band of the opened structure. This band varied with the number of layers and time of irradiation.

387

Polymeropoulos and Möbius studied monolayers of 1'-octadecyl-3,3'-dimethyl-6-nitrospiro[2H-1-benzopyran-2,2'-indoline] [256]. This spiropyran derivative is an interesting system, since the forward reaction occurs at 366 nm while the back one occurs at 545 nm. Therefore, this system can be used for recording, where two different laser systems can be used, one for writing and the other for erasing. The molecules are probably on their edge in the monolayer, since the area per molecule was ~20 Å^2. The photochromic properties of these molecules were studied at the water–air interface by monitoring surface pressure and surface potential. (The open form is charged and, therefore, a large change in surface potential was expected.) Moreover, reversible ring opening and closure were observed, indicating not only that the molecule is stable, but that the molecular conformation does not change as a result of the first-order C–O bond cleavage reaction (i.e., rotation to form a 90° angle between the two parts of the molecule).

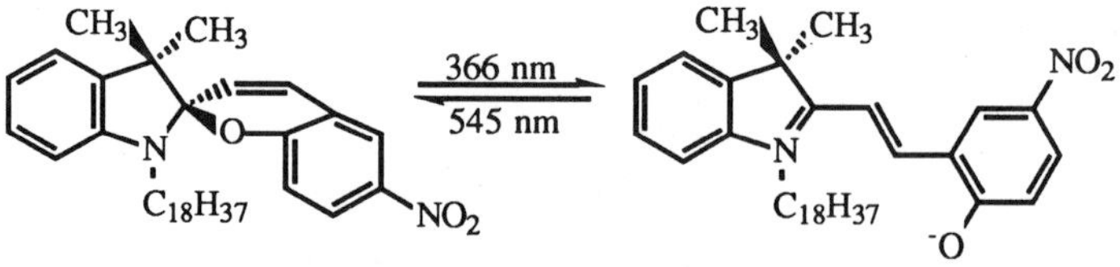

McArdle *et al.* studied the photochromic reaction of some spiropyran derivatives in LB films using FTIR [257]. They reported their results for mixed monolayer films of synthetic polymers and spiropyran [258], and studied LB films of some indolinospiropyrans and reported that ring opening was catalyzed by an acidic subphase [259].

Ando *et al.* compared the stability of LB films of spiropyran molecules with one and two long alkyl chains [260, 261]. They found that while the spiropyran having one alkyl chain is stable only in the opened merocyanine form, the introduction of two alkyl chains stabilizes the closed system, simply because the spiropyran occupies considerably more space than the merocyanine form.

$R = C_{16}H_{33}, R' = H$
$R = C_{16}H_{33}, R' = CH_2OCOC_{21}H_{43}$

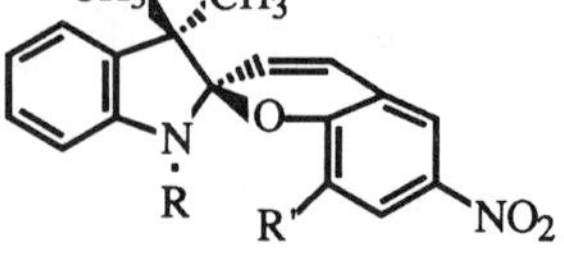

The introduction of *n*-hexadecane into the spiropyran monolayer stabilized it even further, and when irradiated formed a *J*-aggregate with a half-decay time which was 10^4 times longer than that of conventional spiropyran. Blair and

McArdle studied photochromism of LB films of monomeric, a side-chain, and a main-chain azobenzene polymers [262]. The most interesting result in this study is that *cis-trans* photoisomerization was inhibited in both the monomeric and side chain azobenzene LB films, probably because of lack of free volume, while permitted in the main-chain polymer case.

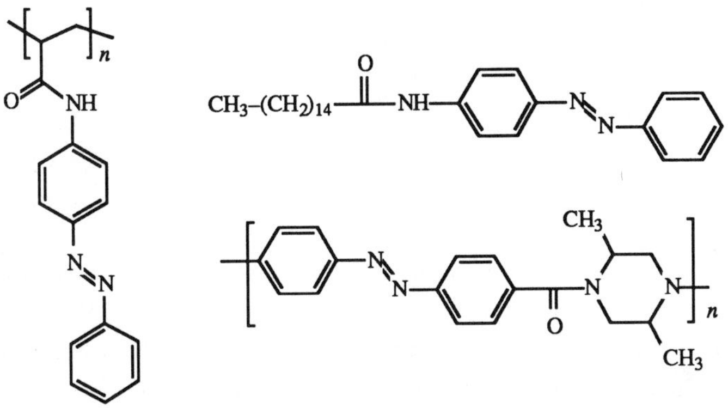

Sandhu *et al.* studied a series of phospholipid molecules, where either one or both chains were substituted by an azobenzene [263]. Both lipids were photochromic and formed good and compact monolayers in the *trans*-form. The *cis*-form, on the other hand, occupied more area, which resulted in considerable hysteresis between compression and expansion cycles before and after photoisomerization.

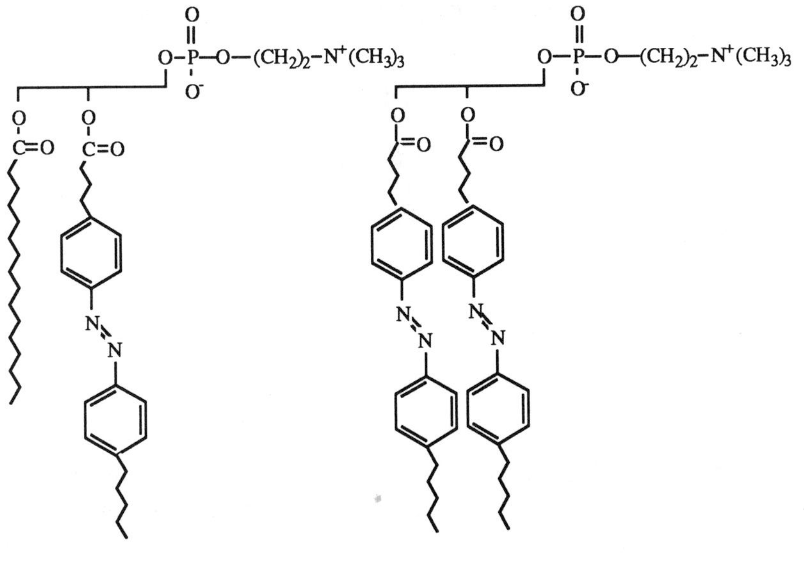

Nishiyama and Fujihira used polyallylamine in the subphase to control the area per molecule in an LB film of an amphiphilic azobenzene [264]. Hence, while the area per molecule for the amphiphile on pure water was 28 Å^2, it was 39 Å^2 on the polyallylamine subphase. This extra free volume allowed a reversible *trans-cis* isomerization of the azobenzene to occur. Thus, a 10 min irradiation with uv light gave the *cis*-form, while a 10 min irradiation with visible light resulted in a recovery of the *trans*-form.

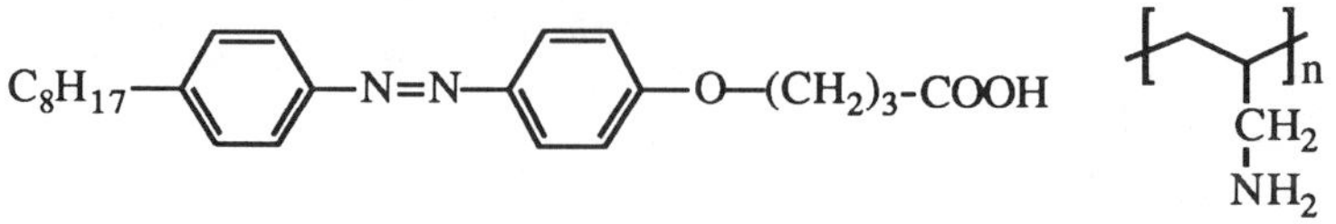

Kawamura *et al.* reported the photochromism of salicylideneanilines in LB multilayer films [265]. The photochromism in this case is a result of tautomerism and not isomerization; therefore, there is no principle difference between the area occupied by the two forms. The molecules formed good monolayers with an area per molecule of 27–28 Å^2, and a Y-type deposition was achieved (on quartz) with a transfer ratio of ~1. The spectrum of the multilayer on quartz was different from that in solution and was considerably blue-shifted. This was a result of the formation of a *H*-aggregates, i.e., an assembly with chromophores arranged in such a way so that their dipoles are parallel to each other. Irradiation with uv light gave the colored quinonic form, which thermally decayed back to the colorless aromatic in a rate constant of 10^{-3} s^{-1} (280–320 K) that was not a simple first-order one. This truly reversible photochromic system demonstrates the importance of molecular design, where reversible chemical reactions, which do not require dramatic changes in molecular shape, can be introduced into an ordered molecular assembly.

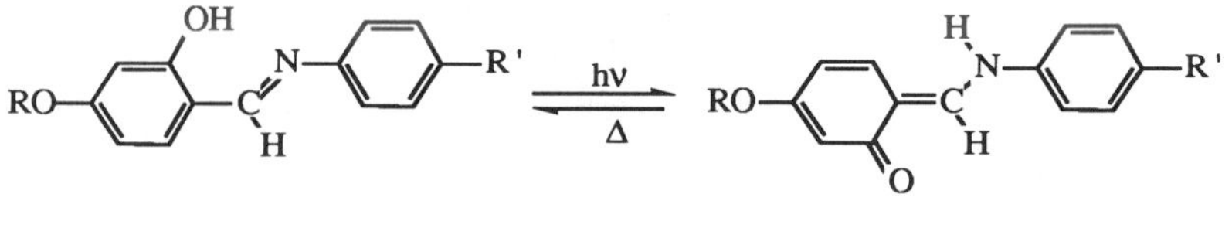

5.6. LB Films as Semiconductors

Very few reports discuss LB films of molecular semiconductors. Most of the work has been done on phthalocyanine (Pc) derivatives, since it has been shown that lutecium bisphthalocyanine (Pc$_2$Lu) and lithium phthalocyanine (PcLi) are intrinsic molecular semiconductors [266–270]. Snow *et al.* reported that LB films

390

of oligomeric Pc complexes are semiconductors [271]. Fujiki *et al.* studied LB films of amide-substituted nickel phthalocyanine (NiPc) [207]. They reported that the PC semiconductor properties were maintained after electron beam irradiation and etching.

Richard *et al.* used glycerol as subphase for monolayers of semi-amphiphilic mixed-valence TCNQ salt, octadecylpyridinium$^+$ (TCNQ$_{1.45}$)$^-$ [272]. The lateral conductivity for this monolayer was 10^{-4} Ω^{-1} cm^{-1}.

5.7. LB Films in Resist Applications

The pursuit of further miniaturization of electronic circuits has made submicron resolution lithography a crucial element in future computer engineering. LB films have long been considered potential candidates for resist applications, since conventional spin-coated photoresist materials have large pinhole densities and variations of thickness. In contrast, LB films are two-dimensional, layered, crystalline solids that provide high control of film thickness and are impermeable to plasma down to thickness of 400 Å [273].

One of the first reports on electron beam (EB) lithography using LB films was that of Barraud *et al.*, who showed that LB films of ω-tricosenoic acid are excellent resist materials capable of a line resolution of 600 Å with a film 1000 Å thick [274, 275]. Although this is by far the best material reported, other materials are also of interest. In an early attempt to prepare good photoresist LB films, Barraud *et al.* used ω-tricosenyl pentadienoate (next structure shown) [276]. The idea was to use polymerization both in the hydrophilic and hydrophobic parts. While the mechanical properties were very good, the polymerized film swelled and the spacial resolution achieved was poor. Fatty acids with diyne groups (diacetylenes) also undergo polymerization [56, 277], but have not been found useful for domain formation, although ω-diynes were better than middle-chain diacetylenes [275].

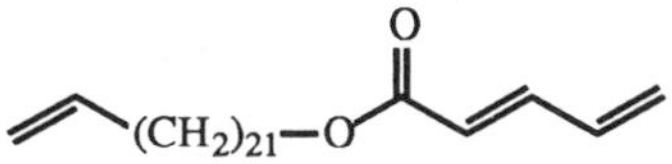

Fariss *et al.* used α-octadecylacrylic acid multilayers both as positive and negative EB resist [278]. They reported resolution of ~500 Å in the positive resist mode. Boothroyd *et al.* discussed two oxirane-containing fatty acids for EB lithography [279, 280]:

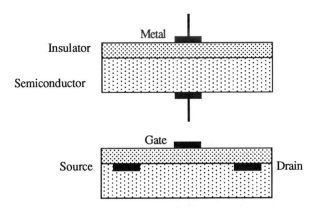

$$\text{H}_2\text{C}\overset{\displaystyle O}{\overset{\displaystyle \diagup\diagdown}{\quad}}\text{CH}-(\text{CH}_2)_{18}-\text{COOH} \qquad \text{CH}_3\text{-}(\text{CH}_2)_{10}-\text{H}_2\text{C}\overset{\displaystyle O}{\overset{\displaystyle \diagup\diagdown}{\quad}}\text{CH}--\text{CH}_2\text{CH}_2\text{COOH} \ .$$

The polymerization reaction in this case may be described as

The reported resolution was not in the region required for advanced microlithography. Miyashita *et al.* used LB films of *N*-octadecyl acrylamide as ultrathin resist materials, and reported resolution of 1000 Å using EB lithography [281, 282]. Jones *et al.* studied LB films of poly(styrene/maleic anhydride) derivatives as EB resist materials [283]. Resolution for ~500 Å thick films was on the order of 2500 Å. Recently, Kuan *et al.* reported the use of ultrathin poly(methyl methacrylate) (PMMA) resist films for microlithography [284].

5.8. LB Films in Sensors

A. Field Effect Devices

We have discussed already the high dielectric constant of LB films and their insulating properties. In the MIS device, an LB film is incorporated between a semiconductor and a metal (Figure 5.15, top).

Figure 5.15. A metal-insulator-semiconductor (MIS) device (top), and a field effect transistor (FET, bottom).

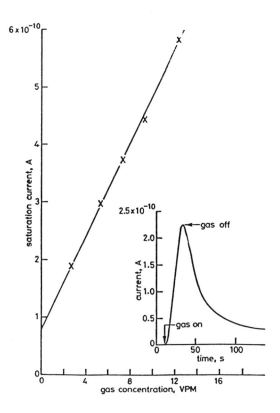

Figure 5.16. Saturation current against NO_2 gas concentration for eight layers of the asymmetrically substituted phthalocyanine deposited onto an Al interdigital electrode pattern. The insert shows the transient response to 120 VPM NO_2 in N_2. (From Baker *et al.* [291], © 1983, Institution of Electrical Engineers)

In such devices, the tunneling current is a function of the dielectric constant of the organic film. Some of the pioneering work in this area was done by Roberts *et al.* [285–290]. The addition of a source and a drain electrode into the MIS device gives a field effect transistor (FET, Figure 5.15, bottom). It was shown by Roberts *et al.* that the conductivity between the source and drain electrodes can be modulated by a gate electrode. In a sense, the FET proposed by Roberts *et al.* is a double dielectric structure, since the LB film is deposited on top of the native oxide layer. If the LB film has semiconducting properties that are sensitive to the structure and oxidation level of the individual organic molecule, this FET can be used as a detector. The use of small organic molecules as LB insulating layers gave encouraging results. It is clear, however, that future more complex structures should include biomolecules, thus providing selectivity and specificity (see the discussion on biosensors in Section 5.8.C).

Baker *et al.* used a FET structure with an eight-layer film of substituted copper phthalocyanine as an NO_2 detector [291]. Figure 5.16 shows the saturation current against NO_2 gas concentration for eight layers of the asymmetrically substituted phthalocyanine deposited onto an Al interdigital electrode pattern. It was found that although the phthalocyanine layer was sensitive to a number of gases, the sensitivity to NO_2 was the largest. The response time to 120 VPM (volume parts per million) NO_2 was 50 s, but the decay to original current level was slow.

Wohltjen *et al.* studied ammonia-sensing by a copper phthalocyanine LB film-based detector [292]. Maximum conductance was achieved in 2 minutes for a 45 layer film exposed to 2 ppm ammonia at 30°C in air. Tredgold *et al.* published a systematic study on the effect of film thickness, NO_2 partial pressure, the central metal ion, and the peripheral substituents on the response of a gas sensor based on a metalloporphyrin LB film [293]. Roberts *et al.* [219] and Henrion *et al.* [294] demonstrated the utility of TCNQ LB films as NO_2 sensors. Recently, Henrion *et al.* described the detection of phosphine with LB films of *N*-octadecyl-pyridinium TCNQ LB films [295].

B. Optical Sensors

Surface plasmons are collective oscillations of the free electrons at the boundary of a metal and a dielectric. The surface plasmons are basically guided waves [296], and their resonance conditions are very sensitive to changes in the thickness and refractive index of the medium adjacent to the metal (see also Section 1.2.J). Indeed, this effect has been used for the study of oxidation of metals and to evaluate optical constants for deposited overlayers [297–303]. Furthermore, it was shown that surface plasmon resonance (SPR) can detect changes of 10^{-5} in the refractive index of the overlayer [304], which makes it comparable with ellipsometry [305].

One of the most promising optical devices is that of the surface plasmon resonance (SPR) [296]. A schematic diagram of the SPR gas detection system is illustrated in Figure 5.17. Lloyd *et al.* used SPR to investigate the interaction between NO_x and LB films of tetra-4-*tert*-butylphthalocyanine containing silicon (see Section 2.7.B for the discussion on LB films containing phthalocyanines and for structure) [306]. Figure 5.18 presents SPR curves for a phthalocyanine sample unexposed and exposed to different concentrations of NO_x. Of course, SPR can be used for other sensor applications such as biosensors (see later).

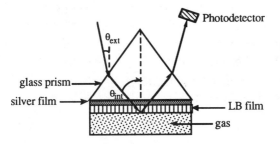

Figure 5.17. A schematic diagram of SPR gas detection system.

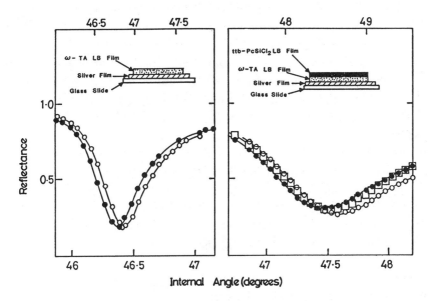

Figure 5.18. SPR curves for a phthalocyanine sample unexposed (•), and after NO_x exposure sequence up to 500 vpm (□) and 900 vpm (o). (From Lloyd *et al.* [306], © 1988, Elsevier Science Publishers.)

Another optical detector of toxic gases was demonstrated using fluorescent porphyrin LB films [307]. Figure 5.19 presents the optical sensor head they developed. They reported changes in fluorescence for NO_2, HCl, and Cl_2 gases. The fluorescence could be quenched quantitatively in the cases of both NO_2 and HCl using NH_3 vapor.

In another work, Beswick and Pitt studied oxygen detection using phosphorescence LB films of tetraphenyl porphyrin palladium(II) (TPPPd) [308]. A

395

surface acoustic wave (SAW) oscillator incorporating LB films was developed by Holcroft and Roberts and had a detection limit of 40 ppb NO_2 in dry air [309]. Schaffar *et al.* studied the effect of layer composition on the response of ion-selective optical electrodes (optrodes) for potassium, based on fluorimetric measurements of membrane potential [310]. The reader may find additional information on LB films as chemical sensors in a review by Moriizumi [311].

C. Biosensors

The quest for microbiosensors has brought the need for highly selective and highly sensitive organic layers, with tailored biological properties that can be incorporated into electronic, optical, or electrochemical devices. The use of LB and SA films in such devices stems from the fact that artificial membranes can be assembled easily at the water–air interface and transferred to different substrates, or assembled at the solution–substrate interface. The molecular dimensions of the LB and SA films make the use of even highly expensive molecules economically attractive, and therefore, the development of sophisticated detection systems possible.

The operation of biosensors can be divided into three main events. In the first, the molecule to be detected is recognized by the sensor. This *recognition* should be both specific, i.e., it is able to recognize the molecule, and selective, i.e., the recognition of the molecule can be done in the presence of other bioactive molecules. The second is *perturbation*, i.e., once the molecule is attached to the recognition site, a detectible physical change is triggered. The last event is *transduction*, which is the ability to recognize the triggered physical change by an instrumental arrangement. The signal then is analyzed and data is provided. This series of steps is illustrated in Figure 5.20.

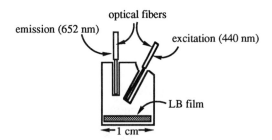

Figure 5.19. Diagram of optical sensor head for porphyrin fluorescence. (Adapted from Beswick and Pitt [307], © 1988, Academic Press.)

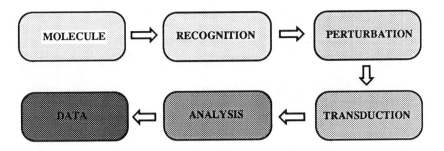

Figure 5.20. A recognition–perturbation–transduction scheme followed by analysis.

The first step, biorecognition, had been reviewed by Aizawa [312]. In this category we include, for example: (*a*) *Enzyme–substrate* — where the enzyme is immobilized on the biosensor. This enzyme then will catalyze a chemical reaction, and the biosensing is done by analyzing either a starting material or a product in this reaction. (*b*) *Antibody–antigen* — where the antigen is immobilized on the biosensor and the formation of an adduct with the antibody is detected. Other systems detect gas molecules (porphyrins) and were discussed earlier.

The second step in the biosensing process, transduction, is summarized in Table 5.2, where a list of transduction modes used in biosensor development is presented [313].

One of the key issues in the design of a successful biosensor is the retainment of full bioactivity of a biomolecule in the adsorbed state in the biosensor. We have elaborated on the subject of surface energy in Section 1.II.1, where the wetting of surfaces was discussed. The reader should bear in mind that in the adsorption or anchoring of a biomolecule to a surface, the molecule will reorganize so that a minimum of the free energy of interaction is achieved. This means, for example, that in the case of a hydrophobic surface, a protein molecule will reorganize so that its hydrophobic part faces the surface. Such reorganization sometimes may result in a complete loss of the bioactivity. An appropriate design of the surface may solve this problem.

Arya *et al.* deposited a lipid membrane onto a polyacrylamide hydrogel, which is an ion-conductive substrate, creating an organized structure capable of electrochemical response [314]. Reichert *et al.* reviewed some of the biosensor issues and discussed the use of LB films and black lipid membranes in these devices [313].

397

Table 5.2. Transduction modes in biosensors. (From Reichert *et al.* [313], © 1987, Elsevier Science Publishers.)

Transduction Mode and Device	Observed Output
Optical Fiber optics and planar devices utilizing absorption, scatter, polarization, reflectivity, and interference of light	Changes in wavelength, intensity, emission profile, reflectivity, fringe patterns, polarization state, and refractive index
Electrochemical Potentiometric devices, amperometric devices (electrode), conductometric devices (chemiresistor)	Changes in voltage, current, impedance, and/or resistance
Galvanometric Piezoelectric microbalances, surface acoustical wave (SAW)	Changes in mass. which produce shifts in frequency or phase of resonance vibrations
Thermal Thermistor devices	Changes in temperature which produce shifts in electrical output

Anzai *et al.* reported on the use of an LB membrane to immobilize penicillinase on the gate surface of an ion-sensitive field effect transistor (ISFET) [315, 316]. Tsuzuki *et al.* immobilized a monolayer of glucose oxidase onto an SnO_2 electrode [317]. They used a mixed monolayer of octadecylamine and octadecanol on silanized SnO_2, connected the enzyme using glutaraldehyde as a spacer bridge, and used amperometry to measure the enzymatically formed H_2O_2. Sriyudthsak *et al.* showed that by a selection of a lipid material, they could enhance enzyme adsorption and incorporation, thus forming an organized system that may serve in biosensors [318]. Okahata used an enzyme–lipid complex and showed that it was possible to form a stable LB film, and to transfer Y-type films onto a platinum electrode [319]. Aizawa *et al.* developed an optical sensor for simultaneous recognition of amino acids in an aqueous solution [320]. Anzai *et al.* used LB films of highly branched polyethyleneimine as a spacer for immobilizing α-chymotrypsin and urease on an ISFET [321].

5.9. Gas Penetration Properties of Monolayers

The use of LB films for evaporation control was attempted as early as 1924 [322–324], and the reader may find accounts for the early work in this area in Gaines' book, and in a review by Blake [325].

The temperature dependence of the transmission coefficient of CO_2 gas across a series of long-chain alcohols from C_{16} to C_{22} monolayers, spread to their equilibrium spreading pressure, was measured by Hawke and White [326]. They used radioactive ^{14}C labeled CO_2 to follow gas transmission, and reported activation energies of ~12 kcal/mol, with an increment of E per CH_2 group of 320 cal/mol. This value is similar to the 400 cal/mol predicted from van der Waals dispersion forces between parallel hydrocarbon chains.

Barnes *et al.* developed a statistical theory of monolayer permeation in which the monolayer was represented as a random array of hard disks and the permeating water molecules as hard spheres [327]. There is a consistent agreement between these calculations and the experimental data for monolayers of long-chain alcohols. This model predicted only a very small temperature dependence for the evaporation, which is not in agreement with experimental data [328]. In another theoretical work, Barnes and Quickenden predicted that there should be no evaporation through a monolayer of long-chain alcohols at areas below 24.3 $Å^2$/molecule [329], again in disagreement with the appreciable permeation at these areas. Turner *et al.* studied permeation of water vapor through lipid monolayers [330]. Dickenson proposed a model in which monolayers are described as hard disks, and the permeation rate is related to the probability of hole formation in the hard disk fluid, and found good agreement with experimental results [331]. In another paper, Dickenson used a two-dimensional model to study the effect of attractive forces between the monolayer molecules. In this model, a dense two-dimensional Lennard–Jones fluid was employed, and the distribution of holes was similar to that in the above hard disk fluid [332]. An energy barrier model for permeation through condensed monolayers was used by Milliken *et al.* to correlate molecular parameters with transport rates across the monolayer [333]. In this model, permeation probabilities, which are directly proportional to the diffusion coefficient through the monolayer, were evaluated as a function of the total intermolecular energy between the permeant and the monolayer. The monolayer was simulated by Monte Carlo configurations of hard-disk particles, and the diffusing species were modeled as permeating hard spheres. The permeation probabilities depended on monolayer density, and on the ratio of permeant radius to monolayer cross-section radius.

Heckmann *et al.* proposed cross-linked LB films for filtration purposes, e.g., for reverse osmosis, and gas separation [334]. Vanderveen and Barnes suggested a new method for measuring the water permeation resistances of monolayers [335]. In this method, a monolayer is deposited onto a cylinder of agar gel by the Langmuir–Blodgett technique, the cylinder is transferred to a vacuum chamber, and the evaporation rate is measured under reduced pressure. In this method, the effect of the monolayer on the evaporation rate is more marked, and lower resistance can be measured. Results for monolayers of octadecanol were in reasonable agreement with published data.

Higashi *et al.* reported on oxygen enrichment by fluorocarbon Langmuir–Blodgett films [336, 337]. They achieved selective permeation of O_2 through films of polyion complex of

$$[CF_3-(CF_2)_7-CH_2-CH_2-CO_2-CH_2-CH_2]_2N-C(O)-CH_2-NMe_3^+Cl^-,$$

with poly(styrenesulfonate). The amphiphile is spread on a subphase containing the polysulfonate, and the monolayer thus formed is transferred to the solid substrate.

Osiander *et al.* used photoacoustic spectroscopy to detect water permeation through cadmium arachidate multilayers deposited on a polymer substrate [338]. They found no dependence on the number of bilayers, and the permeation resistance was 0.35 s cm^{-1} in the temperature region from 30 to 50°C, and 0.7 s cm^{-1} above 50°C. Barnes and Hunter studied evaporation resistance of monolayers of cellulose *tris*(decanoate) [339]. They found that there was no measurable permeation resistance in these monolayers. Since small holes were observed in these monolayers, as well as other monolayers with high permeation resistance, they concluded that there is no correlation between the occurrence of observable holes and permeation resistance.

5.10. Tribology of LB and SA Films

The use of organic thin films as lubricants on solid surfaces has important technological applications (e.g., in magnetic discs). However, to date, the materials used for this purpose are liquids such as fluorinated polyethers. LB and SA films represent an attractive alternative to these lubricants because of their strong adsorption to the surface; i.e., they are expected not to migrate on the surface and not to transfer from one solid surface (e.g., the magnetic disk) to the other (e.g., the magnetic head).

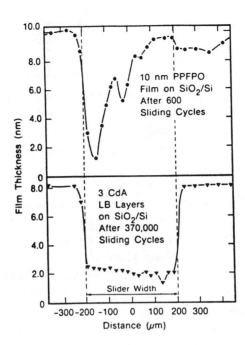

Figure 5.21. Comparison of film thickness profiles before failure for SiO$_2$/Si coated with physisorbed PPFPO and chemisorbed lubricants. (From Novotny *et al.* [349], © 1989, Am. Chem. Soc.)

Early investigations on the tribology of LB films were carried out by Needs [340], Bailey and Courtney-Pratt [341], and Israelachvili and Tabor [342]. The reader may find valuable information on these and other early studies in reviews by Allen and Drauglis [343] and by Fuks [344]. Briscoe *et al.* used curved mica surfaces to study the influence of contact pressure on the sliding behavior of stearic acid monolayers [345]. Briscoe and Evans investigated the shear properties of LB monolayers of fatty acids (C$_{14}$ to C$_{22}$), of fluorinated fatty acids, and of LB multilayers on mica surfaces [346]. They reported that the shear strength varied linearly with pressure, temperature, and the logarithm of sliding velocity. Furthermore, the pressure coefficient of shear strength was increased by the incorporation of fluorine. This is contrary to the low friction properties of polytetrafluoroethylene (PTFE), and was rationalized by differences in energy dissipation processes involved in the ordered monolayers and the polymer.

Seto *et al.* studied the frictional properties of magnetic media coated with LB films [347]. They found that treatment with one monolayer of the barium salt of a

401

fatty acid (C_{16} to C_{22}) reduced the coefficient of kinetic friction from 0.8 in the uncoated surface to 0.2. Additional layers did not reduce the coefficient much further; however, this coefficient was much smaller for the odd number (hydrophobic surface) than for the even number of layers (hydrophilic surface).

Suzuki *et al.* suggested that to lower the coefficient of friction, the surface energy should be small, the lubricant molecule should have a high molecular weight, and it should be bound to the surface [348]. It was concluded, therefore, that a high-molecular-weight fluorinated polymer, which has a functional group to react with the surface, will be a good lubricant.

Novotny *et al.* measured dynamic friction and wear for tribological pairs of silicon substrates coated with LB monolayers and multilayers of cadmium arachidate [349]. Figure 5.21 presents a comparison of film thickness profiles before failure for SiO_2/Si coated with PPFPO (polyperfluoropropylene oxide) and chemisorbed (three CdA monolayers) lubricants. It is clear that while the two upper monolayers were removed after few hundreds of sliding cycles, the first monolayer that apparently is chemisorbed on the surface retained its integrity for hundreds of thousands of sliding cycles.

DePalma and Tillman studied the friction and wear of SA monolayers of alkyltrichlorosilanes on silicon [350]. They reported that the coefficients of kinetic friction (μ_k) were 0.07, 0.09, and 0.16 for monolayers of OTS ($C_{18}H_{37}SiCl_3$), UTS ($C_{11}H_{23}SiCl_3$), and FTS ($CF_3-(CF_2)_8-CH_2-CH_2-SiCl_3$), respectively. These results, which are in agreement with those in literature for SA monolayers of fatty acids [351], suggest that the chemisorption on polar surfaces of carboxylic acids and amine head groups results in bonding comparable in strength to that of a Si–O bond.

References

1. Prasad, P. N.; Williams, D. J. *Introduction to Nonlinear Optical Effects in Organic Molecules and Polymers* , Wiley: New York, 1990.
2. Williams, D. J., Ed., *Nonlinear Optical Properties of Organic Materials,* ACS Symposium Series No. 253; American Chemical Society: Washington, D. C., 1983.
3. Williams, D. J. *Angew. Chem. Intl. Ed. Eng.* **1984**, *23*, 690.
4. Chemla, D. S.; Zyss, J., Eds., *Nonlinear Optical Properties of Organic Molecules and Crystals* , Academic Press: Orlando, 1987.
5. Zernike, F.; Midwinter, J. E. *Applied Nonlinear Optics* , Wiley: New York, 1973, p. 54ff.
6. *ibid*, p 153ff.
7. Ledoux, I.; Zyss, J.; Migus, A.; Etchepare, G.; Grillon, G.; Antonetti, A. *Appl. Phys. Lett.* **1986**, *48*, 1564.

8. Hunsberger, R. G. *Integrated Optics: Theory and Technology*, Springer-Verlag: Berlin, 1982, p. 120ff.
9. Yariv, A. *Quantum Electronics*, Wiley: New York, 1975; p 327ff.
10. Gold, R. *Laser and Applications* **1986**, 67.
11. Sellers, G. J.; Sriram, S. *Laser Focus/Electro-Optics* **1986**, *22*, 74.
12. Tamir, T. *Integrated Optics*, Vol. 7 in Applied Physics, Springer-Verlag: New York, 1975.
13. Gibbs, H. M.; McCall, S. L.; Venkaterson, T. N. C. *Opt. Eng.* **1980**, *19*, 463.
14. Pepper, D. M. *Sci. Amer.* **1986**, *254(1)*, 74.
15. Onsager, L. J. *J. Am. Chem. Soc.* **1936**, *58*, 1486.
16. Zyss, J. *J. Molec. Electron.* **1985**, *1*, 25.
17. Ledoux, I., Josse, D.; Vidakovich, P.; Zyss, J. *Opt. Eng.* **1986**, *25*, 202.
18. Blinov, L. M.; Davydova, N. N.; Lazarev, V. V.; Yudin, S. G. *Sov. Phys. Solid State* **1982**, *24*, 1523.
19. Blinov, L. M.; Dubinin, N. V.; Mikhnev, L. V.; Yudin, S. G. *Thin Solid Films* **1984**, *120*, 161.
20. Aktsipetrov, O. A.; Akhmediev, Mishina, E. D.; Novak *JEPT Lett.* **1983**, *37*, 207.
21. Aktsipetrov, O. A.; Akhmediev, N. N.; Baranova, I. M.; Mishina, E. D.; Novak, V. R. *Sov. Tech. Phys. Lett.* **1985**, *11*, 249.
22. Bloembergen, N.; Pershan, P. S. *Phys. Rev.* **1962**, *128*, 606.
23. Ledoux, I.; Josse, D.; Vidakovich, P.; Zyss, J.; Hann, R. A.; Gordon, P. F.; Bothwell, B. D.; Gupta, S. K.; Allen, S.; Robin, P.; Chastaing, E.; Dubois, J. C. *Europhys. Lett.* **1987**, *3*, 803.
24. Loulerque, J. C.; Dumont, M.; Levy, Y.; Robin, P.; Pocholle, J. P.; Popuchon, M. *Thin Solid Films* **1988**, *160*, 399.
25. Dulcic, A. *Chem. Phys.* **1979**, *37*, 57.
26. Gaines, G. L., Jr. *Anal. Chem.* 1976, 48, 450.
27. Daniel, M. F.; Smith, G. W. *Mol. Cryst. Liq. Cryst.* **1984**, *102*, 193 (Lett.).
28. Kuder, J. E.; Wychick, D. *Chem. Phys. Lett.* **1974**, *102*, 193.
29. Gaines, G. L. Jr., *Nature* **1982**, *298*, 544.
30. Girling, I. R.; Cade, N. A.; Kolinsky, P. V.; Montgomery, C. M. *Elect. Lett* **1985**, *21*, 169.
31. Meredith, G. R., Buchalter, B.; Hanzlik, C. *J. Chem. Phys.* **1983**, *78*, 1533.
32. Marowsky, G.; Chi, L. F.; Möbius, D.; Steinhoff, R.; Shen, Y. R.; Dorsch, D.; Rieger, B. *Chem. Phys. Lett.* **1988**, *147*, 420.
33. Girling, I. R.; Kolinsky, P. V.; Cade, N. A.; Earls, J. D.; Peterson, I. R. *Opt. Comm.* **1985**, *55*, 289.
34. Neal, D. B.; Petty, M. C.; Roberts, G. G.; Ahmad, M. M.; Feast, W. J.; Girling, I. R.; Cade, N. A.; Kolinsky, P. V.; Peterson, I. R. *Electr. Lett.* **1986**, *22*, 460.
35. Earls, J. D.; Peterson, I. R.; Russell, G. J.; Girling, I. R.; Cade, N. A. *J. Mol. Electr.* **1986**, *2*, 85.
36. Mahan, G. D.; Lucas, A. A. *J. Chem. Phys.* **1978**, *68*, 1344.
37. King, F. W.; van Duyne, R. P.; Schatz, G. C. *ibid*, *69*, 4472.
38. Eesley, G. L.; Smith, J. R. *Solid State Comm.* **1979**, *31*, 815.
39. Bagchi, A.; Barrera, R. C.; Dasgupta, B. B. *Phys. Rev. Lett.* **1980**, *44*, 1475.
40. Bagchi, A.; Barrera, R. C.; Fuchs, R. *Phys. Rev. B* **1982**, *25*, 7086.
41. Ye, P.; Shen, Y. R. *Phys. Rev. B* **1983**, *28*, 4288.
42. Hayden, L. M. *Phys. Rev. B* **1988**, *38*, 3718.

403

43. Girling, I. R.; Cade, N. A.; Kolinsky, P. V.; Jones, R. J.; Peterson, I. R.; Ahmad, M. M.; Neal, D. B.; Petty, M. C.; Roberts, G. G.; Feast, W. J. *J. Opt. Soc. Am. B* **1987**, *4*, 950.
44. Schildkraut, J. S.; Penner, T. L.; Willand, C. S.; Ulman, A. *Opt. Lett.* **1988**, *13*, 134.
45. Marowsky, G.; Steinhoff, R. *Opt. Lett.* **1988**, *13*, 707.
46. Cross, G. H.; Peterson, I. R.; Girling, I. R.; Cade, N. A.; Goodwin, M. J.; Carr, N.; Sethi, R. S.; Marsden, R.; Gray, G. W.; Lacey, D.; McRoberts, A. M.; Scrowston, R. M.; Toyne, K. J. *Thin Solid Films* **1988**, *156*, 39.
47. Schildkraut, J. S.; Penner, T. L.; Willand, C. S.; Ulman, A., Unpublished results.
48. Steinhoff, R.; Chi, L. F.; Marowsky, G.; Möbius, D. *J. Opt. Soc. Am. B* **1989**, *6*, 843.
49. Cross, G. H.; Girling, I. R.; Peterson, I. R.; Cade, N. A.; Earls, J. D. *J. Opt. Soc. Am. B* **1987**, *4*, 962.
50. Marowsky, G.; Gieralski, A.; Steinhoff, R.; Dorsch, D.; Eldenshnile, R.; Rieger, B. *J. Opt. Soc. Am. B* **1987**, *4*, 956.
51. Lupo, D.; Prass, W.; Scheunemann, U.; Laschewsky, A.; Ringsdorf. H.; Ledoux, I. *J. Opt. Soc. Am. B* **1988**, *5*, 300.
52. Roberts, G. G.; Holcroft, B.; Richardson, T.; Colbrook, R. *J. de Chim. Phys.* **1988**, *85*, 1093.
53. Popovitz-Biro, R.; Hill, K.; Landou, E. M.; Lahav, M.; Leiserowitz, L.; Sagiv, J.; Hsiung, H.; Meredith, G. R.; Vaherzeele, H. *J. Am. Chem. Soc.* **1988**, *110*, 2672.
54. Carr, N.; Goodwin, M. J. *Makromol. Chem. Rapid Comm.* **1987**, *8*, 487.
55. Anderson, B. L.; Hall, R. C.; Higgins, B. G.; Lindsay, G.; Stroeve, P.; Kowel, S. T. *Synth. Metals* **1989**, *28*, D683.
56. Kajzar, F.; Messier, J. *Thin Solid Films* **1983**, *99*, 109.
57. Kajzar, F.; Messier, J.; Zyss, J.; Ledoux, I. *Opt. Comm.* **1983**, *45*, 133.
58. Kajzar, F.; Messier, J.; Zyss, J. *J. de Phys.* **1983**, *44*, C3-709.
59. Kajzar, F.; Messier, J.; Girling, I. R.; Peterson, I. R. *Electron. Lett.* **1986**, *22*, 1230.
60. Kajzar, F.; Messier, J. *Thin Solid Films* **1985**, *132*, 11.
61. Biegajski, J. E.; Burzynski, R.; Cadenhead, D. A.; Prasad, P. N. *Macromolecules* **1986**, *19*, 2457.
62. Rao, D. N.; Chopra, P.; Ghoshal, S. K.; Swiatkiewicz, J.; Prasad, P. N.; *J. Chem. Phys.* **1986**, *84*, 7049.
63. Prasad, P. N. *Thin Solid Films* **1987**, *152*, 275.
64. Berkovic, G.; Superfine, R.; Guyot-Sionnest, P.; Shen, Y. R.; Prasad, P. N. *J. Opt. Soc. Am.* **1988**, *5*, 668.
65. Logsdon, P. B.; Peleger, J.; Prasad, P. N. *Synth. Met.* **1988**, *26*, 369.
66. Tasaka, S.; Katz, H. E.; Hutton, R. S.; Orenstein, J.; Fredrickson, G. H.; Wang, T. T. *Synth. Met.* **1986**, *16*, 17.
67. Kajzar, F.; Girling, I. R.; Peterson, I. R. *Thin Solid Films* **1988**, *160*, 209.
68. Ogawa, K.; Mino, N.; Tamura, H.; Sonoda, N. *Langmuir*, **1989**, *5*, 1415.
69. Novak, V. R.; Myagkov, I. V. *Pis'ma Zh. Tekh. Fiz.* **1985**, *11*, 385.
70. Snow, A. W.; Barger, W. R.; Klusky, M.; Wohltjen, H.; Jarvis, N. L. *Langmuir* **1986**, *2*, 513.
71. Okahata, Y.; Ariga, K. *J. Chem. Soc. Chem. Commun.* **1987**, 1535.
72. Biddle, M. B.; Rickert, S. E. *Ferroelectrics* **1987**, *76*, 133.
73. Li, S. C. *Ferroelectrics* **1983**, *46*, 209.

74. Christie, P.; Jones, C. A.; Petty, M. C.; Roberts, G. G. *J. Phys. D* **1986**, *19*, L167.
75. Christie, P.; Petty, M. C.; Roberts, G. G.; Richards, D. H.; Service, D.; Stewart, M. J. *Thin Solid Films* **1985**, *134*, 75.
76. Blinov, L. M.; Mikhnev, L. V.; Sobotka, E. B.; Yudin, S. G. *Sov. Tech. Phys. Lett.* **1983**, *9*, 640.
77. Blinov, L. M.; Mikhnev, L. V.; Yudin, S. G. *Phys. Chem. Mech. Surfaces* **1985**, *3*, 2919.
78. Smith, G. W.; Daniel, M. F.; Barton, J. W.; Ratcliffe, N. *Thin Solid Films* **1985**, *132*, 125.
79. Christie, P.; Roberts, G. G.; Petty, M. C. *Appl. Phys. Lett.* **1986**, *48*, 1101.
80. Jones, C. A.; Petty, M. C.; Roberts, G. G. *IEE Trans. on Ultrasonic Ferroelectrics, and Frequency Control.*
81. Agarwal, V. K. *Electrocomponent Sci. Tech.* **1975**, *2*, 1.
82. Forlani, F.; Minnaja, N. *Phys. Status Solidi* **1966**, *4*, 311.
83. Forlani, F.; Minnaja, N. *J. Vac. Sci. Technol.* **1969**, *6*, 518.
84. Klein, N. *Advanc. Phys.* **1972**, *21*, 605.
85. Agarwal, V. K.; Srivastava, V. K. *Thin Solid Films* **1975**, *27*, 49.
86. Kapur, U.; Srivastava, V. K. *Phys. Status Solidi A* **1976**, *38*, K77.
87. Honig, E. P. *Thin Solid Films* **1975**, *33*, 231.
88. Gupta, S. K.; Singal, C. M.; Srivastava, V. K. *Proc. Nucl. Phys. Solid State Symp.* **1976**, *19C*, 88.
89. Srivastava, V. K. *Electrocomponenet Sci. Technol.* **1976**, *3*, 117.
90. Taylor, D. M.; Mahboubian-Jones, M. G. B. *Thin Solid Films* **1982**, *87*, 167.
91. Iwamoto, M.; Kang, D. Y. *Coll. Eng. Tokyo Inst. Tech.* **1989**, *38*, 35.
92. Handy, R. M.; Scala, L. C. *J. Electrochem. Soc.* **1966**, *113*, 109.
93. Miles, J.; McMahon, H. P. *J. Appl. Phys.* **1961**, *32*, 1176.
94. Mann, B.; Kuhn, H. *J. Appl. Phys.* **1971**, *42*, 4398.
95. Polymeropoulos, E. E. *Solid State Commun.* **1978**, *28*, 883.
96. Polymeropoulos, E. E.; Sagiv, J. *J. Chem. Phys.* **1978**, *69*, 1836.
97. Sugi, M.; Fukui, T.; Iizima, S. *Appl. Phys. Lett.* **1975**, *27*, 559.
98. Sugi, M.; Nembach, K.; Möbius, D.; Kuhn, H. *Solid State Commun.* **1974**, *15*, 1867.
99. Sugi, M.; Nembach, K.; Möbius, D. *Thin Solid Films* **1975**, *27*, 205.
100. Iizima, S.; Sugi, M. *Appl. Phys. Lett.* **1976**, *28*, 548.
101. Sugi, M.; Fukui, T.; Iizima, S. *Chem. Phys. Lett.* **1977**, *45*, 163.
102. Kuhn, H. *J. Photochem.* **1979**, *10*, 111.
103. Yamamoto, N.; Ohnishi, T.; Hatakeyama, M.; Tsubomura, H. *Bull. Chem. Soc. Jpn.* **1976**, *51*, 3462.
104. Peterson, I. R. *Aust. J. Chem.* **1980**, *33*, 1713.
105. Peterson, I. R.; Russell, G. J.; Roberts, G. G. *Thin Solid Films* **1983**, *109*, 371.
106. Peterson, I. R.; Girling, I. R. *Sci. Prog. Oxf.* **1985**, *69*, 533.
107. Veale, G.; Girling, I. R.; Peterson, I. *Thin Solid Films* **1985**, *127*, 293.
108. Peterson, I. R.; Veale, G.; Montgomery, C. M. *J. Colloid Interface Sci.* **1986**, *109*, 527.
109. Peterson, I. R. *Br. Polym. J.* **1987**, *19*, 391.
110. Peterson, I. R.; Earls, J. D.; Girling, I. R.; Russell, G. J. *Mol. Cryst. Liq. Cryst.* **1987**, *147*, 141.
111. Tredgold, R. H.; Winter, C. S. *J. Phys. D: Appl. Phys.* **1981**, *14*, L185.

112. Tredgold, R. H.; Vickers, A. J.; Allen, R. A. *J. Phys. D: Appl. Phys.* **1984**, *17*, L5.
113. Hao, S.; Blott, B. H.; Melville, D. *Thin Solid Films* **1985**, *132*, 63.
114. Couch, N. R.; Montgomery, C. M.; Jones, R. *Thin Solid Films* **1986**, *135*, 173.
115. Couch, N. R.; Movaghar, B.; Girling, I. R. *Solid State Commun.* **1986**, *59*, 7.
116. Gemma, N.; Mizushima, K.; Miura, A.; Azuma, M. *Synth. Met.* **1987**, *18*, 809.
117. Iwamoto, M.; Shidoh, S.; Kubota, T.; Sekinne, M. *Jpn. J. Appl. Phys.* **1988**, *27*, 1825.
118. Matsuda, H.; Sakai, K.; Kawada, H.; Eguchi, K.; Nakagiri, T. *J. Mol. Electron.* **1989**, *5*, 107.
119. Gerischer, H.; Willig, F. *Top. Curr. Chem.* **1976**, *61*, 31.
120. Schumacher, E. *Chimia* **1978**, *32*, 193.
121. Whitten, D. G.; Delaire, P. J.; Foreman, T. K.; Mercer-Smith, J. A.; Schmehle, R. H.; Giannotti, C. in *Solar Energy: Chemical Conversion and Storage* , Hautala, R. R.; King, R. B.; Kutal, C., Eds.; Humana Press: Clinton, NJ, 1979, p. 117.
122. Hong, F. T.; Mauzerall, D. *Proc. Natl. Acad. Sci. USA* **1974**, *71*, 1564.
123. Katusin-Razem, B.; Wong, M.; Thomas, J. K. *J. Am. Chem. Soc.* **1978**, *100*, 1679.
124. Schmehl, R. H.; Whitten, D. G. *J. Am. Chem. Soc.* **1980**, *102*, 1938.
125. Takuma, K.; Kajiwara, M.; Matsuo, T. *Chem. Lett.* **1977**, 1199.
126. Kalyanasundaram, K.; Kiwiw, J.; Grätzel, M. *Helv. Chim. Acta.* **1978**, *61*, 2720.
127. Gohn, M.; Getoff, N. Z. *Naturforsch. A* **1979**, *34*, 1135.
128. Moroi, Y.; Braun, A.; Grätzel, M. *Ber. Bunsenges, Phys. Chem.* **1978**, *82*, 950.
129. Willner, I.; Ford, W. E.; Otvos, J.; Calvin, M. *Nature* **1979**, *280*, 823.
130. Tributsch, H.; Gerischer, K. *Ber. Bunsenges, Phys. Chem.* **1969**, *73*, 251.
131. Memming, R. *Photochem. Photobiol.* **1972**, *16*, 325.
132. Gerischer, H., in *Light-Induced Charge Separation in Biology and Chemistry* , Gerischer, H.; Katz, J. J., Eds.; Dahlem Konferenzen: West Berlin, 1979, p. 61.
133. Gundlach, K. H.; Kadlech, J. *Chem. Phys. Lett.* **1974**, *25*, 293.
134. Polymeropoulos, E. E. *J. Appl. Phys.* **1977**, *48*, 2404.
135. Tredgold, R. H. *Rep. Prog. Phys.* **1987**, *50*, 1609.
136. Penner, T. L.; Möbius, D. *J. Am. Chem. Soc.* **1982**, *104*, 7407.
137. Fugihira, M.; Nishiyama, K.; Aoki, K. *Thin Solid Films* **1988**, *160*, 317.
138. Fugihira, M.; Akoi, K. *Proc. Electrochem. Soc.* **1988**, *88-14*, 280.
139. Marcus, R. A. *J. Chem. Phys.* **1965**, *63*, 2654.
140. Marcus, R. A.; Sutin, N. *Biochim. Biophys. Acta* **1985**, *811*, 265.
141. Siders, P.; Marcus, R. A. *J. Am. Chem. Soc.* **1981**, *103*, 748.
142. Ulstrup, J.; Jortner, J. *J. Chem. Phys.* **1975**, *63*, 4358.
143. Murakata, T.; Miyashita, T.; Matsuda, M. *Macromolecules* **1988**, *21*, 2730.
144. Fonash, S. J. *Solar Cell Device Physics*, Academic Press: New York, 1981.
145. Fujihira, M.; Ohnishi, N.; Osa, T. *Nature* **1977**, *268*, 226.
146. Fujihira, M.; Aoki, K.; Imoue, S.; Takemura, H.; Muraki, H.; Aoyagui, S. *Thin Solid Films* **1985**, *132*, 221.
147. Miyasaka, T.; Watanabe, T.; Fujishima, A.; Honda, K. *J. Am. Chem. Soc.* **1978**, *100*, 6657.

148. Miyasaka, T.; Watanabe, T.; Fujishima, A.; Honda, K. *Nature* **1979**, *277*, 638.
149. Miyasaka, T.; Watanabe, T.; Fujishima, A.; Honda, K. *Photochem. Photobioi* **1980**, *32*, 217.
150. Petty, M. C.; Roberts, G. G. *Electron. Lett.* **1979**, *15*, 335.
151. Dharmadasa, I. M.; Roberts, G. G.; Petty, M. C. *Electron. Lett.* **1980**, *1* 201.
152. Roberts, G. G.; Petty, M. C.; Dharmadasa, I. M. *IEE Proc.* **1981**, *128*, 197.
153. Yoneyama, M.; Sugi, M.; Saito, M.; Ikegami, K.; Kuroda, S.; Iizima, S. *Jpn J. Appl. Phys.* **1986**, *25*, 961.
154. Willig, F.; Charlé, K. -P.; van der Auwerfaer, M.; Bitterling, K. *Mol. Crys. Liq. Cryst.* **1986**, *137*, 329.
155. van der Auweraer, M.; Willig, F. *Isr. J. Chem.* **1985**, *25*, 274.
156. Willig, F.; Eichberger, R.; Bitterling, K.; Durfee, W. S.; Storck, W. *Ber. Bunsenges. Phys. Chem.* **1987**, *91*, 869.
157. Yoneyama, M.; Nagao, T.; Miura, K. *Thin Solid Films* **1988**, *160*, 165.
158. Furuno, T.; Takimoto, K.; Kouyama, T.; Ikegami, A.; Sasabe, H. *Thin Solid Films* **1988**, *160*, 145.
159. Sasabe, M.; Furuno, T.; Takimoto, K. *Synth. Met.* **1989**, *28*, C787.
160. Terenin, A.; Putzeiko, E.; Akimov, I. *Disc. Faraday Soc.* **1959**, *27*, 83.
161. Rosenberg, G.; Camiscoli, J. F. *J. Chem. Phys.* **1961**, *35*, 982.
162. Tang, C. W.; Albrecht, A. C. *J. Chem. Phys.* **1975**, *63*, 2139.
163. Corker, G. A.; Lundstrom, I. *J. Appl. Phys.* **1978**, *49*, 686.
164. Ghosh, A. K.; Morel, D. L.; Feng, T.; Shaw, R. F.; Rowe, C. A. *J. Appl. Phys.* **1974**, *45*, 230.
165. Fan, F. R.; Faulkner, L. R. *J. Chem. Phys.* **1978**, *69*, 3341.
166. Gaines, G. L., Jr. *Insoluble Monolayers at Liquid–Gas Interfaces*, Interscience: New York, 1966.
167 Ke, B. in *The Chlorophylls*, Vernon, L. P.; Seely, G. R., Eds.; Academic Press: New York, 1966, pp. 253–279.
168. De Costa, B. S. M.; Froines, J. R.; Harris, J. M.; Leblanc, R. M.; Orger, B. H.; Porter, G. *Proc. R. Soc.* **1972**, *A326*, 503.
169. Aghion, J.; Leblanc, R. M. *J. Membr. Biol.* **1978**, *42*, 189.
170. Jones, R.; Tredgold, R. H.; O'Mullane, J. E. *Photochem. Photobiol.* **1980**, *32*, 223.
171. Reucroft, P. J.; Simpson, W. H. *Photochem. Photobiol.* **1969**, *10*, 79.
172. Simpson, W. H.; Reucroft, P. J. *Thin Solid Films* **1970**, *6*, 167.
173. Janzen, A. F.; Bolton, J. R. *J. Am. Chem. Soc.* **1979**, *101*, 6342.
174. Lawrence, M. F.; Dodelet, J. P.; Ringuet, M. *Photochem. Photobiol.* **1981**, *34*, 393.
175. Blinov, L. M.; Davydova, N. N.; Mikhnev, L. V.; Yudin, S. G. *Poverhost* **1985**, *6*, 141.
176. Holoyda, J.; Kannewurf, C. R.; Kauffman, J. W. *Thin Solid Films* **1980**, *68*, 205.
177. Saito, M.; Sugi, M.; Fukui, T.; Iizima, S. *Thin Solid Films* **1983**, *100*, 117.
178. Iriyama, K.; Yoshiura, M.; Ozaki, Y.; Ishii, T.; Yasui, S. *Thin Solid Films* **1985**, *132*, 229.
179. Sugi, M.; Sakai, K.; Saito, M.; Kawabata, Y.; Iizima, S. *Thin Solid Films* **1985**, *132*, 65.
180. Saito, M.; Sugi, M.; Iizima, S. *Jpn. J. Appl. Phys.* **1985**, *24*, 379.

181. Nagamura, T.; Matano, K.; Ogawa, T. *Ber. Bunsenges. Phys. Chem.* **1987**, *91*, 759.
182. Polymeropoulos, E. E.; Möbius, D.; Kuhn, H. *J. Chem. Phys.* **1978**, *68*, 3918.
183. Polymeropoulos, E. E.; Möbius, D.; Kuhn, H. *Thin Solid Films* **1980**, *68*, 173.
184. Sugi, M.; Fukui, T.; Iizima, S. *Mol. Cryst. Liq. Cryst.* **1980**, *62*, 165.
185. Sugi, M.; Iizima, S. *Thin Solid Films* **1980**, *68*, 199.
186. Sugi, M.; Saito, M.; Fukui, T.; Iizima, S. *Thin Solid Films* **1983**, *133*, 17.
187. Yamamoto, N.; Ohnishi, T.; Hatakeyama, M.; Tsubomura, H. *Thin Solid Films* **1980**, *68*, 191.
188. Nagamura, T.; Matano, K.; Ogawa, T. *J. Phys. Chem.* **1987**, *91*, 2019.
189. Steven, J. K.; Sudiwala, R. V.; Wilson, E. G. *Thin Solid Films* **1988**, *160*, 171.
190. Tran-Thi, T.; Palacin, S.; Clergeot, B. *Chem. Phys. Lett.* **1989**, *157*, 92.
191. Barlow, W. A.; Finney, J. A.; McGinnity, T. M.; Roberts, G. G.; Vincett, P. S. *Physics of Semiconductors 1978 (Inst. Phys. Conf. Ser. 43)*, 1979, Ch. 22.
192. Vincett, P. S.; Barlow, W. A.; Boyle, F. T.; Finney, J. A.; Roberts, G. G. *Thin Solid Films* **1979**, *60*, 265.
193. Roberts, G. G.; McGinnity, T. M.; Barlow, W. A.; Vincett, P. S. *Solid State Commun.* **1979**, *32*, 683.
194. Roberts, G. G.; McGinnity, T. M.; Barlow, W. A.; Vincett, P. S. *Thin Solid Films* **1980**, *68*, 223.
195. Vincett, P. S.; Barlow, W. A. *Thin Solid Films* **1980**, *71*, 305.
196. Day, D. R.; Lando, J. B. *J. Appl. Polym. Sci.* **1981**, *26*, 1605.
197. Jones, R.; Tredgold, R. H.; Hodge, P. *Thin Solid Films* **1983**, *99*, 25.
198. Jones, R.; Tredgold, R. H.; Hoorfar, A. *Thin Solid Films* **1984**, *113*, 115.
199. Jones, R.; Tredgold, R. H.; Hoorfar, A. *Thin Solid Films* **1985**, *123*, 307.
200. Tredgold, R. H.; Winter, C. S.; El-Badawy, Z. I. *Electron. Lett.* **1985**, *21*, 554.
201. Tredgold, R. H.; Evans, S. D.; Hodge, P.; Hoorfar, A. *Thin Solid Films* **1988**, *160*, 99.
202. McArdle, C. B.; Ruaudel-Teixier, A. *Thin Solid Films* **1985**, *133*, 93.
203. Baker, S.; Petty, M. C.; Roberts, G. G.; Twigg, M. V. *Thin Solid Films* **1983**, *99*, 53.
204. Hann, R. A.; Gupta, S. K.; Fryer, J. R.; Eyres, B. L. *Thin Solid Films* **1985**, *134*, 35.
205. Hua, Y. L.; Roberts, G. G.; Ahmad, M. M.; Petty, M. C. *Phil. Mag. B* **1986**, *53*, 105.
206. Kutzler, F. W.; Barger, W. R.; Snow, A. W.; Wohltjen, H. *Thin Solid Films* **1987**, *155*, 1.
207. Fujiki, M.; Tabei, H.; Imamura, S. *Jpn. J. Appl. Phys.* **987**, *26*, 1224.
208. Fujiki, M.; Tabei, H. *Langmuir* **1988**, *4*, 320.
209. Nakamura, T.; Tanaka, H.; Matsumoto, M. *Chem. Lett.* **1988**, 1667.
210. Nakamura, T.; Tanaka, H.; Matsumoto, M.; Tachibana, H.; Manda, E.; Kawabata, T. *Synth. Met.* **1988**, *27*, B601.
211. Vandevyver, M.; Lesieur, J. R.; Ruaudel-Teixier, A.; Barraud, A. *Mol. Cryst. Liq. Cryst.* **1986**, *134*, 337.
212. Rieutord, F.; Benattar, J. J.; Bosio, L. *J. Physique* **1986**, *47*, 1249.
213. Vandevyver, M.; Richard, J.; Barraud, A.; Ruaudel-Teixier, A.; Lequan, M.; Lequan, R. M. *J. Chem. Phys.* **1987**, *87*, 6754.

214. Richard, J.; Vandevyver, M.; Lesieur, P.; Ruaudel-Teixier, A.; Barraud, A. *J. Chem. Phys.* **1987**, *86*, 2428.
215. Dhindsa, A. S.; Bryce, M. R.; Lloyd, M. P.; Petty, M. C. *Synth. Met.* **1987**, *22*, 185.
216. Richard, J.; Barraud, A.; Vandevyver, M.; Ruaudel-Teixier, A.*Thin Solid Films* **1988**, *159*, 207.
217. Barraud, A.; Flörsheimer, M.; Möhwald, H.; Richard, J.; Ruaudel-Teixier, A.; Vandevyver, M. *J. Colloid Interface Sci.* **1988**, *121*, 491.
218. Richard, J.; Vandevyver, M.; Barraud, A.; Delhaes, P.*Thin Solid Films* **1988**, *161*, L73.
219. Roberts, G. G.; Holcroft, B.; Barraud, A.; Richard, J.*Thin Solid Films* **1988**, *160*, 53.
220. Nakamura, T.; Tanaka, M.; Sekiguchi, T.; Kawabata, Y. *J. Am. Chem. Soc.* **1986**, *108*, 1302.
221. Nakamura, T.; Takei, F.; Tanaka, M.; Matsumoto, M.; Sekiguchi, T.; Manda, E.; Kawabata, Y.; Saito, G. *Chem. Lett.* **1986**, 323.
222. Nakamura, T.; Matsumoto, M.; Takei, F.; Tanaka, M.; Sekiguchi, T.; Manda, E.; Kawabata, Y. *Chem Lett.* **1986**, 709.
223. Matsumoto, M.; Nakamura, T.; Takei, F.; Tanaka, M.; Sekiguchi, T.; Mizuno, M.; Manda, E.; Kawabata, Y. *Synth. Met.* **1987**, *19*, 675.
224. Barraud, A.; Lesieur, P.; Richard, J.; Ruaudel-Teixier, A.; Vandevyver. M; Lequan, M.; Lequan, R. M. *Thin Solid Films* **1988**, *160*, 81.
225. Barraud, A.; Lequan, M.; Lequan, R. M.; Lesieur, P.; Richard, J.; Ruaudel-Teixier, A.; Vandevyver, M. *J. Chem. Soc. Chem. Commun.* **1987**, 797.
226. Saito, G. *Pure Appl. Chem.* **987**, *59*, 999.
227. Dhindsa, A. S.; Bryce, M. R.; Lloyd, J. P.; Petty, M. C. *Thin Solid Films* **1988**, *165*, L97.
228. Dhindsa, A. S.; Bryce, M. R.; Lloyd, J. P.; Petty, M. C. *Synth. Met.* **1988**, *27*, B563.
229. Richard, J.; Vandevyver, M.; Barraud, A.; Morand, J. P.; Delhaes, P.*J. Colloid Interface Sci.* **1989**, *129*, 254.
230. Bertho, F.; Talham, D.; Robert, A.; Batail, P.; Megtert, S.; Robin, P. *Mol. Cryst. Liq. Cryst.* **1988**, *156*, 339.
231. Nakamura, T.; Takei, F.; Tanaka, M.; Matsumoto, M.; Sekiguchi, T.; Manda, E.; Kawabata, Y.; Saito, G. *Chem. Lett.* **1986**, 323.
232. Nakamura, T.; Takei, F.; Matsumoto, M.; Tanaka, M.; Sekiguchi, T.; Manda, E.; Kawabata, Y.; Saito, G. *Synth. Met.* **1987**, *19*, 681.
233. Ikegami, K.; Kuroda, S.; Sugi, M.; Saito, M.; Iizima, S.; Nakamura, T.; Matsumoto, M.; Kawabata, Y. *Synth. Met.* **1987**, *19*, 669.
234. Ikegami, K.; Kuroda, S.; Saito, K.; Saito, M.; Sugi, M.; Nakamura, T.; Matsumoto, M.; Kawabata, Y.; Saito, G. *Synth. Met.* **1988**, *27*, 587.
235. Kawabata, Y.; Nakamura, T.; Matsumoto, M.; Tanaka, M.; Sekiguchi, T.; Komizu, H.; Manda, E.; Saito, G. *Synth. Met.* **1987**, *19*, 663.
236. Matsumoto, M.; Nakamura, T.; Manda, E.; Kawabata, Y.; Ikegami, K.; Kuroda, S.; Sugi, M.; Sairo, G. *Thin Solid Films* **1988**, *160*, 61.
237. Fujiki, M.; Tabei, H. *Synth. Met.* **1987**, *18*, 815.
238. Richard, J.; Vandevyver, M.; Barraud, A.; Morand, J. P.; Lapouyade, R.; Delhaés, P.; Jacquinot, J. F.; Roulliay, M. *J. Chem. Soc. Chem. Commun.* **1988**, 754.
239. Morand, J. P.; Lapouyade, R.; Delhaés, P.; Vandevyver, M.; Richard, J.; Barraud, A.*Synth. Met.* **1988**, *27*, B569.

240. Iyoda, T.; Ando, M.; Kaneko, A.; Ohtani, A.; Shimidzu, T.; Honda, K. *Tet, Lett.* **1986**, *27*, 5633.
241. Iyoda, T.; Ando, M.; Kaneko, T.; Ohtani, A.; Shimidzu, T.; Honda, K. *Langmuir* **1987**, *3*, 1169.
242. Shimidzu, T.; Iyoda, T.; Ando, M.; Ohtani, A.; Kaneko, T.; Honda, K. *Thin Solid Films* **1988**, *160*, 67.
243. Hong, K.; Rubner, M. F. *Thin Solid Films* **1988**, *160*, 187.
244. Nakahara, H.; Nakayama, J.; Hoshino, M.; Fukuda, K. *Thin Solid Films* **1988**, *160*, 87.
245. Watanabe, I.; Hong, K.; Rubner, M. F. *J. Chem. Soc. Chem. Commun.* **1989**, 123.
246. Watanabe, I.; Hong, K.; Rubner, M. F.; Loh, I. H. *Synth. Met.* **1988**, *28*, C473.
247. Nishikata, Y.; Kakimoto, M.; Imai, Y. *J. Chem. Soc. Chem Commun.* **1988**, 1040.
248. Schweig, A. *Z. Naturforsch* **1970**, *22a*, 724.
249. Bücher, H.; Kuhn, H. *Z. Naturforsch* **1970**, *25b*, 1323.
250. Yamamoto, H.; Sugiyama, T.; Tanaka, M. *Jpn. J. Appl. Phys.* **1985**, *24*, L305.
251. Reich, R.; Scheerer, R. *Ber. Bunsenges. Phys. Chem.* **1976**, *80*, 542.
252. Fukui, T.; Saito, M.; Sugi, M.; Iizima, S. *Thin Solid Films* **1983**, *109*, 247.
253. Decher, G.; Tieke, B. *Thin Solid Films* **1988**, *160*, 407.
254. Gruda, I.; Leblanc, R. M. *Can. J. Chem.* **1976**, *54*, 576.
255. Morin, M.; Leblanc, R. M.; Gruda, I. *Can. J. Chem.* **1980**, *58*, 2038.
256. Polymeropoulos, E. E.; Möbius, D. *Ber. Bunsenges. Phys. Chem.* **1979**, *83*, 1215.
257. McArdle, C. B.; Blair, H. S.; Barraud, A.; Ruaudel-Teisier, A. *Thin Solid Films* **1983**, *99*, 181.
258. Blair, H. S.; McArdle, C. B. *Polymer* **1984**, *25*, 999.
259. McArdle C. B.; Blair, H. S. *Colloid. Polym. Sci.* **1984**, *262*, 481.
260. Ando, E.; Miyazaki, J.; Morimoto, K.; Nakahara, H.; Fukuda, K. *Thin Solid Films* **1985**, *133*, 21.
261. Ando, E.; Hibino, J.; Hashida, T.; Morimoto, K. *Thin Solid Films* **1988**, *160*, 279.
262. Blair, H. S.; McArdle, C. B. *Polymer* **1984**, *25*, 1347.
263. Sandhu, S. S.; Yianni, Y. P.; Morgan, C. G.; Taylor, D. M.; Zaba, B. *Biochim. Biophys. Acta* **1986** , *860*, 253.
264. Nishiyama, K.; Fujihira, M. *Chem Lett.* **1988**, 1257.
265. Kawamura, S.; Tsutsui, T.; Saito, S.; Murao, Y.; Kina, K. *J. Am. Chem. Soc.* **1988**, *110*, 509.
266. André, J. -J.; Holezer, K.; Petit, P.; Riou, M. -T.; Clarrisse, C.; Even, R.; Fourmigué, M.; Simon, J. *Chem. Phys. Lett.* **1985**, *115*, 463.
267. Maitrot, M.; Guillaud, G.; Boudjema, B.; André, J. -J.; Strzelecka, H.; Simon, J.; Even, R. *Chem. Phys. Lett.* **1987**, *133*, 59.
268. Turek, P. ; Petit, P.; André, J. -J.; Simon, J.; Even, R.; Boudjema, B.; Guillaud, G.; Maitrot, M. *J. Am. Chem. Soc.* **1987**, *109*, 5119.
269. Even, R.; Simon, J.; Markovitsi, D. *Chem. Phys. Lett.* **1989**, *156*, 609.
270. Robinet, S.; Clarisse, C. *Thin Solid Films* **1989**, *170*, L51.
271. Snow, A.; Barger, W.; Jarvis, N.; Wahltjen, H. *16th Natl. SAMPE Tech. Conf.*, October 9-11, 1984, p 388.

272. Richard, J.; Barraud, A.; Vandevyver, M.; Ruaudel-Teixier, A. *J. Mol. Electron.* **1986**, *2*, 193.
273. Barraud, A.; Rosilio, C.; Ruaudel-Teixier, A. *Microcircuit Engineering 79* , Institut of Semiconductors and Electronics, Aachen, 1979, p. 127.
274. Barraud, A.; Rosilio, C.; Ruaudel-Teixier, A., *Thin Solid Films* **1980**, *68*, 91.
275. Barraud, A. *Thin Solid Films* **1983**, *99*, 317.
276. Barraud, A.; Rosilio, C.; Ruaudel-Teixier, A. *Polym. Prepr., Am. Chem. Soc., Div. Polym. Chem.* **1978**, *19*, 179.
277. Teike, B.; Lieser, G.; Wegner, G. *J. Polym. Sci.* **1979**, *17*, 1631.
278. Fariss, G.; Lando, J.; Rickert, S. *Thin Solid Films* **1983**, *99*, 305.
279. Boothroyd, B.; Delaney, P. A.; Hann, R. A.; Johnstone, R. A. W.; Ledwith, A. *Br. Polym. J.* **1985**, *17*, 360.
280. Boothroyd, B.; Delaney, P. A.; Hann, R. A.; Johnstone, R. A. W.; *Polym. Deg. Stab.* **1987**, *17*, 253.
281. Miyashita, T.; Yoshida, H.; Matsuda, M. *Thin Solid Films* **1987**, *155*, L11.
282. Miyashita, T.; Matsuda, M. *Thin Solid Films* **1989**, *168*, L47.
283. Jones, R.; Winter, C. S.; Tredgold, R. H.; Hodge, P.; Hoorfar, A. *Polymer* **1987**, *28*, 1619.
284. Kuan, S. W. J.; Frank, C. W.; Yen Lee, Y. H.; Eimori, T.; Allee, D. R.; Pease, R. F. W.; Browning, R. *J. Vac. Sci. Tech.* **1989**, *B7*, 1745.
285. Roberts, G. G.; Pande, K. P.; Barlow, W. A. *Proc. IEE Part 1* **1978**, *2*, 169.
286. Petty, M. C.; Roberts, G. G. *Proc. INFOS 79*, Roberts, G. G., Ed.; London: Institute of Physics, 1980, p 186.
287. Kan K. K.; Petty, M. C.; Roberts, G. G. *Thin Solid Films* **1983**, *99*, 291.
288. Lloyd, J. P.; Petty, M. C.; Roberts, G. G.; LeComber, P. G.; Spear, W. E. *Thin Solid Films* **1982**, *89*, 4.
289. Lloyd, J. P.; Petty, M. C.; Roberts, G. G.; LeComber, P. G.; Spear, W. E. *Thin Solid Films* **1982**, *89*, 2974.
290. Roberts, G. G.; Pande, K. P.; Barlow, W. A. *Electron. Lett.* **1977**, *13*, 581.
291. Baker, S.; Roberts, G. G.; Petty, M. C. *Proc. IEE Part 1* **1983**, *130*, 260.
292. Wohltjes, H.; Barger, W.; Snow, A. W.; Jarvis, N. L. *IEE Trans. Electron. Dev.* **1985**, *ED-32*, 1170.
293. Tredgold, R. H.; Young, M C. J.; Hodge, P.; Hoorfar, A. *IEE Proc.* **1985**, *132*, 151.
294. Henrion, L.; Derost, G.; Ruaudel-Teixier, A.; Barrayd, A. *Sensors Actuators* **1988**, *14*, 251.
295. Henrion, L.; Derost, G.; Barraud, A.; Ruaudel-Teixier, A. *Sensors Actuators* **1989**, *17*, 493.
296. Raether, H. *Phys. Thin Films* **1977**, *9*, 145.
297. Weber, W. H. *Phys. Rev. Lett.* **1977**, *39*, 153.
298. Pockrand, I.; Swalen, J. D.; Gordon II, J. G.; Philpott, M. R. *Surf. Sci.* **1977**, *74*, 237.
299. Pockrand, I. *Surf. Sci.* **1978**, *72*, 577.
300. Chu, K. C.; Chen, C. K.; Shen, Y. R. *Mol. Cryst. Liq. Cryst.* **1980**, *59*, 97.
301. Pockrand, I.; Swalen, J. D.; Santo, R.; Brillante, A.; Philpott, M. R. *J. Chem. Phys.* **1978**, *69*, 4001.
302. Barnes, W. L.; Sambles, J. R. *Thin Solid Films* **1986**, *143*, 237.
303. Welford, K. R.; Sambles, J. R.; Clark, M. G. *Liq. Cryst.* **1987**, *2*, 91.
304. Liedberg, B.; Nylander, C.; Lundstrom, I. *Sens. Actuators* **1983**, *4*, 299.
305. Gordon II, J. G.; Swalen, J. D. *Opt. Commun.* **1977**, *22*, 374.

306. Lloyd, J. P.; Pearson, C.; Petty, M. C. *Thin Solid Films* **1988**, *160*, 431.
307. Beswick, R. B.; Pitt, C. W. *J. Colloid Interface Sci.* **1988**, *124*, 146.
308. Beswick, R. B.; Pitt, C. W. *Chem. Phys. Lett.* **1988**, *143*, 589.
309. Holcroft, B.; Roberts, G. G. *Thin Solid Films* **1988**, *160*, 445.
310. Schaffar, B. P. H.; Wolfbeis, O. S.; Leitner, A. *Analyst* **1988**, *113*, 693.
311. Moriizumi, T. *Thin Solid Films* **1988**, *160*, 413.
312. Aizawa, M., in *Analytical Chemistry Series, Chemical Sensors* , Seiyama, T.; Fueki, K.; Shiokama, J.; Suzuki, S., Eds., Elsevier: Fukuoka, 1983, pp. 683–692.
313. Reichert, W. M.; Bruckner, C. J.; Joseph, J. *Thin Solid Films* **1987**, *152*, 345.
314. Arya, A.; Krull, U. J.; Thompson, M.; Wong, H. E. *Anal. Chim. Acta* **1985**, *173*, 331.
315. Anzai, J.; Furuya, K.; Chen, C.; Osa, T.; Matsuo, T. *Anal.. Sci.* **1987**, *3*, 271.
316. Anzai, J.; Hashinoto, J.; Osa, T.; Matsuo, T. *Anal. Sci.* **1988**, *4*, 247.
317. Tsuzuki, H.; Watanabe, T.; Okawa, Y.; Yoshida, S.; Yano, S.; Koumoto, K.; Komiyana, M.; Nihei, Y. *Chem. Lett.* **1988**, 1265.
318. Sriyudthsak, M.; Yamagishi, H.; Moriizumi, T. *Thin Solid Films* **1988**, *160*, 463.
319. Okahata, Y.; Tsuruta, T.; Ijiro, K.; Ariga, K. *Langmuir* **1988**, *4*, 1373.
320. Aizawa, M.; Matsuzawa, M.; Shinohara, H. *Thin Solid Films* **1988**, *160*, 477.
321. Anzai, J.; Lee, S.; Osa, T. *Makromol. Chem., Rapid Commun.* **1989**, *10*, 167.
322. Hedestrand, G. *J. Phys. Chem.* **1924**, *28*, 1244.
323. Adam, N. K. *J. Phys. Chem.* **1925**, *29*, 610.
324. Rideal, E. K. *J. Phys. Chem.* **1925**, *29*, 1585.
325. Blake, M. *Tech. Surface Colloid Chem. Phys.* **1972**, *1*, 41.
326. Hawke, J. G.; White, I. *J. Phys. Chem.* **1970**, *74*, 2788.
327. Barnes, G. T.; Quickenden, T. I.; Saylor, J. E. *J. Colloid Interface Sci.* **1970**, *33*, 236.
328. LaMer, V. K.; Healy, T. W.; Aylmore, L. A. G. *J. Colloid Sci.* **1964**, *19*, 673.
329. Barnes, G. T.; Quickenden, T. I. *J. Collois Sci.* **1971**, *37*, 581.
330. Turner, S. R.; Litt, M. L.; William, S. *J. Colloid Interface Sci.* **1975**, *50*, 181.
331. Dickenson, E. *J. Colloid Interface Sci.* **1978**, *63*, 461.
332. Dickenson, E. *J. Chem. Soc. Faraday Trans. 2* **1978**, *74*, 821.
333. Milliken, J. O.; Zollweg, J. A.; Bobalek, E. G. *J. Colloid Interface Sci.* **1980**, *77*, 41.
334. Heckmann, K.; Strobl, C.; Bauer, S. *Thin Solid Films* **1983**, *99*, 265.
335. Vanderveen, R. J.; Barnes, G. T. *Thin Solid Films* **1985**, *134*, 227.
336. Higashi, N.; Kunitake, T.; Kajiyama, T. *Kobunshi Ronbunshu* **1986**, *43*, 761.
337. Higashi, N.; Kunitake, T.; Kajiyama, T. *Polym. J. (Tokyo)* **1987**, *19*, 289.
338. Osiander, R.; Korpium, P.; Duschl, C.; Knoll, W. *Thin Solid Films* **1988**, *160*, 501.
339. Barnes, G. T.; Hunter, D. S. *J. Colloid Interface Sci.* **1989**, *129*, 585.
340. Needs, S. J. *Trans. ASME* **1940**, *62*, 331.
341. Bailey, A. I.; Courtney-Pratt, J. S. *Proc. R. Soc. London* **1955**, *A227*, 500.
342. Israelachvili, J. N.; Tabor, D. *Wear* **1973**, *24*, 385.

343. Allen, C. M.; Drauglis, E. *Wear* **1969**, *14*, 363.
344. Fuks, G. I. in *Research in Surface Forces,* Deryagin, B. V., Ed.; Vol. I, 1960; Translation by Consultants Bureau, New York, 1964.
345. Briscoe, B. J.; Evans, D. C. B.; Tabor, D. *J. Colloid Interf. Sci.* **1977**, *61*, 9.
346. Briscoe, B. J.; Evans, D. C. B. *Proc. R. Soc. London* **1982**, *A380*, 389.
347. Seto, J.; Nagai, T.; Ishimoto, C.; Watanabe, H. *Thin Solid Films* **1985**, *134*, 101.
348. Suzuki, M.; Saotome, Y.; Yanagisawa, M. *Thin Solid Films* **1988**, *160*, 453.
349. Novotny, V.; Swalen, J. D.; Rabe, J. P. *Langmuir* **1989**, *5*, 485.
350. DePalma, V.; Tillman, N. *Langmuir* **1989**, *5*, 868.
351. Levine, O.; Zisman, W. A. *J. Phys. Chem.* **1957**, *61*, 1068.

AUTHOR'S FINAL NOTE

We almost have come to the end of this book on ultrathin organic films, and at this point, I would like to share with the reader some of my views and thoughts on the future of this area. Many of us are asked frequently if LB and SA films ever will become a viable technology. While being a legitimate concern, let me emphasize that the areas of LB and SA films still need many years of research in physics, chemistry, and especially in chemical engineering before they become technologies for large-scale production of molecular devices. Thus, the purpose of this note is to try to define some of the future directions as I see them, and highlight their significance.

LB and SA films are important junctions where biology, chemistry, physics, and engineering meet. This uniqueness, i.e., being at the interface of disciplines, defines the role of LB and SA research in the science and technology of the 1990s and beyond. Kuhn has compared biology to engineering [1]. He wrote that in biology, living objects are considered as organisms consisting of organs, i.e., "purposefully interacting parts, which cooperate and act as functional units." Likewise, "engineering considers machines as purposefully designed entities of interlocked and interacting parts, functional units, i.e., systems with basically new properties which result from the cooperation of the parts." On the other hand, chemistry deals with the synthesis of new molecules and the study of their properties. Thus, taking individual molecules and assembling them into two- and three-dimensional systems provides a clear pathway from the molecular to the supermolecular, from chemistry to materials science and solid-state physics.

We mentioned in Part Four that the understanding of molecular organization in two dimensions is vital to the engineering of materials at the molecular level. It

415

is clear, however, that in the fabrication of supermolecular structures where more than one type of molecule is involved, one also has to worry about how these different molecules interlock to form the two-dimensional assembly. Indeed, it was Kuhn who introduced us to the concept of molecular organizates [2]. I believe that in the evaluation of more complex molecular organizates it is essential to use computer modeling at the most detailed and accurate level. I will write more on this subject when discussing self-assembly (later in this section).

Most of the physics and chemistry studies were carried out on LB films. However, in most cases, the actual quality of the film under study has not been disclosed. Questions such as the film's mechanical strength, its integrity, the size of its molecular clusters, and the orientational and positional order in these clusters are of great importance. There are many types of materials that tend to crystallize (e.g., phthalocyanines, cyanine dyes, etc.) and, as a result, an LB film of such materials, sometimes, is nothing but a bunch of small crystallites. Fighting this tendency is not easy, and one solution to this problem is in polymeric amphiphiles.

In trying to define the long-term goals for LB film research, one has to be careful in using terms such as molecular electronics. It was suggested in the past that individual molecules could perform electronic functions and that molecular wires could transfer information. This is wrong, and, in fact, talk about molecular memories has had a damaging effect on the credibility of this area. I will write more about this issue toward the end of this note.

When assessing the role of LB film research in materials science in the 1990s and beyond, one should not consider only potential applications, but first and foremost the understanding of the molecular assembly in two dimensions. There is no doubt that the potential utility of one's research has to be considered very early on, and that the impact of the research should be understood and put forward in any proposal. However, one of the great mistakes is to ask the question: What is the *specific* device that you have in mind? I believe that such a question is a mistake since it confines scientific research to very narrow, existing niches, thus reducing innovation and enthusiasm. I suggest that once a better understanding of monolayer assembly is achieved, there will be no shortage of applications. One of the most powerful tools in the understanding of two-dimensional organization is the π-A isotherm. We have elaborated on the structure of LB films at the water–air interface, and the reader may turn back to Part Two for further details. The role of LB film research is in:

(*a*) The study of mixed monolayers, where mechanisms such as H-bonding and dipole–dipole interactions are incorporated to facilitate self-aggregation.

(b) The study of supermolecular structures, i.e., deposition of different kinds of monolayers in a planned sequence on top of each other, thus creating desired physical phenomena or devices [2–4].

(c) The study of chemical manipulations, i.e., highly specific, quantitative reactions, either intra- or interplane [5, 6]. One highly desired reaction is a polymerization that does not result in shrinkage of the molecular film.

(d) The study of two-dimensional organization of prefabricated supermolecular elements. One example may be using amino acids as folding directors for the design of polymeric operation units, and subsequently organizing these units at the water–air interface.

(e) The study of the physics and chemistry of complex systems at reduced dimensionality. Here, for example, one can study problems such as chemical reactions in confined geometries, physical processes of polymers, etc.

More direct goals in LB film research are in the area of applications [7–9]:

(a) NLO is the first application that comes to mind, mainly due to the high inherent order parameter in these two-dimensional systems and the ångstrom-level control on film thickness. However, for an LB film to be a good NLO material, it must have a very high linear optical quality, too. Thus, any light scattering due to grain boundaries, for example, will cause losses in the NLO device. (An acceptable loss is < 1 dB/cm.)

(b) Development of novel materials for the control of friction and wear (tribology).

(c) Development of microsensors. Here, the uniqueness of the LB technique in the production of lipid bilayers is clear. Thus, once bioactive molecules are incorporated into such films, biosensors that mimic cell membranes can be fabricated. Chemical sensors also are a promising application. Indeed, a chlorine sensor based on LB films of phthalocyanine has been developed recently and may become a product [10, 11]. Other sensors using the piezoelectric effect also may become viable products.

(d) Development of novel materials for solar energy conversion.

We note that some the aforementioned goals clearly can be adapted to SA monolayers. For example, thiols on gold can be used for the formation of surfaces with novel chemical and biological properties.

In the discussion of LB films, we elaborated on film transfer and deposition and talked about reactive transfer. This, together with the accounts on SA monolayers of fatty acids, suggests that a good LB deposition has a strong element of self-assembly in it. This is true for a deposition, for example, of fatty acids on AgO/Ag, Al_2O_3/Al, and even on clean glass (of course, only for the first monolayer). The preceding observation, together with the frequent difficulties in preparing high quality LB films from complex materials, brings me to the obvious conclusion that self-assembly or self-organization should be an important direction of future research in ultrathin organic films. (If one thinks about it in a philosophical way, just by looking at the mirror one should see the advantages of self-organization, since self-organization of lipid bilayers is an important part of our body.) Therefore, the reader is encouraged to think about self-assembly in a more general way, and to separate it from the existing techniques of SA monolayers. For example, the controlled folding and self-organization of proteins and of polymers containing oligopeptides may be an exciting route for the fabrication of three-dimensional objects at the nanometer-size level.

Future science and technology developments in self-assembled thin organic films will require (*a*) a much better control on the structure of two-dimensional molecular assemblies, and (*b*) the development of quantitative surface reactions. These two areas present scientific challenges in organic chemistry, computer modeling, and materials science. We shall discuss these two issues in more detail.

A better control on the structure of two-dimensional molecular assemblies can be achieved with the aid of computer modeling, from the level of individual molecules to that of three-dimensional assemblies. There is a need to develop more accurate parameters and force fields, faster algorithms, and faster and bigger computers. There is no doubt that parallel processing will make it possible to simulate large molecular assemblies in great detail, and, therefore, will make computer modeling an essential part of the future of materials science. These advances in computer science will help in understanding the relationships between molecular structures and the structure of their assemblies, perhaps to the degree where prediction will be possible. Such prediction capabilities, together with the flexibility offered by organic synthesis, will enable the construction of highly ordered monomolecular layers with interesting electronic, magnetic, and electro-optic properties.

In trying to assess the future potential of SA films of trichlorosilanes, we should consider both the advantages and disadvantages. The advantages of trichlorosilane chemistry are clear. First and foremost are the strong chemisorption to the surface via Si–O bonds and the intra- and interlayer bonding through the formation of polysiloxane backbones. Second, and vital for any fabrication of a multilayer structure, is the healing mechanism, i.e., the *in situ* polymer formation at the solution–substrate interface that overcomes defects. Third are the chemical and thermal stabilities of these SA monolayers, and fourth is the noncentrosymmetry of the monolayer formed. The disadvantages, on the other hand, are not conceptual, but rather practical. First, and in my view the most important issue when electronic and electro-optic materials are considered, is the fact that once a bulky, planar π-system has been introduced into the alkyltrichlorosilane, it cannot be purified. Second, and important from an environmental standpoint, is the use of harsh chemicals such as $LiAlH_4$ or B_2H_6. Thus, while SA monolayers of alkyltrichlorosilane may have importance in surface engineering (e.g., controlled wetting) and as adhesion layers (e.g., in biosensors), multilayers of such materials will not become a viable technology unless the serious problems alluded to are solved.

The development of new self-assembly chemistries is of importance, and at the time of writing this final note, there are several groups that are working in this area. Two recent communications have indicated that ionic interactions may serve as a useful vehicle for the production of multilayer films via self-assembly [12, 13]. The first by Lee et al. reported the fabrication of SA multilayers using ionic interactions, in which ordered zirconium 1,10-decanebiphosphonate multilayers were formed on silicon surfaces [12]. In the second, Smotkin et al. exposed cadmium arachidate multilayers to H_2S, thereby preparing CdS [13]. Utilizing this idea of forming a metal salt on a carboxylic acid surface, one could envisage that with a suitable choice of metal ions an alkanethiol monolayer could be chemisorbed on the carboxylic metal–salt surface. Works done both in Sagiv's group at the Weizmann Institute and in our group, indicate that this indeed is that case, and that specific ionic interactions may have an important role as a self-assembly mechanism.

Copper sulfide is highly insoluble in water ($K_{sp} = 1 \times 10^{-44}$) [14], while copper acetate is soluble [15]. Thus, one could presume (rather naively) that ω-mercaptoalkanoic acids ($HS-(CH_2)_n-COOH$) would interact with ionic copper surfaces to give *specifically* the CuS rather than the –COOCu bonding mode.

The formation of multilayers was achieved by the following process [16]: In the first step, the gold substrate is exposed to a solution of the ω-mercaptoalkanoic acid in ethanol, producing a monolayer. After washing with ethanol and water, the thickness of the first monolayers was 14 Å and 18 Å for $HS-(CH_2)_{10}-COOH$ (I) and $HS-(CH_2)_{15}-COOH$ (II), respectively, and both monolayers were completely wet by water ($\theta = 0°$). The acid-surface was then exposed to a solution of $Cu(ClO_4)_2$ in absolute ethanol, and then exposed to the thiol solution again. Repetition of the second and third steps led to a multilayer of the ω-mercaptoalkanoic acid. The addition of a new layer resulted in a completely water-wettable surface.

The formation of multilayers was monitored by both ellipsometry and grazing-angle FTIR spectroscopy. Figure 6.1 shows a plot of film thickness vs. number of layers for multilayers of I and II. Figure 6.2 presents the FTIR reflection absorption spectrum vs. number of layers for multilayers of I.

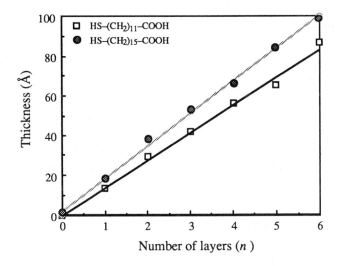

Figure 6.1. Film thickness vs. number of layers.

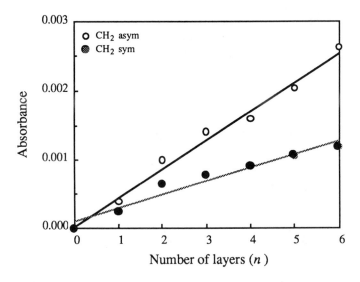

Figure 6.2. FTIR reflection absorption spectrum vs. number of layers for multilayers of I.

420

Although films of only six layers have been shown here, films of greater numbers of layers have now been prepared in our laboratory from II with the same linear behavior as shown in Figures 6.1 and 6.2. It is believed, therefore, that there is, in principle, no limitation to the total number of layers that may be deposited.

In trying to understand the structure of these multilayers, careful XPS and FTIR studies were carried out. It was found that upon copper salt formation, the peaks at 1719 cm^{-1} and 1628 cm^{-1} associated with the –COOH group were replaced by a very strong peak at 1628 cm^{-1}, associated with the COO^{-} group. The addition of a second layer resulted in the replacement of the COO^{-} peak by a broad peak at 1716 cm^{-1}. Further exposure of the two-layer system to the Cu^{2+} solution resulted in the reappearance of the strong 1628 cm^{-1} peak. The methylene stretch vibrations occurred at *ca.* 2919 and 2851 cm^{-1} for the asymmetric and symmetric vibrations, respectively, in a sample of six monolayers of I. These showed decreasing FWHM values as the number of layers increased, going from 30 to 21.5 cm^{-1} for $v_a(CH_2)$, and from 19 to 12 cm^{-1} for $v_s(CH_2)$, between two and six layers of I, respectively. These shifts indicate a transformation from a "liquid-like" structure in the two-layer system to a "solid-like" structure in the six-monolayer sample. Both these peaks displayed high asymmetry. XPS showed a shift of the C=O carbon peak from 289.4 eV in the free acid to 288.5 eV in the copper salt, and indicated the formation of a dimeric structure, $(–COO^{-})_2Cu^{2+}$. Further XPS studies were ambiguous, possibly due to monolayer desorption.

A detailed literature search gave very few structures that contain both thiolate and carboxylate groups around a divalent metal ion [17–19]. Usually, a thiolate is the preferred ligand, even when a carboxylate group is present [17, 20]. Furthermore, weakly bound short chain carboxylic acid species have been used in self-assembly to "protect" gold surfaces from unwanted adventitious contaminants. The acid is removed easily during the subsequent thiol adsorption. From angle-resolved XPS studies to determine whether there is any competitive adsorption between the carboxylate and the thiol functionalities on gold surfaces, we have found that there is, within the detectable limits, no carboxylate-substrate binding and that the thiolate-substrate reaction is dominant [21]. Finally, it has been found that the binding energies of alkanethiols to copper and to gold surfaces are similar [22].

The structure emerging from the data presented so far is that multilayer formation occurs via the formation of copper (II) thiolate adsorbed on a free-acid surface. It is clear from wetting data that the ω-mercaptoalkanoic acid is attached to the copper carboxylate surface through the thiol-end (the advancing contact angle on an SH surface is 71 ± 3° [23]. It is not clear, however, why complete monolayers from simple alkanethiols (e.g., octadecanethiol $CH_3–(CH_2)_{17}–SH$) on the copper carboxylate surface could not be produced. The only speculation we can provide is that since there is only one Cu^{2+} ion for two alkyl chains, the carboxylic groups play an important role in the adsorption, probably through H-bond stabilization of molecular dimers. While this work is by no means complete, I have brought it here as a trigger for new ideas, with the hope that it will initiate further studies in this area.

The conclusion from the recent preliminary works cited here is that our chemical knowledge is like a Lego® box, with many organic and inorganic reactions. For example, from the work of Mallouk and co-workers it is clear that a phosphate (OPO_3^-) or a phosphonate ($–PO_3^-$) group should self-assemble on an ionic surface. Indeed, this was recently reported by Sagiv [24]. Clearly, the rules of the self-assembly game are different from those in LB, and have not yet been fully defined. The very recent results mentioned above are undoubtedly just the beginning. Much work is needed in both the development of new self-assembly reactions and the understanding of the limits of specific ionic interactions. Furthermore, the incorporation of inorganic ions in the self-assembly process is not limited to one layer, but should, in principle, be extended to thicker inorganic layers. This should make possible the construction of organic–inorganic superlattices, and in fact, the demonstration of quantum effects, similar to those reported by Smotkin *et al.*

Another important issue for both LB and SA films is the development of a precise positioning technology for communication with molecular assemblies. The development of scanning tunneling microscopy (STM) offers not only surface analysis at the atomic level, but also precise positioning possibilities. In fact, different research groups in this country and in Japan are trying to develop the STM as a production tool. Considering an area per molecule of ~40 Å^2, a metal dot with a diameter of ~200 Å, when placed on a monolayer surface, covers an assembly of ~800 molecules.

It is important to emphasize that it is unreasonable, because of the uncertainty principle, that electronic operations will be conducted by individual molecules, and that a ratio of ~10^4–10^5 molecules per bit of information will probably be needed. A simple calculation shows that 10^4 molecules occupy an area of ~4 x 10^5 Å^2, or a ~600 x 600 Å square. Such dimensions (~0.06 μm), may be possible to achieve with x-ray lithography. The reader is referred to recent papers by Miller on the general topic of molecular materials [25–27]. It therefore can be suggested that hybrids between lithographically produced substrates and self-assembly multilayers will be the more immediate future devices. This will allow us to combine high resolution substrate preparation with high speed π-system-based electronics and electrooptics. It is thus apparent that self-assembly should make important contributions to these areas of ultrathin organic films and of molecular devices.

We finally come to the conclusion of this book, with the vision of the 1990s being the decade of molecular engineering and self-organization. Langmuir and Blodgett made it possible for us to engineer molecular films at the water–air interface and to transfer them onto solid substrates. However, self-assembly will slowly become more important when nanometer-size, complex devices are considered. With this in mind, the game has really just started.

References

1. Kuhn H. *Thin Organic Films* **1989**, *178*, 1.
2. Kuhn, H.; Möbius, D.; Bücher, H., in *Physical Methods of Chemistry;* Weissberger, A.; Rossiter, B., eds., Wiley: New York, 1972, Vol. I, Part 3B, p 577.
3. Blinov, L. M. *Russ. Chem. Rev.* **1983**, *52*, 713.
4. Zwick, M. M.; Kuhn, H. *Z. Naturforsch* **1962**, *17a*, 411.
5. Ruaudel-Teixier, A. *J. Chim. Phys.* **1988**, *85*, 1067.
6. Ringsdorf, H.; Schlarb, B.; Venzmer, J. *Angew. Chem. Int. Ed. Engl.* **1988**, *27*, 113.
7. Roberts, G. G. *Adv. Phys.* **1985**, *34*, 475.
8. Swalen, J. D.; Allara, D. L.; Andrade, J. D.; Chandross, E. A.; Garoff, S.; Israelachvili, J.; McCarthy, T. J.; Murray, R.; Pease, P. F.; Rabolt, J. F.; Wynne, K. J.; Yu, K. *Langmuir* **1987**, *3*, 932.
9. Barraud, A. *J. Chim. Phys.* **1988**, *85*, 1121.
10. Wang, H. Y.; Ko, W. H.; Batzel, D. A.; Kenney, M. E.; Lando, J. B. *Sensors and Actuators* **1990**, *B1*, 138.
11. Ko, W. H.; Fu, C. W.; Wang, H. Y.; Batzel, D. A.; Kenney, M. E.; Lando, J. B. *Sensors and Mat.* **1990**, *2*, 39.
12. Lee, H.; Kepley, L. J.; Hong, H. -G.; Mallouk, T. E. *J. Am. Chem. Soc.* **1988**, *110*, 618.
13. Smotkin, E. S.; Lee, Chongmok, Bard, A. J.; Campion, A.; Fox, M. A.; Mallouk T. E.; Webber, S. E.; White, J. M. *Chem. Phys. Lett.* **1988**, *152*, 265.
14. Vogel, A. I. *Macro and Semimicro Qualitative Inorganic Analysis;* Longmans: London, 1964, p. 37.
15. *CRC Handbook of Chemistry and Physics,* Weast R. C., Ed., 66th ed., CRC Press: Boca Raton, 1985.
16. Evans, S. D.; Ulman, A. *J. Am. Chem. Soc.,* submitted.
17. Chow, S. T.; McAuliffe, C. A.; Sayle, B. J. *J. Inorg. Nucl. Chem.* **1974**, *37*, 451.
18. Salinas, F.; Martinez-Vidal, J. L.; Martinez Galera, M.; Perez-Alvarez, I. J. *Thermocim. Acta* **1989**, *153*, 87.

19. Khosla, M. M. L.; Rao, S. P. *Microchem. J.* **1972**, *17*, 388.
20. Nigam, H. L.; Srivastava, U. C. *Inorg. Chim. Acta* **1971**, *5*, 338.
21. Ulman, A.; Evans, S. D., unpublished results.
22. Nuzzo, R. G., private communication.
23. Balachander, N.; Sukenik, C. N. *Langmuir* **1990**, *6*, 1621.
24. Sagiv, J., in a lecture presented at the American Chemical Society Meeting, Boston, March 1990.
25. Miller, J. S. *Adv. Materials* **1990**, *2*, 378.
26. Miller, J. S. *Adv. Materials* **1990**, *2*, 495.
27. Miller, J. S. *Adv. Materials* **1990**, *2*, 601.

INDEX

A

438

X

Y

Z

About the Author

Abraham Ulman was born in Haifa, Israel, in 1946. He studied chemistry at Bar-Ilan University in Ramat-Gan, Israel, and received his B.Sc. in 1969. He received his M.Sc. in phosphorus chemistry in 1971. After a brief period in industry he moved to the Weizmann Institute in Rehovot, Israel, and received his Ph.D. in 1978 for work on heterosubstituted porphyrins. He then spent two years at Northwestern University in Evanston, Illinois, where his main interest was one-dimensional organic conductors. Since 1985 he has been with the Corporate Research Laboratories of the Eastman Kodak Company, in Rochester, New York, where he is currently a research associate. His research interests encompass nonlinear optics, self-assembled monolayers, and surface engineering.